W9-CGZ-476

Andreas Hirsch
Michael Brettreich
Fullerenes

Further Reading from WILEY-VCH

S. Reich, C. Thomsen, J. Maultzsch

Carbon Nanotubes

2004
ISBN 3-527-40386-8

C. N. R. Rao, A. Müller, A. K. Cheetham (Eds.)

The Chemistry of Nanomaterials
2 Volumes

2004
ISBN 3-527-30686-2

A. de Meijere, F. Diederich (Eds.)

Metal-Catalyzed Cross-Coupling Reactions
Second Edition, 2 Volumes

2004
ISBN 3-527-30518-1

R. Mahrwald (Ed.)

Modern Aldol Reactions
2 Volumes

2004
ISBN 3-527-30714-1

Andreas Hirsch, Michael Brettreich

Fullerenes

Chemistry and Reactions

WILEY-VCH Verlag GmbH & Co. KGaA

Authors:

Prof. Dr. Andreas Hirsch
Dr. Michael Brettreich

Department of Organic Chemistry
Friedrich Alexander University
of Erlangen-Nuremberg
Henkestrasse 42
91054 Erlangen
Germany

Library of Congress Card No.: Applied for

British Library Cataloging-in-Publication Data:
A catalogue record for this book is available from the British Library.

Bibliographic information published by Die Deutsche Bibliothek
Die Deutsche Bibliothek lists this publication in the Deutsche Nationalbibliografie; detailed bibliographic data is available in the internet at http://dnb.ddb.de.

Printed in the Federal Republic of Germany
Printed on acid-free paper

Typesetting Manuela Treindl, Laaber
Printing Strauss GmbH, Mörlenbach
Bookbinding Litges & Dopf Buchbinderei GmbH, Heppenheim

ISBN 3-527-30820-2

Foreword

Ten years ago Prof. Andreas Hirsch, then finishing his "Habilitation" (an advanced junior research academic rank used in some European countries) in Prof. Hanack's laboratories in Tübingen, wrote THE seminal book about fullerene chemistry: "*The Chemistry of the Fullerenes*". This book, in our group, has many yellowed pages and an abundance of fingerprints, attesting to its outstanding usefulness as a reference book, or, better, a manual. The book has nine chapters covered in 203 pages, including an index.

In the intervening decade, though it would appear to some chemists in the world that the field of fullerene science had "matured", the current version of the book proves otherwise. In "*Fullerenes – Chemistry and Reactions*" Professor Hirsch has expanded the original 9 chapters to 14. The new chapters cover halogenation (Chap. 9), regiochemistry (Chap. 10), cluster modified fullerenes (Chap. 11), heterofullerenes (Chap. 12) and higher fullerenes (Chap. 13). The book is now 419 pages long, over double the number of the original book! The last three chapters give a strong impression that the field of fullerene chemistry is still vibrant and replete with challenges. Upon perusing its content, it is obvious that this book will be equally, or more, useful than its antecedent.

The statement I made on the back cover of "*The Chemistry of the Fullerenes*" applies today to "*Fullerenes – Chemistry and Reactions*" with only very minor changes that took place with time:

> "Though the synthetic fullerenes have been with us for only three years, the scientific articles based on them number in the thousands. It is therefore important and necessary to have a source of information which summarizes the most important and fundamental aspects of the *organic and organometallic chemistry* of the fullerenes. Dr. Hirsch is already well known for his important contributions to the field and is uniquely qualified to write what will become the *primary source* of information for the practicing organic and organometallic chemist.
> This book is logically arranged and information is easy to retrieve. The style lends itself to effortless reading and to learning more about the chemical properties of a family of molecules that constitute new building blocks for novel architectures in the still expanding universe of synthetic chemistry.
> This book will not only be found in university libraries but also on bookshelves of chemists interested in the art and science of structure and property manipulation by synthesis. Dr. Hirsch's "*The Chemistry of the Fullerenes*" is a

Fullerenes: Chemistry and Reactions. Andreas Hirsch and Michael Brettreich

ISBN: 3-527-30820-2

genuine, single author book. It is the first, and so far the best, monograph in the field. It stands out because its quality surpasses that of other multi-author compendia, that preceded it."

September 2004 *Fred Wudl*

Preface of "The Chemistry of the Fullerenes" by Andreas Hirsch (1994)

"Buckminsterfullerene: Is it a real thing?" This question was in our heads that evening in November 1990, when Fred Wudl came into the lab and showed us the first 50 mg sample of C_{60}. Although at that time there was evidence for the geometrical structures of this soccer ball shaped molecule and his bigger brother the "American football" C_{70}, no one knew what the chemical and physical properties of these fascinating molecular allotropes of carbon would be. On that same night we started to work on C_{60} and three reversible one electron reductions of the carbon sphere were found – one more than already detected by the Rice group. We were all enthusiastic and Fred projected possible chemical transformations and proposed remarkable electronic properties of fulleride salts – a prediction that shortly thereafter became reality by the discovery of the superconductivity of K_3C_{60} in the Bell labs. It was one of the greatest opportunities in my scientific life, that Fred asked me to participate in the ball game and to investigate organic fullerene chemistry. He also encouraged me to carry on this work in Germany, once I had finished my postdoctoral time in his group in Santa Barbara. The pioneering research of these early days is still a basis of my present work. Fullerene chemistry, which is unique in many respects, has meanwhile exploded. In a very short period of time a huge number of chemical transformations of the "real thing" C_{60} and outstanding properties of fullerene derivatives have been discovered. Many principles of fullerene chemistry are understood. The fullerenes are now an established compound class in organic chemistry.

It is therefore the right time to give a first comprehensive overview of fullerene chemistry, which is the aim of this book. This summary addresses chemists, material scientists and a broad readership in industry and the scientific community. The number of publications in this field meanwhile gains such dimensions that for nonspecialists it is very difficult to obtain a facile access to the topics of interest. In this book, which contains the complete important literature, the reader will find all aspects of fullerene chemistry as well as the properties of fullerene derivatives. After a short description of the discovery of the fullerenes all methods of the production and isolation of the parent fullerenes and endohedrals are discussed in detail (Chapter 1). In this first chapter the mechanism of the fullerene formation, the physical properties, for example the molecular structure, the thermodynamic, electronic and spectroscopic properties as well as solubilities are also summarized. This knowledge is necessary to understand the chemical behavior of the fullerenes.

Fullerenes: Chemistry and Reactions. Andreas Hirsch and Michael Brettreich

ISBN: 3-527-30820-2

The description of the chemistry of the fullerenes themselves is organized according to the type of chemical transformation, starting from the reduction (Chapter 2), nucleophilic additions (Chapter 3), cycloadditions (Chapter 4), hydrogenation (Chapter 5), radical additions (Chapter 6), transition metal complex formation (Chapter 7) through oxidation and reactions with electrophiles (Chapter 8). Most of the examples presented in these chapters are reactions with C_{60}, since only very little work has been published on C_{70} and the higher fullerenes. It is the aim to provide an understanding of the basic characteristics of fullerene chemistry. This is achieved by a comparative description of both experimental and theoretical investigations in each of these chapters. It is also emphasized in each chapter wherever a reaction type leads to a fullerene derivative with a potential for practical application. In the last chapter (Chapter 9) the emerging principles of fullerene chemistry, such as reactivity and regiochemistry, are evaluated and summarized. The fullerene chemistry is still in an early stage of development. For synthetic chemists a lot of challenging work remains to be done. A prediction of the future directions of fullerene chemistry is therefore also given in Chapter 9 and finally, since from the beginning of the fullerene era many practical uses have been proposed, perspectives for applications are evaluated.

Writing this book prevented me for some time from carrying out practical work with my own hands. Nevertheless, since I have had co-workers like Thomas Grösser, Iris Lamparth, Andreas Skiebe and Antonio Soi, the experimental work on fullerenes in my lab has proceeded, even with much success. I am also indebted to my co-workers for their assistance in preparing figures and illustrations presented in this book. I thank Dr. L. R. Subramanian for reading the entire manuscript and for useful suggestions.

I thank Dr. J. P. Richmond for the excellent co-operation, which enabled the fast realization of this book project.

I am very grateful to my teacher Prof. Dr. Dr. h. c. M. Hanack, who has been supporting me for many years and who provided the starting conditions for writing this book.

It was Prof. Dr. F. Wudl who introduced me to the art of fullerene chemistry and who inspired me to continue with this work, for which I want to thank him very much. On reviewing the manuscript he came up with cogent comments and suggestions.

My special thanks go to my wife Almut, who, despite the fact that for several months she was very often deprived of the company of her husband, responded with warmth and understanding.

Tübingen, May 1994 *Andreas Hirsch*

Contents

Fullerenes: Chemistry and Reactions. Andreas Hirsch and Michael Brettreich

ISBN: 3-527-30820-2

Abbreviations

a.u.	arbitrary units
AIBN	2,2′-azobisisobutyronitril
ALS	amyotrophic lateral sclerosis
APCI	atmospheric pressure chemical ionization
ATP	adenosintriphosphate
BN	boron nitride
BP	biphenyl
BtOH	1H-benzotriazol
CD	circular dichroism
CI	chemical ionization
CSA	camphor sulfonic acid
CT	charge transfer
CV	cyclic voltammetry
CVM	chemical vapor modification
DABCO	1,4-diazabicyclo[2.2.2]octane
DBU	1,8-diazabicyclo[5.4.0]undec-7-ene
DCC	N,N′-dicyclohexylcarbodiimide
DCI	desorptive chemical ionization
DDQ	2,3-dichloro-5,6-dicyanobenzoquinone
DFT	density functional theory
DIBAL-H	diisobutylaluminium-hydride
DIOP	2,3-O-isopropylidene-2,3-dihydroxy-1,4-bis(diphenylphosphanyl)butane
DMA	9,10-dimethylanthracene
DMAD	dimethylacetylenedicarboxylate
DMAP	4-(dimethylamino)pyridine
DMB	dimethoxybenzene
DMF	dimethylformamide
DMSO	dimethylsulfoxide
DNA	desoxyribonucleid acid
DOS	density of states
DPIF	1,3-diphenylisobenzofurane
dppb	1,2-bis(diphenylphosphino)benzene
DPPC	dipalmitoylphosphatidylcholine

Fullerenes: Chemistry and Reactions. Andreas Hirsch and Michael Brettreich

ISBN: 3-527-30820-2

dppe	1,2-bis(diphenylphosphino)ethane
dppf	1,1′-bis(diphenylphosphino)-ferrocene
DSC	digital scanning calorimetry
EDCI	N′-(3′-dimethylaminopropyl)-N-ethylcarbodiimide hydrochloride
EI	electron impact
EPR	electron paramagnetic resonance
ESCA	electron spectroscopy for chemical analysis
ESR	electron spin resonance
FAB	fast atom bombardment
FD	field desorption
FT-ICR	fourier transform ion cyclotron resonance
FVP	flash-vacuum pyrolysis
GPC	gel permeation chromatography
HETCOR	heteronuclear chemical shift correlation
HIV	human immunodeficiency virus
HMPA	hexamethyl phosphoric acid
HOMO	highest occupied molecular orbital
HPLC	high pressure liquid chromatography
IC	ion chromatography
ICR	ion cyclotron resonance
INADEQUATE	incredible natural abundance double quantum transition experiment
IPR	isolated pentagon rule
IR	infrared
ITO	indium-tin-oxide
LB	Langmuir-Blodgett
LDA	Lithium diisopropylamide
LESR	light-induced ESR measurement
LUMO	lowest unoccupied molecular orbital
MALDI	matrix-assisted laser desorption ionization
MCPBA	m-chloroperoxybenzoic acid
MEM	methoxy ethoxy methyl
MO	molecular orbital
MRI	magnetic resonance imaging
MS	mass spectrometry
NCS	N-chlorosuccinimide
NICS	nucleus-independent chemical shift
NIR	near infrared
NMA^+	N-methylacridinium-hexafluorophosphate
NMR	nuclear magnetic resonance
ODCB	ortho-dichlorobenzene
OL	optical limiting
PAH	polycyclic aromatic hydrocarbon
PCBA	[6,6]-phenyl-C_{61}-butyric acid

PCC	pyridinium chlorochromate
PET	photoinduced electron transfer
POAV	π orbital axis vector
POM	polarizing optical microscopy
PPV	poly-para-phenylen-vinylene
PVK	poly(N-vinylcarbazole)
QCM	quartz crystal microbalance
RE	resonance energy
RETOF	reflectron time of flight
SAM	self-assembled monolayer
SEC	size-exclusion chromatography
SET	single electron transfer
SWNT	single-walled carbon nanotube
TBA	tetrabutylammonium
TCE	trichloroethylene
TCNE	tetracyanoethylene
TCNQ	tetracyano-p-quinodimethane
TDAE	tetrakis(dimethylamino)ethylene
TEMPO	2,2,6,6-tetramethylpiperidin-1-yloxyl
TFA	trifluoroacetic acid
TGA	thermal gravimetric analysis
THA	tetrahexylammonium
THF	tetrahydrofuran
TMEDA	N,N,N',N'-tetramethylene diamine
TMM	trimethylenemethane
TOF	time of flight
TosOH	4-toluenesulfonic acid
TPP	tetraphenylporphyrin
TPP^+	triphenylpyriliumtetrafluoroborate
TTF	tetrathiafulvalene
UHV-STM	ultra-high vacuum scanning tunneling microscopy
UV	ultraviolet
UV/Vis	ultraviolet/visible
VB	valence bond
Vis	visible
XPS	X-ray photoelectron spectroscopy

1
Parent Fullerenes

1.1
Fullerenes: Molecular Allotropes of Carbon

For synthetic chemists, who are interested in the transformation of known and the creation of new matter, elemental carbon as starting material once played a minor role. This situation changed dramatically when the family of carbon allotropes consisting of the classical forms graphite and diamond became enriched by the fullerenes. In contrast to graphite and diamond, with extended solid state structures, fullerenes are spherical molecules and are soluble in various organic solvents, an important requirement for chemical manipulations.

Fullerenes are built up of fused pentagons and hexagons. The pentagons, absent in graphite, provide curvature. The smallest stable, and also the most abundant fullerene, obtained by usual preparation methods is the I_h-symmetrical Buckminsterfullerene C_{60} (Figure 1.1). Buckminsterfullerene has the shape of a soccer ball. The next stable homologue is C_{70} (Figure 1.2) followed by the higher fullerenes C_{74}, C_{76}, C_{78}, C_{80}, C_{82}, C_{84}, and so on. The building principle of the fullerenes is a consequence of the Euler theorem, which says that for the closure of each spherical network of n hexagons, 12 pentagons are required, with the exception of $n = 1$.

Compared to small two-dimensional molecules, for example the planar benzene, the structures of these three-dimensional systems are aesthetically appealing. The beauty and the unprecedented spherical architecture of these molecular cages immediately attracted the attention of many scientists. Indeed, Buckminsterfullerene C_{60} rapidly became one of the most intensively investigated molecules. For synthetic chemists the challenge arose to synthesize exohedrally modified derivatives, in which the properties of fullerenes can be combined with those of other classes of materials. The following initial questions concerned the derivatization of fullerenes: What kind of reactivity do the fullerenes have? Do they behave like a three-dimensional "superbenzene"? What are the structures of exohedral fullerene derivatives and how stable are they?

The IUPAC method of naming Buckminsterfullerene given below is too lengthy and complicated for general use [1]:

Hentriacontacyclo[$29.29.0.0.^{2,14}.0^{3,12}.0^{4,59}.0^{5,10}.0^{6,58}.0^{7,55}.0^{8,53}.0^{9,21}.0^{11,20}.0^{13,18}.0^{15,30}.0^{16,28}.0^{17,25}.0^{19,24}.0^{22,52}.0^{23,50}.0^{26,49}.0^{27,47}.0^{29,45}.0^{32,44}.0^{33,60}.0^{34,57}.0^{35,43}.0^{36,56}.0^{37,41}.0^{38,54}.0^{39,51}.0^{40,48}.0^{42,46}$]hexaconta-1,3,5(10),6,8,11,13(18),14,16,19,21,23,25,27,

Fullerenes: Chemistry and Reactions. Andreas Hirsch and Michael Brettreich

ISBN: 3-527-30820-2

29(45),30,32,(44),33,35(43),36,38(54),39(51),40(48),41,46,49,52,55,57,59-triacontaene.

Furthermore, the enormous number of derivatives, including the multitude of possible regioisomers, available by chemical modifications requires the introduction of a simple nomenclature. According to the latest recommendation, the icosahedral Buckminsterfullerene C_{60} was named as $(C_{60}\text{-}I_h)$[5,6]fullerene and its higher homologue C_{70} as $(C_{70}\text{-}D_{5h})$[5,6]fullerene [2, 3]. The parenthetical prefix gives the number of C-atoms and the point group symbol; the numbers in brackets indicate the ring sizes in the fullerenes. Fullerenes involving rings other then pentagons and hexagons are conceptually possible (*quasi*-fullerenes [4]). The identification of a well defined and preferably contiguous helical numbering pathway is the basis for the numbering of C-atoms within a fullerene. Such a numbering system is important for the unambiguous description of the multitude of possible regioisomeric derivatives formed by exohedral addition reactions. A set of rules for the atom numbering in fullerenes has been adopted [2, 3]. The leading rule (Fu-3.1.1) is:

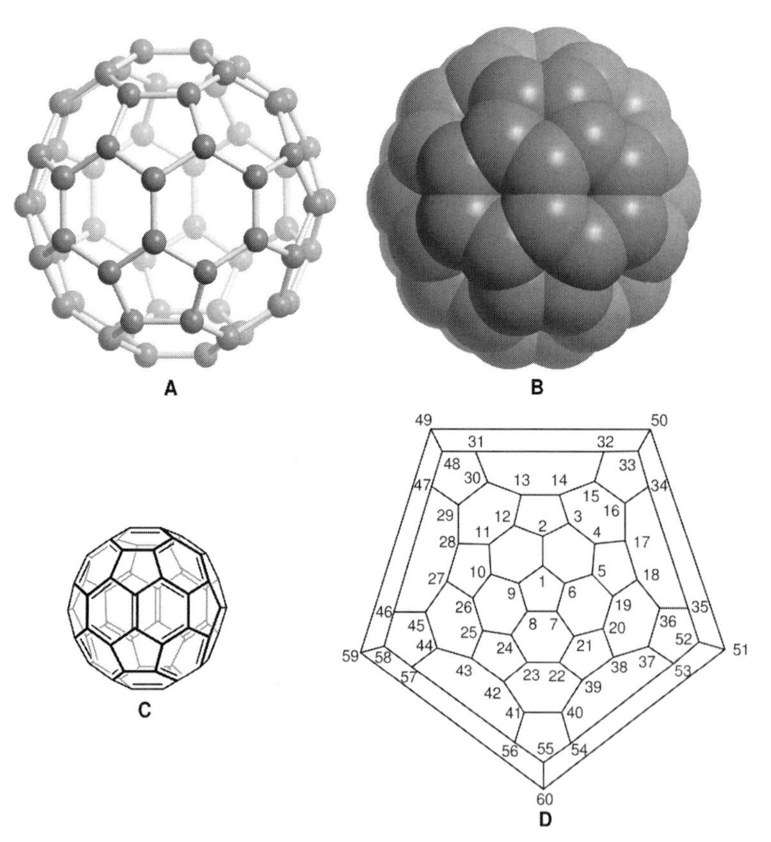

Figure 1.1 Schematic representations of C_{60}. (A) ball and stick model, (B) space filling model, (C) VB formula, (D) Schlegel diagram with numbering of the C-atoms (according to [4]).

Proper rotation axes (C_n) are examined in sequence from the highest-order to the lowest-order axis, until at least one contiguous helical pathway is found that begins in a ring through which a proper rotation axis passes, at the end a bond bisected by a proper rotation axis, or at an atom through which a proper rotation axis passes. Numbering begins at the end of such a contiguous helical pathway, and the corresponding axis is called the "reference axis".

This system allows also for the indication of the absolute configuration of inherently chiral fullerenes by introducing the stereodescriptors ($^{f,s}C$) and ($^{f,s}A$) ("f" = fullerene; "s" = systematic numbering; "C" = clockwise; "A" = anti-clockwise).

In another nomenclature recommendation it was suggested that fullerenes be named in the same way as annulenes, for which the number of C-atoms is indicated in square brackets in front of the word [4]. For fullerenes the number of C-atoms is accompanied by the point group symmetry and by the number of the isomer (using capital Roman) in cases were there are more than one. This is especially important for higher fullerenes. Thus, for Buckminsterfullerene the full description is

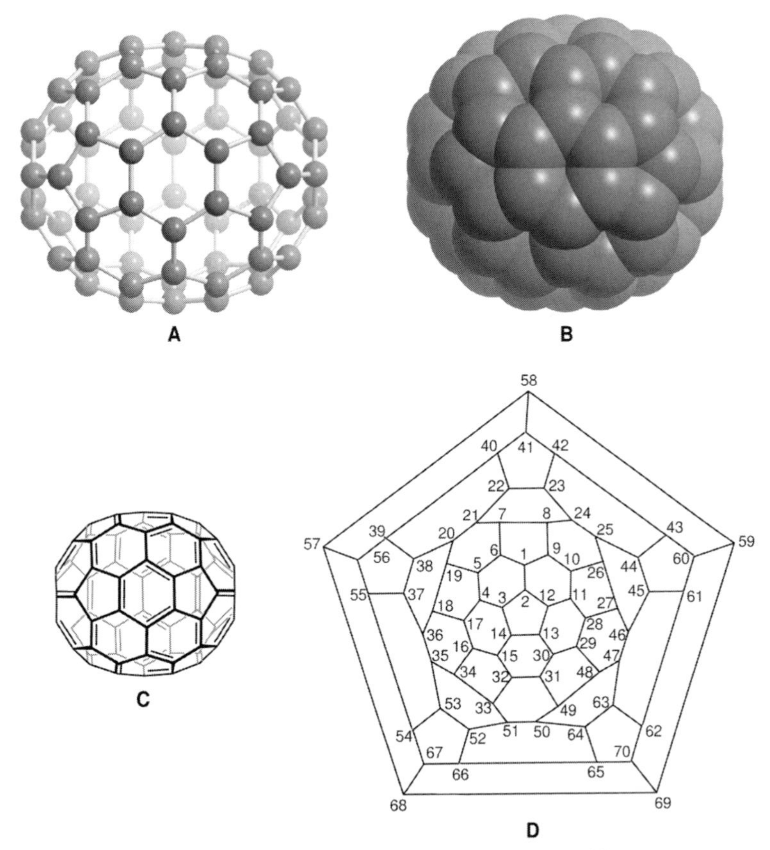

Figure 1.2 Schematic representations of C_{70}. (A) ball and stick model, (B) space filling model, (C) VB formula, (D) Schlegel diagram with numbering of the C-atoms (according to [4]).

[60-I_h]fullerene and for C_{70} (Figure 1.2) [70-D_{5h}]fullerene. In most cases further simplification to [60]fullerene and [70]fullerene or even C_{60} and C_{70} is made, since there are no other stable isomers of these fullerenes. An alternative numbering of C-atoms to that pointed out above [2, 3] is also based on a contiguous spiral fashion but numbers the bond of highest reactivity as 1,2 (Figure 1.1) [4]. For [60]fullerene these are the bonds at the junction of two hexagons ([6,6]-bonds). Since the chemistry of [70]fullerene has many similarities to that of [60]fullerene, it is advantageous if the numbering scheme for [70]fullerene parallels that of [60]fullerenes, which is indeed possible (Figure 1.2) [4].

Valence bond (VB) formulas or Schlegel diagrams are useful for simple schematic representations of fullerenes and their derivatives. VB formulas are mostly used for parent fullerenes or for derivatives with a few modifications of the cage structure only. A Schlegel diagram shows each C-atom of the fullerene, which is flattened out in two dimensions. This model is suitable for considering polyadducts, for example, polyhydrofullerenes.

The main type of chemical fullerene derivatizations are addition reactions. Regardless of the relatively many possible reaction sites, addition reactions show a remarkable regioselectivity, especially when the number of addends is small. This is another fulfilled requirement, which makes these molecular spheres exciting objects for synthetic chemists.

1.2 Discovery of the Fullerenes

In 1966 Deadalus alias D. E. H. Jones considered the possibility of making large hollow carbon cages, structures now called giant fullerenes [5, 6]. This suggestion elicited no reaction from the scientific community. Four years later, in 1970, simulated by the synthesis of the bowl shaped corannulene **1** [7], Osawa first proposed the spherical I_h-symmetric football structure for the C_{60} molecule (**2**) [8, 9]. During his efforts to find new three-dimensional superaromatic π-systems, he recognized corannulene to be a part of the football framework. Subsequently, some theoretical papers of other groups appeared, in which inter alia Hückel calculations on C_{60} were reported [10–13].

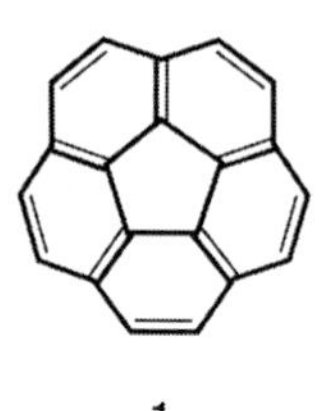

1

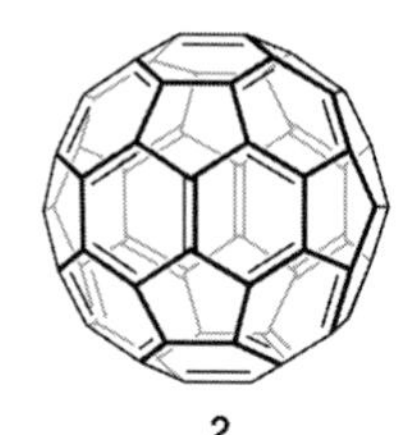

2

In 1984 it was observed that, upon laser vaporization of graphite, large carbon-only clusters C_n with $n = 30–190$ can be produced [14]. The mass distribution of these clusters was determined by time-of-flight mass spectrometry. Only ions with even numbers of carbon atoms were observable in the spectra of large carbon clusters ($n \geq 30$). Although C_{60} and C_{70} were among these clusters, their identity was not recognized. The breakthrough in the experimental discovery of the fullerenes came in 1985 [15] when Kroto visited the Rice University in Houston. Here, Smalley and co-workers developed a technique [16] for studying refractory clusters by mass spectrometry, generated in a plasma by focusing a pulsed laser on a solid, in this case graphite. Kroto and Smalley's original goal was to simulate the conditions under which carbon nucleates in the atmospheres of red giant stars. Indeed, the cluster beam studies showed that the formation of species such as the cyanopolyynes HC_7N and HC_9N, which have been detected in space [17, 18], can be simulated by laboratory experiments [19]. These studies found that, under specific clustering conditions, the 720 mass peak attributed to C_{60}, and to a lesser extent the peak attributed to C_{70}, exhibits a pronounced intensity in the spectra (Figure 1.3). Conditions could be found for which the mass spectra were completely dominated by the C_{60} signal. Kroto and Smalley immediately drew the right conclusion of these experimental findings. The extra stability of C_{60} is due to its spherical structure, which is that of a truncated icosahedron with I_h symmetry [15]. This molecule was named after the architect Buckminster Fuller, whose geodesic domes obey similar building principles. Retrospectively, the enhanced intensity of the peak of C_{70}, which is also a stable fullerene, became understandable as well. Although Buckminsterfullerene (C_{60}) was discovered, a method for its synthesis in macroscopic amounts was needed.

This second breakthrough in fullerene research was achieved by Krätschmer and Huffman [20]. Their intention was to produce laboratory analogues of interstellar dust by vaporization of graphite rods in a helium atmosphere [21]. They observed that, upon choosing the right helium pressure, the IR-spectrum of the soot, generated by the graphite vaporization, shows four sharp stronger absorptions,

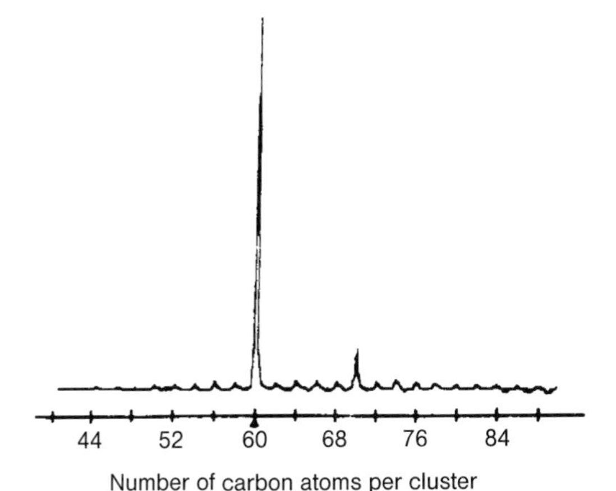

Figure 1.3 Time-of-flight mass spectrum of carbon clusters produced by laser vaporization of graphite under the optimum conditions for observation of a dominant C_{60} signal [15].

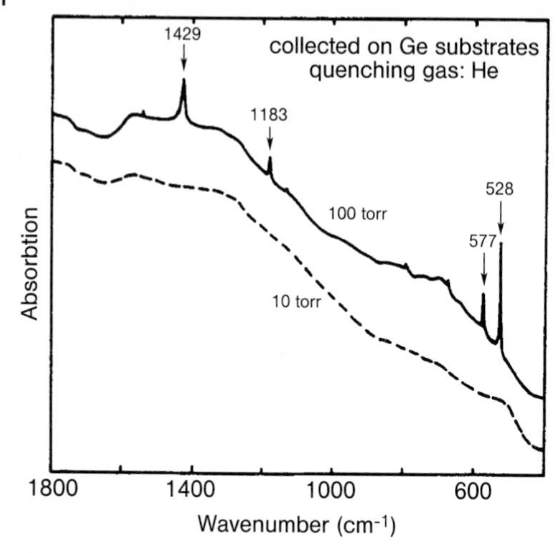

Figure 1.4 IR-spectra of soot particles produced by evaporation of graphite under different helium quenching gas pressures. The occurrence of the four additional sharp peaks at elevated helium pressures turned out to originate from [60-I_h]fullerene (C_{60}) [20].

together with those of the continuum of regular soot (Figure 1.4) [22]. These absorptions were close to the positions predicted by theory for Buckminsterfullerene [23]. The fullerenes were then isolated from the soot by sublimation or extraction with benzene. This allowed the verification of their identity by spectroscopic and crystallographic methods as well as by control experiments with ^{13}C-enriched material. Along with Buckminsterfullerene C_{60}, higher homologues are also obtained by this technique. Fullerenes were then available for the scientific community.

1.3 Fullerene Production

1.3.1 Fullerene Generation by Vaporization of Graphite

1.3.1.1 Resistive Heating of Graphite

Macroscopic quantities of fullerenes were first generated by resistive heating of graphite [20]. This method is based on the technique for the production of amorphous carbon films in a vacuum evaporator [24]. The apparatus (Figure 1.5) that Krätschmer and Fostiropoulos used for the first production of fullerenes consisted of a bell jar as recipient, connected to a pump system and a gas inlet. In the interior of the recipient two graphite rods are kept in contact by a soft spring. Thereby, one graphite rod is sharpened to a conical point, whereas the end of the other is flat. The graphite rods are connected to copper electrodes.

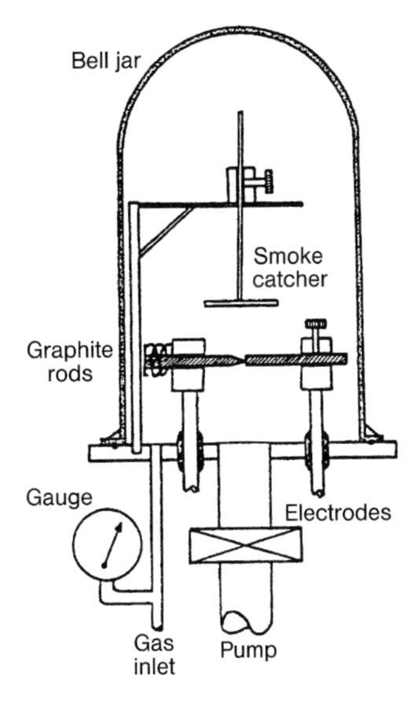

Figure 1.5 Fullerene generator originally used by Krätschmer [20].

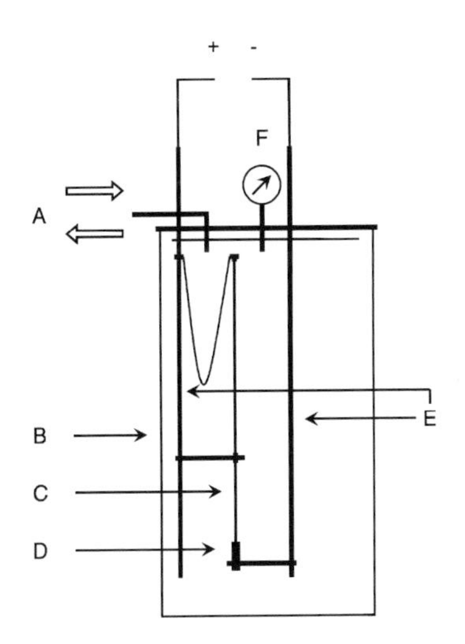

Figure 1.6 Simple benchtop reactor developed by Wudl [27]. Helium supply and connection to a vacuum system (A), Pyrex bell jar (B), graphite rod (3 mm) (C), graphite rod (12 mm) (D), copper electrode (E), manometer (F).

To produce soot, the apparatus is repeatedly evacuated and purged with helium and finally filled with about 140 mbar of helium. After applying a voltage, the electric current passing through the rods dissipates most of its Ohmic power heating at the narrow point of contact. This leads to a bright glowing in this area at 2500–3000 °C. Simultaneously, smoke develops at the contact zone, being transported away by convection and collected on the cooler areas (bell jar and smoke catcher) of the apparatus. The evaporation of the graphite is most efficient at the sharpened end of the rod. After the reaction is over, fullerenes are extracted from the soot, for example with toluene, in about 10–15% yield.

Modifications of this type of fullerene reactor are gravity feed generators [25–27]. The advantage of these generators is their simple construction principle. This, together with their low costs, makes them attractive for synthetic chemists. A schematic representation of such a simple benchtop reactor, developed by Wudl [27] is given in Figure 1.6. A thin graphite rod (3 mm), guided by a copper sleeve, with a sharpened tip is placed on a thick rod (12 mm). A commercially available arc welder serves as power supply. After applying a current (AC or DC) of about 40–60 A, only the material of the thin rod evaporates, whereupon it slips downward, guided by the copper sleeve that keeps the electrical contact. After a few minutes the rod is consumed to the point that it can not any longer make contact with the 12 mm rod. The power is then shut off. Based on evaporated graphite, fullerene yields of 5–10% are obtained [27, 28].

The buffer gas cools the plasma by collisions with the carbon vapor. The gas has to be inert, to prevent reactions with smaller carbon clusters or atoms, initially formed by the evaporation. Using N_2 dramatically reduces the yield of fullerenes, presumably due to nitrogen atoms, formed in the hot zone of the generator, reacting with the carbon fragments [28]. The highest yields of fullerenes are obtained if helium is used as buffer gas. Also, the concentration of the buffer gas is important (Figure 1.7), with maximum yields obtained between 140 and 160 mbar [28]. With a very low buffer gas pressure the carbon radicals diffuse far from the hot zone and the clusters continue to grow in an area that is too cool to allow an annealing to spherical carbon molecules. Conversely, if the pressure of the buffer gas is too high, a very high concentration of carbon radical results in the hot reaction zone. This leads to a fast growth of particles far beyond 60 C-atoms and the annealing process to fullerenes cannot compete [29].

During these resistive heating procedures the formation of slag, depositing on the thicker graphite rod, can be observed after some time of evaporation. As long as this vapor-deposited boundary layer remains between the two electrodes in a sufficiently thick and resistive form, the electrical power continues to be dissipated just in this small zone, and carbon vaporization from the end of the thin graphite rod proceeds efficiently [30]. Thus, the formation of such a resistive layer may be an important requirement for the continuation of smoke production. In the beginning of the reaction this was guaranteed by the sharpened thin graphite rod (heat dissipation in this small resistive zone). For graphite rods, with diameters of 6 mm or greater, the resistive layer does not remain sufficiently resistive and the entire length of the graphite rod eventually begins to glow. This causes inefficient evaporation of carbon from the center of the rod. Therefore, only comparatively thin graphite rods can be used for efficient fullerene production by the resistive heating technique.

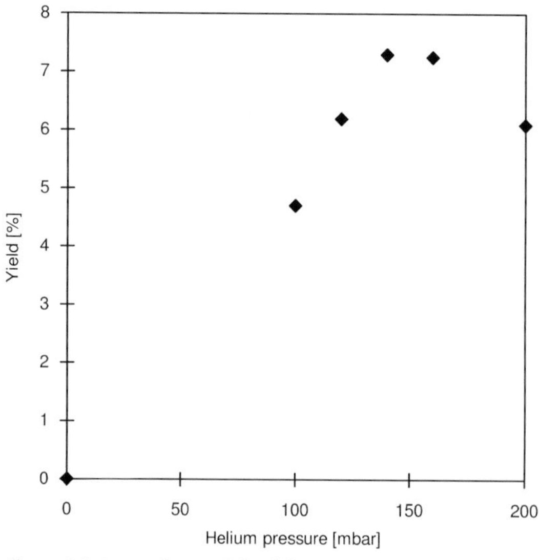

Figure 1.7 Dependence of the fullerene yield on the helium gas pressure in the fullerene generator.

1.3.1.2 Arc Heating of Graphite

An alternative to resistive heating is arc vaporization [29, 31–36] of graphite, first developed by Smalley [31]. If the tips of two sharpened graphite rods are kept in close proximity, but not in a direct contact, the bulk of the electrical power is dissipated in an arc and not in Ohmic heating. In an original generator a spring tension was adjusted to maintain the arc between the nearly contacting graphite electrodes. The most efficient operation occurs when the electrodes are barely touching, which lead to the term "contact-arcing" [31]. This method also allows an efficient evaporation of carbon with somewhat thicker, for example, 6 mm rods. The yield of fullerenes obtained by this technique was found to be about 15%. However, by increasing the rod diameter the yield decreases almost linearly [31], which also prevents an upscaling to very large rod sizes. The reason for the low yields observed by using larger rod-sizes is the fullerenes sensitivity towards UV-radiation. Very intense UV-radiation originates from the central portion of the arc plasma. Newly formed fullerenes moving from the region around the arc are exposed to this intense light flux. The absorption of UV-light produces a triplet state (T_1), which lives for a few microseconds (Scheme 1.1) [37].

$$C_{60}\ (S_0) + h\upsilon \rightarrow C_{60}^{*}\ (T_1)$$

Scheme 1.1

In this T_1 state the fullerene is an open shell system and very susceptible to other carbon species C_n. As a result of such a reaction a non-vaporizable insoluble product may be formed (Scheme 1.2) [30].

$$C_{60}^{*}\ (T_1) + C_n \rightarrow \text{insoluble carbon}$$

Scheme 1.2

The effect of increased rod sizes is a larger photochemically dangerous zone. The rate of migration of the newly formed fullerenes through this zone, however, remains constant. Therefore, the yield of fullerenes that migrate through this region without reacting with other carbon species linearly decreases with the rod diameter [30]. A mathematical model for an arc reactor has taken into account (a) cooling and mixing of carbon vapor with buffer gas, (b) non-isothermal kinetics of carbon cluster growth and (c) formation of soot particles and heterogeneous reactions at their surface. This model provided good coincidence of experimental and calculated values both for the fullerene yields and the C_{60}/C_{70} ratio in the reaction products obtained under widely varied conditions [38].

The ratio of C_{60} to higher fullerenes is typically about 8 : 2. The relative yields of higher fullerenes were improved when graphite containing light elements such as B, Si or Al was used and the buffer gas He was mixed with a small amount of N_2 [39, 40]. Fullerenes have also been synthesized by a pulse arc discharge of 50 Hz–10 kHz and 150–500 A, with graphite electrodes and ambient helium (about 80 torr). Instead of graphite, coal was also used as carbon source [41]. Extraction of the corresponding soot with toluene resulted in a 4–6% yield of fullerenes.

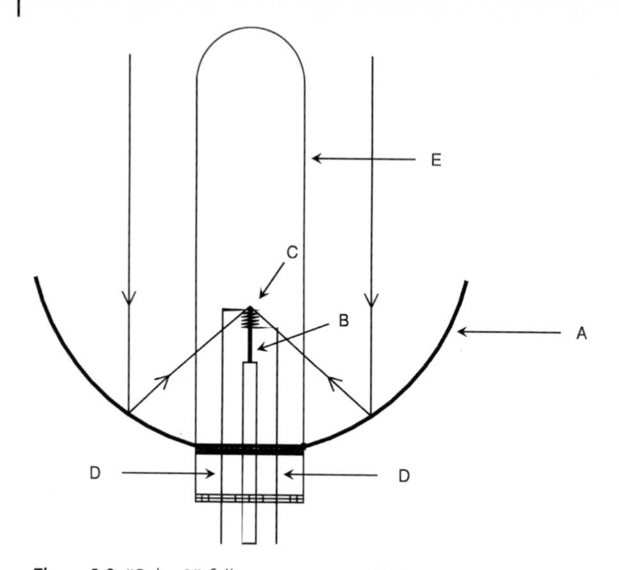

Figure 1.8 "Solar 1" fullerene generator [30].
(A) Parabolic mirror, (B) graphite target, (C) preheater,
(D) insulated preheater connectors and (E) glass tube.

1.3.1.3 **Solar Generators**

The problem of intense UV-radiation is avoided by the use of solar furnaces as fullerene generators [30, 42]. Although sun light is used to evaporate graphite the exposure of generated fullerenes to radiation is far less extensive than with resistive heating or arc vaporization techniques. As an example solar generator, "Solar 1" developed by Smalley [30] will be discussed (Figure 1.8). Sunlight is collected by parabolic mirrors and focused onto a tip of a graphite rod. This rod is mounted inside a Pyrex tube. To minimize conductive heat loss and to provide suitable conditions for the annealing process of the carbon clusters, the graphite rod was enclosed by a helical tungsten preheater. After degassing the system with the preheater, it is filled with about 50 Torr of argon and sealed off. To run the reaction the apparatus is adjusted so that the sunlight is focused directly onto the tip of the graphite target. The argon gas heated by the tungsten preheater is efficiently carried up over the solar-irradiated carbon tip by convection (solar flux: 800–900 W m^{-2}). The condensing carbon vapor quickly moves from the intensive sunlight, cools in the upper regions of the Pyrex tube and subsequently deposits on the upper walls. Although fullerenes can be obtained this way, the efficiency of the prototype "Solar-1" generator is not very high.

1.3.1.4 **Inductive Heating of Graphite and Other Carbon Sources**

Fullerenes can also be produced by direct inductive heating of a carbon sample held in a boron nitride support [43]. Evaporation at 2700 °C in a helium atmosphere affords fullerene-containing soot that is collected on the cold Pyrex glass of the reaction tube. This method allows a continuous operation by keeping the graphite sample in the heating zone. Upon evaporating 1 g of graphite, 80 to 120 mg of fullerene extract can be obtained in 10 min.

Continuous production of fullerenes was possible by pyrolysis of acetylene vapor in a radio-frequency induction heated cylinder of glassy polymeric carbon having multiple holes through which the gas mixture passes [44]. Fullerene production is seen at temperatures not exceeding 1500 K. The yield of fullerenes, however, generated by this method is less than 1%. A more efficient synthesis (up to 4.1% yield) was carried out in an inductively coupled radio-frequency thermal plasma reactor [45].

1.3.2 Fullerene Synthesis in Combustion

The existence of fullerenes in sooting flames was first revealed by mass spectrometry studies [46, 47]. Also, the production of fullerenes in optimized sooting flames is possible [48–52]. For this purpose premixed laminar benzene–oxygen–argon flames have been operated under a range of conditions, including different pressures, temperatures and carbon-to-oxygen ratios. Along with fullerenes and soot, polyaromatic hydrocarbons (PAHs) are formed simultaneously. The yield of fullerenes, as well as the C_{70}:C_{60} ratio, strongly depends on the operation mode. The amount of C_{60} and C_{70}, produced under different sooting flame conditions is in the range 0.003–9% of the soot mass. Expressed as percentage of fuel carbon, the yields varies from $2 \cdot 10^{-4}$ to 0.3% for a non-sooting flame, obtained at optimum conditions, at a pressure of 20 Torr, a carbon-to-oxygen ratio of 0.995 with 10% argon and a flame temperature of about 1800 K. The C_{70}:C_{60} ratio varies from 0.26 to 5.7, which is much larger than that observed for graphite vaporization methods (0.02–0.18). This ratio tends to increase with increasing pressure [48].

Further optimization of the formation of fullerenes in combustion lead to the development of efficient pilot plants [53–56]. Currently, 400 kg of fullerenes per year (Mitsubishi's Frontier Carbon Corporation) are obtained by these methods. Ton scale production is expected in the near future. This remarkable development has allowed fullerenes to be sold for less than \$300 kg^{-1}, a sharp improvement on the \$ 40 000 kg^{-1} rate that prevailed not long ago [57].

1.3.3 Formation of Fullerenes by Pyrolysis of Hydrocarbons

Fullerenes can also be obtained by pyrolysis of hydrocarbons, preferably aromatics. The first example was the pyrolysis of naphthalene at 1000 °C in an argon stream [58, 59]. The naphthalene skeleton is a monomer of the C_{60} structure. Fullerenes are formed by dehydrogenative coupling reactions. Primary reaction products are polynaphthyls with up to seven naphthalene moieties joined together. Full dehydrogenation leads to both C_{60} as well as C_{70} in yields less than 0.5%. As side products, hydrofullerenes, for example $C_{60}H_{36}$, have also been observed by mass spectrometry. Next to naphthalene, the bowl-shaped corannulene and benzo[*k*]fluoranthene were also used as precursors to C_{60} [60]. Fullerene synthesis by laser pyrolysis is possible using benzene and acetylene as carbon sources [61]. Soot-free C_{60} has been produced in the liquid phase of an aerosol precursor of soot at 700 °C [62]. The precursor soot aerosol, a high temperature stable form of hydrocarbon, was produced by pyrolysis of acetylene at atmospheric pressure in a flow tube reactor. Further pyrolysis-based methods for the generation of fullerenes include CO_2-laser pyrolysis of small hydrocarbons such as butadiene and thermal plasma dissociation of hydrocarbons [63].

1.3.4
Generation of Endohedral Fullerenes

Since fullerenes are hollow molecules it should be possible to trap atoms inside the cage. Indeed, one week after the initial discovery of C_{60}, evidence for an endohedral lanthanum complex of C_{60} was obtained [64]. Laser vaporization of a graphite disk soaked in $LaCl_3$ solution produced an additional peak in the time-of-flight (TOF) mass spectrum due to La encapsulated by C_{60}. Evidence that endohedral complexes are so-formed came from "shrink-wrap" experiments showing that these complexes can lose, successively, C_2 fragments without bursting the cluster or losing the incorporated metal (Scheme 1.3) [65]. This is valid up to a certain limit, dictated by the ionic radius of the internal atom. For example, it was difficult to fragment past LaC_{44}^+ and impossible to go past LaC_{36}^+ without bursting the cluster [66].

$$M@C_x^+ \rightarrow [M@C_{x-m}] + C_m \; (m = 0\ 2,4,6\ ...)$$
$$M@C_x^+ \not\rightarrow M + C_x^+$$
$$M@C_x^+ \not\rightarrow M^+ + C_x$$

Scheme 1.3

To facilitate discussion of these somewhat more complicated fullerenes with one or more atoms inside the cage, a special symbolism and nomenclature was introduced [66]. Thereby the symbol @ is used to indicate the atoms in the interior of the fullerene. All atoms listed to the left of the @ symbol are located inside the cage and all atoms to the right are a part of the cage structure, which includes heterofullerenes, e.g. $C_{59}B$. A C_{60}-caged metal species is then written as $M@C_{60}$, expanded as "metal at C_{60}". The corresponding IUPAC nomenclature is different from the conventional $M@C_n$ representation. IUPAC recommend that $M@C_n$ be called [*n*]fullerene-*incar*-lanthanum and should be written *i*MC_n [4].

The production of endohedral fullerene complexes in visible amounts was first accomplished by a pulsed laser vaporization of a lanthanum oxide–graphite composite rod in a flow of argon gas at 1200 °C [66]. In this procedure, the newly formed endohedrals, together with empty fullerenes, sublime readily and are carried away in the flowing gas, depositing on the cool surfaces of the apparatus. This sublimate contains the complexes $La@C_{60}$, $La@C_{74}$ and $La@C_{82}$ (Figure 1.9). Among these, the endohedral molecule $La@C_{82}$ exhibits an extra stability. It can

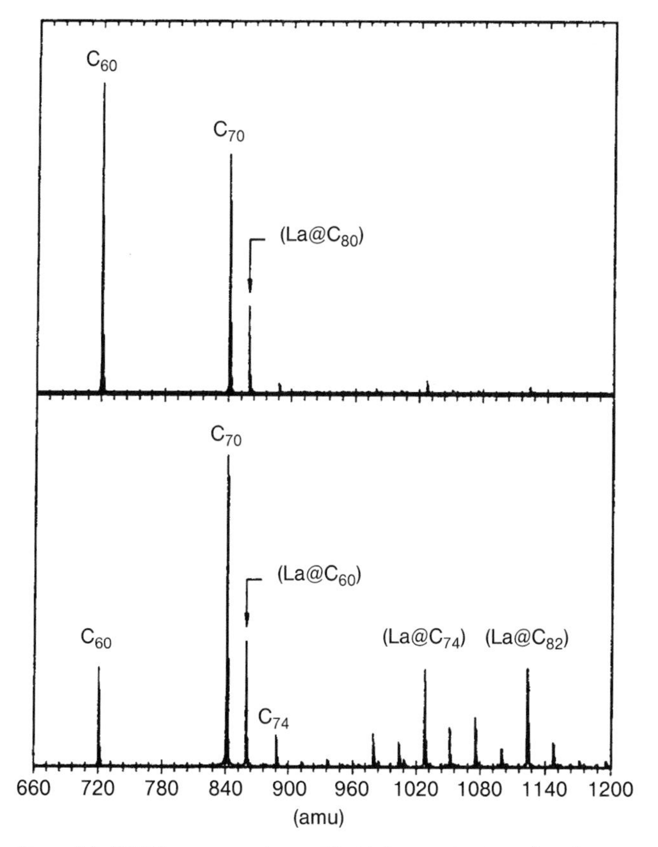

Figure 1.9 FT-ICR mass spectrum of hot toluene extract produced by laser vaporization of a lanthanum oxide–graphite composite rod [66].

be dissolved in toluene or carbon disulfide and exposed to air. The smaller lanthanum fullerenes are less stable and do not survive a hot extraction with toluene on air. The laser vaporization technique was also successful for the production of yttrium fullerenes such as $Y@C_{60}$, $Y@C_{70}$ and $Y_2@C_{82}$ in addition to $Y@C_{82}$ [67].

Along with laser vaporization methods, endohedral fullerene complexes can also be produced by arc vaporization of graphite impregnated with various metal oxides [66, 68–73] or rare earth metal carbides [74]. The yield of the lanthanofullerene $La@C_{82}$, for example, increases by a factor of 10 when, instead of metal-oxide-, metal-carbide-enriched composite graphite rods are used to generate soot [74]. More recent developments in the generation and characterization of endohedral metallofullerenes have been extensively reviewed by Shinohara [75, 76] and Nagase et al. [77]. In general, the endohedral complexes $M_n@C_{82}$ ($n = 1$–3) are the most abundant in the toluene or carbon disulfide extracts. However, metals, encapsulated in C_{60}, C_{74}, C_{76}, C_{80}, C_{84} and even in heterofullerenes such as $C_{79}N$ [78] have also been generated [75–77]. Endohedral complexes of inter alia La, Y, Sc, Ce, Pr, Nd, Sm, Eu, Gd, Tb, Dy, Ho, Er Tm, Yb and Lu, and also of Ca, Sr, Ba and Ti, can be prepared

[75–77]. The interior of C_{82}, approximately 8 Å in diameter, can readily accommodate three rare earth trivalent ions, which have average diameters of 2.06 to 2.34 Å [69]. The yield of dimetallo- or trimetallofullerenes depends on the metal-to-carbon ratio, whereas higher yields of multiple metal species are obtained by a higher metal content in the composite. Significantly, metallofullerenes encaged by C_{60} generally exhibit a very low abundance within most solvent-extracts, although empty C_{60} is the most abundant fullerene generated by the usual production methods. Possible reasons for this phenomenon are [72]: (1) $M@C_{60}$ fullerenes are unstable and destroyed upon exposure to air or moisture; (2) $M@C_{60}$ fullerenes are insoluble in organic solvents; and (3) the $M@C_{60}$ fullerenes are not preferably formed by this production method.

Interestingly, $M_n@C_{82}$ fullerenes are the most abundant, although empty C_{82} is not a dominant species formed under normal conditions. The fact that C_{80}, which in the empty case is an open shell system and therefore very unstable, also efficiently encapsulates rare earth metals, to form stable endohedrals $M_2@C_{80}$, suggests, that the electronic structure of the fullerene shell is dramatically influenced by the central metals. This, conversely, may be one reason for the possible instability of $M@C_{60}$. ESR studies on $M@C_{82}$ (M = Sc, Y, La) demonstrate that the metals are in the +3 oxidation state, [68, 73, 79] which leaves the fullerene in the trianionic state. This finding is corroborated by cyclic voltammetry studies on HPLC purified $La@C_{82}$ [80]. The cyclic voltammogram of dark green dichlorobenzene solutions of $La@C_{82}$ shows one reversible oxidation peak, the oxidation potential of which is approximately equal to that of ferrocene, implying that the complex is a moderate electron donor and therefore an oxygen-stable molecule. In addition, five reversible reductions are observed, showing that $La@C_{82}$ is a stronger electron acceptor than empty fullerenes, with the first reduction potential being especially low lying. These finding can be interpreted with the proposed molecular orbital diagram (Figure 1.10).

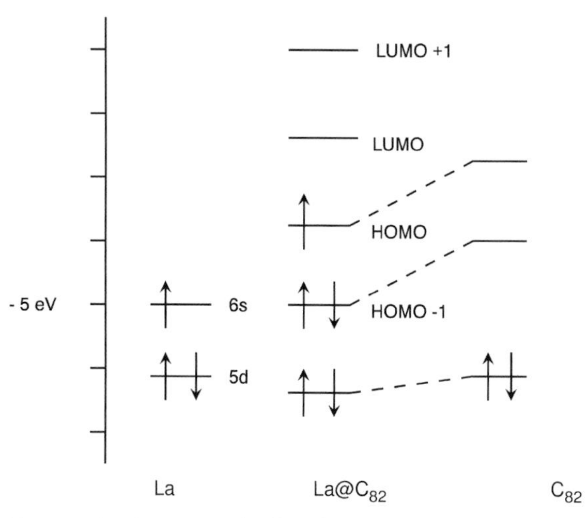

Figure 1.10 Schematic molecular orbital diagram of $La@C_{82}$ [80].

The removal of the radical electron corresponds to the first oxidation process. The resulting cation should be diamagnetic. The first reduction is relatively easy, because filling of the HOMO leads to the closed shell species $La@C_{82}^{-}$. Theoretical calculations predicted that the location of the lanthanum within the cage is off-center, which allows a stronger interaction with carbon atoms of the fullerene sphere [81–83].

Dynamic circular motion of the metal atoms within $La_2@C_{80}$ the cage has been investigated by NMR spectroscopy [77, 84, 85]. Synchrotron X-ray diffraction, ^{13}C NMR and ultra-high vacuum scanning tunneling microscopy (UHV-STM) studies also demonstrate the encapsulation of the metal atoms by the carbon cage [75, 76]. The metal atoms are not in the center of the fullerene cage but very close to the carbon cage, indicating a strong metal–cage interaction. Further electronic properties of endohedral metallofullerenes, such as redox behavior, have been investigated by cyclic voltammetry [75, 76, 86]. Various exohedral chemical functionalizations of endohedral metallofullerenes such as $La@C_{82}$, $Pr@C_{82}$ and $Gd@C_{82}$ have been carried out [87, 88]. The endohedral metallofullerene $Gd@C_{82}$ is considered to be a promising agent for magnetic resonance imaging (MRI) due to the high spin state of the encapsulated ion [89].

Major drawbacks of the above synthesis methods are their low yields, typically $< 0.5\%$, and the formation of multiple endohedral fullerene isomers. This makes it difficult to perform detailed studies of their properties and to use them for practical applications. In the regard the preparation of endohedral fullerenes such as $Sc_3N@C_{80}$ stands out [90]. The presence of small quantities of N_2 in a fullerene reactor allowed the synthesis of new endohedral fullerenes, such as the trigonal-planar Sc_3N unit encased in the high-symmetry, icosahedral C_{80}. The exact structure of $Sc_3N@C_{80}$ was determined by X-ray crystal structure analysis. ^{13}C NMR spectroscopic investigations revealed that, at room temperature, the Sc_3N unit is free to move in the C_{80} cage and so only two signals for the C-atoms in C_{80} are observed. This C_{80} isomer (Figure 1.11) is a third icosahedral fullerene, alongside C_{20} and C_{60}, and has an unstable and antiaromatic ground-state open-shell structure as an

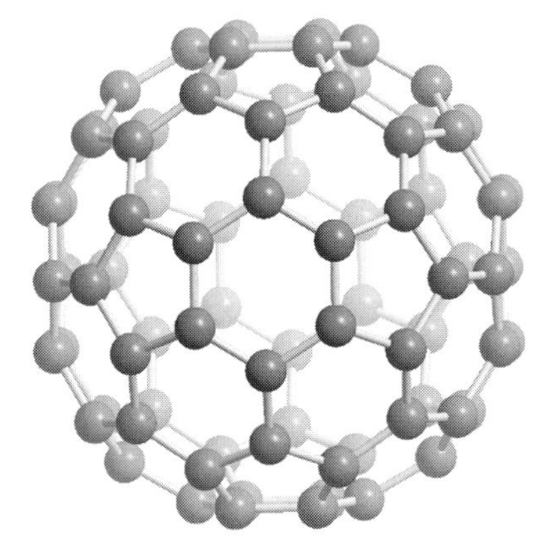

Figure 1.11 Icosahedral C_{80} cage of $Sc_3N@C_{80}$.

empty cage and has not yet been isolated. For $Sc_3N@C_{80}$, the cage is stabilized by six negative charges, which results in aromaticity [91]. Next to $Sc_3N@C_{80}$ a whole new family of stable endohedral fullerenes encapsulating trimetallic nitride clusters, $Er_xSc_{3-x}N@C_{80}$ ($x = 0–3$), has been generated [90, 92]. This 'trimetallic nitride template' process generates significant quantities of product, containing 3–5% $Sc_3N@C_{80}$. Also, the endohedral fullerene $Sc_3N@C_{78}$ was prepared and completely characterized [93]. Exohedral modification of $Sc_3N@C_{80}$ using Diels–Alder chemistry has been carried out [94]. Encapsulation has even been achieved of Sc_3N by the fullerene C_{68}, which does not obey the isolated pentagon rule [95], and by a C_{80} with D_{5h} symmetry [96].

Although electropositive metals afford endohedral complexes with fullerenes during their formation in macroscopic quantities, this is not the case with helium, the buffer gas used for the fullerene synthesis. Of 880 000 fullerene molecules generated by the arc-vaporization method only one was shown to contain a helium atom [97]. Conversely, mass spectrometry studies showed that an intact empty fullerene can be penetrated by the noble gas atoms He and Ne [98–104]. For this purpose singly and multiply charged C_x^{n+} fullerene molecules ($x = 60, 70$; $n = 1–3$) are shot through a stationary noble gas atmosphere in a molecular beam experiment. The resulting $M@C_{60}^+$ species show the "shrink-wrap" behavior with retention of M and successive elimination of C_2 (Scheme 1.3) that is typical for endohedral complexes. Upon reduction of $He@C_{60}^+$ the neutral complex $He@C_{60}$ can be obtained, which has a finite lifetime of > 90 μs [101]. This is further evidence that the noble gas is physically trapped inside the fullerene cage. The endohedral complex $He@C_{60}$ is the first noble gas–carbon compound.

To investigate the properties of endohedral noble gas fullerene compounds it is necessary to increase the fraction of the molecules occupied. This has been achieved by heating C_{60} or C_{70} in a noble gas atmosphere at high pressures (e.g. 2700 bar) and temperatures (e.g. 600 °C) [105, 106]. Next to helium, neon, argon, krypton and xenon have also been incorporated into fullerenes by this method. Mole fractions of $X@C_{60}$ and $X@C_{70}$ (X = Ne, Ar, Kr, Xe) in the range 0.04–0.3% have been estimated by mass spectrometry [106, 107]. In addition, beam implantation methods have been used for the generation of endohedral noble gas fullerenes [108]. Since the abundance of C_{60} and C_{70} with four ^{13}C atoms is high enough to interfere with the peaks of $He@C_{60}^+$ and $He@C_{70}^+$ it is difficult to measure the extent of helium incorporation by mass spectrometric methods. The use of ^3He, however, allows one to record ^3He NMR spectra since ^3He has spin = ½. The ^3He NMR spectra of $^3He@C_{60}$ and $^3He@C_{70}$ show that the incorporation fraction is about 0.1% [105, 106]. ^3He NMR spectroscopy can also be used to measure the shielding environment inside the fullerene cavity. The ^3He nuclei encapsulated in C_{60} and C_{70} are shielded by 6.3 and 28.8 ppm respectively relative to free ^3He [105, 106]. These shieldings indicate significant diamagnetic ring currents in C_{60} and very large ones in C_{70}. ^3He NMR spectroscopy of endohedral He complexes of a whole series of fullerenes [109–111], including exohedral fullerenes adducts [112–119], charged fullerenes [120, 121] heterofullerenes [122] and other cluster modified fullerenes [123], is a very powerful method for investigating the magnetic properties of carbon cages

[124]. It can be also very important for determining the number of isomers of higher fullerenes of regioisomers of exohedral fullerene adducts and for carrying out mechanistic investigations on the escape of the endohedral guest and reversible addition reactions [125]. The ^{129}Xe NMR spectrum of Xe@C_{60} has also been reported [126].

Since the penetration of a noble gas atom through a six-membered ring would afford a very high activation energy, the formation of X@C_{60} or X@C_{70} (X = noble gas) must be accompanied by the breaking of at least one bond of the fullerene core. A window mechanism has been proposed to explain the penetration of noble gases into the fullerene cages [127]. The energetics of opening both the [6,6]-bond and the [5,6]-bond have been calculated [127]. A comparison of the reaction coordinates for the processes of breaking a [6,6]-bond in the C_{60} singlet ground state and a [5,6]-bond in the triplet state reveals an energetic preference for the latter process. The opening of one [5,6]-bond leads to the formation of a nine-membered ring, which is expected to be large enough for atoms to pass in or out. After the thermal breaking of a bond the fullerene, either filled with a noble gas atom or empty, can reform by closing the opened bond or it can react further, via another irreversible pathway, to form degraded fullerenes. Degradation products have indeed been found by the synthesis of X@C_{60} or X@C_{70} [105, 106].

Another striking development with endohedral fullerenes was the synthesis of N@C_{60}, the first example of encapsulation of a reactive nonmetal atom [128–132]. It is particularly surprising that the enclosed nitrogen exists as a single atom and no bonding to the fullerene framework is apparent. The fullerene therefore represents a trap for extremely reactive atomic nitrogen. This occurs because curvature results in the inner surface of the fullerene cage being inert, whilst the outer surface is distinguished by high reactivity [133]. The formation of a covalent bond with the enclosed N-atom, which has a half-filled p-shell (three unpaired electrons) and so displays minimal electron affinity, would lead to a distinct rise in the strain energy of the total system. Thanks to the absence of relaxation mechanisms, the lowest ESR linewidths observed occur in the spectra of N@C_{60} [134]. The wavefunction of the enclosed N atom is influenced by subsequent exohedral adduct formation with one or more addends [134]. This is caused by the altered cage structure of the adducts. The analogous complexes N@C_{70} and P@C_{60} are made similarly to N@C_{60}, by bombarding a thin layer of fullerene on a cathode with energy-rich N- or P-ions, respectively [135, 136].

1.3.5
Total Synthesis Approaches

Fullerene generation by vaporization of graphite or by combustion of hydrocarbons is very effective and certainly unbeatable what facile production in large quantities is concerned. However, total synthesis approaches are attractive because (a) specific fullerenes could be made selectively and exclusively, (b) new endohedral fullerenes could be formed, (c) heterofullerenes and (d) other cluster modified fullerenes could be generated using related synthesis protocols.

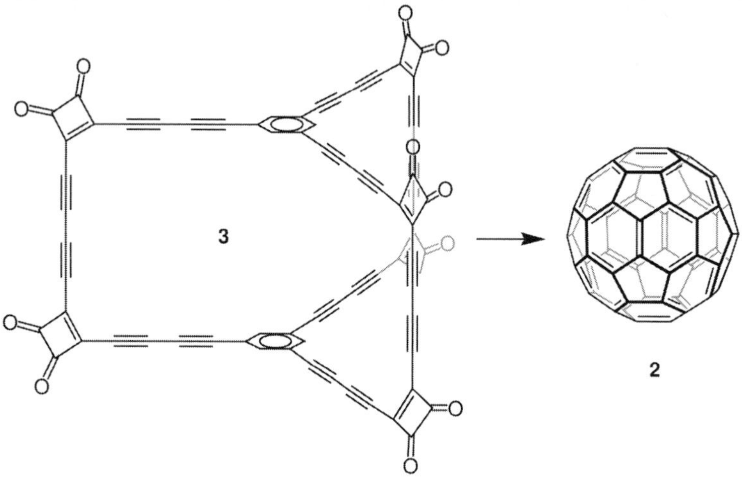

3

2

Scheme 1.4 Conversion of cyclophane **3** into C_{60} in the gas phase in laser desorption mass spectrometry.

Three approaches for rational syntheses of fullerenes have been developed. The first approach is the zipping up of fullerene precursors [137–144]. Carbon-rich acetylenic spherical macrocycles such as **3** (Scheme 1.4) may function as precursors of C_{60} and even its endohedral metal complexes in a process analogous to the coalescence annealing of mono- and polycycles with sp-hybridized C-atoms [139] during gas-phase formation of fullerenes from evaporated graphite [137]. Indeed **3** is effectively converted into $C_{60}H_6$ and C_{60} ions in the gas phase in laser desorption mass spectrometry experiments [138]. Related cyclophanes involving pyridine – instead of benzene moieties – are precursors for the conversion into the heterofullerene $C_{58}N_2$ in a mass spectrometer [143].

The second approach is based on the idea of synthesizing bowl-shaped hydrocarbons in which curved networks of trigonal C-atoms map out the same patterns of five- and six-membered rings as those found on the surfaces of C_{60} and/or the higher fullerenes [145–152]. An example for such an "open geodesic polyarene" is circumtrindene (**5**), generated by flash-vacuum pyrolysis (FVP) of trichlorodecacyclene **4** (Scheme 1.5) [152]. Circumtrindene represents 60% of the framework of C_{60}.

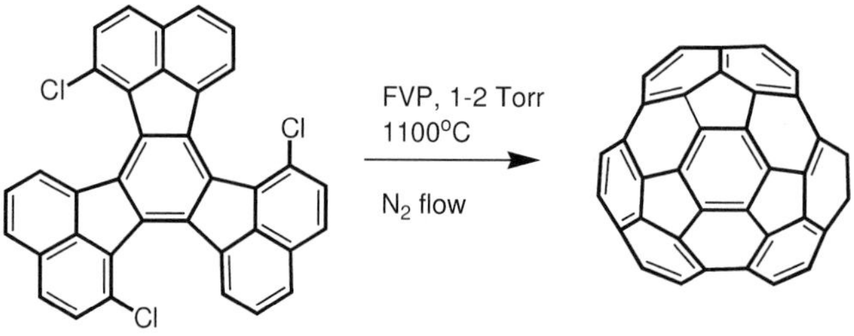

Cl
Cl
Cl
FVP, 1-2 Torr
1100°C
N_2 flow

4

5

Scheme 1.5 Synthesis of circumtrindene (**5**), representing 60% of C_{60}.

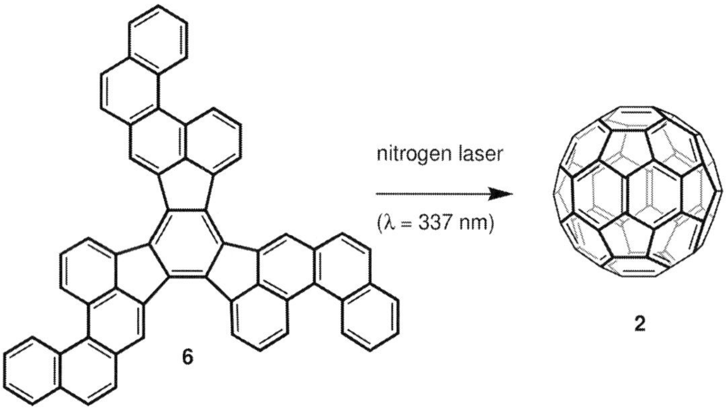

nitrogen laser
(λ = 337 nm)
6
2

Scheme 1.6 Generation of C_{60} by cyclodehydrogenation of polyarene **6**.

All 60 C-atoms of C_{60} are incorporated in the $C_{60}H_{30}$ polycyclic aromatic hydrocarbon (PAH) **6**, for which an efficient synthesis was developed [153]. Laser irradiation of **6** at 337 nm induces hydrogen loss and the formation of C_{60}, as detected by mass spectrometry (Scheme 1.6). Control experiments with ^{13}C-labeled material and with the $C_{48}H_{24}$ homologue of **6** verified that the C_{60} is formed by a molecular transformation directly from the $C_{60}H_{30}$ PAH and not by fragmentation and recombination in the gas phase.

In a third approach substituents were removed from a preformed dodecahedrane cage. In this way the synthesis of [20-I_h]fullerene was possible in the gas phase [154]. However, extensions of this approach to syntheses of fullerenes consisting of 60 or more C-atoms are conceptually very difficult.

1.3.6 Formation Process

Even though fullerenes are significantly destabilized with respect to graphite [155], they are readily formed out of a chaotic carbon plasma at about 3000 K. A thermodynamically controlled pathway leading to highly symmetric fullerenes can be ruled out. If the fullerene formation would be thermodynamically controlled, then, for example, the yield of C_{70}, which is more stable than C_{60}, [156, 157] should be much higher than found in extracts from graphitic soot. The toluene extract of soot, obtained by carbon vaporization, contains predominantly C_{60} followed by C_{70}. The ratio C_{60}:C_{70} is about 85 : 15 [25]. Therefore kinetic factors must govern the fullerene generation. [60-I_h]Fullerene is the most stable C_{60} isomer. Furthermore, the energy per carbon of [60-I_h]fullerene is much lower than any isomer of C_{58} and C_{62}. Locally, [60-I_h]fullerene is in a deep potential well and, if once formed from clustering carbon, it is chemically inert. This explains the much lower abundance of the higher fullerenes. One reason for the stability of C_{60} is that it obeys the "isolated pentagon rule" (IPR), which allows only fullerenes in which all pentagons are completely separated from each other by hexagons [158, 159]. [60-I_h]Fullerene is at the same time the smallest possible fullerene that obeys the IPR. The next most stable isomer of C_{60} has two pairs of adjacent pentagons and has been calculated to be 2 eV higher in energy [160]. These stability considerations already imply that if clusters smaller than C_{60} are formed they will undergo further reactions in the plasma and if those being a little bit larger than C_{60} are formed, for example C_{62}, they can be stabilized by loss of small fragments (C_2) and rearrangements, leading to the survivor [60-I_h]fullerene. Analogous processes should be valid for the higher fullerenes.

A mechanism for fullerene generation [38, 88, 161–167] by vaporization of graphite has to consider three major stages: (1) the vaporization itself and the nature of the initially formed intermediates; (2) the structure of the growing clusters and (3) the annealing to the fullerenes. The first step in the fullerene generation by evaporation of graphite is the formation of carbon atoms, as shown experimentally by $^{12}C/^{13}C$ isotope scrambling measurements [168–170]. The distribution of ^{13}C among the C_{60} molecules follows exactly Poisson statistics. The next step is the clustering of the carbon atoms. The smaller clusters are linear carbon chains or carbon rings. This was concluded by mass spectrometry studies of carbon clusters generated by laser desorption [171–174] and is supported by calculations [175, 176]. If reactive components are added to the buffer gas during the graphite vaporization such early intermediates can be quenched and the fullerene formation is suppressed (Scheme 1.7) [19, 177–179]. Laser desorption of graphite in the presence of H_2 allowed the mass spectrometric observation of polyynes [19]. Similar rod-shaped molecules, the dicyanopolyynes NC-$(C{\equiv}C)_n$-CN ($n = 3$–7), are obtained in high yields upon vaporization of graphite in an He/$(CN)_2$ atmosphere (Scheme 1.7) [178]. Obviously, in both cases irreversible additions of H atoms and CN-radicals to the ends of linear carbon chain intermediates are occurring. If Cl_2 is added to the

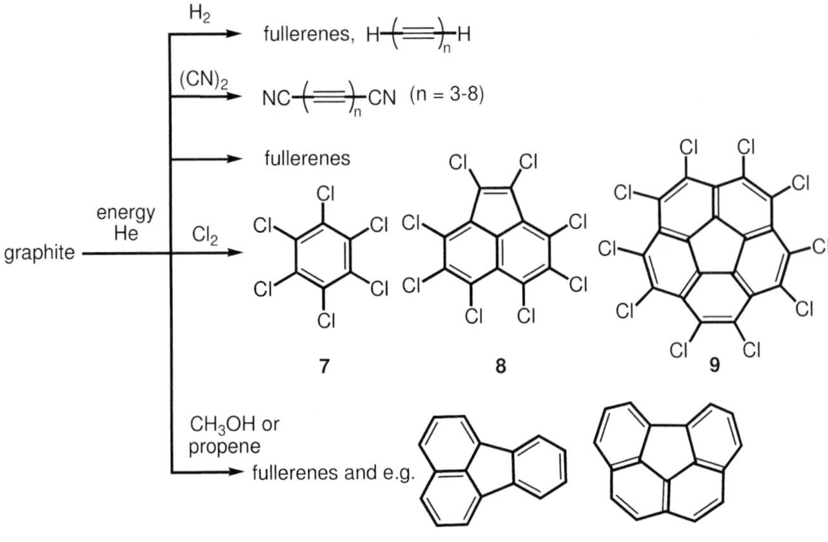

Scheme 1.7

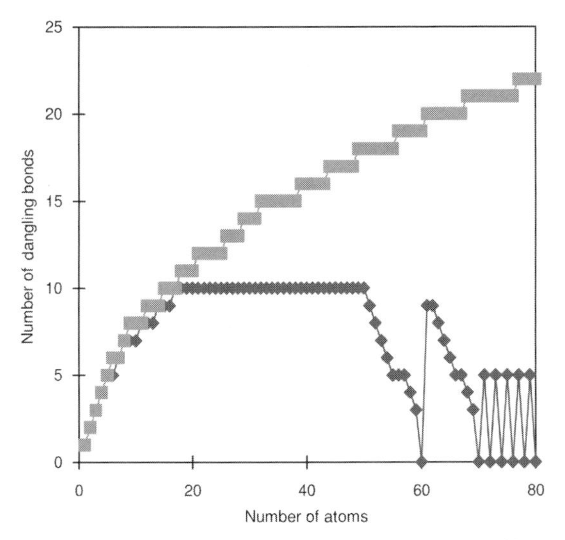

Figure 1.12 Number of dangling bonds in the best possible structure for a graphitic sheet with (*top*) all hexagons, compared to (*bottom*) those which obey the IPR, as a function of the number of atoms in the sheet [181].

buffer gas, instead of linear chains, perchlorinated cyclic compounds are found in the toluene extract of the soot (Scheme 1.7). The amount of fullerenes in the extract is < 5%. Each carbon framework of the cyclic compounds **7–9** represents a part of the C_{60}-structure. Thereby, it is remarkable that, already, structures containing pentagons, for example, the decachlorocorannulene **9** are formed. Polyaromatic hydrocarbons (PAHs) such as **10** and **11** can be isolated, if the vaporization of graphite is carried out in an atmosphere of He/propene or He/methanol (Scheme 1.7). Systems with a similar carbon framework are suggested to be present as intermediates during the formation of fullerenes in benzene flames [51]. However, it can not be concluded from the quenching experiments alone whether clusters with structures like **7–11** are indeed predominant intermediates in the fullerene formation, since the presence of reactive species influences the growth process. The mechanisms of fullerene formation by graphite evaporation and by combustion or pyrolysis of naphthalene are thought to be different.

Graphitic sheets as intermediates for cluster sizes of 30 or more C-atoms are proposed in the "pentagon road" model [162, 180, 181]. Thereby it is emphasized that the growth process of the graphitic sheet follows a low energy path by minimizing the number of dangling bonds. This can be achieved by the introduction of pentagons accompanied by a curling process (Figure 1.12). A complete closure to a fullerene reduces the number of dangling bonds to zero. This model also takes into account that high energy adjacent pentagon structures are avoided and also explains (1) the need for an elevated temperature for effective fullerene formation, because prior to a further growth an annealing process leads to an isolated pentagon network and (2) the role of the buffer gas as well as the pressure dependence on the fullerene yield, because the helium concentration controls the diffusion of the C_n species from the hot into cooler zones of the plasma. Higher buffer gas pressures lead to an increase in concentration of reactive C_n, which causes the annealing process to be not competitive with the cluster growth. During annealing, the formation of a low energy structure obeying the IPR could be accomplished inter alia by the Stone–Wales rearrangement (Scheme 1.8) [182]. This is a concerted process, involving a Hückel four-center transition state.

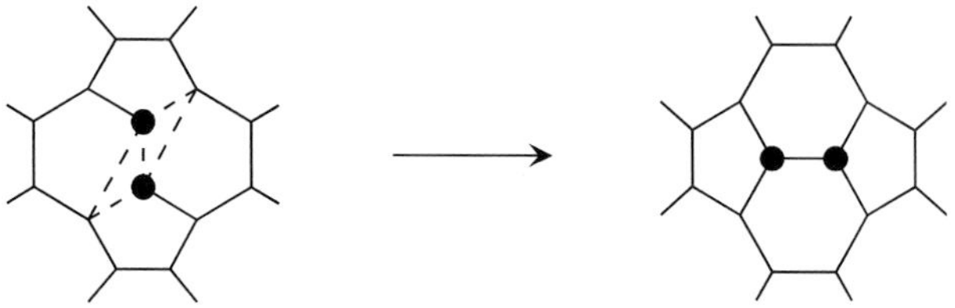

Scheme 1.8

Graphitic sheets, however, are not detectable in the carbon-plasma by ion chromatography (IC) [162, 174, 183–185]. This method provides a means for separating carbon cluster ions with different structures because the reciprocal of the ion mobility is proportional to the collision cross-section. Several species C_n^+ with different structures coexist and their relative amount depends on the cluster size. Small clusters C_n^+ ($n < 7$) are linear. In the range $n = 7$–10 chains as well as monocycles coexist. The clusters C_{11}^+–C_{20}^+ are exclusively monocycles and the range between C_{21}^+ and C_{28}^+ is characterized by planar mono- and bicyclic systems. The first three-dimensional structure is detected at C_{29}^+ and the first fullerenes appear at C_{30}^+ and dominate from C_{50}^+. Therefore, the growth pattern for carbon in the plasma starting from atoms is linear → monocyclic rings → polycyclic rings and finally → fullerenes.

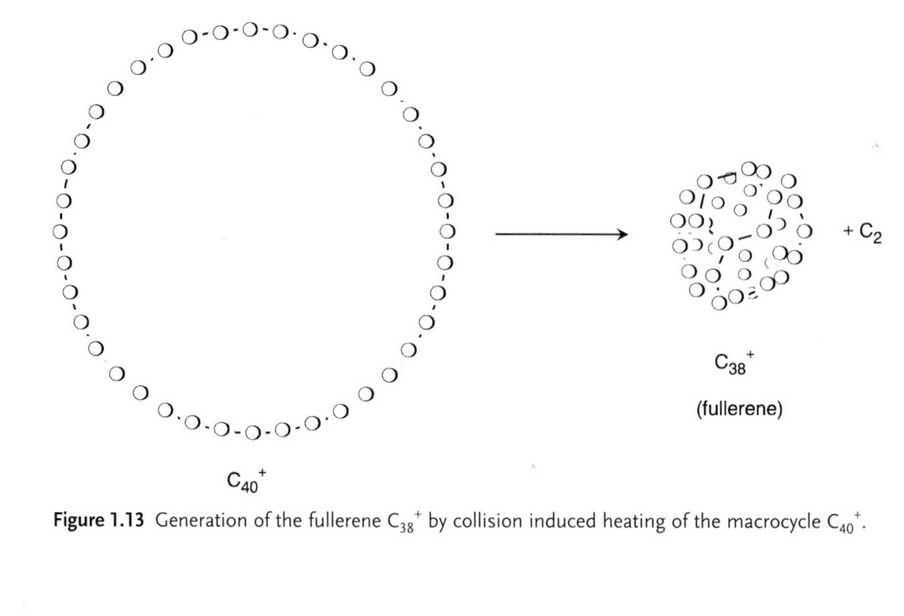

Figure 1.13 Generation of the fullerene C_{38}^+ by collision induced heating of the macrocycle C_{40}^+.

If such mixtures of ions, generated by laser desorption of graphite, are annealed by collisions, predominantly monocyclic rings are formed upon isomerization of the initial structures. At very high collision energies, isomerization to fullerenes occurs. The fullerenes C_n^+ so-obtained carry enormous internal energy, which can be dissipated by the cleavage of C_1- or C_3- (for odd n) or C_2-fragments (for even n) (Figure 1.13). From drift experiments, the more rings the initial structure contains the smaller the ring barrier is to rearrangement to a fullerene [162]. Tricyclic or tetracyclic isomers rearrange readily to form fullerenes, while bicyclic and monocyclic rings are significantly slower.

Another important series of experiments has shown, by ion-cyclotron-resonance (ICR) studies, that monocyclic carbon rings can coalesce very efficiently to fullerenes (Scheme 1.9) [139]. The carbon rings are obtained by laser desorption of carbon oxide $C_n(CO)_{n/3}$ precursors, out of which, upon loss of CO, the cyclo[n]carbons C_{18}, C_{24} and C_{30} are generated. Fullerene formation proceeds via collisions of positively charged cluster ions C_n^+ with the corresponding neutrals. In the positive ion mode, reactions between cations and neutral molecules of the cyclo[n]carbons lead to fullerene ions. Remarkably, the cyclocarbons C_{18} and C_{24} predominantly lead, via

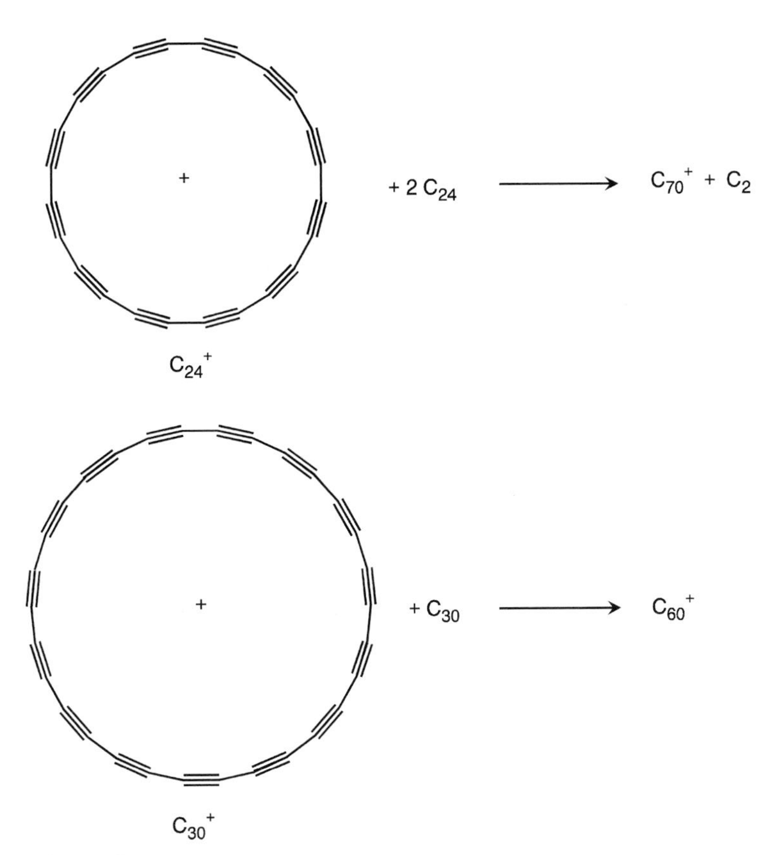

Scheme 1.9

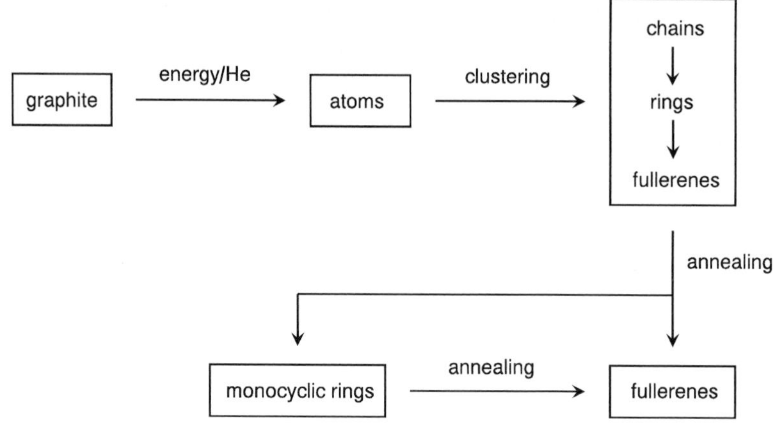

Scheme 1.10

C_{72}^+, to C_{70}^+ and no signal for C_{60}^+ is observed. Conversely, C_{30} precursors coalesce almost exclusively to C_{60}. Subsequent losses of C_2 fragments demonstrate the fullerene nature of the products. The distribution of fullerenes can therefore be directly affected by the size of the precursors [163, 186].

These investigations impressively demonstrate that fullerenes can form through excitation or coalescence of relatively large monocyclic precursors. A growth mechanism of smaller fullerenes by subsequent incorporation of C_2- or C_3-units is not necessarily required. In the excitation experiments of C_n^+ clusters as well as the coalescence studies of cyclocarbons, a scenario is investigated that comes close to the situation in a carbon plasma generated by Krätschmer–Huffman graphite evaporation. Based on the experimental data available, fullerene generation is schematically summarized in Scheme 1.10. Upon vaporization of graphite carbon atoms are formed, which subsequently cluster to a mixture of chains, monocyclic rings, polycyclic rings and fullerenes. In an annealing process these clusters rearrange, especially efficiently via monocyclic rings, to fullerenes. During this annealing, stable fullerenes can be formed by loss of small fragments (e.g. C_2, C_3) and by ring rearrangements (e.g. Stone–Wales rearrangement). However, the availability of fullerenes by several methods implies the existence of different mechanisms, which may also operate simultaneously.

1.4 Separation and Purification

The raw product obtained by the evaporation of graphite is soot and slag. Next to soluble fullerenes the soot and slag contain other kinds of closed carbon structures, e.g. giant fullerenes [187] and nanotubes [188, 189]; the rest is amorphous carbon. Fullerenes can be isolated from the soot either by sublimation or by extraction. The first isolation of fullerenes was achieved by a simple sublimation with a Bunsen burner as heat source [190]. In a more controlled procedure, sublimation from soot was achieved by gradient sublimation [191]. Thereby, the fullerenes are deposited on a quartz rod along a temperature gradient. With this method a partial separation or enrichment of the different fullerenes is possible. In addition, endohedrals can be enriched and C_{74}, not obtainable by extraction, sublimes out of the soot. However, these methods put thermal stress on the fullerenes, leading to partial decomposition. This has been overcome, at least partly, by using microwave-aided sublimation [192].

The most common method for isolating the fullerenes from soot is by extraction with organic solvents [20, 25, 187, 193, 194]. In general, toluene is used since it provides a sufficient solubility and is less toxic than benzene or carbon disulfide. Either a hot extraction of the soot followed by filtration or a Soxhlet extraction is possible. Longer extraction times lead to higher yields of fullerenes [187]. Alternatively, hexane or heptane can also be used as solvents for this method. Fullerene yields of up to 26% from the soot have been reported for Soxhlet extractions [187]. The toluene extract is typically red to red brown. It contains, along with C_{60} and C_{70}, higher fullerenes such as C_{76}, C_{78}, C_{84}, C_{90} and C_{96}; [60-I_h]fullerene is the most abundant species. Upon further extractions of soot with other solvents, for example pyridine or 1,2,3,5-tetramethylbenzene, additional soluble material is accessible. In such fractions clusters with very high masses (giant fullerenes) up to C_{466} are detectable by mass spectrometry [187]. However, under these more drastic extraction conditions, chemical reactions between fullerenes and the solvent have been observed [187]. The preferred method of isolating fullerenes from the soot is Soxhlet extraction with toluene.

To separate fullerenes predominantly chromatographic methods are used. A clean separation of C_{60} as the first fullerene fraction is obtained by column chromatography with alumina as stationary phase with either pure hexane or hexane–toluene (95 : 5) as eluent [25, 195, 196]. However, since fullerenes are not very soluble in hexane, enormous amounts of alumina and solvents are needed, rendering this method inefficient. Also, a one-step separation of the higher fullerenes is impossible, because of the pronounced tailing effect of the preceding fullerenes. Solvents with a higher solubility, for example toluene, cause the fullerenes to elute together without retention on alumina. The efficiency of the separation of fullerenes on alumina with hexane as mobile phase has been dramatically improved by using Soxhlet-chromatography (Figure 1.14) [197, 198].

This method combines distillation or Soxhlet extraction with chromatography and therefore does not require either large quantities of solvent or constant monitoring. Gram quantities of pure C_{60} can be obtained in reasonable periods of time. After C_{60} is eluted, a new hexane-containing flask is used to collect eluting C_{70}.

With chromatography on graphite, higher amounts of toluene in the hexane–toluene mixture can be used to improve solubility [199, 200]. The major breakthrough, however, allowing the use of pure toluene as mobile phase was achieved by chromatography on mixtures of charcoal and silica gel [201, 202]. This is the most inexpensive and efficient method for a fast separation of C_{60}. A flash

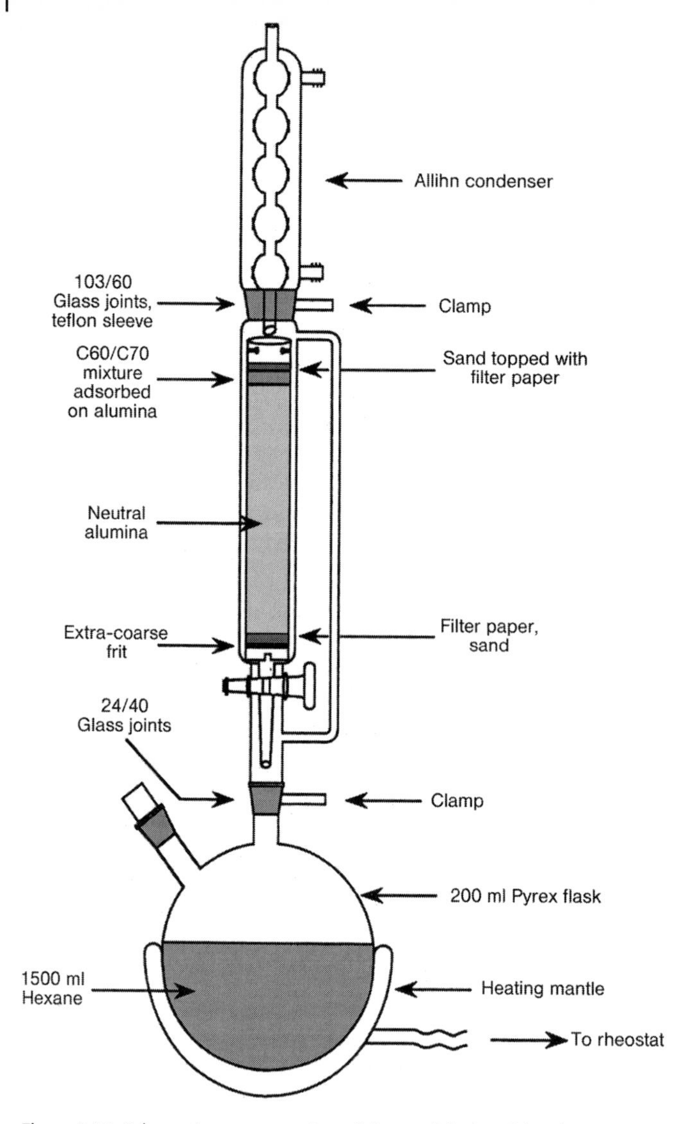

Figure 1.14 Schematic representation of the modified Soxhlet chromatography apparatus [198].

chromatography – or even simpler a plug filtration – setup can be used. For example [202], a fritted funnel (10 cm diameter) is covered with a slurry of 63 g charcoal and 125 g silica gel, which gives a plug of 5.5 cm height. A concentrated toluene solution of fullerene extract is loaded and eluted by the application of a slight vacuum at the filter flask. After a few minutes a dark purple solution of 1.5 g of C_{60} can be obtained from the initial 2.5 g extract.

In a related and even more efficient method the purification was accomplished by filtration of a fullerene extract through a thin layer of activated carbon [203].

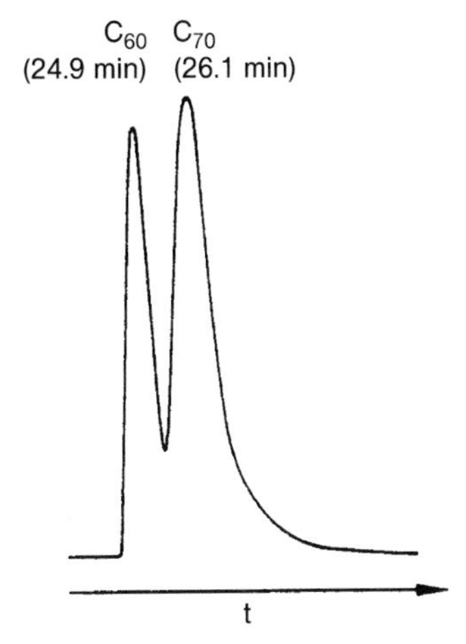

Figure 1.15 Separation of C_{60} and C_{70} on polystyrene gel column with toluene as mobile phase [207].

Under the optimum conditions, chlorobenzene was used as eluent to diminish the volume of the solvent. The carbon layer was only 3–17 mm thick, enough to obtain pure C_{60} and C_{70} in good yields.

Although Soxhlet chromatography on alumina and chromatography on charcoal are very simple and efficient methods, a certain amount of the fullerenes decomposes or irreversibly adsorbs on these stationary phases. Especially, higher fullerenes hardly elute from charcoal columns. Upon chromatography on alumina, light-catalyzed degradation of the fullerenes was observed [204]. Polystyrene gel, however, is completely unreactive towards fullerenes, and a good separation of fullerenes can be obtained on this stationary phase with toluene as mobile phase (Figure 1.15) [205–207].

A completely automated system with reinjection/sample collection and solvent recovery allows the separation of up to 10 g of extract a day. The C_{60} is obtained in very high purity and the recovery is nearly 100%. A simple benchtop method for the enrichment of preparative amounts of C_{60}, C_{70} and higher fullerenes (up to C_{100}) from a crude fullerene mixture is based on a single elution through a column of poly(dibromostyrene)–divinylbenzene using chlorobenzene as mobile phase [208].

Various other packing materials have been used for the separation of fullerenes both by analytical and preparative HPLC, including C_{18} reversed phase [209–211], Pirkle-type phases [212–214], γ-cyclodextrin phases [215], and reversed phases containing polycyclic aromatics [216], cyclopentadiene moieties [217] or multi-legged phenyl groups [218]. The eluent most commonly used is hexane, either pure or in combination with more polar components such as CH_2Cl_2, Et_2O, THF or toluene. For HPLC on C_{18}-reversed phase silica, toluene/MeCN 75 : 25 or $CHCl_3$/MeCN 60 : 40, for example, are suitable mobile phases. Although the solubility of parent fullerenes is very low in these combinations of solvents, such HPLC methods are

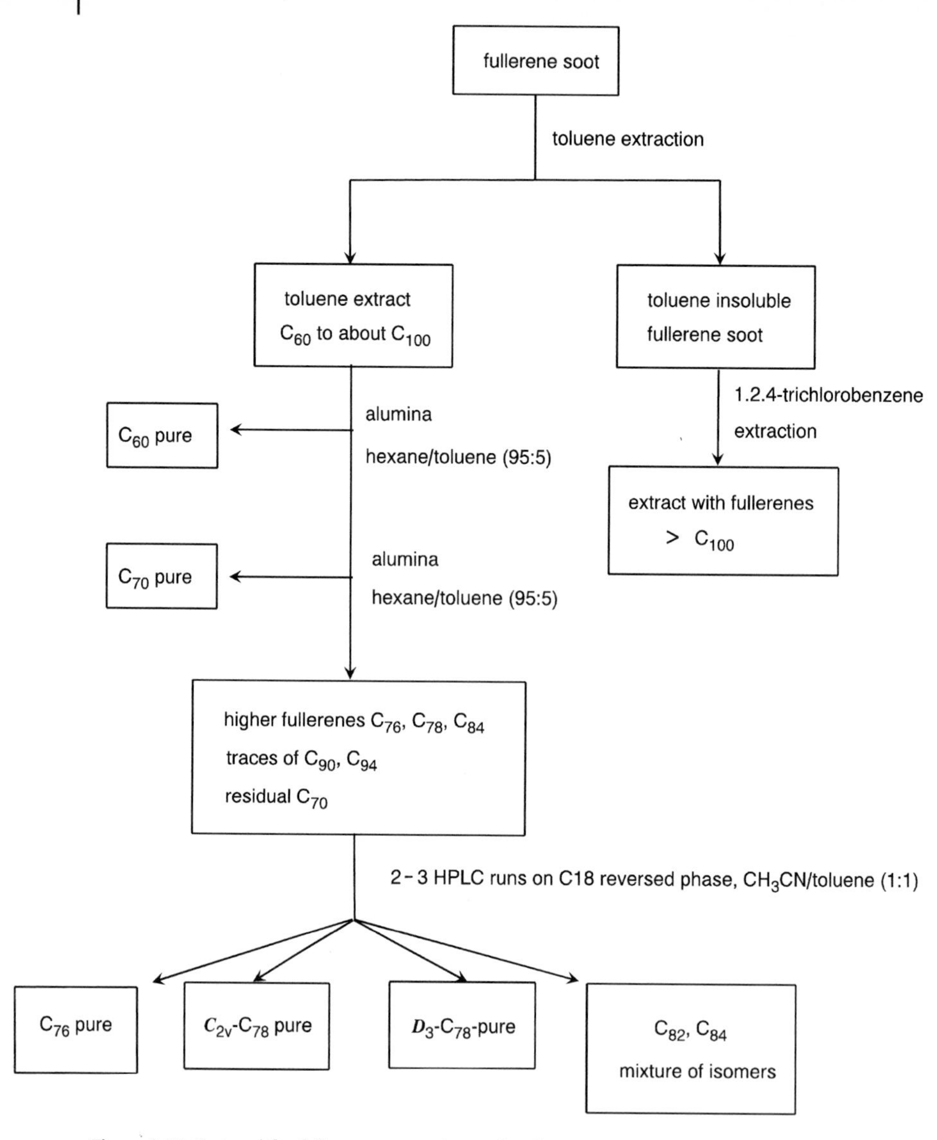

Figure 1.16 Protocol for fullerene separation and isolation [219].

useful for analytical purposes or for the separation of higher fullerenes [210, 211]. The currently most efficient columns for the HPLC separation of parent fullerenes, including endohedral fullerenes and heterofullerenes, are the commercially available COSMOSIL columns (Nacalai Tesque, Japan), which allow the use of toluene as mobile phase. A complete protocol for the isolation of fullerenes from soot and their separation by column chromatography and HPLC elaborated by Diederich is represented in Figure 1.16 [219].

To separate fullerene derivatives containing covalently bound groups, HPLC methods are also very important. Addends on the fullerene core have a dramatic influence on the solubility properties and the retention behavior. Often, more polar eluents in mixtures or in a pure form can be used and efficient separations on silica gel or several reversed phases (medium polarity), even of different regioisomers of addition products, are possible [220–228]. A separation of C_{60} from C_{70} has also been achieved on the basis of the small difference in their chemical reactivity [229]. The treatment of a carbon disulfide solution of a C_{60}/C_{70} mixture with $AlCl_3$ leads to a preferred precipitation of a $C_{70}[AlCl_3]_n$ complex, whereas the purple C_{60} is mostly unaffected and its solutions can be separated by filtration from the precipitate. The parent fullerenes can be recovered from the C_{70} enriched precipitate by treatment with water.

Purification of C_{60} by fractional crystallization in 1,3-diphenylacetone represents another inexpensive method [230]. After three steps, 99.5% pure C_{60} was obtained with a total yield of 69%. The purity of C_{60} can be increased up to 99.99% by adsorption of the residual C_{70} fullerene on charcoal. Preferential precipitation of C_{70} over C_{60} has been obtained by the addition of *p*-trihalohomocalix[3]arenes [231]. In this way C_{70} with up to 92% purity was obtained.

1.5 Properties

1.5.1 Structures

Each fullerene contains $2(10 + M)$ carbon atoms corresponding to exactly 12 pentagons and M hexagons. This building principle is a simple consequence of Euler's theorem. Hence, the smallest fullerene that can be imagined is C_{20}. Starting at C_{20} any even-membered carbon cluster, except C_{22}, can form at least one fullerene structure. With increasing M the number of possible fullerene isomers rises dramatically, from only 1 for $M = 0$ to over 20 000 for $M = 29$ and so on [232, 233]. The soccer-ball shaped C_{60} isomer [60-I_h]fullerene is the smallest stable fullerene. The structure of [60-I_h]fullerene was determined theoretically [234–238] and experimentally (Table 1.1) [239–242]. These investigations confirm the icosahedral structure of [60-I_h]fullerene. Two features of this C_{60} structure are of special significance: (1) all twelve pentagons are isolated by hexagons and (2) the bonds at the junctions of two hexagons ([6,6] bonds) are shorter than the bonds at the junctions of a hexagon and a pentagon ([5,6] bonds) (Table 1.1 and Figure 1.17).

The pentagons within fullerenes are needed to introduce curvature, since a network consisting of hexagons only is planar. [60-I_h]fullerene is the only C_{60} isomer and at the same time the smallest possible fullerene to obey the "isolated pentagon rule" (IPR) [158, 159]. The IPR predicts fullerene structures with all the pentagons isolated by hexagons to be stabilized against structures with adjacent pentagons. A destabilization caused by adjacent pentagons is due to (1) pentalene-type

Table 1.1 Calculated and measured bond distances in [60-I_h]fullerene in Å.

Method	*[5,6]-bonds*	*[6,6]-bonds*	*Reference*
HF(STO-3G)	1.465	1.376	234
HF(7*s*3*p*/4*s*2*p*)	1.453	1.369	235
LDF(11*s*6*p*)	1.43	1.39	236
HF	1.448	1.37	237
MP2	1.446	1.406	238
NMR	1.448	1.370	239
Neutron diffraction	1.444	1.391	240
Electron diffraction	1.458	1.401	241
X-ray	1.467	1.355	242

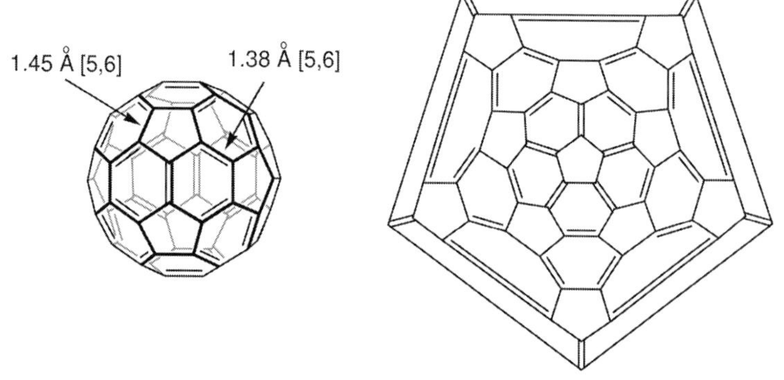

Figure 1.17 Schematic representation of [60-I_h]fullerene with the lengths of the two different bonds in the molecule and Schlegel diagram of the lowest energy Kekulé structure.

8π-electron systems, leading to resonance destabilization and (2) an increase of strain energy, as a consequence of the enforced bond angles. The formation of IPR structures is accompanied with an enhancement of the sphericity of the molecules. A spherical shape distributes the strain as evenly as possible and minimizes the anisotropic contribution to the strain energy [243]. The bond length alternation in [60-I_h]fullerene shows that, in the lowest energy Kekulé structure, the double bonds are located at the junctions of the hexagons ([6,6] double bonds) and there are no double bonds in the pentagonal rings. Each hexagon in [60-I_h]fullerene exhibits cyclohexatriene character and each pentagon [5]radialene character. The diameter of [60-I_h]fullerene has been determined by NMR measurements to be 7.10 ± 0.07 Å [239, 244]. When taking into account of the size of the π-electron cloud associated with the C-atoms, the outer diameter of the C_{60} molecule can be estimated as 7.10 + 3.35 = 10.34 Å, where 3.35 Å is an estimate of the thickness of the π-electron cloud surrounding the C-atoms on the C_{60} framework [244]. The volume per C_{60} molecule is estimated to be $1.87 \cdot 10^{-22}$ cm^3 [244].

The next higher stable and IPR satisfying fullerene is [70-D_{5h}]fullerene [25]. Its structure was calculated [245] and determined by X-ray crystallography (Figure

Table 1.2 Calculated and experimental structure of [70-D_{5h}]fullerene.

Bond type	Bond length (Å) Experiment		Bond length (Å) Theory (dzp/SCF) [245]
	a)	b)	
1	1.462	1.462	1.475
2	1.423	1.414	1.407
3	1.441	1.426	1.415
4	1.432	1.447	1.457
5	1.372	1.378	1.361
6	1.453	1.445	1.446
7	1.381	1.387	1.375
8	1.463	1.543	1.451

a) Data taken from the crystal structure of the complex $(\eta^2\text{-}C_{70})Ir(CO)Cl(PPh_3)_2$ [246].
b) Data taken from the crystal structure of $C_{70} \cdot 6\,(S_8)$ [247].

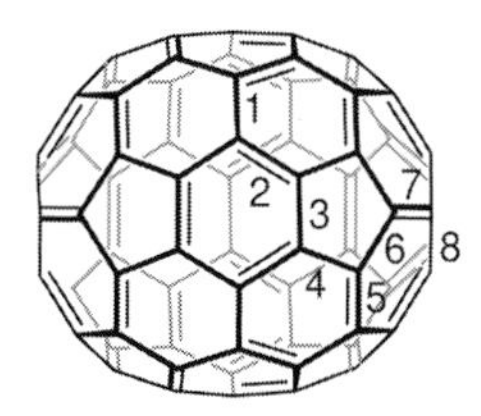

Figure 1.18 Bond type definitions for the [70-D_{5h}]fullerene.

1.18, Table 1.2) [246, 247]. The geometry at the poles (highest curvature) of [70-D_{5h}]fullerene is very similar to that of [60-I_h]fullerene. The corannulene subunits (bonds 5,6,7,8) have the same type of bond length alternation. In contrast to [60-I_h]fullerene, this fullerene has an equatorial belt consisting of fused hexagons. The analysis of bond lengths within the equatorial phenylene belt of [70-D_{5h}]fullerene as well as its D_{5h}-symmetry per se makes clear that two equivalent lowest-energy Kekulé structures per equatorial hexagon are required to describe its structure and reactivity properties (see Chapter 14).

The number of possible fullerene isomers without any constraints reduces considerably if only IPR structures are allowed. Taking into account that open-shell structures are avoided [233, 248, 249] and that the number of double bonds in pentagons is minimized [250], which favors a meta over a para relationship of the pentagons (Figure 1.19) [251], the number of allowed isomers is further reduced.

With these constraints, magic numbers n for stable fullerenes C_n can be predicted, which are $n = 60$, 70, 72, 76, 78, 84 [233]. With the exception of C_{72} at least one representative for all of these fullerenes, including C_{74} [252] and C_{80} [253], has now been produced, isolated and characterized. The number of allowed fullerene isomers are, for example, one for C_{60}, one for C_{70}, one for C_{76} (D_2), five for C_{78} [D_{3h} (1), D_{3h} (2), D_3, C_{2v} (1), C_{2v} (2)], 24 for C_{84} and 46 for C_{90} [233, 254]. The interplay between theoretical predictions and ^{13}C NMR experiments was the key to structure elucida-

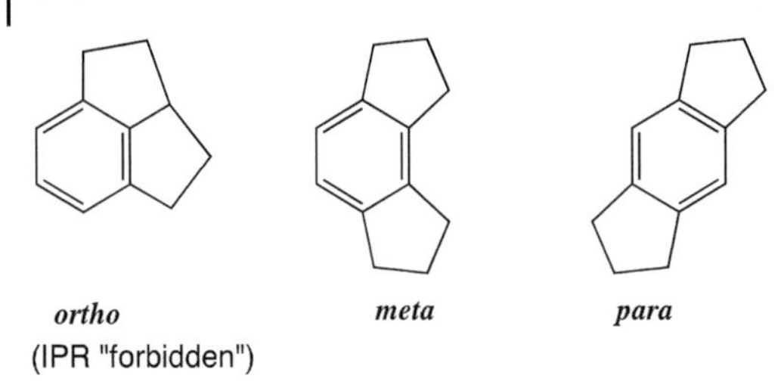

Figure 1.19 Ortho, meta, and para relationships of fused pentagons in a fullerene.

tion of some principle isomers of the higher fullerenes C_{76}, and C_{78}, C_{80}, C_{82} and C_{84} [210, 211, 219, 253, 255–262]. Structure assignments have been accomplished by comparing the ^{13}C NMR spectra (number and intensity of lines) of given isomers, isolated or enriched by HPLC, with the theoretical possible structures suggested by Manolopoulos and Fowler [233]. In addition, the number of lines in different chemical shift regions provide further useful information for the structure determination. In some cases, even structure determination by X-ray crystallography was possible. Certain to confident structure assignment [263] was possible for [74-D_{3h}]-fullerene [264], [76-D_2]fullerene, [210] [78-$C_{2v(1)}$]fullerene [211, 256, 258], [78-$C_{2v(2)}$]fullerene [256, 258], [78-D_3]fullerene [211, 256, 258], [80-D_2]fullerene [253], [82-C_2]fullerene, [84-D_{2d}]fullerene, [260, 261] and [84-D_2]fullerene [256, 258, 261, 262]. A few examples are shown in Figure 1.20. Since only one isomer of each C_{60}, C_{70} and C_{76} is formed, in the following the indication of symmetry will be omitted.

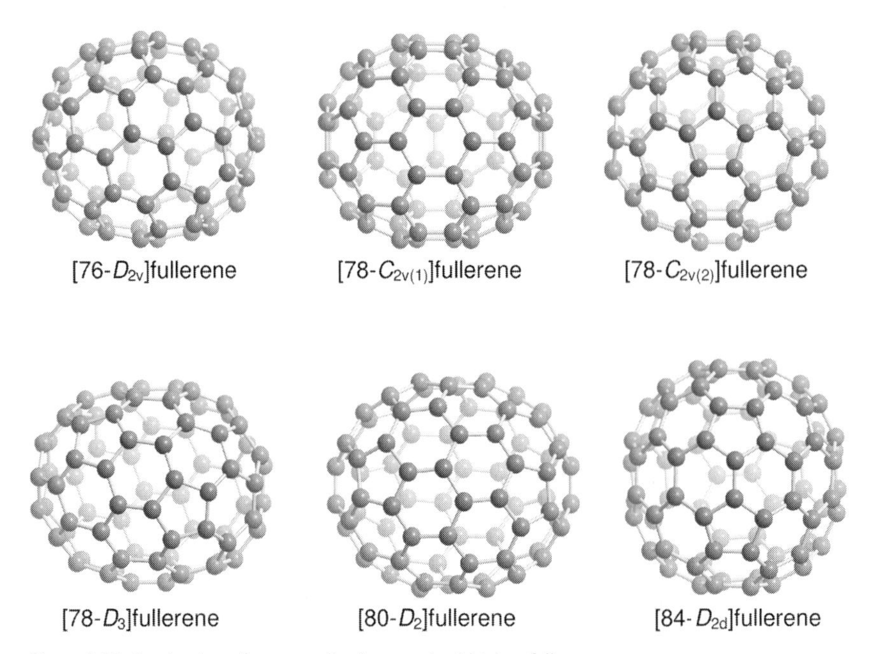

Figure 1.20 A selection of structurally characterized higher fullerenes.

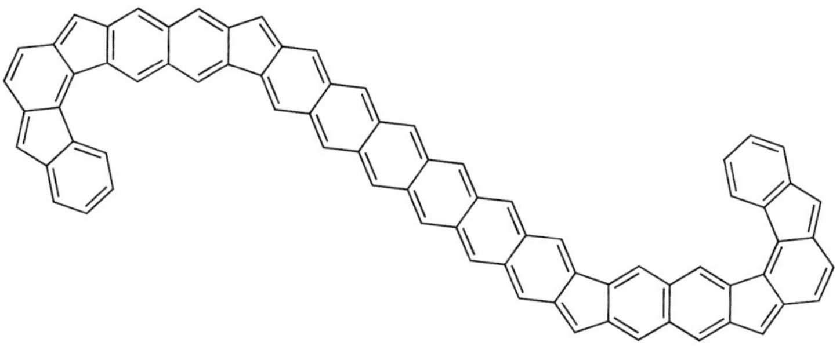

Figure 1.21 Helix motif in double-helical [76-D_2]fullerene.

The fact that for C_{78} the most symmetrical D_{3h} isomer is not favored implies that an important stabilizing factor is the maximization of the sphericity rather than symmetry. Interestingly, some of the higher fullerenes are chiral, for example [76-D_2]fullerene, [80-D_2]fullerene, [82-C_2]fullerene, and [84-D_2]fullerene. The chirality of [76-D_2]fullerene is based on the helical arrangement of the sp^2-atoms in space (Figure 1.21). To describe the absolute configuration of a chiral fullerene the stereodescriptors ($^{f,s}C$) and ($^{f,s}A$) introduced by Diederich et al. (see also Section 1.1) can be used [265]. Optically pure isomers of [76-D_2]fullerene were obtained by functionalization of (±)-[76-D_2]fullerene with an enantiopure chiral addend, separation of the resulting diastereoisomers, and subsequent removal of the functionality [266]. In the same way optical resolution of [84-D_2]fullerene was possible [267]. Comparison of the obtained circular dichroism (CD) spectra with calculated spectra allowed the assignment of the absolute configuration of the enantiomers of [76-D_2]fullerene [263, 268].

1.5.2 Physical and Spectroscopic Properties

Investigations of physical properties in solution as well as in the solid state have been predominantly carried out on C_{60}, and to a minor extent on C_{70}, since these are the most abundant fullerenes. Currently, little material is available for the higher fullerenes or endohedrals, because it takes about 250 h to produce 1 mg of any of these [269].

The heat of formation of C_{60} and C_{70} have been determined theoretically and experimentally by calorimetry to be 10.16 kcal mol^{-1} per C-atom for C_{60} and 9.65 kcal mol^{-1} per C-atom for C_{70} [127, 155]. The fullerenes are therefore thermodynamically less stable than graphite and diamond, which exhibit heats of formation of zero and 0.4 kcal mol^{-1} respectively. It is expected that upon increasing the size of the fullerene the energy content of the spheres asymptotically reaches that of graphite [270, 271]. The binding energy per C-atom of C_{60} was calculated to be about 0.4–0.7 eV smaller than that for graphite [244]. The binding energy of C_{70} is about 0.2 eV greater than that of C_{60} [244]. Further physical constants of C_{60} and C_{70} are summarized in Tables 1.3 and 1.4. Measurements of the electron affinity

Table 1.3 Physical constants for C_{60} molecules [244].

Quantity	*Value*	*Reference*
Moment of inertia I	$1.0 \cdot 10^{-43}$ kg m^2	272
Volume per C_{60}	$1.87 \cdot 10^{-22}$ cm^{-3}	–
Number of distinct C sites	1	–
Number of distinct C–C bonds	2	–
Binding energy per atom	7.40 eV	272
Heat of formation (per g C atom)	1.16 kcal	155
Electron affinity	2.65 ± 0.05 eV	273
Cohesive energy per C atom	1.4 eV atom^{-1}	274
Spin–orbit splitting of C($2p$)	0.00022 eV	275
First ionization potential	7.58 eV	276
Second ionization potential	11.5 eV	272
Optical absorption edge	1.65 eV	272

Table 1.4 Physical constants for C_{70} molecules [244].

Quantity	*Value*	*Reference*
Average C–C distance (Å)	1.43	155, 277
C_{70} c-axis diameter (Å)	7.96	277
C_{70} a-b-axis diameter (Å)	7.12	277
Moment of inertia $I_{\parallel}$ (kg m^2)	$1.24 \cdot 10^{-43}$	272
Moment of inertia $I_{\perp}$ (kg m^2)	$1.24 \cdot 10^{-43}$	72
Volume per C_{70} (cm^{-3})	$1.56 \cdot 10^{-22}$	–
Number of distinct C sites	5	278
Number of distinct C–C bonds	8	278
Binding energy per C atom (eV)	7.42	278
Heat of formation (kcal g^{-1} °C^{-1})	9.65	157, 279
Electron affinity (eV)	2.72	280
Ionization potential (1st) (eV)	7.61	281
Optical absorption edge (eV)	1.7	278, 282–284
Atomic zero point motion (Å)	0.07	285

E_A have been reported also for the higher fullerenes C_{76} (2.88 ± 0.05 eV), C_{78} (3.01 ± 0.07 eV) and C_{84} (3.05 ± 0.08 eV), showing the general increase in E_A with increasing cluster size [244].

To chemically modify fullerenes, in most cases it is necessary that they are in the solution. For extractions or chromatographic separations the solubility also plays a crucial role. The solubility of C_{60} in various organic solvents has been investigated systematically (Table 1.5) [286–289].

In polar and H-bonding solvents such as acetone, tetrahydrofuran or methanol C_{60} is essentially insoluble. It is sparingly soluble in alkanes, with the solubility increasing with the number of atoms. In aromatic solvents and in carbon disulfide, in general appreciable solubilities are observed. A significant increase of the solubility takes place on going from benzenes to naphthalenes. Although there are trends for the solution behavior of C_{60}, there is no direct dependence of the solubility on a certain solvent parameter like the index of refraction n. When the solubility is

Table 1.5 Solubility of C_{60} in various solvents [287].

Solvent	***[C_{60}] (mg mL^{-1})***	***Mole fraction (· 10^4)***	***n***
n-Pentane	0.005	0.008	1.36
n-Hexane	0.043	0.073	1.38
Cyclohexane	0.036	0.059	1.43
n-Decane	0.071	0.19	1.41
Decalines	4.6	9.8	1.48
Dichloromethane	0.26	0.27	1.42
Carbon disulfide	7.9	6.6	1.63
Dichloromethane	0.26	0.27	1.42
Chloroform	0.16	0.22	1.45
Tetrachloromethane	0.32	0.40	1.46
Tetrahydrofuran	0.000	0.000	1.41
Benzene	1.7	2.1	1.50
Toluene	2.8	4.0	1.50
Tetraline	16	31	1.54
Benzonitrile	0.41	0.71	1.53
Anisole	5.6	8.4	1.52
Chlorobenzene	7.0	9.9	1.52
1,2-Dichlorobenzene	27	53	1.55
1-Methylnaphthalene	33	68	1.62
1-Chloronaphthalene	51	97	1.63
Acetone	0.001	0.001	1.36
Methanol	0.000	0.000	1.33

expressed in mole fraction units it is evident that C_{60} is not very soluble even in the best solvents listed in Table 1.5. The solubility of C_{70} follows qualitatively similar trends.

The electronic absorption spectra [25] of C_{60} and C_{70} are characterized by several stronger absorptions between 190 and 410 nm as well as by some forbidden transitions in the visible part of the spectrum (Figures 1.22 and 1.23). For C_{60}, the assignment of the transitions has been carried out using the results of theoretical calculations [290, 291]. The absorptions between 190 and 410 nm are due to allowed $^1T_{1u}$–^1A_g-transitions, whereas those between 410 and 620 nm are due to orbital forbidden singlet–singlet transitions. These latter absorptions in the visible are responsible for the purple color of C_{60} and the red color of C_{70}.

The fullerenes, in particular C_{60}, exhibit a variety of remarkable photophysical properties, making them very attractive building blocks for the construction of photosynthetic antenna and reaction center models (Table 1.6) [292–295].

The singlet excited state of C_{60} decays very efficiently via intersystem crossing to the energetically lower-lying triplet excited state (Scheme 1.11). The triplet quantum yields are very high (Table 1.6). The triplet excited states are susceptible to various deactivation processes, including ground state quenching, triplet–triplet annihilation, quenching by molecular oxygen leading to $^1O_2$ and electron transfer to donor molecules [295]. The reorganization energy of the rigid fullerene core in electron transfer reaction is exceptionally low. The covalent binding of fullerene to electroactive donor molecules such as porphyrins allows for the development of molecular

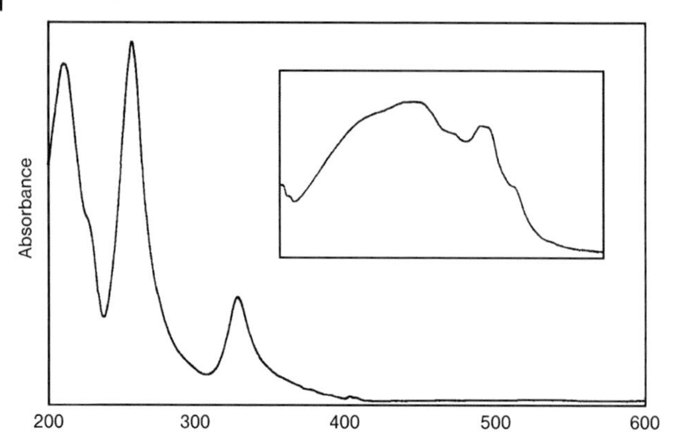

Figure 1.22 Electronic absorption spectrum of C_{60} in hexane. Inset: 420–470 nm region.

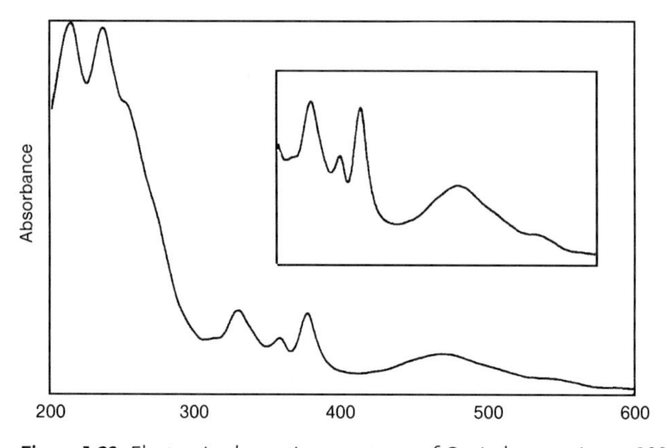

Figure 1.23 Electronic absorption spectrum of C_{70} in hexane. Inset: 300–600 nm region.

Table 1.6 Photophysical properties of C_{60} and C_{70} [295].

Property	C_{60}	C_{70}
E (singlet)	1.99 eV	1.90 eV
E (triplet)	1.57 eV	1.60 eV
λ_{max} (singlet)	920 nm	660 nm
λ_{max} (triplet)	747 nm	400 nm
ε (triplet)	20 000 M^{-1} cm^{-1}	14 000 M^{-1} cm^{-1}
τ (singlet)	1.3 ns	0.7 ns
τ (triplet)	135 μs	11.8 ms
$E_{1/2}$ ($^{1*}C_{60}/C_{60}^{\bullet-}$)	1.44 V vs SCE	1.39 vs SCE
$E_{1/2}$ ($^{3*}C_{60}/C_{60}^{\bullet-}$)	1.01 vs SCE	1.09 vs SCE
Φ (fluorescence)	$1.0 \cdot 10^{-4}$	$3.7 \cdot 10^{-4}$
Φ (triplet)	0.96	0.9
k_q (oxygen)	$1.6 \cdot 10^9$ M^{-1} s^{-1}	$1.9 \cdot 10^9$ M^{-1} s^{-1}
k_q (DABCO)	$2.5 \cdot 10^9$ M^{-1} s^{-1}	
k_{ET} (biphenyl)	$1.7 \cdot 10^{10}$ M^{-1} s^{-1}	$2.0 \cdot 10^{10}$ M^{-1} s^{-1}

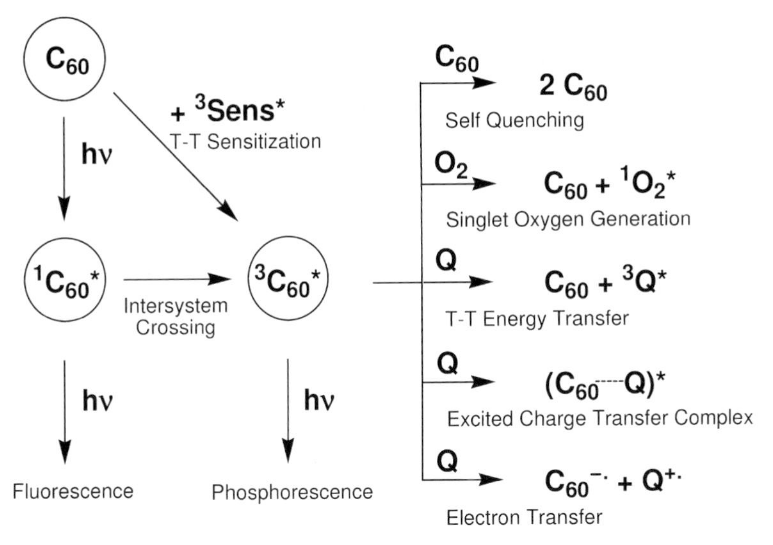

Scheme 1.11

dyads in which photoinduced energy- and/or electron transfer takes place. In many of such architectures C_{60} accelerates photoinduced charge separation and retards charge recombination in the dark [292–295]. As a consequence, long-lived charge separated states can be generated, which can be used for subsequent energy conversion processes.

As expected for the icosahedral symmetry, the ^{13}C NMR spectrum of C_{60} (Figure 1.24) shows one signal, at $\delta = 143.2$ [25]. Since the amount of fullerene in the NMR samples due to low solubility is small and the spin–lattice relaxation

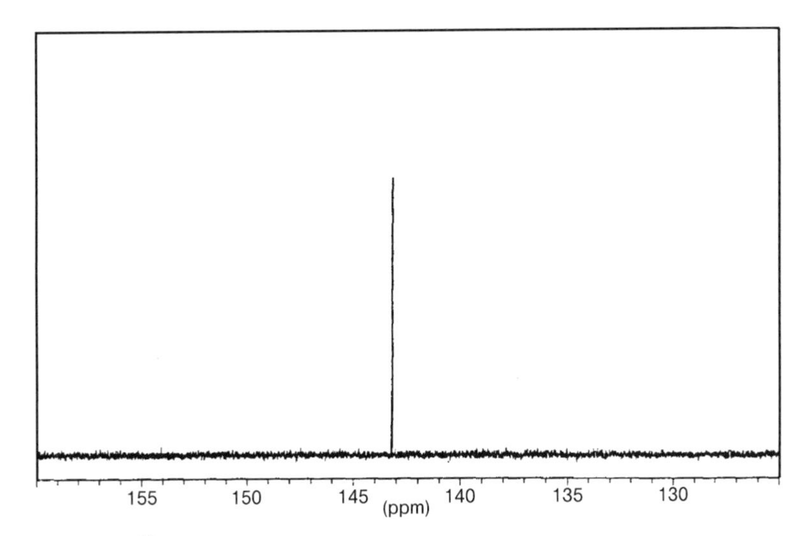

Figure 1.24 ^{13}C NMR spectrum of C_{60}.

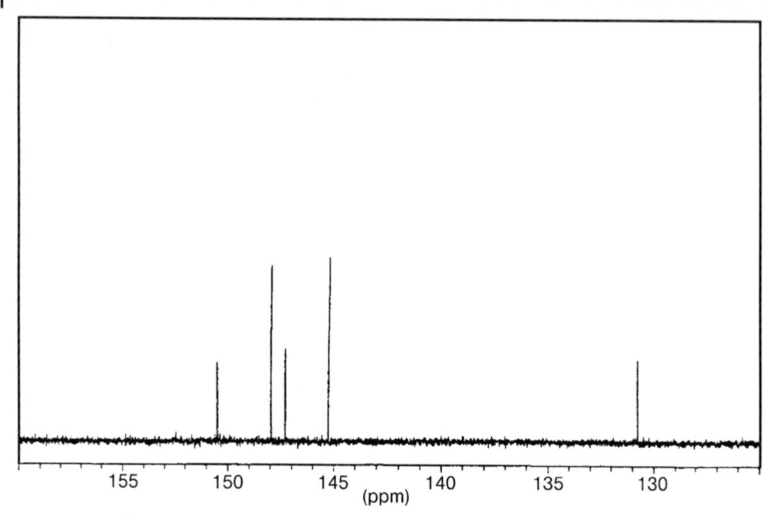

Figure 1.25 ^{13}C NMR spectrum of C_{70}.

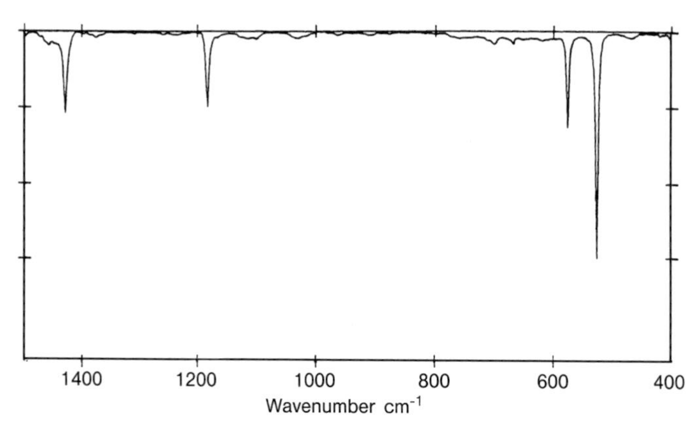

Figure 1.26 IR spectrum (KBr) of C_{60}.

times are quite long (≥ 20 s for C_{60}) [25] a comparatively large number of accumulations is necessary. The ^{13}C NMR spectrum of C_{70} (Figure 1.25) shows five peaks [25] in the intensity ratio of 1 : 2 : 1 : 2 : 1, which have been attributed to the five different sets of C-atoms by 2D NMR analysis [296]. The center of gravity of the ^{13}C resonances shifts continuously towards higher field on going from C_{70} to the higher fullerenes C_{76}, C_{78}, C_{82} and C_{84} [219]. This can be interpreted in terms of an increase of the more benzoid pyrene-type carbons with increasing size of the carbon spheres.

Out of the 174 vibrational modes [23] for C_{60} giving rise to 42 fundamentals of various symmetries, four have t_{1u} symmetry and are IR active (Figure 1.26). This fact has historical significance in the first detection of C_{60} [20].

Solid C_{60} forms a face-centered-cubic (FCC) structure at room temperature [244, 297, 298]. The density in the solid state d_{ffc} is 1.72 g cm^{-1} [299]. Four equivalent molecules are contained in a unit cube with edge length a = 14.17 Å, at the origin

122 A. Weitz, K. Holczer, C. Bellavia-Lund, F. Wudl, M. Saunders, *Proc. – Electrochem. Soc.* **1998**, *98-8*, 1039.

123 Y. Rubin, T. Jarrosson, G.-W. Wang, M. D. Bartberger, K. N. Houk, G. Schick, M. Saunders, R. J. Cross, *Angew. Chem.* **2001**, *113*, 1591; *Angew. Chem. Int. Ed. Engl.* **2001**, *40*, 1543.

124 M. Bühl, A. Hirsch, *Chem. Rev.* **2001**, *101*, 1153.

125 G. W. Wang, M. Saunders, R. J. Cross, *J. Am. Chem. Soc.* **2001**, *123*, 256.

126 M. S. Syamala, R. J. Cross, M. Saunders, *J. Am. Chem. Soc.* **2002**, *124*, 6216.

127 R. L. Murry, G. E. Scuseria, *Science* **1994**, *263*, 791.

128 T. A. Murphy, T. Pawlik, A. Weidinger, M. Höhne, R. Alcala, J. M. Späth, *Phys. Rev. Lett.* **1996**, *77*, 1075.

129 B. Pietzak, M. Waiblinger, T. Almeida Murphy, A. Weidinger, M. Höhne, E. Dietel, A. Hirsch, *Chem. Phys. Lett.* **1997**, *279*, 259.

130 B. Pietzak, M. Waiblinger, T. A. Murphy, A. Weidinger, M. Höhne, E. Dietel, A. Hirsch, *Carbon* **1998**, *36*, 613.

131 T. Suetsuna, N. Dragoe, W. Harneit, A. Weidinger, H. Shimotani, S. Ito, H. Takagi, K. Kitazawa, *Chem. Eur. J.* **2002**, *8*, 5079.

132 A. Hirsch, *Angew. Chem.* **2001**, *113*, 1235; *Angew. Chem. Int. Ed. Engl.* **2001**, *40*, 1195.

133 H. Mauser, N. J. R. van Eikema Hommes, T. Clark, A. Hirsch, B. Pietzak, A. Weidinger, L. Dunsch, *Angew. Chem.* **1997**, *109*, 2858; *Angew. Chem. Int. Ed. Engl.* **1997**, *36*, 2835.

134 E. Dietel, A. Hirsch, B. Pietzak, M. Waiblinger, K. Lips, A. Weidinger, A. Gruss, K.-P. Dinse, *J. Am. Chem. Soc.* **1999**, *121*, 2432.

135 B. Pietzak, K. Lips, M. Waiblinger, A. Weidinger, E. Dietel, A. Hirsch, *Proc. – Electrochem. Soc.* **2000**, *2000-12*, 353.

136 M. Waiblinger, K. Lips, W. Harneit, A. Weidinger, E. Dietel, A. Hirsch, *Phys. Rev. B: Condens. Matter Mater. Phys.* **2001**, *63*, 045421/1.

137 Y. Rubin, *Chem. Eur. J.* **1997**, *3*, 1009.

138 Y. Rubin, T. C. Parker, S. J. Pastor, S. Jalisatgi, C. Boulle, C. L. Wilkins, *Angew. Chem.* **1998**, *110*, 1353; *Angew. Chem. Int. Ed. Engl.* **1998**, *37*, 1226.

139 S. W. McElvany, M. M. Ross, N. S. Goroff, F. Diederich, *Science* **1993**, *259*, 1594.

140 Y. Tobe, N. Nakagawa, K. Naemura, T. Wakabayashi, T. Shida, Y. Achiba, *J. Am. Chem. Soc.* **1998**, *120*, 4544.

141 Y. Tobe, N. Nakagawa, J. y. Kishi, M. Sonoda, K. Naemura, T. Wakabayashi, T. Shida, Y. Achiba, *Tetrahedron* **2001**, *57*, 3629.

142 Y. Tobe, H. Nakanishi, N. Nakagawa, R. Furukawa, *Proc. – Electrochem. Soc.* **1999**, *99-12*, 146.

143 Y. Tobe, H. Nakanishi, M. Sonoda, T. Wakabayashi, Y. Achiba, *Chem. Commun.* **1999**, 1625.

144 U. H. F. Bunz, Y. Rubin, Y. Tobe, *Chem. Soc. Rev.* **1999**, *28*, 107.

145 L. T. Scott, M. M. Hashemi, D. T. Meyer, H. B. Warren, *J. Am. Chem. Soc.* **1991**, *113*, 7082.

146 L. T. Scott, M. M. Hashemi, M. S. Bratcher, *J. Am. Chem. Soc.* **1992**, *114*, 1920.

147 L. T. Scott, *Pure Appl. Chem.* **1996**, *68*, 291.

148 P. W. Rabideau, A. Sygula, *Acc. Chem. Res.* **1996**, *29*, 235.

149 A. Borchardt, A. Fuchicello, K. V. Kilway, K. K. Baldridge, J. S. Siegel, *J. Am. Chem. Soc.* **1992**, *114*, 1921.

150 S. Hagen, M. S. Bratcher, M. S. Erickson, G. Zimmermann, L. T. Scott, *Angew. Chem.* **1997**, *109*, 407; *Angew. Chem. Int. Ed. Engl.* **1997**, *36*, 406.

151 L. T. Scott, H. E. Bronstein, D. V. Preda, R. B. M. Ansems, M. S. Bratcher, S. Hagen, *Pure Appl. Chem.* **1999**, *71*, 209.

152 R. B. M. Ansems, L. T. Scott, *J. Am. Chem. Soc.* **2000**, *122*, 2719.

153 M. M. Boorum, Y. V. Vasil'ev, T. Drewello, L. T. Scott, *Science* **2001**, *294*, 828.

154 H. Prinzbach, A. Weller, P. Landenberger, F. Wahl, J. Worth, L. T. Scott, M. Gelmont, D. Olevano, B. v. Issendorff, *Nature* **2000**, *407*, 60.

155 H. D. Beckhaus, C. Rüchardt, M. Kao, F. Diederich, C. S. Foote, *Angew. Chem.* **1992**, *104*, 69; *Angew. Chem. Int. Ed. Engl.* **1992**, *31*, 63.

156 R. F. Curl, *Phil. Trans. Royal Soc. London, Ser. A: Math., Phys. Eng. Sci.* **1993**, *343*, 19.

157 H. D. Beckhaus, S. Verevkin, C. Rüchardt, F. Diederich, C. Thilgen, H. U. ter Meer, H. Mohn, W. Müller, *Angew. Chem.* **1994**, *106*, 1033; *Angew. Chem. Int. Ed. Engl.* **1994**, *33*, 996.
158 T. G. Schmalz, W. A. Seitz, D. J. Klein, G. E. Hite, *Chem. Phys. Lett.* **1986**, *130*, 203.
159 H. W. Kroto, *Nature* **1987**, *329*, 529.
160 K. Raghavachari, C. M. Rohlfing, *J. Phys. Chem.* **1992**, *96*, 2463.
161 H. Schwarz, *Angew. Chem.* **1993**, *105*, 1475; *Angew. Chem. Int. Ed. Engl.* **1993**, *32*, 1412.
162 N. S. Goroff, *Acc. Chem. Res.* **1996**, *29*, 77.
163 J. L. Martins, F. A. Reuse, *Condensed Matter Theor.* **1998**, *13*, 355.
164 G. N. Churilov, P. V. Novikov, V. E. Tarabanko, V. A. Lopatin, N. G. Vnukova, N. V. Bulina, *Carbon* **2002**, *40*, 891.
165 V. A. Schweigert, A. L. Alexandrov, Y. N. Morokov, V. M. Bedanov, *Chem. Phys. Lett.* **1995**, *235*, 221.
166 T. Belz, H. Werner, F. Zemlin, U. Klengler, M. Wesemann, B. Tesche, E. Zeitler, A. Reller, R. Schloegl, *Angew. Chem.* **1994**, *106*, 1919; *Angew. Chem. Int. Ed. Engl.* **1994**, *33*, 1866.
167 P. T. A. Reilly, R. A. Gieray, W. B. Whitten, J. M. Ramsey, *J. Am. Chem. Soc.* **2000**, *122*, 11596.
168 G. Meijer, D. S. Bethune, *J. Chem. Phys.* **1990**, *93*, 7800.
169 J. M. Hawkins, A. Meyer, S. Loren, R. Nunlist, *J. Am. Chem. Soc.* **1991**, *113*, 9394.
170 T. W. Ebbesen, J. Tabuchi, K. Tanigaki, *Chem. Phys. Lett.* **1992**, *191*, 336.
171 D. C. Parent, S. W. McElvany, *J. Am. Chem. Soc.* **1989**, *111*, 2393.
172 S. W. McElvany, B. I. Dunlap, A. O'Keefe, *J. Chem. Phys.* **1987**, *86*, 715.
173 G. von Helden, M. T. Hsu, N. G. Gotts, P. R. Kemper, M. T. Bowers, *Chem. Phys. Lett.* **1993**, *204*, 15.
174 G. von Helden, M. T. Hsu, N. Gotts, M. T. Bowers, *J. Phys. Chem.* **1993**, *97*, 8182.
175 V. Parasuk, J. Almlof, *Chem. Phys. Lett.* **1991**, *184*, 187.
176 M. Feyereisen, M. Gutowski, J. Simons, J. Almlof, *J. Chem. Phys.* **1992**, *96*, 2926.
177 M. Broyer, A. Goeres, M. Pellarin, E. Sedlmayr, J. L. Vialle, L. Woeste, *Chem. Phys. Lett.* **1992**, *198*, 128.
178 T. Grösser, A. Hirsch, *Angew. Chem.* **1993**, *105*, 1390; *Angew. Chem. Int. Ed. Engl.* **1993**, *32*, 1340.
179 T. M. Chang, A. Naim, S. N. Ahmed, G. Goodloe, P. B. Shevlin, *J. Am. Chem. Soc.* **1992**, *114*, 7603.
180 G. A. Heath, S. C. O'Brien, R. F. Curl, H. W. Kroto, R. E. Smalley, *Comm. Cond. Mater. Phys.* **1987**, *13*, 119.
181 R. E. Smalley, *Acc. Chem. Res.* **1992**, *25*, 98.
182 A. J. Stone, D. J. Wales, *Chem. Phys. Lett.* **1986**, *128*, 501.
183 G. von Helden, N. G. Gotts, M. T. Bowers, *Nature* **1993**, *363*, 60.
184 M. T. Bowers, P. R. Kemper, G. von Helden, P. A. M. van Koppen, *Science* **1993**, *260*, 1446.
185 G. von Helden, N. G. Gotts, M. T. Bowers, *J. Am. Chem. Soc.* **1993**, *115*, 4363.
186 D. Babic, N. Trinajstic, *J. Mol. Struct.* **1996**, *376*, 507.
187 D. H. Parker, K. Chatterjee, P. Wurz, K. R. Lykke, M. J. Pellin, L. M. Stock, *Carbon* **1992**, *30*, 1167.
188 S. Iijima, *Nature* **1991**, *354*, 56.
189 T. W. Ebbesen, P. M. Ajayan, *Nature* **1992**, *358*, 220.
190 K. Fostiropoulos, Dissertation, University of Heidelberg, **1991**.
191 C. Yeretzian, J. B. Wiley, K. Holczer, T. Su, S. Nguyen, R. B. Kaner, R. L. Whetten, *J. Phys. Chem.* **1993**, *97*, 10097.
192 W. Müller, U. Wirth, H. Mohn, K. Alberti, *Fullerenes '93* **1993**.
193 J. Theobald, M. Perrut, J.-V. Weber, E. Millon, J.-F. Muller, *Separation Sci. Technol.* **1995**, *30*, 2783.
194 T. Jovanovic, D. Koruga, B. Jovancicevic, J. Simic-Krstic, *Fullerenes, Nanotubes, Carbon Nanostruct.* **2003**, *11*, 383.
195 R. Taylor, J. P. Hare, A. a. K. Abdul-Sada, H. W. Kroto, *J. Chem. Soc., Chem. Commun.* **1990**, 1423.
196 R. D. Johnson, G. Meijer, D. S. Bethune, *J. Am. Chem. Soc.* **1990**, *112*, 8983.
197 K. Chatterjee, D. H. Parker, P. Wurz, K. R. Lykke, D. M. Gruen, L. M. Stock, *J. Org. Chem.* **1992**, *57*, 3253.

198 K. C. Khemani, M. Prato, F. Wudl, *J. Org. Chem.* **1992**, *57*, 3254.
199 A. M. Vassallo, A. J. Palmisano, L. S. K. Pang, M. A. Wilson, *J. Chem. Soc., Chem. Commun.* **1992**, 60.
200 I. N. Kremenskaya, M. A. Nudelman, I. G. Shlyamina, V. I. Shlyamin, *Mendeleev Commun.* **1993**, 9.
201 W. A. Scrivens, P. V. Bedworth, J. M. Tour, *J. Am. Chem. Soc.* **1992**, *114*, 7917.
202 L. Isaacs, A. Wehrsig, F. Diederich, *Helv. Chim. Acta* **1993**, *76*, 1231.
203 N. Komatsu, T. Ohe, K. Matsushige, *Carbon* **2004**, *42*, 163.
204 R. Taylor, J. P. Parsons, A. G. Avent, S. P. Rannard, T. J. Dennis, J. P. Hare, H. W. Kroto, D. R. M. Walton, *Nature* **1991**, *351*, 277.
205 M. S. Meier, J. P. Selegue, *J. Org. Chem.* **1992**, *57*, 1924.
206 A. Gügel, M. Becker, D. Hammel, L. Mindach, J. Raeder, T. Simon, M. Wagner, K. Müllen, *Angew. Chem.* **1992**, *104*, 666; *Angew. Chem. Int. Ed. Engl.* **1992**, *31*, 644.
207 A. Gügel, K. Müllen, *J. Chromatogr.* **1993**, *628*, 23.
208 W. A. Scrivens, A. M. Rawlett, J. M. Tour, *J. Org. Chem.* **1997**, *62*, 2310.
209 A. Mittelbach, W. Hoenle, H. G. von Schnering, J. Carlsen, R. Janiak, H. Quast, *Angew. Chem.* **1992**, *104*, 1681; *Angew. Chem. Int. Ed. Engl.* **1992**, *31*, 1642.
210 R. Ettl, I. Chao, F. Diederich, R. L. Whetten, *Nature* **1991**, *353*, 149.
211 F. Diederich, R. L. Whetten, C. Thilgen, R. Ettl, I. Chao, M. M. Alvarez, *Science* **1991**, *254*, 1768.
212 J. M. Hawkins, T. A. Lewis, S. D. Loren, A. Meyer, J. R. Heath, Y. Shibato, R. J. Saykally, *J. Org. Chem.* **1990**, *55*, 6250.
213 W. H. Pirkle, C. J. Welch, *J. Org. Chem.* **1991**, *56*, 6973.
214 C. J. Welch, W. H. Pirkle, *J. Chromatogr.* **1992**, *609*, 89.
215 K. Cabrera, G. Wieland, M. Schaefer, *J. Chromatogr.* **1993**, *644*, 396.
216 K. Kimata, K. Hosoya, T. Araki, N. Tanaka, *J. Org. Chem.* **1993**, *58*, 282.
217 B. Nie, V. M. Rotello, *J. Org. Chem.* **1996**, *61*, 1870.
218 K. Jinno, K. Yamamoto, T. Ueda, H. Nagashima, K. Itoh, J. C. Fetzer, W. R. Biggs, *J. Chromatogr.* **1992**, *594*, 105.
219 C. Thilgen, F. Diederich, R. L. Whetten, *Buckminsterfullerenes* **1993**, 59.
220 A. Hirsch, I. Lamparth, H. R. Karfunkel, *Angew. Chem.* **1994**, *106*, 453; *Angew. Chem. Int. Ed. Engl.* **1994**, *33*, 437.
221 A. Hirsch, T. Grösser, A. Skiebe, A. Soi, *Chem. Ber.* **1993**, *126*, 1061.
222 A. Hirsch, O. Vostrowsky, *Eur. J. Org. Chem.* **2001**, 829.
223 A. Hirsch, *Top. Curr. Chem.* **1999**, *199*, 1.
224 F. Djojo, A. Hirsch, *Chem. Eur. J.* **1998**, *4*, 344.
225 F. Djojo, A. Hirsch, S. Grimme, *Eur. J. Org. Chem.* **1999**, 3027.
226 B. Gross, V. Schurig, I. Lamparth, A. Hirsch, *J. Chromatogr., A* **1997**, *791*, 65.
227 B. Gross, V. Schurig, I. Lamparth, A. Herzog, F. Djojo, A. Hirsch, *Chem. Commun.* **1997**, 1117.
228 U. Reuther, T. Brandmüller, W. Donaubauer, F. Hampel, A. Hirsch, *Chem. Eur. J.* **2002**, *8*, 2833.
229 I Bucsi, R. Aniszfeld, T. Shamma, G. K. S. Prakash, G. A. Olah, *Proc. Natl. Acad. Sci.* **1994**, *91*, 9019.
230 R. J. Doome, A. Fonseca, H. Richter, J. B. Nagy, P. A. Thiry, A. A. Lucas, *J. Phys. Chem. Solids* **1997**, *58*, 1839.
231 N. Komatsu, *Org. Biomol. Chem.* **2003**, *1*, 204.
232 D. E. Manolopoulos, J. C. May, S. E. Down, *Chem. Phys. Lett.* **1991**, *181*, 105.
233 P. W. Fowler, R. C. Batten, D. E. Manolopoulos, *J. Chem. Soc., Faraday Trans.* **1991**, *87*, 3103.
234 J. M. Schulman, R. L. Disch, M. A. Miller, R. C. Peck, *Chem. Phys. Lett.* **1987**, *141*, 45.
235 H. P. Lüthi, J. Almlof, *Chem. Phys. Lett.* **1987**, *135*, 357.
236 D. Dunlap, B. Samori, C. Bustamante, *NATO ASI Ser., Ser. C: Math. Phys. Sci.* **1988**, *242*, 275.s
237 G. E. Scuseria, *Chem. Phys. Lett.* **1991**, *176*, 423.
238 M. Haser, J. Almlof, G. E. Scuseria, *Chem. Phys. Lett.* **1991**, *181*, 497.
239 C. S. Yannoni, P. P. Bernier, D. S. Bethune, G. Meijer, J. R. Salem, *J. Am. Chem. Soc.* **1991**, *113*, 3190.
240 W. I. F. David, R. M. Ibberson, J. C. Matthewman, K. Prassides, T. J. S. Dennis, J. P. Hare, H. W. Kroto,

R. Taylor, D. R. M. Walton, *Nature* **1991**, *353*, 147.
241 K. Hedberg, L. Hedberg, D. S. Bethune, C. A. Brown, H. C. Dorn, R. D. Johnson, M. De Vries, *Science* **1991**, *254*, 410.
242 S. Liu, Y. J. Lu, M. M. Kappes, J. A. Ibers, *Science* **1991**, *254*, 408.
243 T. G. Schmalz, D. J. Klein, *Buckminsterfullerenes* **1993**, 83.
244 M. S. Dresselhaus, G. Dresselhaus, P. C. Eklund, Sience of Fullerenes and Carbon Nanotubes. Academic Press, San Diego, **1996**.
245 G. E. Scuseria, *Chem. Phys. Lett.* **1991**, *180*, 451.
246 A. L. Balch, V. J. Catalano, J. W. Lee, M. M. Olmstead, S. R. Parkin, *J. Am. Chem. Soc.* **1991**, *113*, 8953.
247 H. B. Bürgi, P. Venugopalan, D. Schwarzenbach, F. Diederich, C. Thilgen, *Helv. Chim. Acta* **1993**, *76*, 2155.
248 D. E. Manolopoulos, *J. Chem. Soc., Faraday Trans.* **1991**, *87*, 2861.
249 D. E. Manolopoulos, P. W. Fowler, *Chem. Phys. Lett.* **1991**, *187*, 1.
250 R. Taylor, *Tetrahedron Lett.* **1991**, *32*, 3731.
251 R. Taylor, *J. Chem. Soc., Perkin Trans. 2: Phys. Org. Chem. (1972–1999)* **1992**, 3.
252 M. D. Diener, J. M. Alford, *Nature* **1998**, *393*, 668.
253 F. H. Hennrich, R. H. Michel, A. Fischer, S. Richard-Schneider, S. Gilb, M. M. Kappes, D. Fuchs, M. Bürk, K. Kobayashi, S. Nagase, *Angew. Chem.* **1996**, *108*, 1839; *Angew. Chem. Int. Ed. Engl.* **1996**, *35*, 1732.
254 K. Balasubramanian, *Chem. Phys. Lett.* **1993**, *206*, 210.
255 F. Diederich, R. Ettl, Y. Rubin, R. L. Whetten, R. Beck, M. Alvarez, S. Anz, D. Sensharma, F. Wudl, et al., *Science* **1991**, *252*, 548.
256 K. Kikuchi, N. Nakahara, T. Wakabayashi, S. Suzuki, H. Shiromaru, Y. Miyake, K. Saito, I. Ikemoto, M. Kainosho, Y. Achiba, *Nature* **1992**, *357*, 142.
257 R. Taylor, G. J. Langley, T. J. S. Dennis, H. W. Kroto, D. R. M. Walton, *J. Chem. Soc., Chem. Commun.* **1992**, 1043.
258 R. Taylor, G. J. Langley, A. G. Avent, J. S. Dennis, H. W. Kroto, D. R. M. Walton, *J. Chem. Soc., Perkin Trans. 2: Phys. Org. Chem.* **1993**, 1029.
259 R. H. Michel, M. M. Kappes, P. Adelmann, G. Roth, *Angew. Chem.* **1994**, *106*, 1742; *Angew. Chem. Int. Ed. Engl.* **1994**, *33*, 1651.
260 A. L. Balch, A. S. Ginwalla, J. W. Lee, B. C. Noll, M. M. Olmstead, *J. Am. Chem. Soc.* **1994**, *116*, 2227.
261 T. J. S. Dennis, T. Kai, T. Tomiyama, H. Shinohara, *Chem. Commun.* **1998**, 619.
262 A. G. Avent, D. Dubois, A. Penicaud, R. Taylor, *J. Chem. Soc., Perkin Trans. 2: Phys. Org. Chem.* **1997**, 1907.
263 C. Thilgen, F. Diederich, *Top. Curr. Chem.* **1999**, *199*, 135.
264 A. Goryunkov, V. Y. Markov, I. N. Ioffe, D. Bolskar Robert, M. D. Diener, I. Kuvytchko, S. H. Strauss, V. Boltalina Olga, *Angew. Chem.* **2004**, *116*, 1015; *Angew. Chem. Int. Ed. Engl.* **2004**, *43*, 997.
265 C. Thilgen, A. Herrmann, F. Diederich, *Helv. Chim. Acta* **1997**, *80*, 183.
266 R. Kessinger, J. Crassous, A. Hermann, M. Rüttimann, L. Echegoyen, F. Diederich, *Angew. Chem.* **1998**, *110*, 2022; *Angew. Chem. Int. Ed. Engl.* **1998**, *37*, 1919.
267 J. Crassous, J. Rivera, N. S. Fender, L. Shu, L. Echegoyen, C. Thilgen, A. Herrmann, F. Diederich, *Angew. Chem.* **1999**, *111*, 1716; *Angew. Chem. Int. Ed. Engl.* **1999**, *38*, 1613.
268 H. Goto, N. Harada, J. Crassous, F. Diederich, *J. Chem. Soc., Perkin Trans. 2* **1998**, 1719.
269 R. Taylor, D. R. Walton, *Nature* **1993**, *363*, 685.
270 B. L. Zhang, C. H. Xu, C. Z. Wang, C. T. Chan, K. M. Ho, *Phys. Rev. B: Condensed Matter Mater. Phys.* **1992**, *46*, 7333.
271 J. Tersoff, *Phys. Rev. B: Condensed Matter Mater. Phys.* **1992**, *46*, 15546.
272 C. Christides, T. J. S. Dennis, K. Prassides, R. L. Cappelletti, D. A. Neumann, J. R. D. Copley, *Phys. Rev. B: Condensed Matter Mater. Phys.* **1994**, *49*, 2897.
273 D. L. Lichtenberger, K. W. Nebesny, C. D. Ray, D. R. Huffman, L. D. Lamb, *Chem. Phys. Lett.* **1991**, *176*, 203.
274 A. Tokmakoff, D. R. Haynes, S. M. George, *Chem. Phys. Lett.* **1991**, *186*, 450.

275 G. Dresselhaus, M. S. Dresselhaus, J. G. Mavroides, *Carbon* **1966**, *4*, 433.

276 J. De Vries, H. Steger, B. Kamke, C. Menzel, B. Weisser, W. Kamke, I. V. Hertel, *Chem. Phys. Lett.* **1992**, *188*, 159.

277 A. V. Nikolaev, T. J. S. Dennis, K. Prassides, A. K. Soper, *Chem. Phys. Lett.* **1994**, *223*, 143.

278 S. Saito, A. Oshiyama, *Phys. Rev. B: Condensed Matter Mater. Phys.* **1991**, *44*, 11532.

279 T. Kiyobayashi, M. Sakiyama, *Fullerene Sci. Technol.* **1993**, *1*, 269.

280 O. V. Boltalina, L. N. Sidorov, A. Y. Borshchevsky, E. V. Sukhanova, E. V. Skokan, *Rapid Commun. Mass Spectrom.* **1993**, *7*, 1009.

281 P. Wurz, K. R. Lykke, M. J. Pellin, D. M. Gruen, *J. Appl. Phys.* **1991**, *70*, 6647.

282 K. Harigaya, *Chem. Phys. Lett.* **1992**, *189*, 79.

283 W. Andreoni, F. Gygi, M. Parrinello, *Phys. Rev. Lett.* **1992**, *68*, 823.

284 K. Nakao, N. Kurita, M. Fujita, *Phys. Rev. B: Condensed Matter Mater. Phys.* **1994**, *49*, 11415.

285 G. Onida, W. Andreoni, J. Kohanoff, M. Parrinello, *Chem. Phys. Lett.* **1994**, *219*, 1.

286 N. Sivaraman, R. Dhamodaran, I. Kaliappan, T. G. Srinivasan, P. R. V. Rao, C. K. Mathews, *J. Org. Chem.* **1992**, *57*, 6077.

287 R. S. Ruoff, D. S. Tse, R. Malhotra, D. C. Lorents, *J. Phys. Chem.* **1993**, *97*, 3379.

288 W. A. Scrivens, J. M. Tour, *J. Chem. Soc., Chem. Commun.* **1993**, 1207.

289 M. V. Korobov, A. L. Smith, *Fullerene: Chem. Phys. Technol.* **2000**, 53.

290 Z. Gasyna, P. N. Schatz, J. P. Hare, T. J. Dennis, H. W. Kroto, R. Taylor, D. R. M. Walton, *Chem. Phys. Lett.* **1991**, *183*, 283.

291 S. Leach, M. Vervloet, A. Despres, E. Breheret, J. P. Hare, T. J. Dennis, H. W. Kroto, R. Taylor, D. R. M. Walton, *Chem. Phys.* **1992**, *160*, 451.

292 H. Imahori, Y. Sakata, *Adv. Mater.* **1997**, *9*, 537.

293 H. Imahori, Y. Sakata, *Eur. J. Org. Chem.* **1999**, 2445.

294 D. M. Guldi, *Chem. Soc. Rev.* **2002**, *31*, 22.

295 D. M. Guldi, P. V. Kamat, *Fullerenes: Chem. Phys. Technol.* **2000**, 225.

296 R. D. Johnson, G. Meijer, J. R. Salem, D. S. Bethune, *J. Am. Chem. Soc.* **1991**, *113*, 3619.

297 R. M. Fleming, B. Hessen, T. Siegrist, A. R. Kortan, P. Marsh, R. Tycko, G. Dabbagh, R. C. Haddon, *ACS Symp. Ser.* **1992**, *481*, 25.

298 A. R. Kortan, N. Kopylov, S. Glarum, E. M. Gyorgy, A. P. Ramirez, R. M. Fleming, F. A. Thiel, R. C. Haddon, *Nature* **1992**, *355*, 529.

299 P. W. Stephens, L. Mihaly, P. L. Lee, R. L. Whetten, S. M. Huang, R. Kaner, F. Deiderich, K. Holczer, *Nature* **1991**, *351*, 632.

300 C. S. Yannoni, R. D. Johnson, G. Meijer, D. S. Bethune, J. R. Salem, *J. Phys. Chem.* **1991**, *95*, 9.

301 R. Tycko, R. C. Haddon, G. Dabbagh, S. H. Glarum, D. C. Douglass, A. M. Mujsce, *J. Phys. Chem.* **1991**, *95*, 518.

302 R. D. Johnson, C. S. Yannoni, H. C. Dorn, J. R. Salem, D. S. Bethune, *Science* **1992**, *255*, 1235.

303 P. A. Heiney, J. E. Fischer, A. R. McGhie, W. J. Romanow, A. M. Denenstein, J. P. McCauley, Jr., A. B. Smith, III, D. E. Cox, *Phys. Rev. Lett.* **1991**, *66*, 2911.

304 A. Dworkin, H. Szwarc, S. Leach, J. P. Hare, T. J. Dennis, H. W. Kroto, R. Taylor, D. R. M. Walton, *Comp. Rend. Acad. Sci., Ser. II: Mecan., Phys., Chim., Sci. la Terre et de l'Univers* **1991**, *312*, 979.

305 J. S. Tse, D. D. Klug, D. A. Wilkinson, Y. P. Handa, *Chem. Phys. Lett.* **1991**, *183*, 387.

306 P. A. Heiney, *J. Phys. Chem. Solids* **1992**, *53*, 1333.

307 M. F. Meidine, P. B. Hitchcock, H. W. Kroto, R. Taylor, D. R. M. Walton, *J. Chem. Soc., Chem. Commun.* **1992**, 1534.

308 A. L. Balch, J. W. Lee, B. C. Noll, M. M. Olmstead, *J. Chem. Soc., Chem. Commun.* **1993**, 56.

309 S. M. Gorun, K. M. Creegan, R. D. Sherwood, D. M. Cox, V. W. Day, C. S. Day, R. M. Upton, C. E. Briant, *J. Chem. Soc., Chem. Commun.* **1991**, 1556.

310 J. D. Crane, P. B. Hitchcock, H. W. Kroto, R. Taylor, D. R. M. Walton, *J. Chem. Soc., Chem. Commun.* **1992**, 1764.

311 A. Izuoka, T. Tachikawa, T. Sugawara, Y. Suzuki, M. Konno, Y. Saito, H. Shinohara, *J. Chem. Soc., Chem. Commun.* **1992**, 1472.

312 D. V. Konarev, R. N. Lyubovskaya, A. Y. Kovalevsky, P. Coppens, *Chem. Commun.* **2000**, 2357.

313 J. D. Crane, P. B. Hitchcock, *J. Chem. Soc., Dalton Trans.: Inorg. Chem.* **1993**, 2537.

314 V. Konarev Dmitri, S. Khasanov Salavat, G. Saito, I. Vorontsov Ivan, A. Otsuka, N. Lyubovskaya Rimma, M. Antipin Yury, *Inorg. Chem.* **2003**, *42*, 3706.

315 P. D. W. Boyd, M. C. Hodgson, C. E. F. Rickard, A. G. Oliver, L. Chaker, P. J. Brothers, R. D. Bolskar, F. S. Tham, C. A. Reed, *J. Am. Chem. Soc.* **1999**, *121*, 10487.

316 D. R. Evans, N. L. P. Fackler, Z. Xie, C. E. F. Rickard, P. D. W. Boyd, C. A. Reed, *J. Am. Chem. Soc.* **1999**, *121*, 8466.

317 D. Sun, F. S. Tham, C. A. Reed, L. Chaker, M. Burgess, P. D. W. Boyd, *J. Am. Chem. Soc.* **2000**, *122*, 10704.

318 A. L. Balch, *Fullerenes: Chem. Phys. Technol.* **2000**, 177.

319 D. Sun, F. S. Tham, C. A. Reed, L. Chaker, P. D. W. Boyd, *J. Am. Chem. Soc.* **2002**, *124*, 6604.

320 F. Michaud, M. Barrio, D. O. Lopez, J. L. Tamarit, V. Agafonov, S. Toscani, H. Szwarc, R. Ceolin, *Chem. Mater.* **2000**, *12*, 3595.

321 P. R. Birkett, C. Christides, P. B. Hitchcock, H. W. Kroto, K. Prassides, R. Taylor, D. R. M. Walton, *J. Chem. Soc., Perkin Trans. 2: Phys. Org. Chem.* **1993**, 1407.

322 O. Ermer, *Helv. Chim. Acta* **1991**, *74*, 1339.

323 G. Roth, P. Adelmann, *J. Phys. I* **1992**, *2*, 1541.

324 T. Andersson, K. Nilsson, M. Sundahl, G. Westman, O. Wennerstroem, *J. Chem. Soc., Chem. Commun.* **1992**, 604.

325 L. J. Barbour, G. W. Orr, J. L. Atwood, *Chem. Commun.* **1998**, 1901.

326 J. L. Atwood, L. J. Barbour, M. W. Heaven, C. L. Raston, *Angew. Chem.* **2003**, *115*, 3376; *Angew. Chem. Int. Ed. Engl.* **2003**, *42*, 3254.

2 Reduction

2.1 Introduction

The first chemical transformations carried out with C_{60} were reductions. After the pronounced electrophilicity of the fullerenes was recognized, electron transfer reactions with electropositive metals, organometallic compounds, strong organic donor molecules as well as electrochemical and photochemical reductions have been used to prepare fulleride salts respectively fulleride anions. Functionalized fulleride anions and salts have been mostly prepared by reactions with carbanions or by removing the proton from hydrofullerenes. Some of these systems, either functionalized or derived from pristine C_{60}, exhibit extraordinary solid-state properties such as superconductivity and molecular ferromagnetism. Fullerides are promising candidates for nonlinear optical materials and may be used for enhanced photoluminescence material.

These phenomena are related to internal properties of the C_{60} molecule. Among organic molecules, the behavior of C_{60} and its chemically modified forms is unique. Reductive transformations of fullerenes have not only been carried out to prepare fulleride salts. The fulleride ions themselves are reactive species and easily undergo subsequent reactions, for example, with electrophiles. Therefore, the anions provide a valuable synthetic potential for fullerene chemistry.

2.2 Fulleride Anions

Theoretical calculations of the molecular orbital levels of C_{60} show that the lowest unoccupied molecular orbitals LUMO (t_{1u}-symmetry) and the LUMO+1 (t_{1g}-symmetry) exhibit a comparatively low energy and are triply degenerated (Figure 2.1) [1–7].

Therefore, C_{60} was predicted to be a fairly electronegative molecule, being reducible up to the hexaanion [8]. Indeed, this was supported by very early investigations carried out with C_{60} in solution, namely cyclic voltammetry studies, which showed its facile and stepwise reduction. In these first investigations, C_{60} was reduced by two [9] and three [10] electrons.

Fullerenes: Chemistry and Reactions. Andreas Hirsch and Michael Brettreich

ISBN: 3-527-30820-2

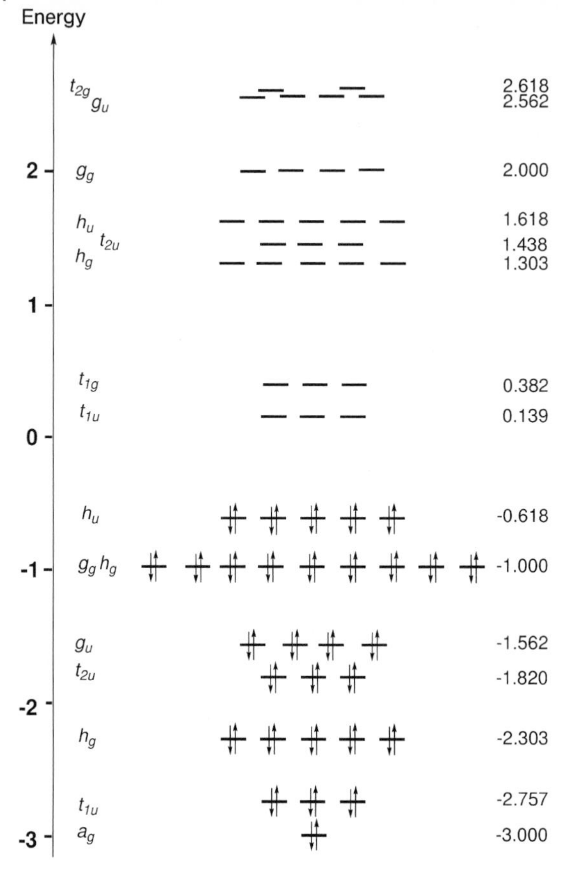

Figure 2.1 Schematic representation of the Hückel molecular orbital diagram for C_{60}.

Shortly thereafter, by increasing the expansion of the available potential window, the formation of the tetra- [11], penta- [12] and hexaanions [13–15] (Figure 2.2 and Table 2.1) could be detected by this technique.

The reductions were also found to be reversible at slow scan rates of 100 mV s^{-1}, for example, in a mixture of toluene and acetonitrile as solvent at –10 °C. The different reduction potentials appear almost equidistant with about 450 ± 50 mV between them. Under CV-conditions the penta- and hexaanions are stable and exhibit reversible behavior on the voltammetric time scale [8, 16].

Table 2.1 Reduction potentials of C_{60}.

Solvent	$C_{60}^{0/1-}$	$C_{60}^{1-/2-}$	$C_{60}^{2-/3-}$	$C_{60}^{3-/4-}$	$C_{60}^{4-/5-}$	$C_{60}^{5-/6-}$
PhMe/MeCN (mV vs Fc/Fc$^+$)	–980	–1370	–1870	–2350	–2850	–3260

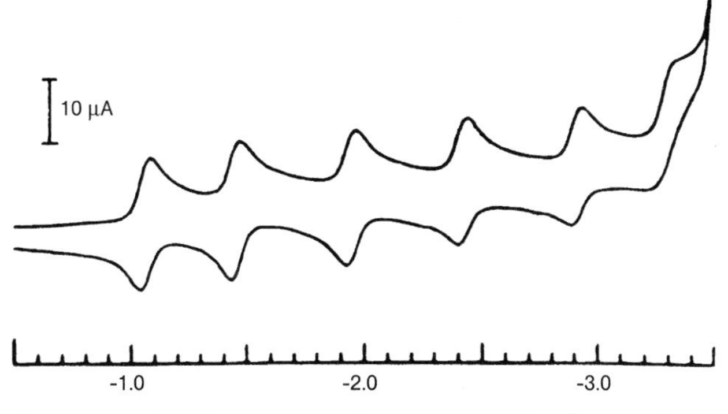

Figure 2.2 Cyclic voltammogram of C_{60} in acetonitrile–toluene with $TBAPF_6$ as supporting electrolyte at −10 °C using CV at a 100 mV s^{-1} scan rate (mV vs Fc/Fc^+) [8].

All the reductions are one-electron transfer processes. This has been demonstrated, for example, for the first four reductions of C_{60} by bulk electrolysis in benzonitrile [11]. The anions up to a charge of 4^- are comparatively stable in solution for up to several days, especially at low temperatures. Despite their stability under CV-conditions, the penta- and hexaanions are not stable under these conditions [8].

The reduction potentials are dependent on the nature of the solvent (Table 2.2), the electrolyte and strongly on the temperature [1, 10, 12, 13, 16–23]. Differences in the values of the first reduction potential of 300 to 400 mV have been observed [22]. This indicates that a large change in solvation occurs with each change of charge [1]. The donor–acceptor properties [18], the Lewis basicity and the strength of the hydrogen bonding interaction [1, 22] of the solvent have a pronounced influence on the reduction potentials. The dependence of potentials on the used electrolyte is diffuse and inconsistent and less substantial than the effect of the solvent.

The higher homologue C_{70} exhibits analogous behavior (Figure 2.3); six reduction waves at comparable potentials have been observed [13].

Table 2.2 Reduction potentials [mV vs $Me_{10}Fc/Me_{10}Fc^+$] of C_{60} in different solvents [1, 8, 18, 22].

Solvent	$C_{60}^{0/1-}$	$C_{60}^{1-/2-}$	$C_{60}^{2-/3-}$	$C_{60}^{3-/4-}$
Acetonitrile		−735	−1225	−1685
DMF	−312	−772	−1362	−1902
Pyridine	−343	−763	−1283	−1813
Aniline	−396	−693	−1158	−1626
Benzonitrile	−397	−817	−1297	−1807
Nitrobenzene	−406			
Benzyl alcohol	−443	−817		
Dichloromethane	−468	−858	−1308	−1758
THF	−473	−1063	−1633	−2133
ODCB	−535	−907	−1360	−1841
Chloroform	−554	−908		

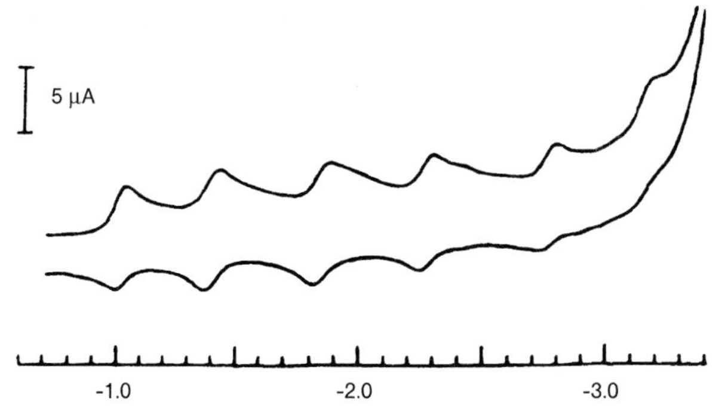

Figure 2.3 Cyclic voltammogram of C_{70} in acetonitrile–toluene with $TBAPF_6$ as supporting electrolyte at −10 °C using CV at a 100 mV s^{-1} scan rate (mV vs Fc/Fc^+) [8].

The electrochemistry of derivatized C_{60} has also been widely investigated [8, 23–28]. As observed by electrochemical reduction, derivatization usually decreases the electron affinity of the C_{60}-sphere. Typically, cathodically (more negative) shifted waves have been observed by cyclovoltammetry and other methods. Depending on the addend, the shifts range from 30 to 350 mV per adduct with respect to those of pure C_{60}. Reduction of some derivatives resulted in the loss of the addend. In some cases, like the retro-Bingel-reaction (Section 3.2.2), this can also be advantageous.

Investigations of the anions C_{60}^{n-} ($n = 1$–6), C_{70}^{n-} ($n = 1$–4) and C_{76}^{n-} ($n = 1$–2) by spectroelectrochemistry have been very valuable for the determination of their individual properties. Each species exhibits a characteristic UV/Vis/NIR and ESR spectrum [1, 11, 29–36]. Further valuable information about the electronic and geometric structure of the C_{60} anions can be obtained by NMR spectroscopy, in some cases by X-ray spectroscopy and by measuring the magnetic susceptibility. This knowledge of the spectroscopic, magnetic and structural properties of the anions allows an unambiguous determination of the oxidation state of reduced C_{60}, obtained by other methods, for example, by chemical reduction.

In contrast to C_{60} itself, the anions C_{60}^{n-} ($n = 1$–6) exhibit strong absorptions in the NIR and the λ_{max} values, which vary from 700 to 1200 nm, significantly depend on n. These low-energy NIR transitions can be rationalized in terms of the molecular orbital diagram of C_{60} (Figure 2.1) [33]. With the parent C_{60} the t_{1u} orbitals are triply degenerated. Placing one electron into the t_{1u} LUMO results in an NIR transition at ca. 1075 nm with an extinction coefficient of about 15 to 20 · 10^3 M^{-1} cm^{-1} [1]. This low energy absorption is assigned to the symmetry allowed t_{1u}–t_{1g} transition [29, 37, 38]; λ_{max} for C_{60}^{n-} with $n = 1$–5 are shown in Table 2.3 (see [1] and references therein).

A broadening and red-shifting was observed for the 330 nm absorption and also for the 255 nm absorption [11], which can also be accounted by a diminution in symmetry [33]. In the visible region, spectral changes produce the colors of the different fulleride anions, which are dark red-purple for C_{60}^{-}, red-orange for C_{60}^{2-} and dark red-brown for C_{60}^{3-} [17].

Table 2.3 NIR transitions of C_{60}^{n-}; approximate values averaged from different references are given in [1].

n	λ_{max} *(nm)*	*Extinction coefficient* (ε) *(*$M^{-1}cm^{-1}$*)*
1	1075	15 000 – 20 000
	925	3 000 – 5 000
2	950	15 000 – 20 000
	830	5 000 – 7 000
3	1350	5 000 – 6 000
	960	6 000 – 9 000
	770	11 000 – 14 000
4	1200	5 000 – 6 000
	730	4 000
5	950	7 000

The four infra-red active modes for neutral C_{60} can also be found in anionic C_{60}. The frequencies of the signal decrease with increasing charge whereas the response of the four signals varies in intensity [1]. The A_g-mode is one of the 10 Raman-active modes, which turned out to be suitable as a diagnostic tool for the charge state. The "pentagonal pinch" mode at 1467 cm^{-1} decreases by 6 cm^{-1} per added charge [1].

ESR spectra [11, 29, 32, 39] of C_{60}^{-} show a broad signal for this $S = ½$ system. The g value is remarkably low, lying between 1.997 and 2.000, which additionally supports the conclusion that the electronic ground state of the monoanion is doubly degenerated [32]. The line width is very temperature dependent and decreases with decreasing temperature [32, 39]. Below 100 K the signal becomes anisotropic due to a distortion from icosahedral symmetry. At room temperature a sharp, weak signal occurs in most of the measurements. This peak has been widely discussed [1, 36, 40, 41] and is assumed to derive from $C_{120}O$, which is often present in C_{60} samples as an impurity. Sharp spikes in ESR spectra were also shown to be caused by C_{60}-derivatives, formed during reduction by reaction of higher anions with, e.g., the solvent. The assignment of the ESR signal of higher anions is (often) controversial [1]. Concerning the dianion C_{60}^{2-} the ESR spectrum seems to be consistent with a triplet ground state ($S = 1$). However, a closer look to the spin-properties with measurements of spin density and magnetic susceptibility indicated an ESR-silent singlet ground-state. The triplet-signal derives from a close-lying triplet-excited state, which is populated at room temperature. C_{60}^{3-} with its $S = ½$ ground state gives an ESR signal very similar to the monoanion. Additional one-electron reductions lead to an ESR-silent singlet ($S = 0$) ground state for the tetraanion, a pentaanion with a spin of ½ and finally to the diamagnetic hexaanion C_{60}^{6-}.

^{13}C NMR spectroscopy [1] is not only a valuable tool to clarify the symmetry of neutral and charged C_{60} and its derivatives but can also reveal more insight into the paramagnetic electronic state and the charge of the C_{60} anions. The paramagnetism, introduced by unpaired electrons and the charge itself, leads to a relatively high downfield-shift of about 15 to 50 ppm. The intensity of this shift depends on the temperature, the charge, the physical condition (solid or solute) and the spin state of the anion salt or intercalated fulleride. It is not dependent on the degree of aromaticity. The paramagnetic or diamagnetic ring-current does not effect the ^{13}C chemical shift, but it does effect the shift of ^{13}C or ^{1}H attached to exohedral addends [42] or the shift of endohedral ^{3}He [43–47].

There are some examples of NMR spectra of discrete $C_{60}{}^{n-}$ for $n = 1$–4 and 6 in solution [1, 48–53]. Despite the useful information about the spin state, solution NMR spectroscopy is not yet a diagnostic tool for unambiguous determination of charge or spin state. The contribution of spin state or charge to the chemical shift is still under examination. In addition, the electron exchange between different charged species is fast on the NMR scale, i.e. it is not possible to detect these species in solution. Only an averaged signal is detectable. The ^{13}C NMR shifts of fullerene anions are usually in the range of 180–200 ppm.

In solid-state NMR [1, 51–64], the magnetic coupling between the fullerene anions has to be taken into account. In the case of metal intercalated fullerides that have metallic properties a contribution from the conduction electrons must be added, a phenomenon called the "Knight shift". Even if this additional "shift" affects the ^{13}C-chemical resonance, the correspondence between extended and discrete systems of comparable $C_{60}{}^{n-}$ oxidation state is quite close [1].

Owing to the sensibility of fullerene anions, the limited range of solvents and counter anions, the growing of single crystals requires patience and luck [65]. Few single crystals are suitable for X-ray crystallography. Crystals of discrete salt structures have mostly been obtained with electrocrystallization (more details see next chapter). Other methods include the growing of crystals on reducing metal surfaces [65] (formation of the monoanionic salt $(Ph_4P)_2C_{60}Br$), reducing with alkaline metals in DMF solution [66] (dianionic $[K([2.2.2]crypt]_2\text{-}C_{60}(C_6H_5CH_3)_4])$, or with alkaline metal in molten crown ether [67] (trianionic $[K([18]\text{-crown-}6)]_3\text{-}[C_{60}](C_6H_5CH_3)_2)$. Other and higher charges located at the fullerene can be obtained by forming crystalline structures from intercalated C_{60}-derivatives. Structures of $C_{60}{}^{n-}$ with n up to 12 are known. Example techniques to produce these crystals are (a) annealing of stoichiometric amounts of alkali metal with C_{60} single crystals at about 450 °C for several days [68, 69]; (b) reaction between alkali metal and C_{60} in liquid ammonia, removal of ammonia and subsequent heating of the solid at 120 to 350 °C under reduced pressure for several days [70]; (c) intercalation of alkali metals into preformed Ba_3C_{60} to form $K_3Ba_3C_{60}$, by loading a piece of potassium metal and Ba_3C_{60} powder in a tantalum cell and heating at 260 °C for 7 days [71]; and (d) slowly heating a pelletized mixture of stoichiometric quantities of purified C_{60} and LiN_3; the crystallinity was improved by a moderate annealing for 6 h at about 800 °C [72].

X-ray structures of C_{60} crystals often show disorder in the arrangement of the C_{60}-anions. The first crystallographically ordered structure of a discrete fulleride was found in the dianionic salt $[PPN^+]_2[C_{60}{}^{2-}]$ [73]. The first example of a highly ordered monoanion was $[Ni(C_5Me_5)_2{}^+][C_{60}{}^-]CS_2$, which was prepared by chemical reduction with decamethylnickelocene [74].

2.3 Reductive Electrosynthesis

2.3.1 Electrocrystallization

The possibility of electrochemically producing C_{60} anions in a defined oxidation state by applying a proper potential can be used to synthesize fulleride salts by electrocrystallization [39, 75–80]. An obvious requirement for this purpose is the insolubility of the salt in the solvent to be used for the electrocrystallization process. This can be achieved by choosing the proper solvent, the oxidation state of C_{60} and the counter cation, which usually comes from the supporting electrolyte.

Fulleride anions are often more soluble, especially in more polar solvents, than the parent fullerenes. For example, in bulk electrolysis experiments with tetra-*n*-butylammonium perchlorate ($TBAClO_4$) as supporting electrolyte, carried out in acetonitrile where C_{60} is completely insoluble, fairly concentrated, dark red-brown solutions of C_{60}^{3-} can be obtained [81]. Upon reoxidation, a quantitative deposition of a neutral C_{60} film on the surface of a gold/quartz crystal working electrode takes place. This C_{60} film can be stepwise reductively doped with TBA^+, leading to $(C_{60}^-)(TBA^+)$ and $(C_{60}^{2-})(TBA^+)_2$ prior to the redissolution of the C_{60}^{3-} species. This process is represented in Scheme 2.1 where (s) and (f) denote "solution" and "film" respectively [81].

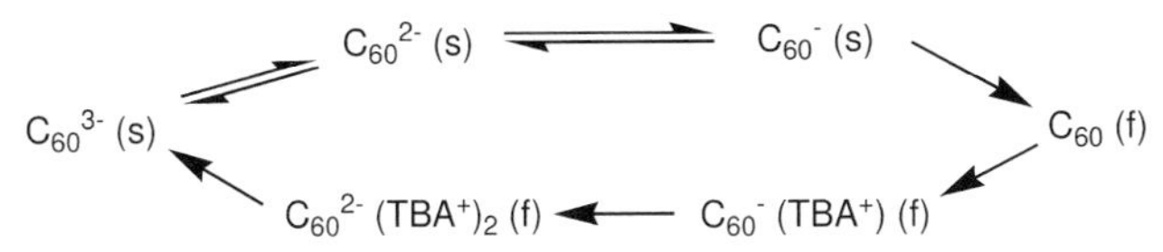

Scheme 2.1

Whereas C_{60} is insoluble and inert in liquid ammonia without any cosolvent, the fulleride anions C_{60}^{n-} ($n = 1$–4), generated electrochemically with KI as supporting electrolyte, dissolve completely in this polar medium [15]. Further reductions lead to the ammonia-insoluble potassium salts of the penta- and hexaanions.

The formation of crystalline fulleride salts at the electrode occurs when less polar solvents and bulky cations are used for the electrosynthesis. The first fulleride salt was synthesized by Wudl by bulk electrolysis of C_{60} in *o*-dichlorobenzene with tetraphenylphosphonium chloride as supporting electrolyte [39, 80]. This black microcrystalline material with the composition $(Ph_4P^+)_3(C_{60}^-)(Cl^-)_2$ exhibits an ESR line with a g-value of 1.9991 and a line width of 45 G at room temperature. Single crystals of the slightly different salts $(Ph_4P^+)_2(C_{60}^-)(Cl^-)$ and $(Ph_4P^+)_2(C_{60}^-)(Br^-)$ could be obtained by electrocrystallization and their crystal structure was determined [82, 83]. Magnetic measurements showed the presence of unpaired spins.

An analogous single crystalline fulleride salt $(Ph_4P^+)_2(C_{60}^-)(I)_x$ was obtained by electrocrystallizing CH_2Cl_2–toluene solutions of C_{60} in the presence of Ph_4PI [78]. The X-ray crystal structure of $(Ph_4P^+)_2(C_{60}^-)(I)_x$ (Figure 2.4) – which was the first

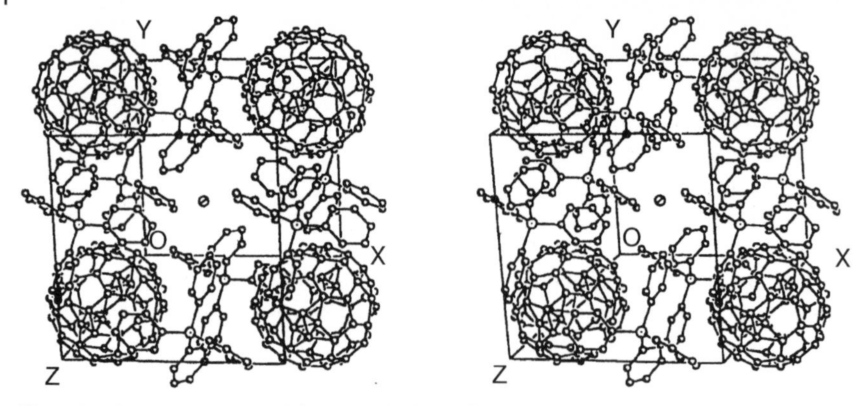

Figure 2.4 Stereoscopic view of the unit cell of $(Ph_4P^+)_2(C_{60}^-)(I^-)_x$ [78]. Iodide anions at the corners of the unit cell have been omitted for clarity.

one solved for a fulleride – shows that the C_{60} anions are surrounded by a tetragonally distorted cube of tetraphenylphosphonium cations. There are no short contacts between the C_{60} anions. The shortest center-to-center distance is 12.6 Å. A deficiency of iodine with x between 0.15 and 0.35 was determined. A repeated examination showed that this salt is a mixed salt rather than a system with a deficiency of iodine [84]. The structure was found to be $(Ph_4P^+)_2(C_{60}^-)Cl_{1-x}I_x$ ($x = 0{,}15$), whereas the chlorine is formed by reduction of methylene chloride. This salt is isostructural and isomorphous with $(Ph_4P^+)_2(C_{60}^-)(Cl^-)$ [82]. Electrocrystallization with tetraphenylarsonium halides gives analogous results [80, 83].

Electrocrystallization in the presence of [bis(triphenylphosphoranylidene)]ammonium chloride (PNPCl) leads to the single crystalline salt $(PNP^+)(C_{60}^-)(C_5H_5Cl)$, which shows no electrical conductivity [77]. C_{60} is surrounded by bulky PPN-cation and crystal solvent and has no short contact between neighboring C_{60} anion radicals [85]. Electrocrystallization with $M^+B(C_6H_5)_4^-$ under the same conditions leads to the single crystals of alkali-metal salts $M_xC_{60}(THF)_y$ ($x \approx 0.4$, $y \approx 2.2$) with M = Na, Li, K, Cs. They are isomorphic and crystallize with hexagonal unit-cells. In contrast to the PPN-salt the Na salt with the structure $Na_{0.39}C_{60}(THF)_{2.2}$ shows metallic properties [85, 86]. The resistivity of this compound was measured to be 50 S cm^{-1} at room temperature and about 1000 S cm^{-1} at 100 K. The lattice spacing of about 9.9 Å at room temperature indicates a direct C_{60}–C_{60} interaction, which could explain the salt's metallic character.

Another semiconducting fulleride salt, $[Ru(bpy)_3](C_{60})_2$ with bpy = 2,2′-bipyridine, crystallizes on the Pt electrode surface out of dichloromethane solutions saturated with $[Ru(bpy)_3]PF_6$ within a few minutes [79]. The NIR spectra of benzonitrile solutions of this salt demonstrate that the only fulleride anion present is C_{60}^-. The temperature dependence of the conductivity is typical for a semiconductor, with the room temperature conductivity being 0.01 S cm^{-1} and the activation energy 0.1 kJ mol^{-1} (0.15 eV). It was postulated that there is an electronic overlap between the two ions of this salt leading to a donation of electron density from the C_{60}^- to the ligand orbitals in the $[Ru(bpy)_3]^{2+}$ ($\Delta E^0 \approx 0.7$) [79].

2.3.2
Electrophilic Additions to Fulleride Anions

The electrochemical generation of fulleride anions can also be used to synthesize covalent organofullerene derivatives by quenching the anions with electrophiles. This was exemplified in the synthesis of dimethyldihydro[60]fullerene, the simplest dialkyl derivative of C_{60} [87]. For this purpose benzonitrile solutions of C_{60} and tetra-*tert*-butylammonium perchlorate ($TBAClO_4$) where exhaustively electroreduced in a dry-box to yield a dark red solution of C_{60}^{2-}. Treatment of this solution with an excess MeI, after the potential is turned off, affords a dark brown solution of $C_{60}(CH_3)_2$ (Scheme 2.2). This addition reaction proceeds comparatively smoothly. After 40 min, about 50% of the dianion is still present, which subsequently can be reoxidized to C_{60}.

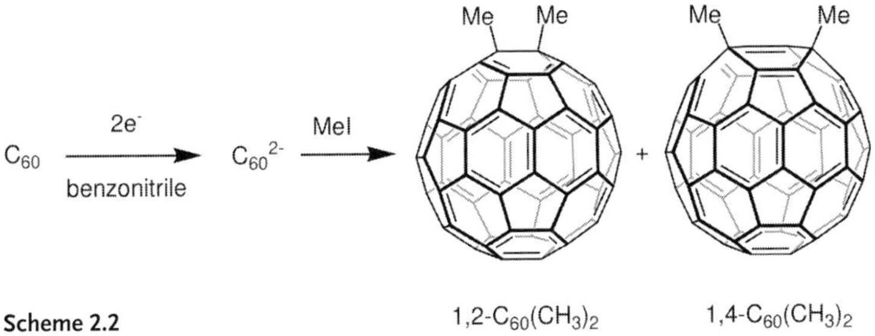

C_{60} —(2e⁻, benzonitrile)→ C_{60}^{2-} —(MeI)→ $1,2\text{-}C_{60}(CH_3)_2$ + $1,4\text{-}C_{60}(CH_3)_2$

Scheme 2.2

The isolated dimethyldihydro[60]fullerenes are a mixture of the 1,2- and 1,4-isomers (1.4 : 1) (Scheme 2.2). Of the 23 possible regioisomers of $C_{60}(CH_3)_2$ the 1,2- and 1,4-isomers are predicted to be the most stable, with the heat of formation of the 1,4-isomer being a little bit larger than that of the 1,2-isomer [87]. The electron density in the intermediate $C_{60}(CH_3)^-$ is calculated to be largest at C-2 (25%) followed by C-4/C-11 (9% each). This also suggests that the second electrophilic attack of MeI should be preferred at C-2, even if it is taken into account that there are two sites available for a 1,4-addition mode.

The preferred formation of 1,2- or 1,4-isomers is also shown with the addition of bulkier $XC_6H_5CH_2$-addends (X = H, Br). Addition of the aryl-addends to the C_{60}^{2-} anion leads to the 1,4-adducts $C_{60}(C_6H_5CH_2)_2$ and $C_{60}(BrC_6H_5CH_2)_2$. The reaction was carried out with $C_6H_5CH_2Br$ and different isomers of $BrC_6H_5CH_2Br$ [88]. Two preferred isomers of $C_{60}(C_6H_5CH_2)_4$ are formed by the reaction of $C_{60}(C_6H_5CH_2)_2^{2-}$ with C_6H_5Br. These isomers show a 1,4-1,4- and a 1,4-1,2-addition pattern [88, 89]. Pd-complex **2** has been synthesized electrochemically by the reaction of electrochemically generated C_{70}^{2-} or C_{60}^{2-} with Pd^{II} and the phosphine ligand **1** (Scheme 2.3) [90].

Other methods that use C_{60} anions as precursor for the synthesis of fullerene-derivatives usually involve chemical formation of the anion. Alkylation of C_{60} has been accomplished, e.g. by reduction with propanethiol and potassium carbonate in DMF [91, 92], sodium methanethiolate in acetonitrile [93], the naphthalene radical anion in benzonitrile[94], potassium naphthalide [95] or simply with zinc [96].

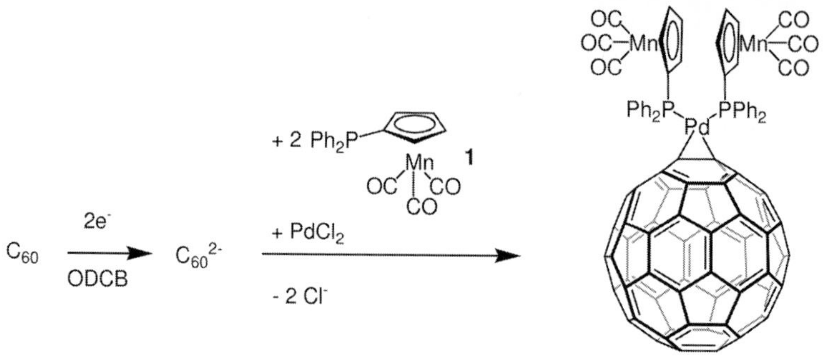

Scheme 2.3

2.4
Reduction with Metals

2.4.1
Alkali Metal Fullerides

2.4.1.1 Generation in Solution and Quenching Experiments

Fullerenes can be easily chemically reduced by the reaction with electropositive metals [1, 97–99], for example, alkali- and alkaline earth metals. The anions C_{60}^{n-} ($n = 1$–5) can be generated in solution by titrating a suspension of C_{60} in liquid ammonia with a solution of Rb in liquid ammonia [100], whereupon the resulting anions dissolve. Monitoring of this titration is possible by detecting the characteristic NIR absorption of each anion by UV/Vis/NIR spectroscopy. The solubility of the alkali metal fullerides in the polar solvent NH_3 demonstrates their salt character.

Ultrasound-aided reduction of C_{60} and C_{70} with Li metal to anionic species can be carried out in THF as the solvent [50]. In contrast to the parent fullerenes, the anions C_{60}^{n-} and C_{70}^{n-} are very soluble in THF. Solutions generated this way are deep red-brown. Alkylation of a C_{60} and C_{70} polyanion mixture with excess MeI leads to polymethylated fullerenes with up to 24 methyl groups covalently bound (see also Chapter 1).

With these methods, the stoichiometry of the generated C_{60}-polyanions is always difficult to adjust. These difficulties in reaching a specific stoichiometry of the alkali fullerides can be overcome by using either the metal in the presence of a crown-ether like dibenzo-18-crown-6 [48, 101, 102] or 2.2.2-cryptand [49, 66] or by using electron carriers such as corranulene [43] or naphthalene [58, 60]. By using a stoichiometric amount of sodium or potassium naphthalenide, KC_{60}, K_3C_{60} or K_6C_{60} can be selectively produced [95, 103]. By the use of 1-methylnaphthalene as electron carrier C_{60}^{n-} salts with $n = 1$ to $n = 3$ have been synthesized [58, 60]. Sonication with excess potassium in tetramethylethylenediamine solution gives, specifically, C_{60}^{3-} [104].

2.4.1.2 Synthesis and Properties of Alkali Metal Fulleride Solids

The discovery of superconductivity [105] in alkali metal doped C_{60} attracted the attention of a broad cross-section of the scientific community. The composition of the first alkali metal doped fullerene was determined to be K_3C_{60} [106, 107]. In this compound, the transition to a superconducting state occurs at $T_c = 19.3$ K [106]. In a short time, a large variety of alkali metal fullerides M_nC_{60} have been synthesized and studied with respect to their solid-state properties [1, 106–114]. Thereby, the stoichiometries of M_nC_{60} as well as the alkali metal with M = Na, K, Rb, Cs have been varied systematically. In addition, mixed alkali compounds $M^1{}_nM^2{}_mC_{60}$ (M^1, M^2 = Li, Na, K, Rb, Cs) [110, 111] and mixed alkali/alkaline earth metal compounds [1, 112–114] can be obtained. All of the alkali metals react directly with C_{60}. Since the heavier homologues of the alkali metals have a higher vapor pressure, their direct reaction with the fullerene proceeds more easily. The synthesis of a defined M_nC_{60} can be achieved by allowing stoichiometric amounts of the metal and C_{60} to react at 200 °C in a sealed tube [106]. In most cases additional annealing steps at elevated temperatures are necessary to complete the reaction. The resulting solids are very air sensitive.

This direct method is a convenient route, especially for saturation-doped fullerides M_6C_{60}. A facile control of the alkali metal fulleride stoichiometry of M_nC_{60} (M = K, Rb, Cs) can be obtained using vapor-transport techniques [115]. Thereby, a weighed amount of C_{60} is treated with a large excess of the alkali metal at 225 °C under vacuum. This procedure leads to the saturation-doped products M_6C_{60}. To these fullerides a specific amount of C_{60} is added to give the desired stoichiometry. The process of the formation of $M_{6m/n}C_{60}$ is completed after a subsequent annealing process under vacuum at 350 °C. Upon heating to 550 °C, K_6C_{60} decomposes to give K_4C_{60}, with the liberated K reacting with the glass [116].

The synthesis of M_nC_{60} (M = Na, $n = 2, 3$; M = K, $n = 3$) has been achieved by the reaction of solid C_{60} with solid MH or MBH_4 [116]. The advantage of these reactions is the easier handling of small quantities of MH or MBH_4 compared with alkali metals. As a source for alkali metals, binary alloys of the type CsM (M = Hg, Tl, Bi) can also be used [109]. However, the heavy metals partly co-intercalate into the C_{60} lattice [115, 117].

Alkali metal doping of C_{60} is also possible by solution-phase techniques [1, 118–121]. K_nC_{60} and Rb_nC_{60} containing small fractions of the superconducting M_3C_{60} phases were prepared by allowing toluene solutions of C_{60} to react with the alkali metal [118, 119]. During the reaction, the alkali metal fullerides form a black precipitate. In another example, sonication of a solution of C_{60} and excess potassium in TMEDA yields $K_3C_{60}(THF)_{14}$ with a defined stoichiometry [104].

Rapid, easy access to high quality alkali metal fulleride superconductors is obtained by the reaction of stoichiometric amounts of C_{60} and alkali metal in liquid ammonia [120]. After the ammonia is removed from the reaction mixture, the products are annealed in sealed evacuated silica tubes at elevated temperatures (225–350 °C). In this way, large amounts of superconducting fractions M_3C_{60} with $M_3 = K_3$, Rb_3, $CsRb_2$, $RbCs_2$ and KRbCs are available. Cs_3C_{60} is obtained via a similar procedure. Heating takes place at 150 °C. With this route, formation of the more stable CsC_{60} and Cs_4C_{60} phases were prevented. Cs_3C_{60} has one of the highest T_c-s, 40 K, even though it was measured under an elevated pressure of 15 kbar. The reaction in liquid ammonia often leads to structures that include a certain percentage of NH_3, as for example in $(NH_3)_xK_3C_{60}$ [122, 123] $(0 < x < 1)$ or in $(NH_3)_xNaA_2C_{60}$ [124] with A = K or Rb and $0.5 < x < 1$. The content of NH_3 (x) in $(NH_3)_xNaA_2C_{60}$ can be controlled by the duration of the drying, which follows the reaction. The presence of NH_3 in the crystal strongly influences the superconductive properties (see below).

Intercalation of C_{60} with lithium has been achieved by solid-state electrochemical doping [125]. In this technique, metallic lithium was used as the negative electrode and a polyethylene oxide lithium perchlorate ($P(EO)_8LiClO_4$) polymer film served as electrolyte. The formation of stoichiometric phases Li_nC_{60} ($n = 0.5$, 2, 3, 4, and 12) has been observed.

The composition of the alkali metal fullerides [112, 114] can be regarded as intercalation compounds of the parent C_{60} fcc (face-centered cubic) solid-state structure, or of hypothetical bcc (body-centered cubic) and bct (body-centered tetragonal) fullerene structures (Figure 2.5). In a fulleride phase M_2C_{60}, for example in the sodium salt Na_2C_{60}, the metal occupies the two tetrahedral interstices of an fcc [126]. The fullerides M_3C_{60} also exhibit fcc structures at room temperature [107]. Here, all the available tetrahedral and octahedral interstices are occupied by the alkali metal. The radius of the tetrahedral sites (1.12 Å) is smaller than that of the octahedral sites (2.06 Å). Thus, ions larger than Na^+ that are occupying tetrahedral sites are lattice expanding [111]. Conversely, all alkali cations are smaller than the octahedral sites; this has a lattice contracting influence. Each alkali metal fulleride

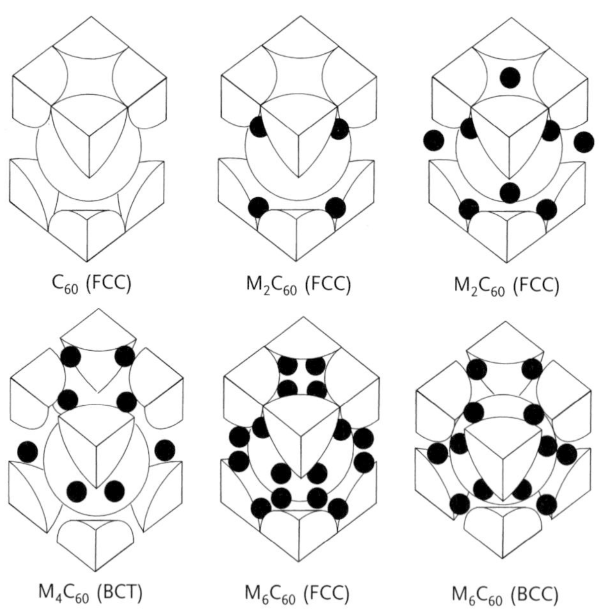

Figure 2.5 Schematic representation of C_{60} and M_nC_{60} structures.

Table 2.4 T_c of different alkali- and alkaline-earth-metal doped C_{60} in dependence of the lattice parameter(s) ([114] and references therein).

Compound	*Structure*	*T_c (K)*	*Lattice parameter (Å)*
Alkali-metal-doped C_{60}			
Li_2CsC_{60}	fcc	10.5	$a = 14.080$
Na_2KC_{60}	fcc	2.5	$a = 14.122$
Na_2RbC_{60}	fcc	3.5	$a = 14.092$
Na_2CsC_{60}	fcc	12	$a = 14.126$
K_3C_{60}	fcc	19	$a = 14.240$
K_2CsC_{60}	fcc	24	$a = 14.292$
Rb_3C_{60}	fcc	29	$a = 14.384$
Rb_2CsC_{60}	fcc	31	$a = 14.431$
$RbCs_2C_{60}$	fcc	33	$a = 14.555$
Cs_3C_{60}	bct	40 (high p)	$a = 12.06$, $c = 11.43$
$(NH_3)_4Na_2CsC_{60}$	fcc	30	$a = 14.473$
$(NH_3)K_3C_{60}$	bco	28 (high p)	$a = 14.971$, $b = 14.895$, $c = 13.678$
Alkaline-earth-metal-doped C_{60}			
Ca_5C_{60}	sc	8.4	$a = 14.01$
Sr_4C_{60}	bco	4	
Ba_4C_{60}	bco	6.7	$a = 11.610$, $b = 11.234$, $c = 10.882$
$K_3Ba_3C_{60}$	bcc	5.6	$a = 11.245$
$Rb_3Ba_3C_{60}$	bcc	2.0	$a = 11.338$
$K_2Ba_4C_{60}$	bcc	3.6	$a = 11.212$

superconductor has an M_3C_{60} stoichiometry and the corresponding fcc structure. Significantly, an expansion of the lattice due to bigger cations is accompanied by an enhancement of the T_c of M_3C_{60} (Table 2.4 and Figure 2.6). Na_3C_{60} is not superconducting, which is attributed to the fact that below 260 K a reversible structural transition takes place that leads to a two-phase mixture of Na_2C_{60} and Na_6C_{60} [126].

The M_4C_{60} and M_6C_{60} (M = K, Rb, Cs) structures are related. The bcc M_6C_{60} structure is based on the bcc packing of the C_{60} molecules. The metal cations occupy four coordinate interstices between rotationally ordered C_{60} molecules [108]. The M_4C_{60} structure consists of a bct packing of the C_{60} molecules [127]. The M_4C_{60} structure can be regarded as a defect M_6C_{60} structure. In M_4C_{60} the six identical sites of a bcc structure are disproportionated into four slightly smaller and two larger ones. The cations M^+ occupy the smaller sites.

The occurrence of superconductivity [112–114] in alkali metal fullerides always requires the trianionic state of C_{60}. In this case, the conduction band derived from the t_{1u} orbitals is half filled. This gives rise to the maximum conductivity of M_3C_{60} [128]. Superconductivity is absent in Li_nC_{60}, Na_nC_{60} and C_{70} based fullerides. Band structure calculations for K_3C_{60} predict the Fermi-level to be close to the lower maximum of the density of states (DOS) of the conduction band [112, 128]. The higher T_cs for those M_3C_{60}s with enhanced unit cell parameters (Table 2.4) can be

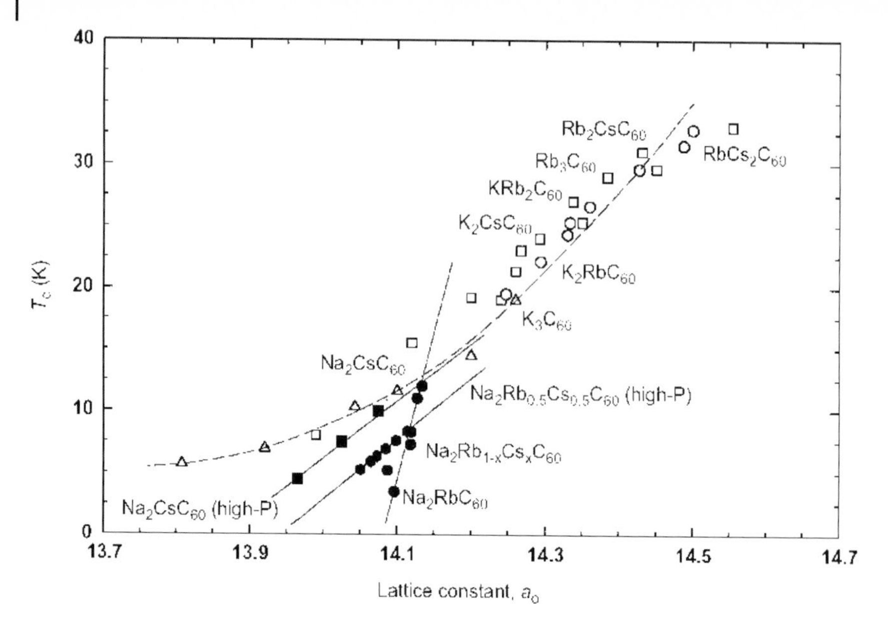

Figure 2.6 Variation of T_c (K) with lattice parameter a_0 (Å) for various compositions of A_3C_{60} and Na_2MC_{60} (M = Rb, Cs) [112].

interpreted with increased densities of states at the Fermi-level. This increase is due to a sharpening of the conduction band caused by a weaker orbital overlap of neighboring C_{60} molecules as the cell volume increases [112–114, 129].

The unit cell volume, and with it the T_c, can also be increased by incorporation of a neutral molecule such as ammonia into the lattice by leaving the fcc-structure and oxidation state of the metal intact (Table 2.4). If the cubic lattice is disordered by this incorporation, a decrease rather than an increase of T_c is observed [113, 122, 123].

In M_6C_{60}, the bands derived from the t_{1u} orbitals are completely filled, making these compounds insulators. Also, the fullerides M_2C_{60} and M_4C_{60} are not conducting, although the t_{1u} bands are partly filled. This has been interpreted by localized Jahn–Teller distortions on $C_{60}{}^{n-}$ [116].

The occurrence of superconductivity in M_3C_{60} can be determined by the disappearance of the resistivity or, more conveniently, by the Meißner–Ochsenfeld–Effect (magnetization measurements). In this method the very air-sensitive fullerides are loaded under inert gas conditions into a capillary tube, which is then sealed under vacuum. Upon the transition to the superconducting state, the magnetic susceptibility sharply decreases, since superconductors in their ground state are diamagnetic.

2.4.2
Alkaline Earth Metal Fullerides

Alkaline earth metals can also be intercalated with C_{60} [1, 112–114, 130–137]. The preparation of Ca_5C_{60} [131], Ba_6C_{60} [133, 138], Ba_xC_{60} (with $x = 3–6$) [134] or Ba_4C_{60} and Sr_4C_{60} [135] has been achieved by the direct reaction of C_{60} with the corresponding alkaline earth metal vapor. Therefore, the alkaline earth-metal is mixed with C_{60} in an inert atmosphere and the powder is then placed into a custom designed high-purity tantalum cell, which allows compression into pellets. The pellets are then loaded into quartz tubes, being subsequently sealed under vacuum. Several heat treatments of these samples, between 550 and 800 °C, for periods ranging from hours to weeks leads to the corresponding fullerides [130–137]. In an alternative preparation method, a Ba layer is placed under a C_{60} layer and the annealing of this sandwich leads to the desired fulleride [138]. In this procedure the amount of Ba in the volume is controlled by photoemission spectroscopy. Other methods to produce fullerides include reaction in THF [139] or in liquid ammonia [140].

Doping of C_{60} with Ca to give Ca_5C_{60} affords a phase transformation, leading to a simple cubic superconducting phase with a T_c of 8.4 K. For Ba_4C_{60} or Sr_4C_{60}, also superconducting ($T_c = 6.7$ and 4 K respectively), a body-centered orthorhombic structure was determined [134, 135]. The fact that Ca_5C_{60} as well as Ba_6C_{60} are superconducting without having a fcc structure implies that the superconductivity mechanism is independent of the coordination number, and that there are no particular rules for the occurrence of superconductivity in earth-alkaline fullerides. Most of the superconductive fullerides have a full t_{1u} conduction band and a partially filled t_{1g} conduction band. Superconductivity is observed in orthorhombic, bcc, and fcc-simple cubic structures [114].

In Ba_3C_{60} the t_{1u} band is completely filled. Higher doping levels are reached in the superconductors Ca_5C_{60} and Ba_6C_{60}. In these compounds, the t_{1g} band is also occupied with charge carriers, as shown by the valence band spectra [131, 138]. Therefore, their superconductivity is associated with the t_{1g} band.

Other valences of the C_{60} fulleride have been obtained by the synthesis of mixed alkaline/alkaline earth compounds [136, 137]. $K_3Ba_3C_{60}$ is formally nonavalent and shows superconductivity with a T_c of 5.6 K. Other mixed fullerides with lower T_cs are $K_2RbBa_3C_{60}$ and $Rb_3Ba_3C_{60}$. All of these compounds were made by intercalation of alkali metal into preformed Ba_3C_{60}. While in alkali fullerides (A_3C_{60}) superconductivity increases with increasing lattice parameter, the opposite is so for the mixed compounds. Fullerides with composition $A_3Ba_3C_{60}$ show decreasing T_cs with increasing cell-volume [112–114].

2.4.3
Reduction with Mercury

A bulk reduction of C_{60} in solution is also possible with less electropositive metals. This was demonstrated by the reduction of C_{60} with mercury, leading to C_{60}^- or C_{60}^{2-} [141]. The addition of a drop of mercury to a 1 mM solution of tetrahexylammonium bromide, (THA)Br, in tetrahydrofuran containing a brownish suspension of C_{60} leads to a rapid disappearance of the suspension and the formation of a clear red solution of the C_{60} monoanion. The solution is stable in the absence of air. As shown by ESR spectroscopy, the amount of dianion is only about 0.4% of the total anion concentration. These results imply that the standard potential of the Hg_2Br_2/Hg couple in THF is more negative than in water and sufficiently negative to reduce C_{60} to the mono- and dianion. As predicted by Nernst's law, the formal potential of the Hg_2Br_2/Hg should be shifted to more negative values upon higher (THA)X concentrations. Indeed, a 0.1 M (THA)Br solution increases the concentration of the dianion to 5%.

The reaction behavior of C_{60} towards the Hg_2X_2/Hg couple strongly depends on the solvent. In a solution of 1 mM (THA)I in benzonitrile, for example, no reaction was observed. This is because the formal potential of the Hg_2I_2/Hg couple under these conditions is not sufficiently negative to allow the reduction of C_{60}. An increase of the (THA)I concentration to 0.1 M, however, leads to the reduction of C_{60} to the monoanion and only traces (0.3%) of the dianion.

In the absence of any solvent no transfer of charges from Hg to the C_{60} cage is observed [142]. The transition-metal fulleride Hg_xC_{60} can be synthesized by mixing C_{60} and Hg and by heating it for several months. This fulleride seems to be an intercalation compound with Hg occupying the octahedral sites of the C_{60} lattice. Lattice expansion, but no reduction of the fullerene, is observed. The high ionization energy of mercury prevents a charge transfer.

2.5 Reduction with Organic Donor Molecules

The synthesis and isolation of fulleride salts is also possible with organic or metal organic donor molecules. Since a large variety of molecular compounds fulfill the requirement of being a donor of sufficient strength, it can be expected to gain access to materials with tunable solid-state properties. The first examples of fullerene based charge-transfer (CT) complexes, containing molecular cationic species already showed a remarkable electronic and magnetic behavior [102, 143–149]. The CT-complex formation is straightforward and usually achieved simply by mixing solutions of the components in a non-polar solvent, for example benzene. This leads to the precipitation of air-sensitive $(D^+)_nC_{60}{}^{n-}$ complexes. The CT-complexes are usually soluble in benzonitrile or THF. If the preparation is carried out in these more polar solvents, the solid CT-complexes can be obtained by precipitation with hexane [102]. With liquid tetrakis(dimethylamino)ethylene (TDAE) as organic donor, a quantitative transformation into $[TDAE]^+C_{60}{}^-$ is achieved by allowing the neat TDAE to react with C_{60} [145].

This early success in forming CT-salts with interesting properties led to the preparation of various donor–C_{60} complexes [1, 26, 150]. Amongst others, complexes with tetrathiafulvalene-derivatives (TTF) [150–153], tetraphenylporphyrin-derivatives (TPP) [143, 150, 154, 155] or metallocenes [74, 102, 149, 156–160] were crystallized

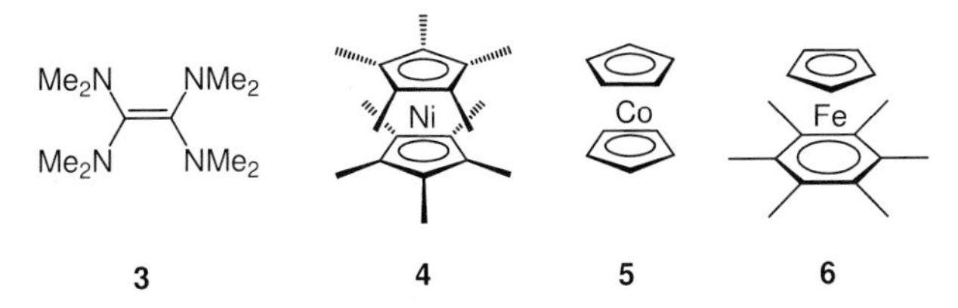

Figure 2.7 Some donors that were used for the formation of charge-transfer-complexes of C_{60}.

and examined. Only a few of these complexes turned out to be ionic. Most of the applied donors can not reduce C_{60} to the anion and the degree of charge transfer is close to zero [26, 150]. However, the use of strong electronic metallocene donors such as cobaltocene ($CoCp_2$) [102, 156, 157], $Fe^I(C_5H_5)(C_6Me_6)$] [149], or decamethylnickelocene [74] (Figure 2.7) has led to the formation of fully ionic salts with promising magnetic or electronic properties.

With the 19-electron complex $[Fe^I(C_5H_5)(C_6Me_6)]$ as donor, the synthesis of a mono-, di- or trianion can be controlled by choosing the right stoichiometry, 1 : 1, 2 : 1 or 3 : 1, respectively [149]. The ease of C_{60} reduction depends on the reduction potential of the donor. These reduction potentials themselves, however, strongly depend on, for example, the polarity of the solvent. Thus, a reliable comparison of the reduction power of the donors in Figure 2.7 would require the knowledge of their reduction potentials determined under identical conditions. The salt formed by combination of C_{60} with decamethylnickelocene $Ni(C_5Me_5)_2$ is a fairly good conductor. $Ni(C_5Me_5)_2$ reduces C_{60} selectively to the monoanion [74]. Also all the other CT-complexes of Figure 2.7 with the exception of **6** contain C_{60} as the monoanion. This can be shown by UV/Vis/NIR and ESR spectroscopy [74, 102, 147, 150, 157]. The characteristic properties of the C_{60} monoanion, such as the 1078 nm NIR absorption or the ESR features, indicating a Jahn–Teller distortion of the t_{1u} orbitals, are found in these chemically prepared systems as well.

A metal-free reductant that is strong enough to produce the C_{60}^--anion was found with the crystal violet radical $\mathbf{7}^\bullet$ [161, 162]; $\mathbf{7}^\bullet$ is a non-dimerizing, persistent radical that can be readily prepared by treatment of crystal violet iodide with zinc. It does not, as expected for a radical, add to C_{60} but reduces it to give $C_{60}^{\bullet-}$ as the brown microcrystalline powder $\mathbf{7}^+C_{60}^{\bullet-}$ (Figure 2.8).

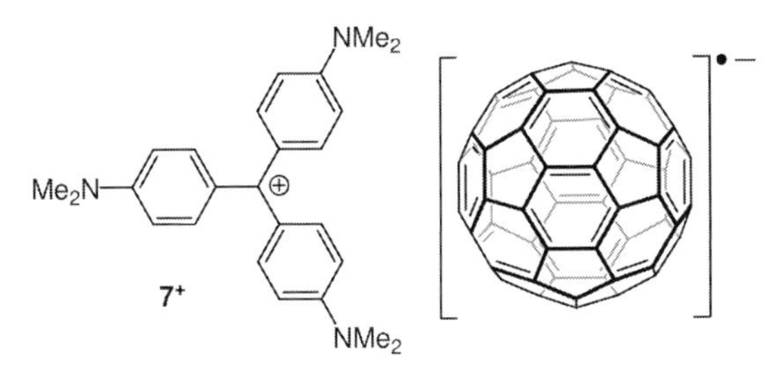

Figure 2.8 Metal-free C_{60}^--salt, prepared by reaction of crystal violet iodide and zinc with C_{60}.

Under certain conditions the CT-complex formation is reversible. If, for example, solid $[TDAE^+]C_{60}^-$ is treated with toluene in the absence of further TDAE, a quantitative reversion to the parent C_{60} takes place upon the formation of a purple solution (Scheme 2.4) [145]. An analogous process occurs with $[TDAE^+]C_{60}^-$ under ultrahigh vacuum at room temperature.

$$C_{60} + TDAE \xleftarrow{\text{ultra high vacuum}} [TDAE^+]C_{60}^- \xrightarrow{\text{toluene}} C_{60} + TDAE$$

Scheme 2.4

Similarly, treatment of $[Cr^{III}(TPP)^+]C_{60}^-$ in toluene reverses the redox reaction [143]. The addition of a few drops of tetrahydrofuran to the toluene solution promotes the electron transfer again. This is consistent with the lower first reduction potential of C_{60} reported in tetrahydrofuran compared with other solvents [10].

Among the CT-complexes represented in Figure 2.7, $[TDAE^+]C_{60}^-$, synthesized by Wudl and coworkers [144], exhibits unique solid-state properties. This material undergoes a transition to a ferromagnetic state at $T_c = 16.1$ K [144, 163–174], a value about 27 times that of previous strictly organic molecular ferromagnets. The ferromagnetic state shows no remanence. This result was interpreted in terms of a soft itinerant ferromagnet [147]. The fact that $[TDAE^+]C_{60}^-$ has the highest transition temperature of all purely organic nonpolymeric ferromagnets has been especially ascribed to a possible Jahn–Teller distortion and a resulting redistribution of the unpaired electronic spin into a belt-like form [166]. Nevertheless, conclusive understanding of the mechanism of the ferromagnetic interactions has not been reached. $[TDAE^+]C_{60}^-$ has a *c*-centered monoclinic unit cell [145], with dimensions $a = 15.85$ Å, $b = 12.99$ Å, $c = 9.97$ Å and $\beta = 93.31°$. The intermolecular contacts along the *c*-axis are only 9.97 Å, a value less than, or comparable to, those of alkali metal doped superconductors M_3C_{60}. This short intermolecular separation suggests the formation of a conduction band, which is partly filled [147]. The pronounced structural anisotropy implies a fairly low-dimensional band structure. The room temperature conductivity of $[TDAE^+]C_{60}^-$ is about 10^{-2} S cm^{-1} [144].

More insight into the reasons for magnetism of C_{60} derivatives may be obtained by a closer look at other recently discovered ferromagnetic C_{60} materials, which include complexes of cobaltocene with a C_{60}-derivative [156, 157], but also metal complexes with europium [175–177] or cerium [178–180] or pure C_{60} in a polymeric modification [181–183].

References

1 C. A. Reed, R. D. Bolskar, *Chem. Rev.* **2000**, *100*, 1075.
2 A. D. J. Haymet, *Chem. Phys. Lett.* **1985**, *122*, 421.
3 R. C. Haddon, L. E. Brus, K. Raghavachari, *Chem. Phys. Lett.* **1986**, *125*, 459.
4 S. Satpathy, *Chem. Phys. Lett.* **1986**, *130*, 545.
5 P. D. Hale, *J. Am. Chem. Soc.* **1986**, *108*, 6087.
6 S. Larsson, A. Volosov, A. Rosen, *Chem. Phys. Lett.* **1987**, *137*, 501.
7 A. Rosen, B. Waestberg, *J. Chem. Phys.* **1989**, *90*, 2525.
8 L. Echegoyen, L. E. Echegoyen, *Acc. Chem. Res.* **1998**, *31*, 593.
9 R. E. Haufler, J. Conceicao, L. P. F. Chibante, Y. Chai, N. E. Byrne, S. Flanagan, M. M. Haley, S. C. O'Brien, C. Pan, Z. Xiao, W. E. Billups, M. A. Ciufolini, R. H. Hauge, J. L. Margrave, L. J. Wilson, R. F. Curl, R. E. Smalley, *J. Phys. Chem.* **1990**, *94*, 8634.
10 P. M. Allemand, A. Koch, F. Wudl, Y. Rubin, F. Diederich, M. M. Alvarez, S. J. Anz, R. L. Whetten, *J. Am. Chem. Soc.* **1991**, *113*, 1050.
11 D. Dubois, K. M. Kadish, S. Flanagan, R. E. Haufler, L. P. F. Chibante, L. J. Wilson, *J. Am. Chem. Soc.* **1991**, *113*, 4364.
12 D. Dubois, K. M. Kadish, S. Flanagan, L. J. Wilson, *J. Am. Chem. Soc.* **1991**, *113*, 7773.
13 Q. Xie, E. Perez-Cordero, L. Echegoyen, *J. Am. Chem. Soc.* **1992**, *114*, 3978.
14 Y. Ohsawa, T. Saji, *J. Chem. Soc., Chem. Commun.* **1992**, 781.
15 F. Zhou, C. Jehoulet, A. J. Bard, *J. Am. Chem. Soc.* **1992**, *114*, 11004.
16 L. Echegoyen, F. Diederich, L. E. Echegoyen, *Full.: Chem. Phys. Technol.* **2000**, 1.
17 L. J. Wilson, S. Flanagan, L. P. F. Chibante, J. M. Alford, *Buckminsterfullerenes* **1993**, 285.
18 D. Dubois, G. Moninot, W. Kutner, M. T. Jones, K. M. Kadish, *J. Phys. Chem.* **1992**, *96*, 7137.
19 R. G. Compton, R. A. Spackman, D. J. Riley, R. G. Wellington, J. C. Eklund, A. C. Fisher, M. L. H. Green, R. E. Doothwaite, A. H. H. Stephens, et al., *J. Electroanal. Chem.* **1993**, *344*, 235.
20 W. R. Fawcett, M. Opallo, M. Fedurco, J. W. Lee, *J. Am. Chem. Soc.* **1993**, *115*, 196.
21 M. V. Mirkin, L. O. S. Bulhoes, A. J. Bard, *J. Am. Chem. Soc.* **1993**, *115*, 201.
22 I. Noviandri, R. D. Bolskar, P. A. Lay, C. A. Reed, *J. Phys. Chem. B* **1997**, *101*, 6350.
23 A. S. Lobach, V. V. Strelets, *Russ. Chem. Bull.* **2001**, *50*, 1593.
24 S. Fukuzumi, H. Mori, T. Suenobu, H. Imahori, X. Gao, K. M. Kadish, *J. Phys. Chem. A* **2000**, *104*, 10688.
25 I. A. Nuretdinov, V. V. Yanilkin, V. I. Morozov, V. P. Gubskaya, V. V. Zverev, N. V. Nastapova, G. M. Fazleeva, *Russ. Chem. Bull.* **2002**, *51*, 263.
26 N. Martin, L. Sanchez, B. Illescas, I. Perez, *Chem. Rev.* **1998**, *98*, 2527.
27 E. Koudoumas, M. Konstantaki, A. Mavromanolakis, S. Couris, M. Fanti, F. Zerbetto, K. Kordatos, M. Prato, *Chem. Eur. J.* **2003**, *9*, 1529.
28 M. Carano, T. Da Ros, M. Fanti, K. Kordatos, M. Marcaccio, F. Paolucci, M. Prato, S. Roffia, F. Zerbetto, *J. Am. Chem. Soc.* **2003**, *125*, 7139.
29 M. A. Greaney, S. M. Gorun, *J. Phys. Chem.* **1991**, *95*, 7142.
30 P. N. Keizer, J. R. Morton, K. F. Preston, A. K. Sugden, *J. Phys. Chem.* **1991**, *95*, 7117.
31 G. A. Heath, J. E. McGrady, R. L. Martin, *J. Chem. Soc., Chem. Commun.* **1992**, 1272.
32 D. Dubois, M. T. Jones, K. M. Kadish, *J. Am. Chem. Soc.* **1992**, *114*, 6446.
33 D. R. Lawson, D. L. Feldheim, C. A. Foss, P. K. Dorhout, C. M. Elliott, C. R. Martin, B. Parkinson, *J. Electrochem. Soc.* **1992**, *139*, L68.
34 D. R. Lawson, D. L. Feldhiem, C. A. Foss, P. K. Dorhout, C. M. Elliott, C. R. Martin, B. Parkinson, *J. Phys. Chem.* **1992**, *96*, 7175.
35 M. Baumgarten, A. Gügel, L. Gherghel, *Adv. Mater.* **1993**, *5*, 458.

36 P. Rapta, A. Bartl, A. Gromov, A. Stasko, L. Dunsch, *ChemPhysChem* **2002**, *3*, 351.
37 T. Kato, T. Kodama, T. Shida, T. Nakagawa, Y. Matsui, S. Suzuki, H. Shiromaru, K. Yamauchi, Y. Achiba, *Chem. Phys. Lett.* **1991**, *180*, 446.
38 T. Kato, T. Kodama, M. Oyama, S. Okazaki, T. Shida, T. Nakagawa, Y. Matsui, S. Suzuki, H. Shiromaru, K. Yamauchi, Y. Achiba, *Chem. Phys. Lett.* **1991**, *186*, 35.
39 P. M. Allemand, G. Srdanov, A. Koch, K. Khemani, F. Wudl, Y. Rubin, F. Diederich, M. M. Alvarez, S. J. Anz, R. L. Whetten, *J. Am. Chem. Soc.* **1991**, *113*, 2780.
40 P. Paul, K.-C. Kim, D. Sun, P. D. W. Boyd, C. A. Reed, *J. Am. Chem. Soc.* **2002**, *124*, 4394.
41 P. Paul, R. D. Bolskar, A. M. Clark, C. A. Reed, *Chem. Commun.* **2000**, 1229.
42 T. Sternfeld, C. Thilgen, R. E. Hoffman, M. d. R. C. Heras, F. Diederich, F. Wudl, L. T. Scott, J. Mack, M. Rabinovitz, *J. Am. Chem. Soc.* **2002**, *124*, 5734.
43 E. Shabtai, A. Weitz, R. C. Haddon, R. E. Hoffman, M. Rabinovitz, A. Khong, R. J. Cross, M. Saunders, P.-C. Cheng, L. T. Scott, *J. Am. Chem. Soc.* **1998**, *120*, 6389.
44 M. Buhl, *Chem. Eur. J.* **1998**, *4*, 734.
45 M. Buhl, A. Hirsch, *Chem. Rev.* **2001**, *101*, 1153.
46 M. Buhl, *Z. Anorg. Allg. Chem.* **2000**, *626*, 332.
47 T. Sternfeld, R. E. Hoffman, M. Saunders, R. J. Cross, M. S. Syamala, M. Rabinovitz, *J. Am. Chem. Soc.* **2002**, *124*, 8786.
48 P. D. W. Boyd, P. Bhyrappa, P. Paul, J. Stinchcombe, R. Bolskar, Y. Sun, C. A. Reed, *J. Am. Chem. Soc.* **1995**, *117*, 2907.
49 Y. Sun, C. A. Reed, *Chem. Commun.* **1997**, 747.
50 J. W. Bausch, G. K. S. Prakash, G. A. Olah, D. S. Tse, D. C. Lorents, Y. K. Bae, R. Malhotra, *J. Am. Chem. Soc.* **1991**, *113*, 3205.
51 M. Wu, X. Wei, L. Qi, Z. Xu, *Tetrahedron Lett.* **1996**, *37*, 7409.
52 J. Chen, Q.-F. Shao, R.-F. Cai, Z.-E. Huang, *Solid State Commun.* **1995**, *96*, 199.
53 J. Chen, F.-F. Cai, Q.-F. Shao, Z.-E. Huang, S.-M. Chen, *Chem. Commun.* **1996**, 1111.
54 R. E. Douthwaite, A. R. Brough, M. L. H. Green, *J. Chem. Soc., Chem. Commun.* **1994**, 267.
55 R. E. Douthwaite, M. A. Green, M. L. H. Green, M. J. Rosseinsky, *J. Mater. Chem.* **1996**, *6*, 1913.
56 Y. Errammach, A. Rezzouk, F. Rachdi, J. L. Sauvajol, *Synth. Methods* **2002**, *129*, 147.
57 A. Rezzouk, F. Rachdi, Y. Errammach, J. L. Sauvajol, *Physica E* **2002**, *15*, 107.
58 J. Chen, Z.-E. Huang, R.-F. Cai, Q.-F. Shao, S.-M. Chen, H.-J. Ye, *J. Chem. Soc., Chem. Commun.* **1994**, 2177.
59 H. Moriyama, H. Kobayashi, A. Kobayashi, T. Watanabe, *Chem. Phys. Lett.* **1995**, *238*, 116.
60 J. Chen, Z.-E. Huang, R.-F. Cai, Q.-F. Shao, H.-J. Ye, *Solid State Commun.* **1995**, 95, 233.
61 F. Rachdi, L. Hajji, M. Galtier, T. Yildirim, J. E. Fischer, C. Goze, M. Mehring, *Phys. Rev. B* **1997**, *56*, 7831.
62 A. J. Fowkes, J. M. Fox, P. F. Henry, S. J. Heyes, M. J. Rosseinsky, *J. Am. Chem. Soc.* **1997**, *119*, 10413.
63 L. Hajji, F. Rachdi, C. Goze, M. Mehrihg, J. E. Fischer, *Solid State Commun.* **1996**, *100*, 493.
64 J. Reichenbach, F. Rachdi, I. Luk'yanchuk, M. Ribet, G. Zimmer, M. Mehring, *J. Chem. Phys.* **1994**, *101*, 4585.
65 A. Perez-Benitez, C. Rovira, J. Veciana, J. Vidal-Gancedo, *Synth. Methods* **2001**, *121*, 1157.
66 T. F. Fässler, A. Spiekermann, M. E. Spahr, R. Nesper, *Angew. Chem.* **1997**, *109*, 1371; *Angew. Chem. Int. Ed. Engl.* **1997**, *36*, 486.
67 T. F. Fässler, R. Hoffmann, S. Hoffmann, M. Worle, *Angew. Chem.* **2000**, *112*, 2170; *Angew. Chem. Int. Ed. Engl.* **2000**, *39*, 2091.
68 C. Coulon, A. Penicaud, R. Clerac, R. Moret, P. Launois, J. Hone, *Phys. Rev. Lett.* **2001**, *86*, 4346.

69 Y. Kubozono, Y. Takabayashi, T. Kambe, S. Fujiki, S. Kashino, S. Emura, *Phys. Rev. B* **2001**, *63*, 045418/1.

70 P. Dahlke, M. J. Rosseinsky, *Chem. Mater.* **2002**, *14*, 1285.

71 S. Margadonna, E. Aslanis, W. Z. Li, K. Prassides, A. N. Fitch, T. C. Hansen, *Chem. Mater.* **2000**, *12*, 2736.

72 L. Cristofolini, M. Ricco, R. De Renzi, *Phys. Rev. B* **1999**, *59*, 8343.

73 P. Paul, Z. Xie, R. Bau, P. D. W. Boyd, C. A. Reed, *J. Am. Chem. Soc.* **1994**, *116*, 4145.

74 W. C. Wan, X. Liu, G. M. Sweeney, W. E. Broderick, *J. Am. Chem. Soc.* **1995**, *117*, 9580.

75 U. Becker, G. Denninger, V. Dyakonov, B. Gotschy, H. Klos, G. Rösler, A. Hirsch, H. Winter, *Europhys. Lett.* **1993**, *21*, 267.

76 V. Dyakonov, G. Rösler, H. Klos, B. Gotschy, G. Denninger, A. Hirsch, *Synth. Methods* **1993**, *56*, 3214.

77 H. Moriyama, H. Kobayashi, A. Kobayashi, T. Watanabe, *J. Am. Chem. Soc.* **1993**, *115*, 1185.

78 A. Penicaud, A. Perez-Benitez, R. Gleason V, E. Munoz P, R. Escudero, *J. Am. Chem. Soc.* **1993**, *115*, 10392.

79 C. A. Foss, Jr., D. L. Feldheim, D. R. Lawson, P. K. Dorhout, C. M. Elliott, C. R. Martin, B. A. Parkinson, *J. Electrochem. Soc.* **1993**, *140*, L84.

80 B. Miller, J. M. Rosamilia, *J. Chem. Soc., Faraday Trans.* **1993**, *89*, 273.

81 W. Koh, D. Dubois, W. Kutner, M. T. Jones, K. M. Kadish, *J. Phys. Chem.* **1992**, *96*, 4163.

82 U. Bilow, M. Jansen, *J. Chem. Soc., Chem. Commun.* **1994**, 403.

83 V. V. Gritsenko, O. A. Dyachenko, G. V. Shilov, N. G. Spitsyna, E. B. Yagubskii, *Russ. Chem. Bull.* **1997**, *46*, 1878.

84 W. Schuetz, J. Gmeiner, A. Schilder, B. Gotschy, V. Enkelmann, *Chem. Commun.* **1996**, 1571.

85 H. Kobayashi, H. Moriyama, A. Kobayashi, T. Watanabe, *Synth. Methods* **1995**, *70*, 1451.

86 H. Kobayashi, H. Tomita, H. Moriyama, A. Kobayashi, T. Watanabe, *J. Am. Chem. Soc.* **1994**, *116*, 3153.

87 C. Caron, R. Subramanian, F. D'Souza, J. Kim, W. Kutner, M. T. Jones, K. M. Kadish, *J. Am. Chem. Soc.* **1993**, *115*, 8505.

88 K. M. Kadish, X. Gao, E. Van Caemelbecke, T. Hirasaka, T. Suenobu, S. Fukuzumi, *J. Phys. Chem. A* **1998**, *102*, 3898.

89 K. M. Kadish, X. Gao, E. V. Caemelbecke, T. Suenobu, S. Fukuzumi, *J. Am. Chem. Soc.* **2000**, *122*, 563.

90 T. V. Magdesieva, D. N. Kravchuk, V. V. Bashilov, I. V. Kusnetsova, V. I. Sokolov, K. P. Butin, *Russ. Chem. Bull.* **2002**, *51*, 1588.

91 E. Allard, L. Riviere, J. Delaunay, D. Rondeau, D. Dubois, J. Cousseau, *Proc. Electrochem. Soc.* **2000**, *2000-10*, 88.

92 E. Allard, J. Delaunay, F. Cheng, J. Cousseau, J. Orduna, J. Garin, *Org. Lett.* **2001**, *3*, 3503.

93 E. Allard, J. Delaunay, J. Cousseau, *Org. Lett.* **2003**, *5*, 2239.

94 S. Fukuzumi, T. Suenobu, T. Hirasaka, R. Arakawa, K. M. Kadish, *J. Am. Chem. Soc.* **1998**, *120*, 9220.

95 Y. Ederle, C. Mathis, *Macromolecules* **1997**, *30*, 4262.

96 Z. Wang, M. S. Meier, *J. Org. Chem.* **2003**, *68*, 3043.

97 T. Yildirim, O. Zhou, J. E. Fischer, *Phys. Chem. Mater.* **2000**, *23*, 23.

98 T. Yildirim, O. Zhou, J. E. Fischer, *Phys. Chem. Mater.* **2000**, *23*, 67.

99 T. Yildirim, O. Zhou, J. E. Fischer, *Phys. Chem. Mater.* **2000**, *23*, 249.

100 W. K. Fullagar, I. R. Gentle, G. A. Heath, J. W. White, *J. Chem. Soc., Chem. Commun.* **1993**, 525.

101 P. Bhyrappa, P. Paul, J. Stinchcombe, P. D. W. Boyd, C. A. Reed, *J. Am. Chem. Soc.* **1993**, *115*, 11004.

102 J. Stinchcombe, A. Penicaud, P. Bhyrappa, P. D. W. Boyd, C. A. Reed, *J. Am. Chem. Soc.* **1993**, *115*, 5212.

103 Y. Chi, J. B. Bhonsle, T. Canteenwala, J.-P. Huang, J. Shiea, B.-J. Chen, L. Y. Chiang, *Chem. Lett.* **1998**, 465.

104 C. Janiak, S. Mühle, H. Hemling, K. Köhler, *Polyhedron* **1996**, *15*, 1559.

105 A. F. Hebard, M. J. Rosseinsky, R. C. Haddon, D. W. Murphy, S. H. Glarum, T. T. M. Palstra, A. P. Ramirez, A. R. Kortan, *Nature* **1991**, *350*, 600.

106 K. Holczer, O. Klein, S. M. Huang, R. B. Kaner, K. J. Fu, R. L. Whetten, F. Diederich, *Science* **1991**, *252*, 1154.
107 P. W. Stephens, L. Mihaly, P. L. Lee, R. L. Whetten, S. M. Huang, R. Kaner, F. Diederich, K. Holczer, *Nature* **1991**, *351*, 632.
108 O. Zhou, J. E. Fischer, N. Coustel, S. Kycia, Q. Zhu, A. R. McGhie, W. J. Romanow, J. P. McCauley, Jr., A. B. Smith, III, D. E. Cox, *Nature* **1991**, *351*, 462.
109 S. P. Kelty, C. C. Chen, C. M. Lieber, *Nature* **1991**, *352*, 223.
110 K. Tanigaki, T. W. Ebbesen, S. Saito, J. Mizuki, J. S. Tsai, Y. Kubo, S. Kuroshima, *Nature* **1991**, *352*, 222.
111 K. Tanigaki, I. Hirosawa, T. W. Ebbesen, J. Mizuki, Y. Shimakawa, Y. Kubo, J. S. Tsai, S. Kuroshima, *Nature* **1992**, *356*, 419.
112 S. Margadonna, K. Prassides, *J. Solid State Chem.* **2002**, *168*, 639.
113 O. Gunnarsson, *Rev. Mod. Phys.* **1997**, *69*, 575.
114 Y. Iwasa, T. Takenobu, *J. Phys. Cond. Mater.* **2003**, *15*, R495.
115 J. P. McCauley, Jr., Q. Zhu, N. Coustel, O. Zhou, G. Vaughan, S. H. J. Idziak, J. E. Fischer, S. W. Tozer, D. M. Groski, et al., *J. Am. Chem. Soc.* **1991**, *113*, 8537.
116 D. W. Murphy, M. J. Rosseinsky, R. M. Fleming, R. Tycko, A. P. Ramirez, R. C. Haddon, T. Siegrist, G. Dabbagh, J. C. Tully, R. E. Walstedt, *J. Phys. Chem. Solids* **1992**, *53*, 1321.
117 Z. Iqbal, R. H. Baughman, B. L. Ramakrishna, S. Khare, N. S. Murthy, H. J. Bornemann, D. E. Morris, *Science* **1991**, *254*, 826.
118 H. H. Wang, A. M. Kini, B. M. Savall, K. D. Carlson, J. M. Williams, K. R. Lykke, P. Wurz, D. H. Parker, M. J. Pellin, D. M. Gruen, U. Welp, W.-K. Kwok, S. Fleshler, G. W. Crabtree, *Inorg. Chem.* **1991**, *30*, 2838.
119 H. H. Wang, A. M. Kini, B. M. Savall, K. D. Carlson, J. M. Williams, M. W. Lathrop, K. R. Lykke, D. H. Parker, P. Wurz, M. I. Pellin, D. M. Gruen, U. Welp, W.-K. Kwok, S. Fleshler, G. W. Crabtree, I. E. Schirber, D. L. Overmyer, *Inorg. Chem.* **1991**, *30*, 2962.
120 D. R. Buffinger, R. P. Ziebarth, V. A. Stenger, C. Recchia, C. H. Pennington, *J. Am. Chem. Soc.* **1993**, *115*, 9267.
121 M. Baumgarten, L. Gherghel, *Appl. Magn. Res.* **1996**, *11*, 171.
122 H. Shimoda, Y. Iwasa, T. Mitani, *Synth. Methods* **1997**, *85*, 1593.
123 O. Zhou, T. T. M. Palstra, Y. Iwasa, R. M. Fleming, A. F. Hebard, P. E. Sulewski, D. W. Murphy, B. R. Zegarski, *Phys. Rev. B* **1995**, *52*, 483.
124 H. Shimoda, Y. Iwasa, Y. Miyamoto, Y. Maniwa, T. Mitani, *Phys. Rev. B* **1996**, *54*, R15653.
125 Y. Chabre, D. Djurado, M. Armand, W. R. Romanow, N. Coustel, J. P. McCauley, Jr., J. E. Fischer, A. B. Smith, III, *J. Am. Chem. Soc.* **1992**, *114*, 764.
126 M. J. Rosseinsky, D. W. Murphy, R. M. Fleming, R. Tycko, A. P. Ramirez, T. Siegrist, G. Dabbagh, S. E. Barrett, *Nature* **1992**, *356*, 416.
127 R. M. Fleming, M. J. Rosseinsky, A. P. Ramirez, D. W. Murphy, J. C. Tully, R. C. Haddon, T. Siegrist, R. Tycko, S. H. Glarum, et al., *Nature* **1991**, *352*, 701.
128 J. Fink, E. Sohmen, *Phys. Bl.* **1992**, *48*, 11.
129 R. M. Fleming, A. P. Ramirez, M. J. Rosseinsky, D. W. Murphy, R. C. Haddon, S. M. Zahurak, A. V. Makhija, *Nature* **1991**, *352*, 787.
130 Y. Chen, F. Stepniak, J. H. Weaver, L. P. F. Chibante, R. E. Smalley, *Phys. Rev. B* **1992**, *45*, 8845.
131 A. R. Kortan, N. Kopylov, S. Glarum, E. M. Gyorgy, A. P. Ramirez, R. M. Fleming, F. A. Thiel, R. C. Haddon, *Nature* **1992**, *355*, 529.
132 A. R. Kortan, N. Kopylov, E. Ozdas, A. P. Ramirez, R. M. Fleming, R. C. Haddon, *Chem. Phys. Lett.* **1994**, *223*, 501.
133 A. R. Kortan, N. Kopylov, S. Glarum, E. M. Gyorgy, A. P. Ramirez, R. M. Fleming, O. Zhou, F. A. Thiel, P. L. Trevor, R. C. Haddon, *Nature* **1992**, *360*, 566.
134 M. Bänitz, M. Heinze, K. Lüders, H. Werner, R. Schlögl, M. Weiden,

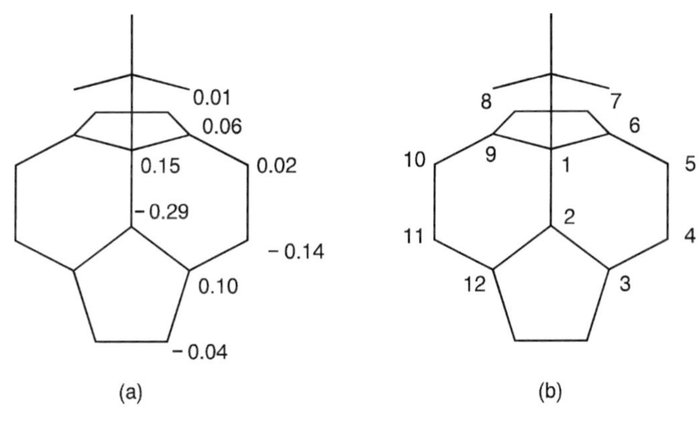

Figure 3.1 Electron densities (Mulliken charges) of the affected pyracylene unit of the intermediate ${}^{t}BuC_{60}^{-}$ obtained from AM1 calculations (a), and numbering of the carbon atoms (b).

situ protonation of the intermediates RC_{60}^{-} [3], protonation of ${}^{t}BuC_{60}^{-}\ Li^{+}{\bullet}4CH_3CN$ afforded the 1,2- as well as the 1,4-isomer [5]. The 1,4-isomer, however, is unstable and, after about 12 h at 25 °C, rearranges to the 1,2-derivative. The anion ${}^{t}BuC_{60}^{-}$ can be oxidized with iodine to give the radical ${}^{t}BuC_{60}^{\bullet}$ [5], which is proposed to dimerize at C-4 [16]. Out of this radical species, 1,2-$C_{60}H^{t}Bu$ can be obtained by treatment with Bu_3SnH upon rearrangement of the initially formed 1,4-$C_{60}H^{t}Bu$ [5]. Whereas the anions RC_{60}^{-} are stable only under inert conditions, solutions of the protonated 1,2-dihydrofullerenes $C_{60}HR$ do not decompose in non-polar solvents in air.

All the expected 32 resonances of the fullerene carbon atoms of the 1-organyl-1,2-dihydro[60]fullerenes, with four of them half the intensity of the others, were observed in the ^{13}C NMR spectra, proving the C_s-symmetry [4, 5]. This experimental finding is corroborated by AM1 calculations for $C_{60}HR$ (R = H, Me, ${}^{t}Bu$), which show that the 1,2-isomers have the lowest heat of formation [3]. The formation of other regioisomers, such as the 1,4- or the 1,6-isomers would require the introduction of at least one [5, 6] double bond, which is energetically unfavorable for the fullerene framework [3, 17, 18]. An enhanced formation of an 1,4-isomer is expected to occur upon an increased steric requirement of the ligands bound to C_{60}, for example, on going from H to ${}^{t}Bu$ [17, 18]. The regioselectivity of these two-step additions is closer to that of an electron deficient alkene than to an aromatic system. A distinct signal for the fullerenyl proton in the ^{1}H NMR spectra of $C_{60}HR$ appears at low field within $\delta = 6$–7 – further direct proof for the electron-withdrawing influence of the fullerene sphere [4]. The chemical shift of the fullerenyl proton also depends on the nature of R. Increased steric requirements, for example, on going from $C_{60}HMe$ to $C_{60}H^{t}Bu$, cause a further down field shift of about 0.7 ppm. These down field positions already imply a pronounced acidity of the fullerenyl proton. Indeed, protonation of the ${}^{t}BuC_{60}^{-}$ anion, monitored quantitatively by electrochemical methods allowed the determination of the pK_a as 5.7 (±0.1) [5].

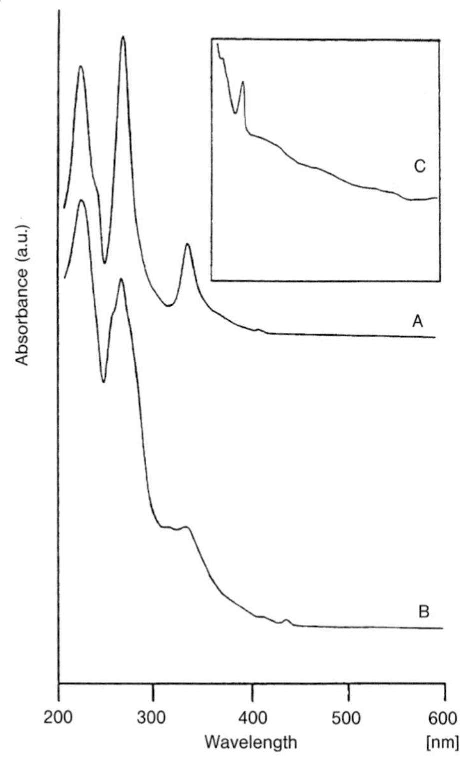

Figure 3.2 Comparative UV/Vis spectra (hexane) of fullerene-60 (A) and $C_{60}H^{t}Bu$ (B); the inset (C) is the 400–700 nm region of $C_{60}H^{t}Bu$ [4].

Therefore, $C_{60}H^{t}Bu$ is one of the strongest acids consisting exclusively of carbon and hydrogen.

The UV/Vis spectra (Figure 3.2) of the chestnut brown solutions of the monoadducts $C_{60}HR$, particularly the intensive bands at $\lambda_{max} = 213$, 257 and 326 nm, are close to those of C_{60}, demonstrating their electronic similarity [4]. The biggest changes in the spectra compared with C_{60} appear in the visible region. The typical features of C_{60} between $\lambda = 400$ and 700 nm are lost, and a new and very characteristic band at $\lambda_{max} = 435$ nm appears, which is independent of the nature of R. Also, the electrochemical properties of $C_{60}HR$ are comparable with those of C_{60} [5, 19]. The first three reversible reduction waves shift about 100 mV to more negative potentials. Therefore, the fullerene core in these monoadducts still exhibits remarkable electron-acceptor properties, which is one reason for almost the identical chemical reactivity compared with C_{60}.

Owing to their stability and low nucleophilicity, metal acetylides are less reactive toward C_{60} than other lithium organyls or Grignard reagents [11]. Though the reaction is slower and higher reaction temperatures are necessary, various acetylene derivatives of C_{60} could be obtained. The first acetylene C_{60} hybrids were (trimethylsilyl)ethynyl- and phenylethynyl-dihydro[60]fullerene, synthesized simultaneously

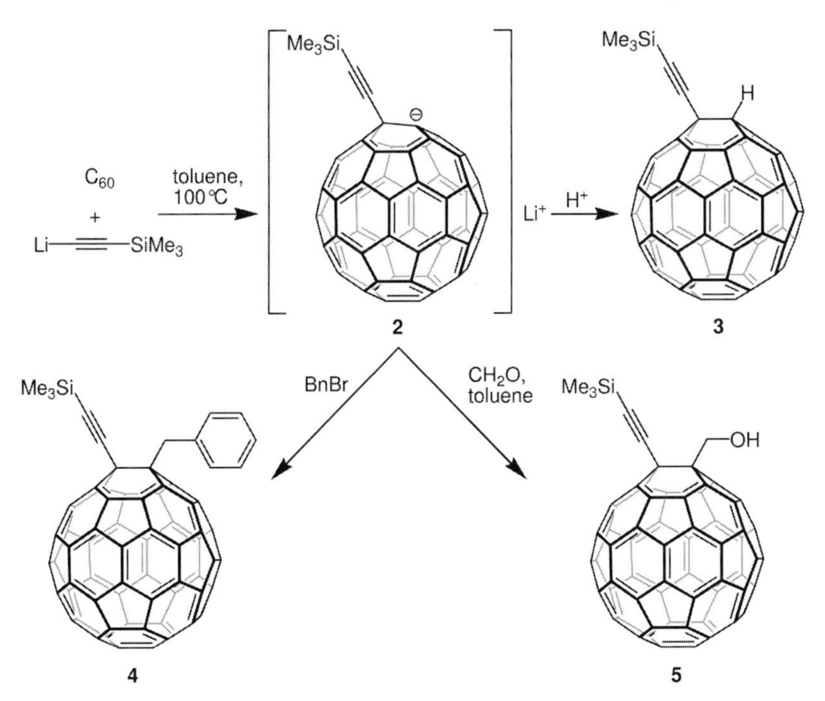

Scheme 3.2

by Diederich [13] and Komatsu [11] (Scheme 3.2). The final products were obtained after quenching with acid. As expected and calculated with the AM1-method [11], protonation takes place at C-2. At the same time, this proton is rather acidic and can be removed easily with K_2CO_3 in dimethyl sulfoxide. The acidity of **3** depends strongly on the solvent and the ability of the solvent to stabilize the corresponding cation of the base [20].

Other Li acetylides Li-C≡C-R with R = hexyl [21] or benzylether dendrons [22, 23] (up to the fourth generation) have also been attached to C_{60} (Figure 3.3), and various different electrophiles have been used to complete the reaction with the intermediate Li-fulleride (Scheme 3.2 and Figure 3.3). Besides the protonation, alkyl-, benzyl-, cycloheptatrienyl-, benzoyl- or vinylether-derivatives or formaldehyde and dichloroacetylene were used as electrophiles [12, 20]. Most of these electrophiles are attached to the anion in the expected C-2 position. The 1,4-adducts are available by quenching the anion with the tropylium cation or benzoyl chloride [12]. The fullerene anion can be stabilized by introduction of benzylether dendrons. The lifetimes of the anions change with the size of the dendrons [22].

The redox properties of the ethynylated dihydrofullerenes depend on the substituents. In benzonitrile three to four reversible reduction peaks and one irreversible oxidation peak can be observed. The reduction potentials for **3** (Scheme 3.2) and **8** (Figure 3.3) have values that are commonly observed for 1,2-dihydro[60]fullerenes, which means a shift of about 0.1 V to negative values.

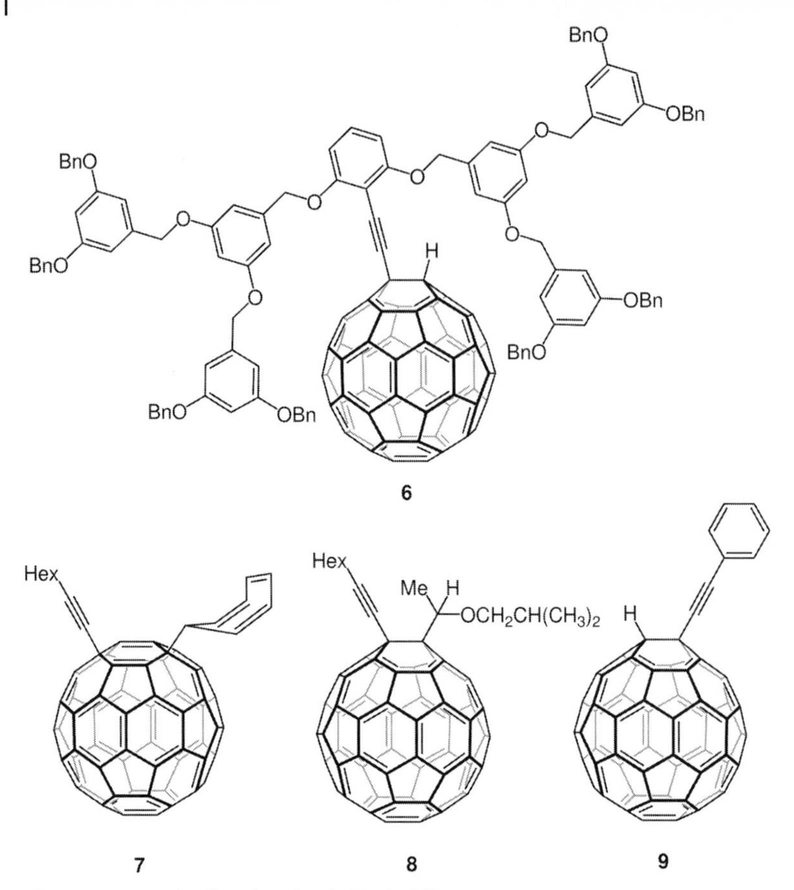

Figure 3.3 Examples for ethynylated dihydrofullerenes.

The reduction potential of **9** (Figure 3.3) is almost identical to the potential of pristine C_{60}, which may be due to the electron-withdrawing effect of the phenylethynyl group canceling out the above-mentioned expected cathodic shift [11].

Ethynylated dihydrofullerenes serve as precursors for buckydumbbells (Scheme 3.3). Coupling of the desilylated compound **10** with CuCl leads to the dimer **11** [24]. Reaction of C_{60} with the acetylide Li-C≡C-Li leads to the dimer with a bridge consisting of one acetylene unit only. Electronic interaction between the fullerene-units in these two buckydumbbells is negligible. Further examples of C_{60}-acetylene-hybrids synthesized by using cyclopropanation reactions are shown in Chapter 4.

Silylmethylation of fullerenes can be achieved by Grignard reaction of C_{60} with $Me_2Si(O^iPr)MgCl$ (Scheme 3.4). Two different types of silylmethylated 1,2-fullerene adducts are formed [8]. The use of different solvents leads either to the 1,2-product **12** or to the unprecedented 1,4-addition product **13**. These compounds react easily with various alcohols or phenols, which makes silylmethylated fullerenes a versatile starting material for regioselectively defined fullerene compounds [10]. Some examples are shown in Scheme 3.4.

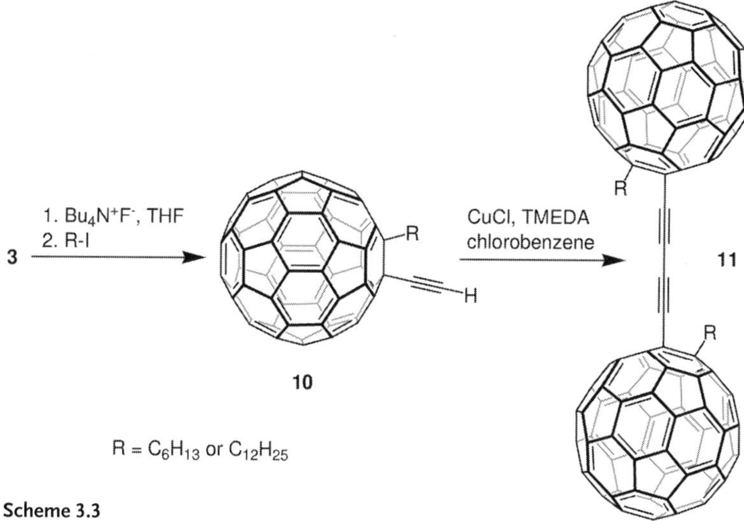

Scheme 3.3

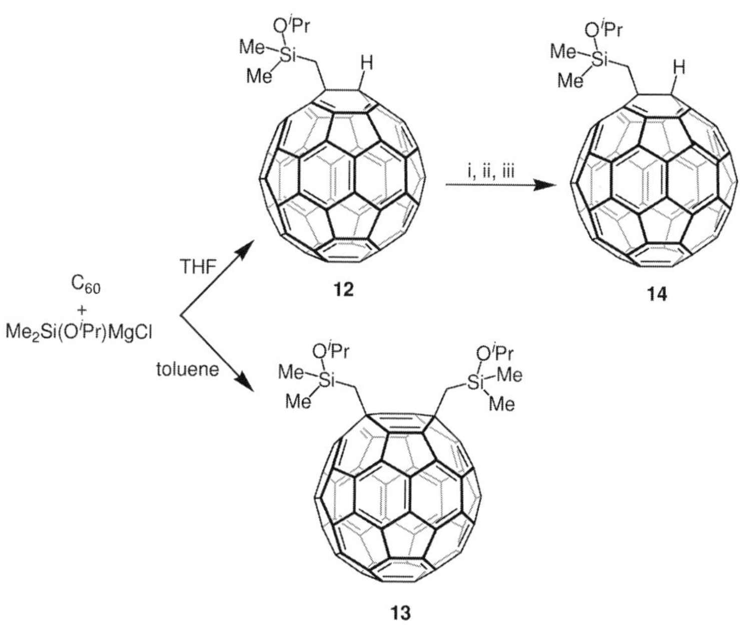

Scheme 3.4

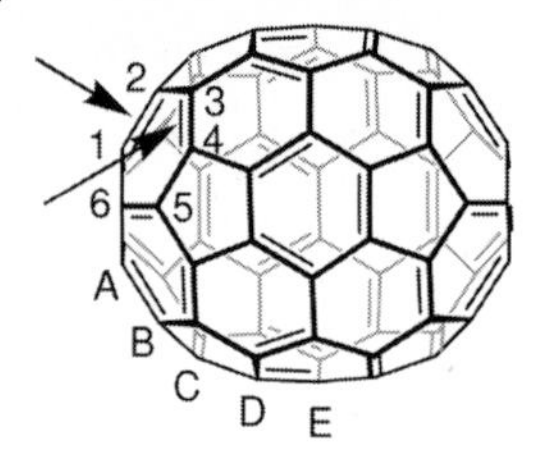

Figure 3.4 Schematic representation of C_{70}, indicating the two types of reactive double bonds, the numbering, and the five different sets of C atoms.

Organocopper compounds prepared by reaction with the corresponding lithium or Grignard-reagents afford, under certain conditions, multiple adducts with complete regioselectivity in very good yields [25, 26]. With this method tris- and pentaadducts are accessible that leave an indenide- or a cyclopentadienide-type π-system, respectively, on the C_{60}-surface [27]. A similar addition pattern can be observed by the reaction of potassium fluorenide with C_{60}, which leads to a tetra- and a pentaadduct with a fulvene- or a cyclopentadienide-type substructure, respectively [28]. All these compounds undergo reactions that are typical for such π-systems, such as deprotonation or complexation. Alkyl-, aryl-, alkenyl- or fluorenyl-addends were used to form these multiple adducts, which are described in more detail in Chapter 10.

Product predictions for nucleophilic additions to C_{70}, based on AM1 calculations, show that among the many possible isomers a few are energetically favored [29]. Two areas within the molecule can be distinguished, which are the inert belt at the equator and the more reactive C_{60}-like double bonds at the poles (Figure 3.4). Experimentally, hydroalkylation and hydroarylation reactions of C_{70} under quantitative HPLC control yield predominantly one isomer of each $C_{70}HR$ [4].

The ^{13}C NMR spectra of $C_{70}HR$ exhibits 37 resonances of the fullerene carbons, with two of them in the sp^3 region, proving C_s-symmetry for the $C_{70}HR$ adducts. This is consistent with an addition to a double bond of a pole corannulene unit (1,2-addition), leading to 1,2-dihydro[70]fullerene derivatives. These particular [6,6] bonds of C_{70}, located between the carbons of the sets A and B, have almost the same bond length as the [6,6] bonds in C_{60} [30], and the pole corannulene unit also exhibits bond alternation with longer [5,6] bonds. ^1H NMR data imply that the initial attack of the nucleophile occurred on C-1 and the protonation on C-2.

3.2.2
Cyclopropanation of C_{60} and C_{70}

The stabilization of reaction intermediates RC_{60}^- and RC_{70}^- to form dihydrofullerene derivatives can also be achieved by intramolecular nucleophilic substitutions (S_Ni), if R contains a leaving group. As shown by Bingel [31], the generation of a carbon nucleophile by deprotonation of α-halo esters or α-halo ketones leads to a clean cyclopropanation of C_{60}.

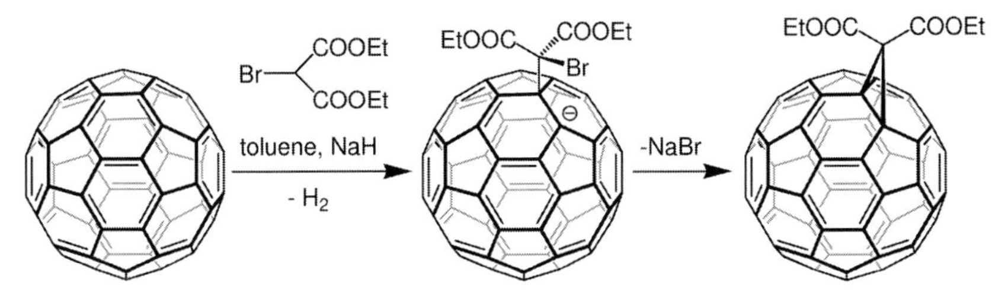

Scheme 3.5

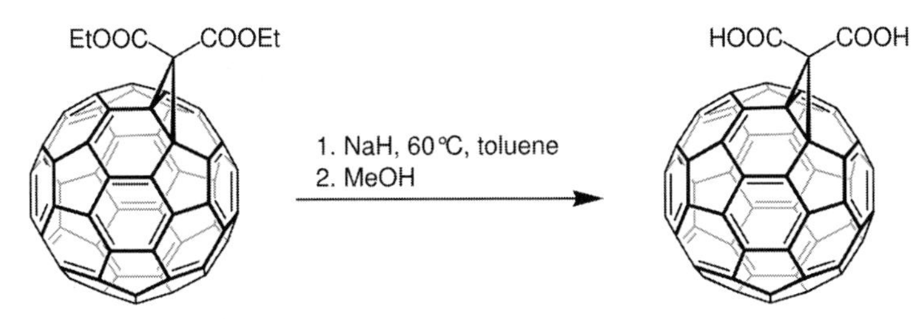

Scheme 3.6

Cyclopropanation of C_{60} with diethyl bromomalonate in toluene with NaH as auxiliary base proceeds smoothly at room temperature (Scheme 3.5). By-products are unreacted C_{60} and higher adducts. The formation of higher adducts is discussed in detail in Chapter 10. The monoadduct can be isolated easily from the reaction mixture by column chromatography. Saponification of such di(ethoxycarbonyl)-methylene adducts of C_{60} is achieved by treatment with NaH in toluene at elevated temperatures and subsequent quenching with methanol (Scheme 3.6) [32]. This method provides easy access to defined water-soluble fullerenes and can also be applied to higher adducts. These malonic acid derivatives of C_{60} are very soluble in polar solvents, for example acetone, THF or basic water, but insoluble in aqueous acids.

Cyclopropanation of C_{60} with methyl-2-chloroacetylacetate, ω-bromoacetophenone and desyl chloride proceeds similarly (Figure 3.5) [31]. In these cases, deprotonation to the stabilized carbanions is accomplished with 1,8-diazabicyclo[5.4.0]undec-7-ene (DBU) as auxiliary base.

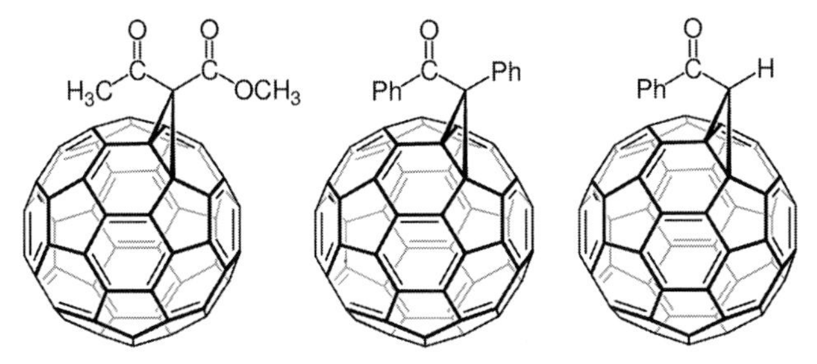

Figure 3.5 Products of the cyclopropanation of C_{60} with methyl-2-chloroacetylacetate, desylchloride and ω-bromoacetophenone.

^1H and ^{13}C NMR investigations demonstrate that the nucleophilic additions followed by S_Ni reactions take place at a [6,6] double bond of the fullerene core. The chemical shifts in the ^{13}C NMR spectra of C-1 and C-2 (66–76 ppm) in all of these 1,2-methano[60]fullerenes prove the cyclopropane structure. Cyclopropanation of C_{70} with diethyl bromomalonate in the presence of DBU yields predominantly one monoadduct (Scheme 3.7) [31]. As with acyclic 1-organyl-1,2-dihydro[70]fullerenes [4], the most C_{60}-like [6,6]-bond at the pole is attacked by the nucleophile, leading to 1,2-di(ethoxycarbonyl)methano[70]hydrofullerene [31].

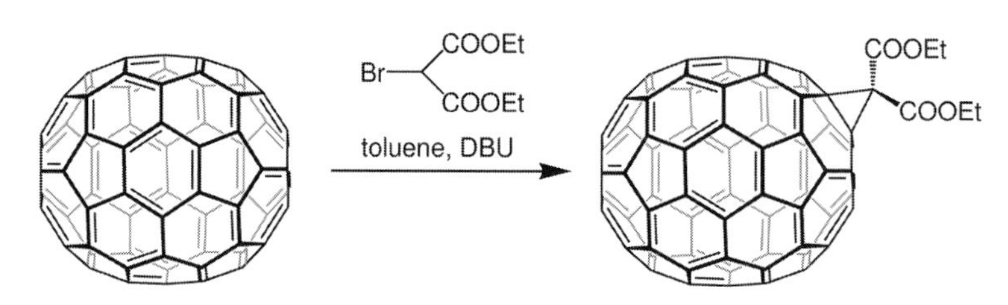

Scheme 3.7

Taking into account that some efficient modifications have been introduced into the original procedure described by Bingel, the Bingel reaction can be considered as one of the most versatile and efficient methodologies to functionalize C_{60}. These modifications work well with esters or ketones. The α-halo esters or α-halo ketones are no longer isolated but generated in situ. Thus, time-consuming purification of the intermediate halogeno-compounds can be omitted. Malonic-derivatives or 1,3-diketones are either treated with iodine [33–35] or with tetrabromomethane [36] and a base, usually DBU. These modified reaction conditions lead generally to good yields (in the range 30–60%). In some cases only one of these methods yields satisfying results.

Recently, a solvent-free variation was found to be equally effective. The inorganic base Na_2CO_3 was utilized to replace the organic base DBU in a mechanochemical reaction under “high-speed vibration milling” conditions [37]. The yields with diethyl bromomalonate are in the range 40–50%.

The Bingel reaction is not restricted to C–H-acidic carbonyl compounds such as malonates. Various other substrates have also been used. Some C–H-acidic compounds, which were successfully subjected to the Bingel reaction, are shown in Table 3.1.

Dipyridylchloromethane **15** (Table 3.1) was used to form cyclophane type Pt-complexes with two C_{60} units (see Chapter 10) [38]. The acetylene derivative **22** represents the precursors toward the synthesis of a new molecular carbon allotrope [20]. Deprotection of the silyl-protected acetylene units and coupling of the acetylene units under Eglinton–Glaser conditions should lead to different C_{60}-oligomers **24** ($n \geq 3$) (Scheme 3.8). In these oligomers the C_{60} units are connected exclusively by acetylene bridges. However, the synthesis of these allotropes has not yet been achieved, but analogous compounds such as **25**, involving functionalized, C_{60} are accessible (Scheme 3.8, see page 84) [43].

Table 3.1 Substrates used in Bingel-like reactions.

Substrate	*Conditions*	*Ref.*	*Substrate*	*Conditions*	*Ref.*
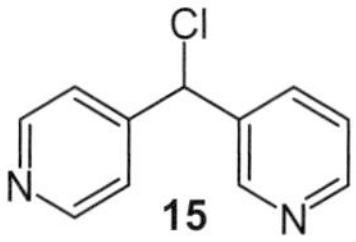 **15**	DBU, toluene, 20 °C, 32%	38	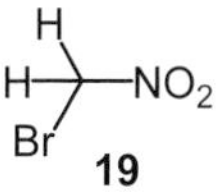 **19**	NEt_3, ODCB, 20 °C, 19%	39
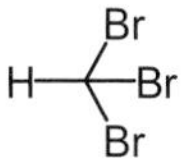 **16**	LDA, toluene, −78 °C, 40%	40	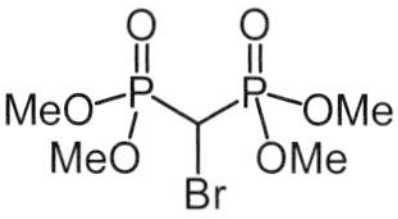 **20**	DBU, toluene, 20 °C, 41%	41
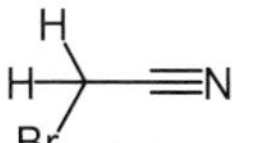 **17**	LDA, toluene, −78 °C, 15%	40	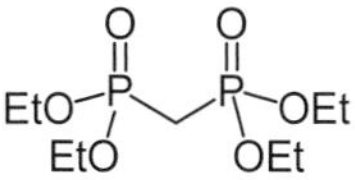 **21**	I_2–NaH, toluene, 20 °C, 38%	42
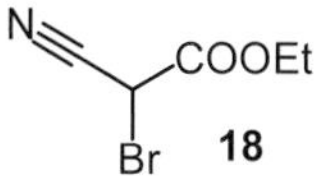 **18**	pyridine, ODCB 20 °C, 31%	39	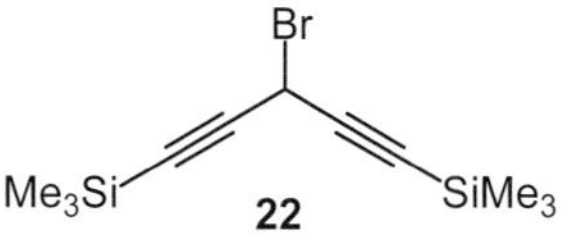 **22**	DBU, toluene, 20 °C, 55%	13, 20

As well as the Bingel reaction and its modifications some more reactions that involve the addition–elimination mechanism have been discovered. 1,2-Methano-[60]fullerenes are obtainable in good yields by reaction with phosphorus- [44] or sulfur-ylides [45, 46] or by fluorine-ion-mediated reaction with silylated nucleophiles [47]. The reaction with ylides requires stabilized sulfur or phosphorus ylides (Scheme 3.9). As well as representing a new route to 1,2-methano[60]fullerenes, the synthesis of methanofullerenes with a formyl group at the bridgehead-carbon is possible. This formyl-group can be easily transformed into imines with various aromatic amines.

The reaction of C_{60} with silylated nucleophiles [47] requires compounds such as silyl ketene acetals, silylketene thioacetals or silyl enol ethers. It proceeds smoothly and in good yields in the presence of fluoride ions (KF/18-crown-6) (Scheme 3.10). The advantage of the latter synthesis is the realization of the cyclopropanation under nearly neutral conditions, which complements the basic conditions that are mandatory for Bingel reactions. Reaction with similar silyl ketene acetals under photochemical conditions and without the use of F^- does not lead to methanofullerenes but to dihydrofullerene acetate [48].

An unusual reaction of electron-deficient acetylenes (usually DMAD is used) and triarylphosphines or -phosphites with C_{60} leads to methanofullerenes that bear an α-ylidic ester (Scheme 3.11). Selective hydrolysis of the phosphite ylides yields phosphonate esters, phosphine oxides or phosphonic acids [49–51].

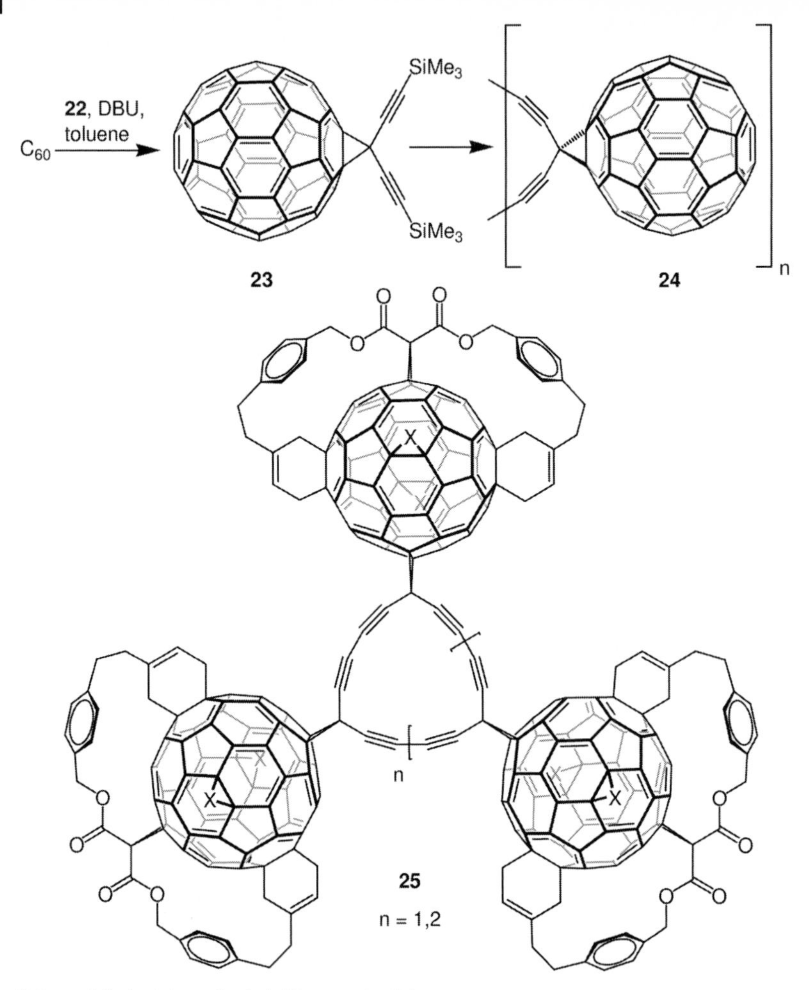

Scheme 3.8 Acetylene–C_{60} hybrids as potential precursors for a new carbon allotrope (X = $(EtO_2CCH_2O_2C)_2C$).

With the above-mentioned variety of addition reactions based on the addition–elimination mechanism almost any functional group or molecule can be attached to C_{60}. Some examples are acetylenes [43, 52], peptides [53], DNA-fragments [53], polymers [54], macrocycles [55, 56], porphyrins [56, 57], dendrimers [58–60] or ligands for complex formation [56]. C_{60} can be turned into hybrids that are biologically active, water soluble, amphiphilic or mixable with polymers [53–55, 58, 61–69].

The reverse reaction to the Bingel cyclopropanation – the so-called retro-Bingel reaction – was developed by Diederich, Echegoyen and coworkers [70] and opens up the possibility to remove the Bingel-addend completely. This removal was successfully done with C_{60} malonates [70, 71], dialkoxyphosphorylmethano[60]-fullerene [72], methano[60]fullerenyl amino acid derivatives [73] and also with

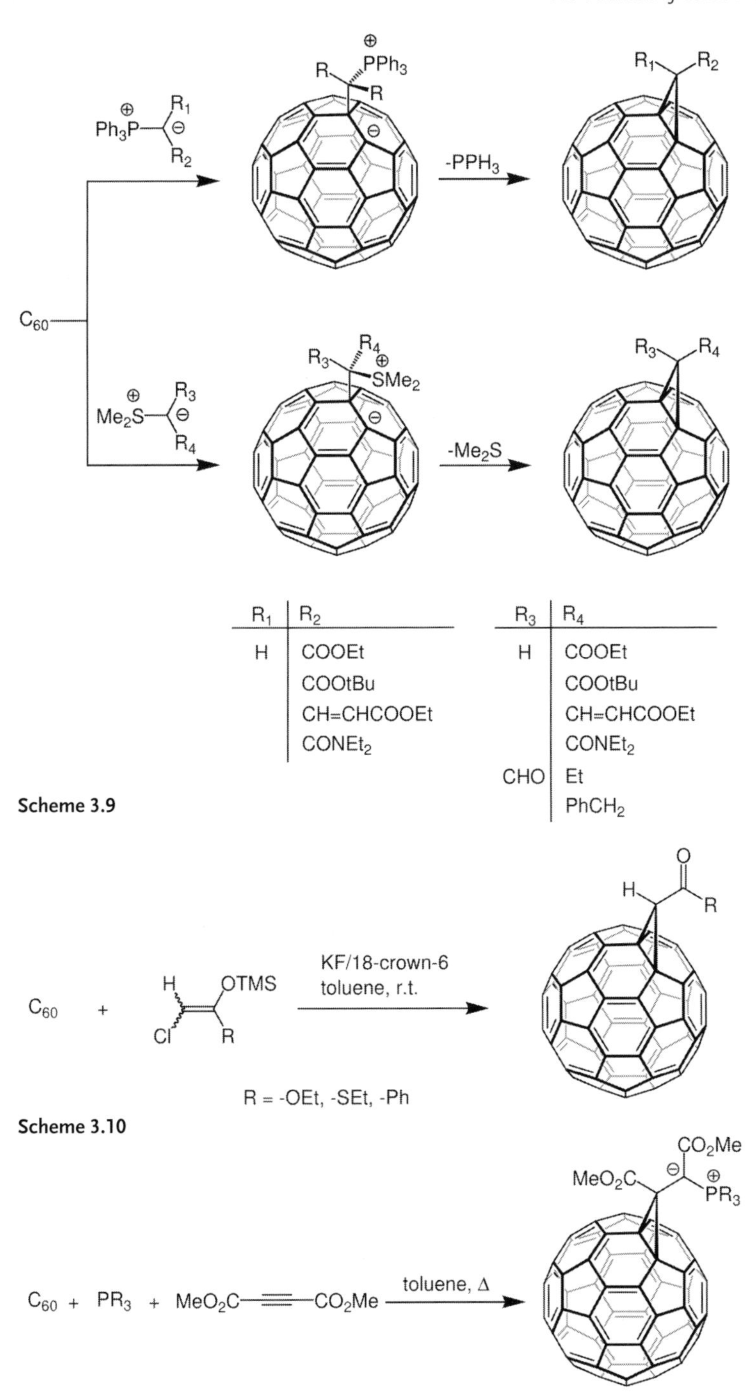

Scheme 3.9

Scheme 3.10

Scheme 3.11 Reaction of DMAD and different phosphines or phosphites with C_{60}.

methano[60]fullerenes derived from reaction with carbenes [74]. The synthetic usefulness was first proven by employing this reaction to the separation of C_{76} enantiomers [70]. The reduction of the methanofullerenes and the removal of the addend at the same time can be done via electrochemistry in a $2e^-$-process or chemically by reaction with amalgamated Mg or Zn/Cu-couple [71]. The yields are satisfying, in a range of about 50 to 70%. In future syntheses it should be possible to use Bingel-addends as protecting groups for fullerenes and as directing groups for further additions to the C_{60} core. Bingel-addends represent an orthogonal protecting group with respect to the pyrrolidine-addends introduced by 1,3 dipolar cycloaddition [71]. Fulleropyrrolidines are stable under the conditions used for removing Bingel-addends.

3.2.3 Addition of Cyanide

Addition of an alkyl nucleophile leads, due to the loss of one double bond, to a decrease of electron affinity and a concomitant negative shift of the reduction potential of about 100 to 150 mV per lost double bond. One possibility to compensate for this negative shift is the introduction of an electron-withdrawing substituent such as cyanide. Reaction of LiCN or NaCN with C_{60} at room temperature generates the monoadduct anion that can be quenched with various electrophiles [6].

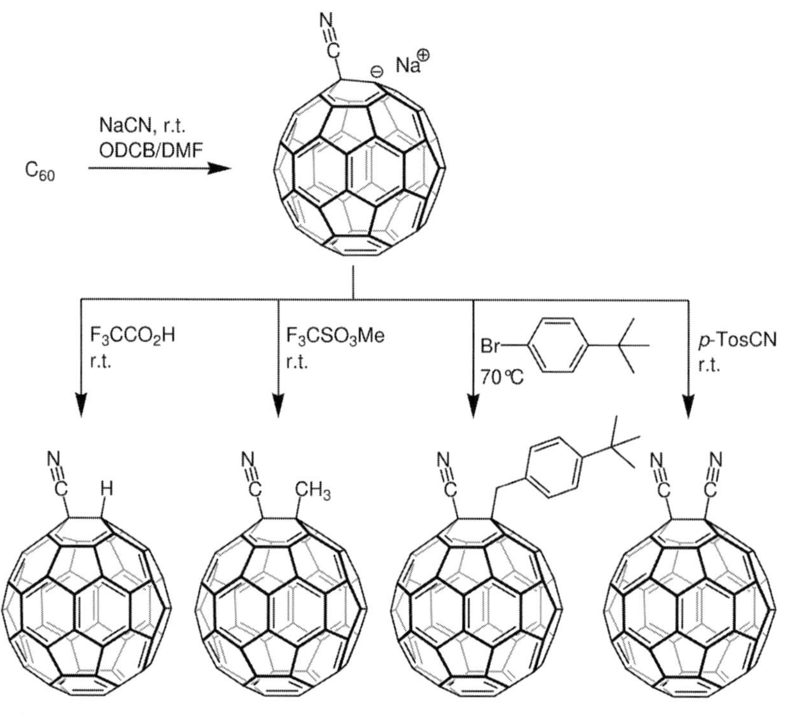

Scheme 3.12

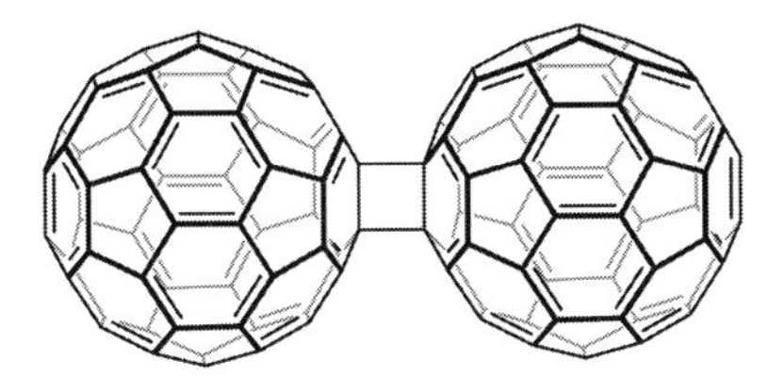

Figure 3.6 C_{60}-dimer C_{120}.

The products of these reactions are shown in Scheme 3.12. The addition of reactive nucleophiles usually yields a complex mixture of multiple adducts. Thus it is remarkable that in this case the monoaddition products can be obtained in very good yields. The reduction potential of these compounds is in the range of pristine C_{60} or even higher.

The solid-state reaction of cyanide with C_{60} using the "high speed vibration mill" technique leads to completely different products [75]. In the first place the same anion is formed, which is very reactive in the solid state and reacts immediately with further C_{60} in a radical mechanism to yield the C_{60}-dimer C_{120} (Figure 3.6) and different C_{180}-trimers. This dimer formation works also with other nucleophiles such as alkali metals or amines such as 4-aminopyridine.

3.3 Addition of Amines

Owing to their high nucleophilicity, primary and secondary aliphatic amines undergo nucleophilic additions with the electron-deficient C_{60} [2, 76–82]. Treating solid C_{60} in neat amines, for example in propylamine or ethylenediamine, leads to the formation of green solutions (fast process) that eventually turn chestnut brown (slow process) [2]. The green color of the intermediates is typical for anion complexes of C_{60} [5, 83]. Careful monitoring of these reactions with ESR and UV/Vis/NIR spectroscopy showed that the first step is a single-electron transfer from the amine to C_{60} to give the C_{60} radical anion. The next step is a radical recombination and the formation of zwitterions, which in principle can be stabilized by proton transfer from the amine to C_{60} or by oxidation followed by deprotonation and radical recombination (Scheme 3.13).

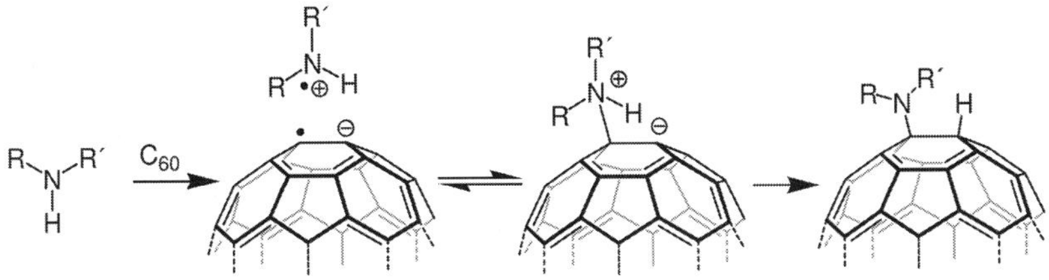

Scheme 3.13

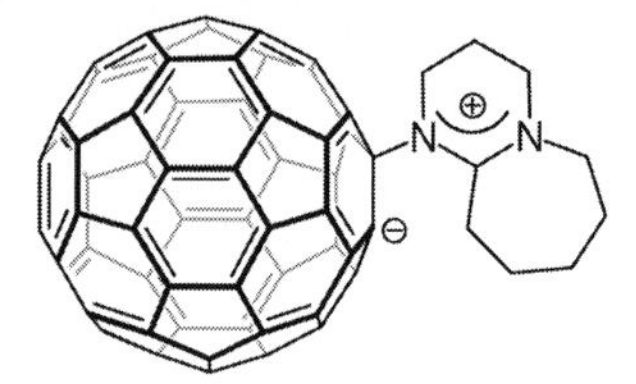

Figure 3.7 Zwitterionic addition product obtained by reaction of DBU with C_{60}.

The first step can be frozen out if a tertiary amine is used which, for steric reasons, is unable to add to C_{60} and to transfer protons. This is valid for the tertiary amine tetrakisdimethylaminoethylene (TDAE). Upon electron transfer the C_{60} radical monoanion salt C_{60}[TDAE] is formed (Chapter 2) [84]. The second step, the formation of the zwitterion, was directly observed by using the less hindered DBU (1,8-diazabicyclo[5.4.0]undec-7-ene) and isolating the green diamagnetic zwitterion complex C_{60}DBU (Figure 3.7) by precipitation in benzene [83]. In solution, this complex stays in equilibrium with the dissociated C_{60}^{-} and DBU^{+} radicals.

Although reactions of [60]fullerenes with primary and secondary amines, either in solution or with the pure amine, proceed very easily, defined aminofullerenes have been isolated in only a few cases because very often complex mixtures of aminoadducts are obtained [85]. Under strict exclusion of oxygen the hydroamination products are formed. Up to twelve propylamino groups on C_{60} were detected by mass spectrometry for $C_{60}(C_3H_9N)_n$. The reaction of excess ethylenediamine [76], excess ethanolamine [86] or even with the amino group in amino acids [87] with C_{60} results in water-soluble multiple adducts. The exact structure of these adducts is often unknown or very poorly defined and only a few examples of defined hydroaminated monoadducts are known. The first defined hydroaminated fullerene derivative was the azacrown-adduct (Figure 3.8) [82, 88]. As an amphiphilic fullerene monoadduct this azacrown-fullerene derivative was used to form stable LB-films, whose second-order susceptibilities were measured [88]. A water-soluble cyclodextrin-fullerene, which seems to be a monohydroaminated product, was obtained in pure form by membrane filtration [89].

In the presence of oxygen the hydroamination products can not be obtained. Instead – especially with secondary amines or diamines – dehydrogenated di- and polyadducts are formed [79]. By reaction of morpholine or piperidine in air-saturated benzene solution the bisadduct, tetraadduct epoxide and the dimer shown in Figure 3.8 could be isolated and characterized. A defined 1,4-addition pattern is found in all these products.

By the reaction of C_{60} with secondary diamines, dehydrogenated 1,2-diaminocycloadducts and polyadducts have been isolated. Isomerically pure monoadducts were obtained by the reaction of secondary diamines, for example *N,N'*-dimethylethylenediamine, piperazine, homopiperazine [81], *N*-ethylethylenediamine and further ethylenediamine derivatives [80], with C_{60} between 0 and 110 °C (Figure 3.9) [80, 81]. In dilute solutions of the reactants, mono- and bisadducts are predominantly formed even if a large excess of the diamine is used.

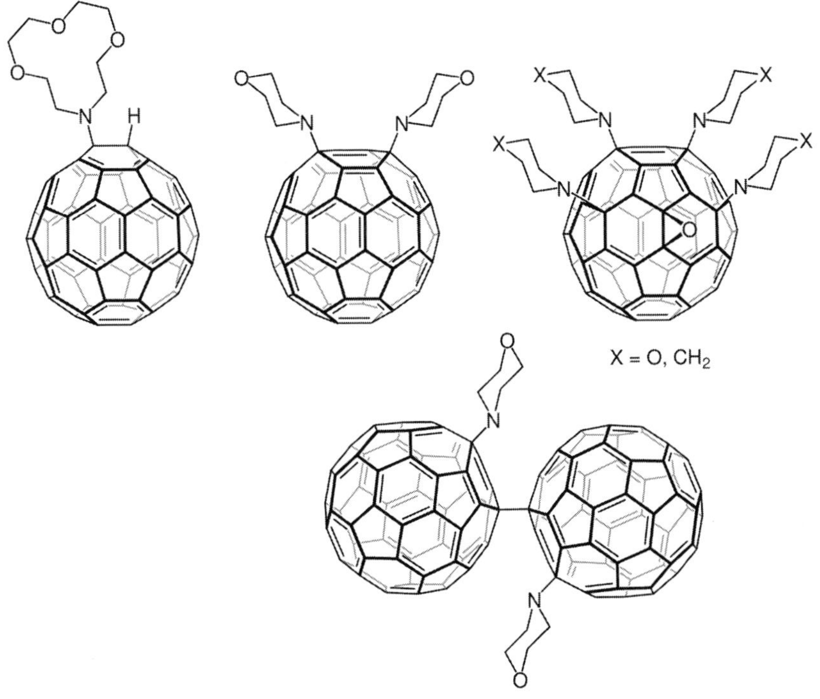

Figure 3.8 Reaction of C_{60} with secondary amines leads to hydroaminated C_{60} or dehydrogenated di- and polyadducts.

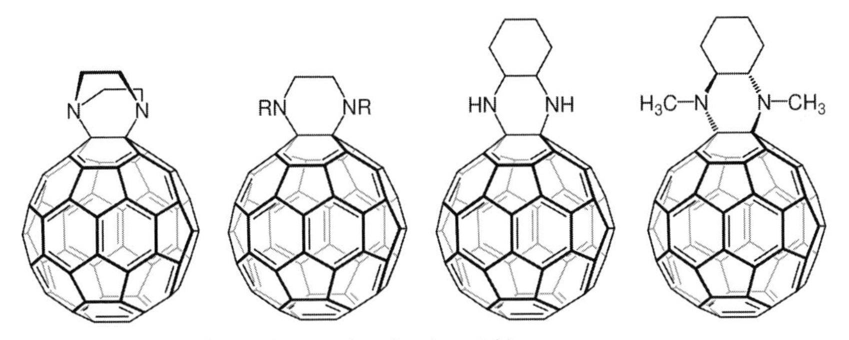

Figure 3.9 Reaction of C_{60} with secondary diamines yields isomerically pure 1,2-diaminocycloadducts.

As shown by mass spectrometry and NMR spectroscopy, these amine derivatives of C_{60}, after work up, are dehydrogenated adducts [81]. After nucleophilic addition, the H atoms brought in by the diamines were oxidatively eliminated. The final adduct formation of the secondary diamines proceeds exclusively at [6,6] bonds. Besides the monoadduct, most of the regioisomeric bisadducts of piperazine or *N,N'*-dimethylethylenediamine and C_{60} could be separated by column chromatography. The structure of the monoadduct and some of these bis-adducts were proven by X-ray crystal-structure analysis [90].

With the addition of both enantiomers of the chiral *trans*-1,2-diaminocyclohexane both enantiomers of a chiral C_{60} amino-monoadducts could be isolated (Figure 3.9). The circular dichroism spectra of the two enantiomers show a very intense chirospectroscopic response [91].

Addition of amines to C_{60} starts, mechanistically, with an electron transfer. Therefore these additions can also be carried out under photochemical conditions, either aerobic or anaerobic. The yields of the addition of *N,N'*-dimethylethylene-diamine or piperazine could be improved and the reaction time was shortened by irradiation with visible light [92, 93]. Aerobic conditions, irradiation and a large excess of a secondary diamine leads almost quantitatively to a well-defined tetraaminoadduct (Chapter 10).

As already mentioned, tertiary amines can not form similar addition products. With stabilized amines such as leuco crystal violet, leucomalachite green and similar dyes neutral and stable charge-transfer complexes were formed [94].

If – dependent on the nature of the tertiary amine – the initially formed CT complex or respective zwitterion is not stable, other reaction pathways such as cycloaddition or insertion of C_{60} into an alkyl–CH bond can be observed [87, 95–99]. More details of these reactions can be found in Chapter 6.

Through the addition reaction with amines, C_{60} can easily be attached to polymers or to self-assembled monolayers (Section 3.7). Self-assembled monolayers (SAMs) [100] of covalently bound C_{60} have been synthesized by the treatment of C_{60} with, for example, $(MeO)_3Si(CH_2)_3NH_2$- [101] or 1,12-diaminododecan- [102] modified indium-tin-oxide (ITO) surfaces (Figure 3.10) as well as, e.g., cysteamine [103] or 8-amino-1-octane thiol [104] modified gold surfaces (Figure 3.10). These cases also take advantage of the fact that primary amino groups easily add to the fullerene double bonds. The use of ITO treated with $(MeO)_3Si(CH_2)_3NH_2$ and the modified

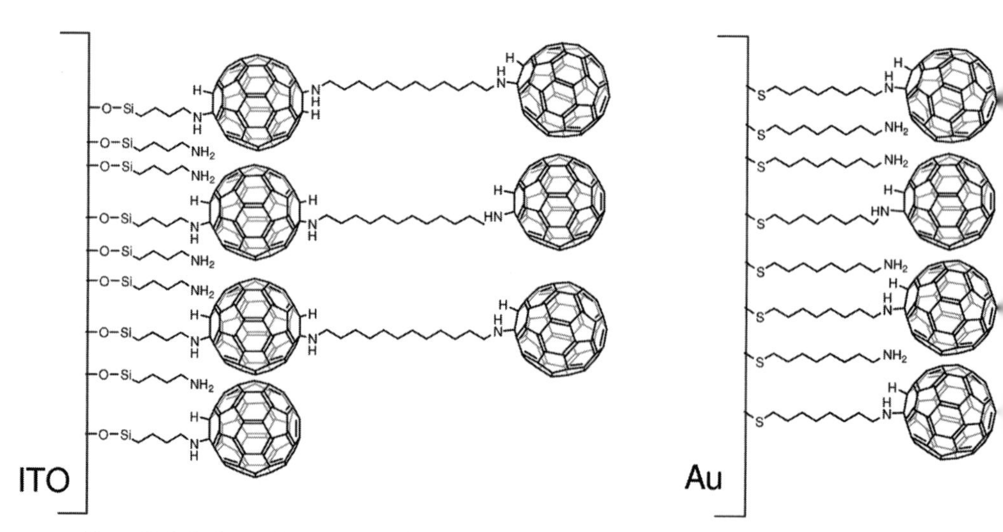

Figure 3.10 Surfaces of e.g. ITO or gold can be functionalized with C_{60} via amine addition to the fullerene.

gold substrates allow the electrochemical characterization of the SAMs of C_{60} by cyclic voltammetry as well as quartz crystal microbalance (QCM) measurements. These investigations showed that monolayers of C_{60} indeed are bound to the surface. The C_{60} surface coverage was determined to be $2.0 \cdot 10^{-10}$ mol cm^{-2}. The monolayer can be further modified with monomeric amine reagents, which demonstrates the potential of the self-assembly process for growing three-dimensional fullerene structures. Different surfaces such as quartz, Si-oxide [105] or ITO [102] were coated with multilayers of fullerene up to stacks of 9 layers. An unidirectional electron transfer is possible across the fullerene multilayers [102]. Not only can multiple layers of fullerenes be connected to a certain surface but amino-functionalized C_{60} can also serve as a linker between two different surfaces. 3-Aminopropyl-tethered glass plates could be linked via a C_{60} layer to 3-aminopropyl covered zeolite crystals [106].

3.4
Addition of Hydroxide and Alkoxides

Heating of C_{60}/C_{70} mixtures in toluene in the presence of an excess of KOH leads to the formation of a precipitate of hydroxylated fullerenes (fullerenols), which are soluble in THF but decompose due to the presence of air [107]. The addition of hydroxide to C_{70} proceeds significantly faster than to C_{60}. This was concluded from time-dependent measurements with UV/Vis spectroscopy, as well as by the observation that during the reaction the red C_{60}/C_{70} mixture turns to purple before all the material precipitates. The formation of fullerenols has also been achieved by the reaction of C_{60} dissolved in benzene with aqueous NaOH in the presence of the phase-transfer catalyst tetrabutylammonium hydroxide in air [108]. This reaction proceeds within a few minutes with the precipitation of a brown solid. In the absence of oxygen the addition of hydroxy groups is very slow. About 24 to 26 hydroxy groups can be attached to the fullerene core by this method. Control of reaction conditions such as reaction time, stoichiometry, temperature and oxidant agent gives access to fullerenols with various degrees of hydroxylation [109]. The low molecular weight fullerenols with less hydroxy groups become less water soluble or even water insoluble but they can form stable Langmuir–Blodgett films. Fullerenols can also be obtained via an electrophilic, oxidative or radical pathway.

The first step in the addition of alkoxides to C_{60} is, consistently, the formation of the alkoxy C_{60} anion. The subsequent process is strongly dependent on the presence of oxygen. In the presence of oxygen, 1,3-dioxolane derivatives of C_{60} are formed [110]. In the absence of oxygen the oligo alkoxy fullerenide anions can be formed [111, 112]. Reaction of alkoxides with C_{60} usually results in complex mixtures. This may be why only a few reactions of C_{60} with alkoxides have been described [113]. Nevertheless, defined alkoxy fullerenes can be obtained by nucleophilic substitution reactions of alkoxides with halogenofullerenes (Chapter 9) [113].

3.5
Addition of Phosphorus Nucleophiles

Compared with the variety of existing carbon or nitrogen nucleophiles that were subjected to nucleophilic addition to C_{60} there are few examples for phosphorus nucleophiles. Neutral trialkylphosphines turn out to be to less reactive for an effective addition to C_{60} even at elevated temperatures [114]. Trialkylphosphine oxides show an increased reactivity. They form stable fullerene-substituted phosphine oxides [115]; it is not yet clear if the reaction proceeds via a nucleophilic mechanism or a cycloaddition mechanism. Phosphine oxide addition takes place in refluxing toluene [115]. At room temperature the charge-transfer complexes of C_{60} with phosphine oxides such as tri-*n*-octylphosphine oxide or tri-*n*-butylphosphine oxide are verifiable and stable in solution [116].

Lithiated phosphines and phosphites also add to the [6,6] double bond of C_{60} [114, 117]. Whereas the phosphite adducts were isolated and purified in good yields, the phosphine adducts could not be separated from unreacted C_{60} and side products. Treatment of C_{60} with lithiated secondary phosphine boranes or phosphinite boranes followed by removal of the BH_3 group afforded 1,2-hydrophosphorylated **27** in good yields (Scheme 3.14) [114, 117]. The phosphorus nucleophiles were generated by deprotonation of the corresponding borane complexes with BuLi in THF–HMPA, and added to toluene solutions of C_{60} at –78 °C. Complexes **26** are stable in air at room temperature for months. In the ^1H NMR spectrum, the proton of **26b** appears as doublet at $\delta = 6.94$. Removal of the BH_3 group by treatment with DABCO proceeded quantitatively to give **27b** as a black powder.

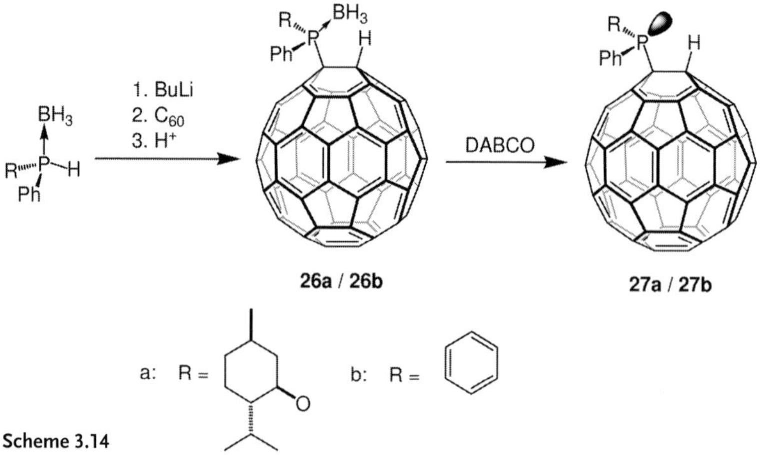

Scheme 3.14

The synthesis of diastereomeric P-chiral fullerenyl phosphine **27a** with a (+)-menthyl group was achieved analogously [114, 117]. No sign of epimerization upon heating **27a** for 14 h at 80 °C in toluene was observed, indicating the remarkable stability of chirality at the phosphorus of **27a**.

3.6
Addition of Silicon and Germanium Nucleophiles

Most of the reactions of silicon or germanium organic compounds proceed photochemically via a radical mechanism or via a cycloaddition mechanism (Chapters 4 and 6). There are few examples of nucleophilic addition of RSi^- or RGe^- to C_{60} [118, 119]. Reaction of silyllithium derivatives R_3SiLi or germyllithium derivatives R_3GeLi with different alkyl- and aryl-substituents R yields mainly the 1,2-adduct **28** or the 1,16-adduct **29**; 1,4-addition and dimerization of two fullerene-units was also found as a minor pathway. One example is given in Scheme 3.15.

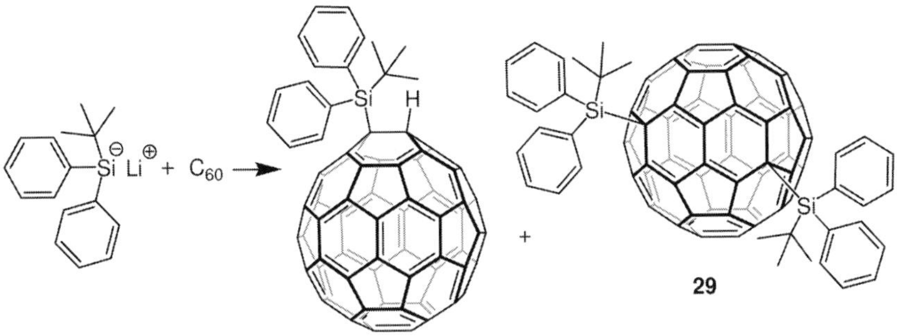

Scheme 3.15

3.7
Addition of Macromolecular Nucleophiles – Fullerene Polymers

The affinity of C_{60} towards carbon nucleophiles has been used to synthesize polymer-bound C_{60} [120] as well as surface-bound C_{60} [121]. Polymers involving C_{60} [54, 68, 69] are of considerable interest as (1) the fullerene properties can be combined with those of specific polymers, (2) suitable fullerene polymers should be spin-coatable, solvent-castable or melt-extrudable and (3) fullerene-containing polymers as well as surface-bound C_{60} layers are expected to have remarkable electronic, magnetic, mechanical, optical or catalytic properties [54]. Some prototypes of polymers or solids containing the covalently bound C_{60} moiety are possible (Figure 3.11) [68, 122]: fullerene pendant systems **Ia** with C_{60} on the side chain of a polymer (on-chain type or "charm bracelet") [123] or on the surface of a solid **Ib** [121], in-chain polymers **II** with the fullerene as a part of the main chain ("pearl necklace") [123], dendritic systems **III**, starburst or cross-link type **IV** or end-chain type polymers **V** that are terminated by a fullerene unit. For **III** and **IV**, one-, two- and three-dimensional variants can be considered. In addition, combinations of all of these types are possible.

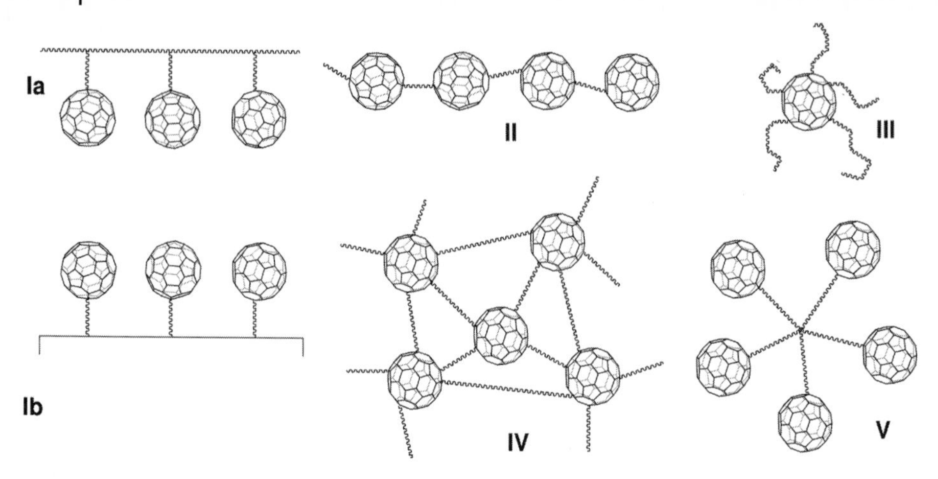

Figure 3.11 Prototypes of polymers involving the C_{60} moiety. Pendant on-chain (**Ia**), pendant on-surface (**Ib**), in-chain (**II**), dendritic (**III**), cross-link (**IV**) and end-chain (**V**) [68, 122].

The addition of living polystyrene to C_{60} leads to the formation of star-shaped polymers with C_{60} in the center (occasionally called "flagellenes") (Scheme 3.16) [120]. These polymers are highly soluble and melt processable [120]. Different "living" anionic polymers such as polystyrene, the block copolymer polystyrene-*b*-poly(phenylvinylsulfoxide) [124] (as a precursor for polyacetylene) or polyisoprene [125] were grafted onto C_{60}. It is possible to control the addition of living polymer anions by stoichiometry and in dependence of the structure of the living polymer. Thus, adducts with a well-defined structure can be prepared. Star-shaped polymers with up to six branches are available when using reactive carbanions such as styryl or isoprenyl [68].

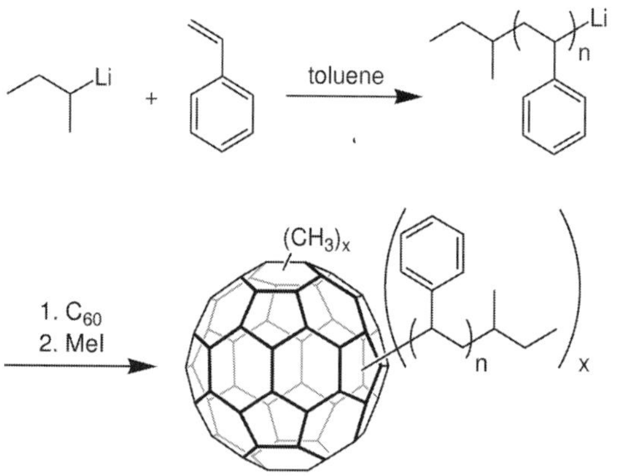

Scheme 3.16

117 S. Yamago, M. Yanagawa, E. Nakamura, *J. Chem. Soc., Chem. Commun.* **1994**, 2093.
118 T. Kusukawa, W. Ando, *Angew. Chem., Int. Ed. Engl.* **1996**, *35*, 1315.
119 T. Kusukawa, W. Ando, *J. Organomet. Chem.* **1998**, *561*, 109.
120 E. T. Samulski, J. M. DeSimone, M. O. Hunt, Jr., Y. Z. Menceloglu, R. C. Jarnagin, G. A. York, K. B. Labat, H. Wang, *Chem. Mater.* **1992**, *4*, 1153.
121 D. E. Bergbreiter, H. N. Gray, *J. Chem. Soc., Chem. Commun.* **1993**, 645.
122 A. Hirsch, *Adv. Mater.* **1993**, *5*, 859.
123 T. Suzuki, Q. Li, K. C. Khemani, F. Wudl, O. Almarsson, *J. Am. Chem. Soc.* **1992**, *114*, 7300.
124 Y. Ederle, D. Reibel, C. Mathis, *Synth. Methods* **1996**, *77*, 139.
125 Y. Ederle, C. Mathis, *Macromolecules* **1997**, *30*, 2546.
126 A. O. Patil, G. W. Schriver, B. Carstensen, R. D. Lundberg, *Polym. Bull.* **1993**, *30*, 187.
127 K. E. Geckeler, A. Hirsch, *J. Am. Chem. Soc.* **1993**, *115*, 3850.
128 C. Weis, C. Friedrich, R. Mülhaupt, H. Frey, *Macromolecules* **1995**, *28*, 403.
129 N. Manolova, I. Rashkov, F. Beguin, H. van Damme, *J. Chem. Soc., Chem. Commun.* **1993**, 1725.
130 Y.-P. Sun, B. Liu, D. K. Moton, *Chem. Commun.* **1996**, 2699.
131 Z. Li, J. Qin, *J. Appl. Polym. Sci.* **2003**, *89*, 2068.
132 Y.-P. Sun, C. E. Bunker, B. Liu, *Chem. Phys. Lett.* **1997**, *272*, 25.
133 S. Samal, B.-J. Choi, K. E. Geckeler, *Chem. Commun.* **2000**, 1373.

4
Cycloadditions

4.1
Introduction

In cycloaddition reactions the [6,6] double bonds of C_{60} exhibit a dienophilic character. A large variety of cycloadditions have carried out with C_{60} and the complete characterization of the products, mainly monoadducts, has greatly increased our knowledge of fullerene chemistry. These chemical transformations also provide a powerful tool for the functionalization of the fullerene sphere. Almost any functional group can be covalently linked to C_{60} by the cycloaddition of suitable addends. Some types of cycloadducts exhibit a remarkable stability; for example, they can be thermally treated up to 400 °C without decomposition. This is an important requirement for further side-chain chemistry as well as for possible applications of the new fullerene derivatives, which may be of interest due to their biological activity or as new materials.

4.2
[4+2] Cycloadditions

The [6,6] double bonds of C_{60} are dienophilic [1], which enables the molecule to undergo various Diels–Alder reactions ([4+2] cycloadditions) [2, 3]. The dienophilic reactivity of C_{60} is comparable to that of maleic anhydride [4] or *N*-phenylmaleimide [5]. The conditions for cycloadduct formation strongly depend on the reactivity of the diene. Most [4+2] reactions with C_{60} are accomplished under thermal conditions, but photochemical reactions have also been reported, and in various additions microwave irradiation can efficiently be used as a source of energy [6]. Equimolar amounts of cyclopentadiene and C_{60} react at room temperature to give the monoadduct in comparatively high yield (Scheme 4.1) [7, 8], while the formation of the cycloadduct with anthracene necessitates an excess of the diene in refluxing toluene [1, 8, 9]. These cycloadducts of C_{60} can be purified by flash-chromatography on SiO_2.

Heating the anthraceno monoadduct in toluene affords the component molecules [1, 8], which demonstrates a facile retro-Diels–Alder reaction. A thermal gravimetric analysis (TGA) of solid C_{60}(anthracene) shows a cleavage of the anthracene moiety

Fullerenes: Chemistry and Reactions. Andreas Hirsch and Michael Brettreich

ISBN: 3-527-30820-2

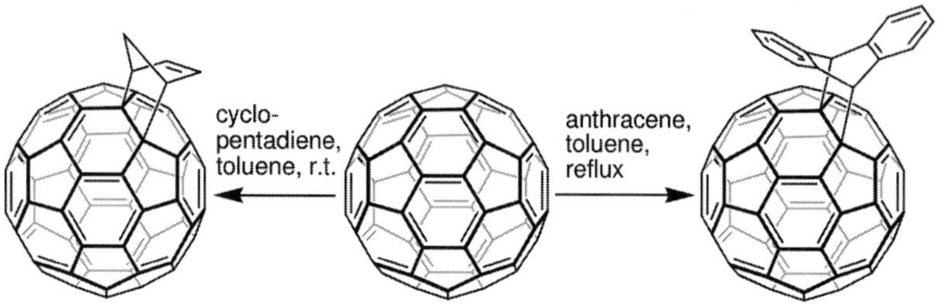

Scheme 4.1

at an onset temperature of 120 °C [9]. These findings clearly show that there is an equilibrium between addition and elimination of anthracene in toluene under reflux conditions, which also explains the comparatively low yields (Scheme 4.1).

An even more pronounced retro-Diels–Alder reaction occurs by using 1,3-diphenylisobenzofuran (DPIF), 9-methylanthracene or 9,10-dimethylanthracene as dienes [8, 10–12]. The monoadduct of DPIF cannot be isolated from the reaction mixture, while the monoadduct of the 9-methyl- or 9,10-dimethyl- derivatives of anthracene can be isolated at temperatures lower than room temperature [10]. Both anthracene derivatives decompose at room temperature, the adduct with one methyl group within hours, the adduct with two methyl groups within minutes. For DPIF and the anthracene compounds the retro-Diels–Alder reaction seems to be facilitated by steric repulsion due to the bulky groups. However, as shown by Wudl and co-workers [13], the cycloadduct of C_{60} with isobenzofuran (Scheme 4.2), which was generated in situ from 1,4-dihydro-1,4-epoxy-3-phenylisoquinoline, is stable in the solid state as well as in solution and shows no tendency to undergo cycloreversion.

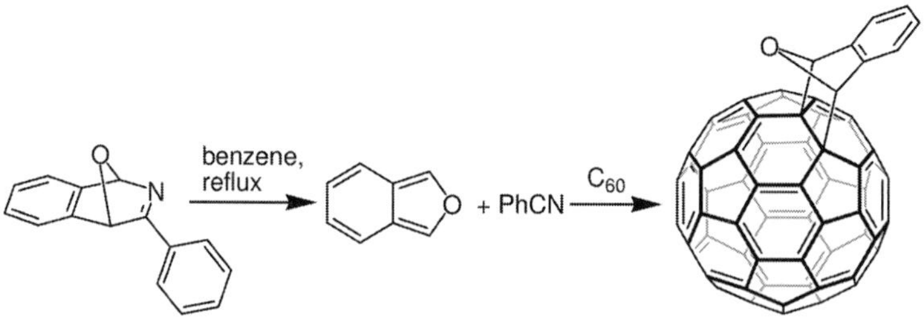

Scheme 4.2

These examples already prove that the potential of such reactions for the synthesis of stable fullerene derivatives is restricted due to the facile cycloreversion to the starting materials. Nevertheless, cycloreversion can also be useful. Reversibility of dimethylanthracene addition was utilized for the selective synthesis of T_h-symmetrical hexakisadducts (see Chapter 10) [12]. In another example, a dendritic polyamidoamine-addend was reversibly attached to C_{60} via an anthracene anchor (Figure 4.1) [14, 15]. The dendrofullerene, which is soluble in polar solvents, can be obtained in 70% yield and the retro-Diels–Alder reaction at 45 °C proceeds with a conversion rate of more than 90%.

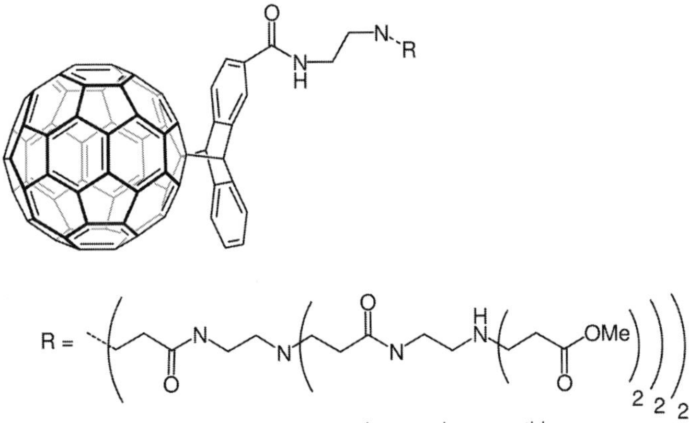

O N H N R

R = N N N N OMe O O O 2 2 2

Figure 4.1 The dendritic anthracene anchor can be reversibly attached to C_{60} via Diels–Alder reaction.

The cycloaddition product of π-extended TTF and C_{60} functions as a fluorescence switch (Scheme 4.3) [16]. The fluorescence can be turned off by attaching the fluorophore to C_{60}, which leads to effective quenching of fluorescence. Release of the addend by heating the solution to 80 °C turns the fluorescence on since intermolecular quenching is not very effective.

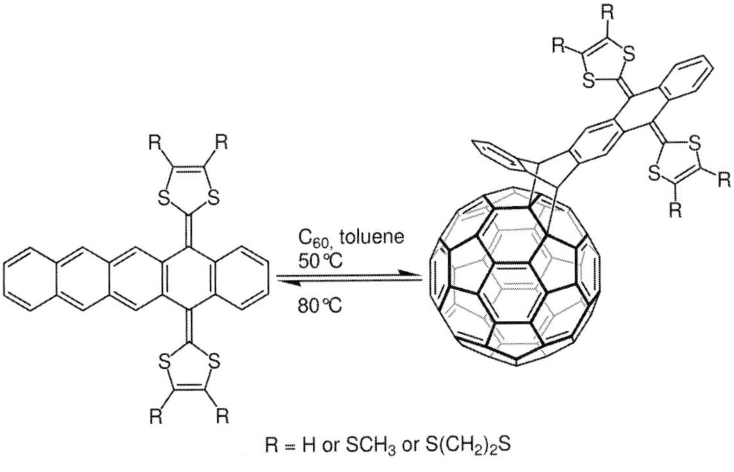

C_{60}, toluene
50 °C
80 °C

R = H or SCH_3 or $S(CH_2)_2S$

Scheme 4.3

Thermal treatment of solid C_{60}(anthracene) was successfully employed for the selective synthesis of the *trans*-1-C_{60}(anthracene)$_2$ bisadduct by the disproportionation of two C_{60}(anthracene) molecules into C_{60}(anthracene)$_2$ and C_{60} [10].

Anthracene as well as its derivatives and also higher acenes can be added to C_{60} either in solution by refluxing in toluene or dichlorobenzene [10, 17, 18] or with solid-state techniques [19]. Anthracene addition may be enhanced by irradiation with light [20]. The high-speed vibration milling technique leads to good yields of mono- and bisadducts of higher acene-homologues such as tetracene, pentacene

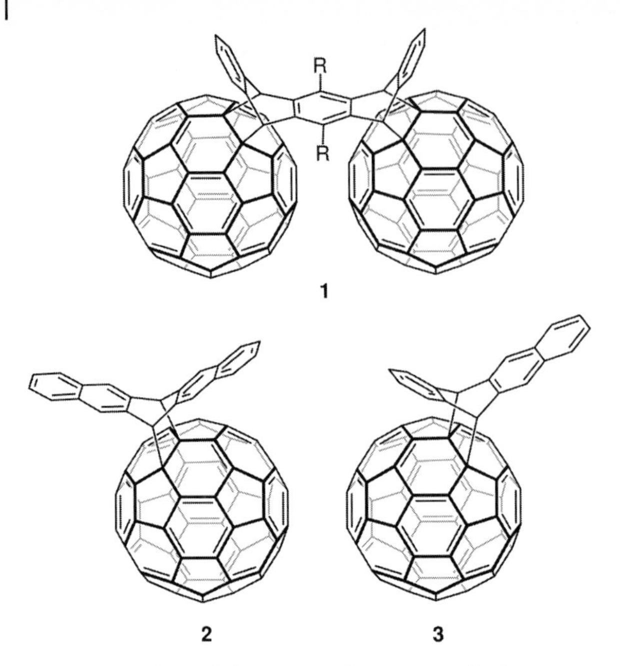

Figure 4.2 Products of the reaction of pentacene with C_{60}.

or naphtho[2,3-*a*]pyrene [19]. With tetracene and pentacene the monoadducts **2** and **3** were isolated as the main products (Figure 4.2) [19]. In the reaction with pentacene, bisfullerene adducts and fullerene bisadducts were also obtained in significant amounts. The *cis* bisfullerene adduct of 6,13-disubstituted pentacene **1** is obtained in solution as the main product [17], whereas the *trans* bisfullerene adduct of unsubstituted pentacene is the major bisfullerene adduct by applying the high speed milling technique [19].

Cyclopentadiene adducts (mono-, tetra- and hexa-adducts) of C_{60} were stabilized against retro-reaction by selective hydrogenation and bromination of the pendant groups [21]. Utilization of Adam's catalyst and dilute bromine solutions exclusively leads to an addition to the cyclopentene double bonds, because C_{60} itself is inert towards these reagents. The increased stability of the reduced cycloadducts can be demonstrated by mass spectrometry [21].

Substituted butadienes form [4+2] cycloadducts with C_{60} that show sufficient stability even without further stabilization. Table 4.1 shows some butadiene derivatives used as substrates. Most of these dienes form stable, isolable products in good yields.

A stable adduct of C_{60} with an electron donor moiety in the side chain (Table 4.1, compound **7**) [24] was synthesized by the cycloaddition of a rigid diene containing a polynorbornyl-bicyclo[2.2.0]hexyl bridge and a dimethoxybenzene (DMB) unit in toluene at reflux in 63% yield. The donor–acceptor interaction of the electron rich DMB unit and C_{60} is indicated by the X-ray crystal structure. The adducts are paired

Table 4.1 [4+2]-Cycloadditions with different substituted butadienes.

Substrate	*Conditions*	*Ref.*
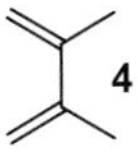 4 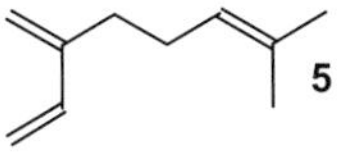 5	Benzene, room temperature (23%; 32%)	4, 22
6 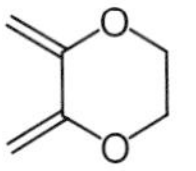	Toluene, room temperature (49%)	23
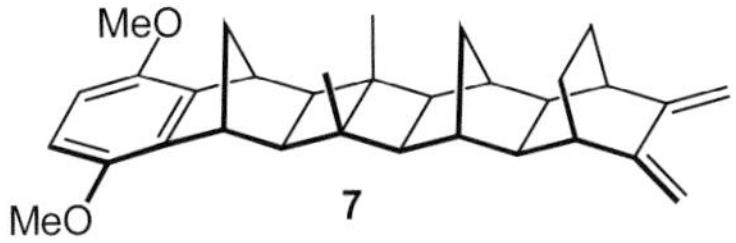 MeO MeO 7	Toluene, reflux (63%)	24
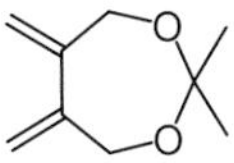 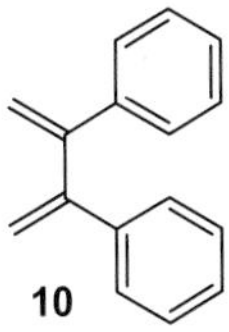 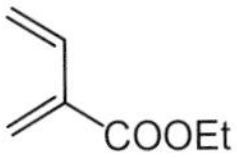 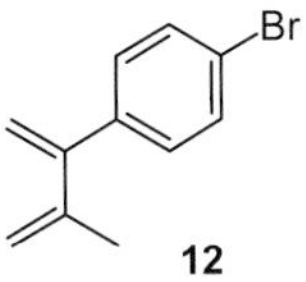 8; 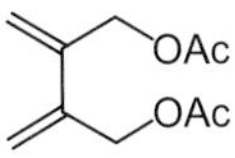 OAc OAc 9; 10; COOEt 11; Br 12	Toluene, reflux (59%; 52%; 49%; 30%; 59%)	25
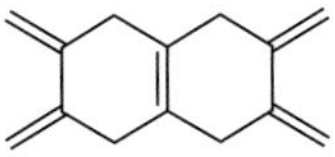 13; 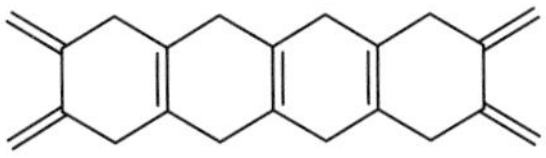 14; 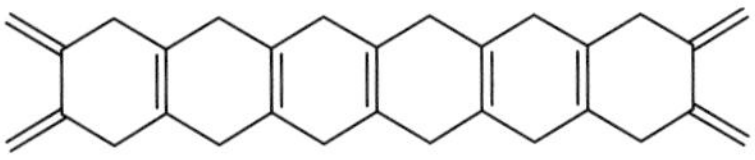 15	Toluene, reflux (70%; 72%; 40%) (bisaddition products)	26
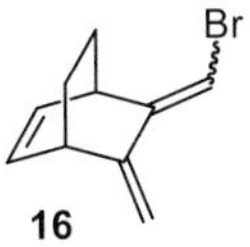 Br 16	Chlorobenzene, 70 °C, product unstable	27
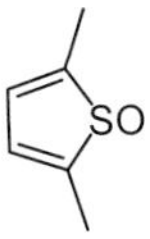 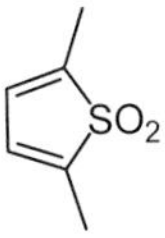 SO 17; SO_2 18	Dichloromethane, room temperature (35%; 26%)	28

Table 4.1 (continued)

Substrate	*Conditions*	*Ref.*
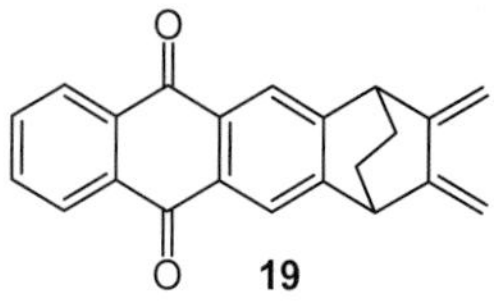 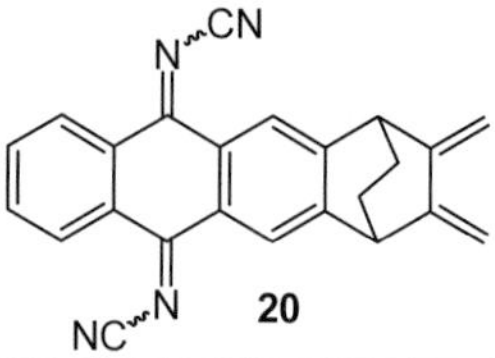 O, O, **19**; N–CN, NC–N, **20**	Toluene, reflux (49%; 42%)	29
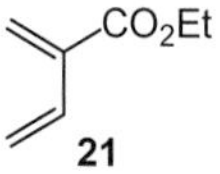 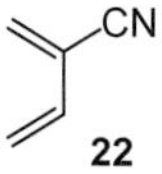 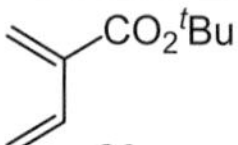 CO_2Et **21**; CN **22**; $CO_2{}^tBu$ **23**	ODCB, 150 °C	30
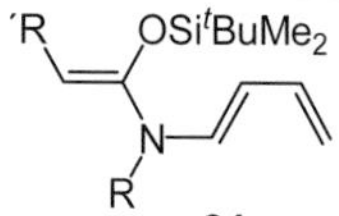 ´R, OSi^tBuMe_2, N, R, **24**	Toluene, room temperature –110 °C	31
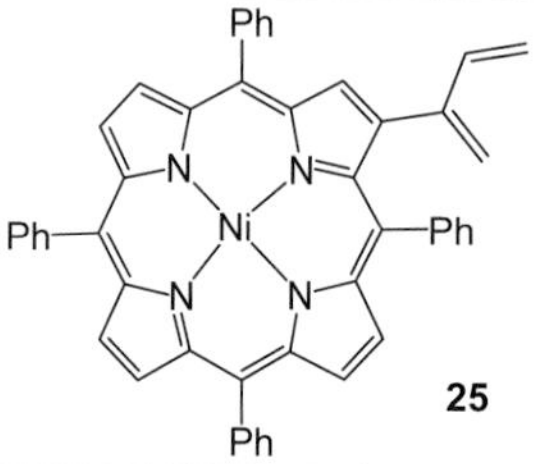 Ph, N, N, Ni, N, N, Ph, Ph, Ph, **25**	Toluene, reflux (95%)	32
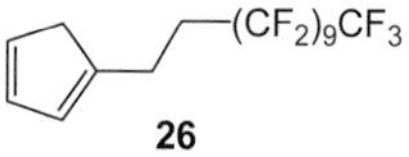 $(CF_2)_9CF_3$ **26**	Toluene–freon 113 (64%)	33
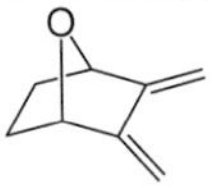 O, **27**	Toluene, 80 °C	34
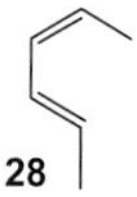 **28**,  **29**	Benzene, 80 °C/25 °C (50%)	35
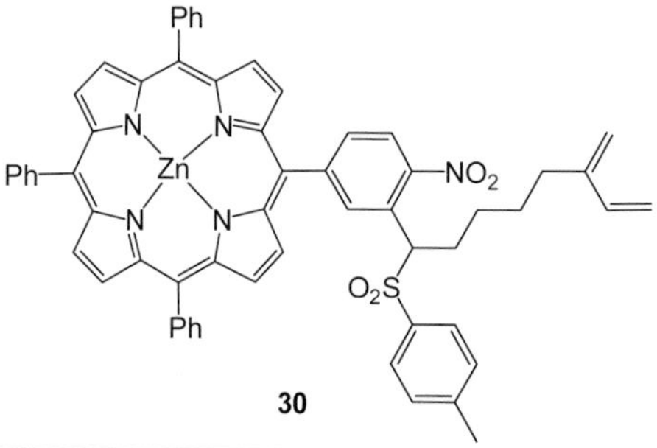 Ph, N, N, Zn, N, N, Ph, Ph, NO_2, O_2S, **30**	Toluene, reflux (31%)	36

up, with the C_{60} unit of one molecule nestling up to the DMB-terminated chain of the other. The cyclohexene ring in the cycloaddition product adopts an extended boat conformation in the crystal. Since a folded boat conformation of the cyclohexene ring was predicted to be more stable, it can be anticipated that crystal-packing forces and π-stacking interactions between the C_{60} and the DMB moiety are responsible for the extended conformation in the solid state.

Regioselective [4+2] cycloadditions to C_{60} are also possible with 2,3-dimethylbuta-1,3-diene (**4**) and with the monoterpene 7-methyl-3-methylideneocta-1,6-diene (**5**, myrcene) [22]. These monoadduct formations proceed under mild and controlled conditions. Most of these addition products of 1,3-butadiene derivatives (e.g. **4**, **5**, **8–12**) are unstable against air and light [25]. The dihydrofullerene moiety in the Diels–Alder adducts act as a $^1O_2$-sensitizer and promotes the oxidation of the cyclohexene moiety to the hydroperoxide. Reduction of the hydroperoxide with PPh_3 yields the corresponding allylic alcohols [25].

The bis-dienes **13–15** form ladder-type bis-fullerene adducts. Cycloadditions proceed in good yields also with electron-deficient dienes such as **21–23**. Porphyrin donor systems such as **25** or **30** or quinoidal acceptor systems such as **19** and **20** were introduced to study the properties of the charge transfer in C_{60} donor–acceptor systems.

Mechanistic studies of the addition of **27–29** to C_{60} [34, 35] support clearly a concerted mechanism with a symmetrical transition state. All three mono-addition products are stable against cycloreversion at least up to 80 °C. Interestingly, the two hexadiene isomers show different reactivity [35]. The *trans-trans*-2,4-hexadiene **29** reacts smoothly at room temperature whereas the *cis-trans*-2,4-hexadiene **28** does not react at all at ambient temperature. The *cis-trans*-isomer **28** cycloadds to C_{60} at 80 °C, while isomerization of **28** into **29** occurs. Thus, reaction of **28** leads to a product mixture.

Cyclohexadiene derivatives are less reactive than butadiene derivatives, thus only a few examples of cycloadditions with these compounds are known (Figure 4.3) [37–40]. The cyclohexadiene bicyclic derivative **32** was synthesized by rhodium-catalyzed reaction of toluene with *tert*-butyldiazoacetate and cycloadds in about 40% yield to C_{60} [39]. The product has anti-cyclopropane orientation relative to the entering dienophile C_{60}. Valence isomerization of **33** (Scheme 4.4) leads to the cyclobutene-fused cyclohexene **35** that adds in good yields (50%) at moderate temperatures (110 °C) to C_{60} [40]. The reaction of C_{60} with the electron-deficient cyclohexene **34** is also possible in moderate yields [38].

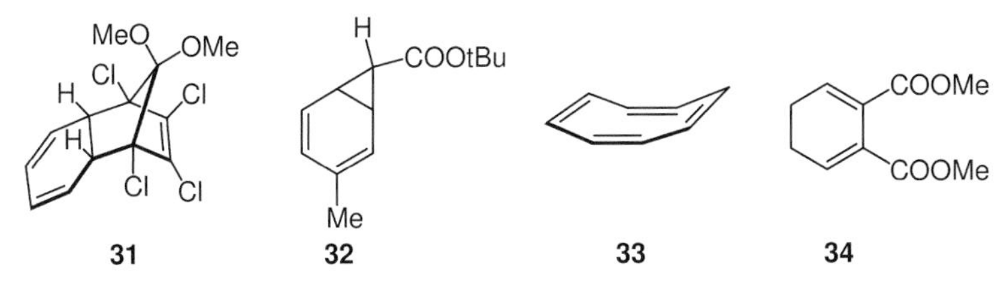

Figure 4.3 Cyclohexadiene-substrates for [4+2]-cycloadditions to C_{60}.

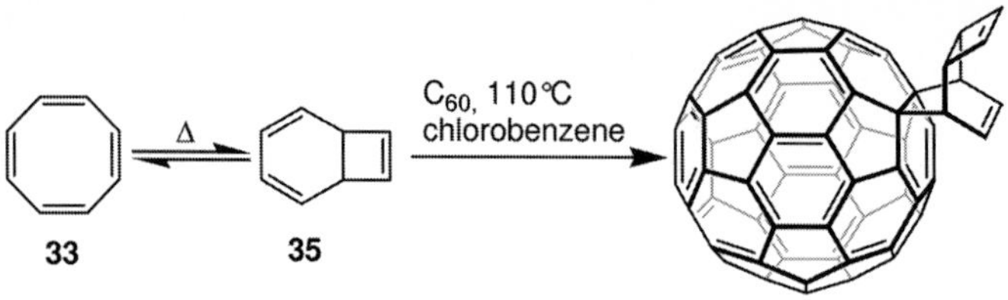

Scheme 4.4

Another approach to synthesize stable Diels–Alder adducts of C_{60} was introduced by Müllen and co-workers [41–43]. The use of *o*-quinodimethane derivatives as dienes, prepared in situ, leads to the formation of thermally stable cycloadducts (Scheme 4.5). As with the isobenzofuran addition product [13], a cycloreversion of these adducts would need to overcome the stabilization provided by the aromatic system and would also give the unstable *o*-quinodimethane intermediate. A fast ring inversion, at elevated temperatures, of the cyclohexane moiety causes a C_{2v}-symmetry of the cycloadduct, leading to 17 lines for the fullerenyl carbons in the ^{13}C NMR spectra [41].

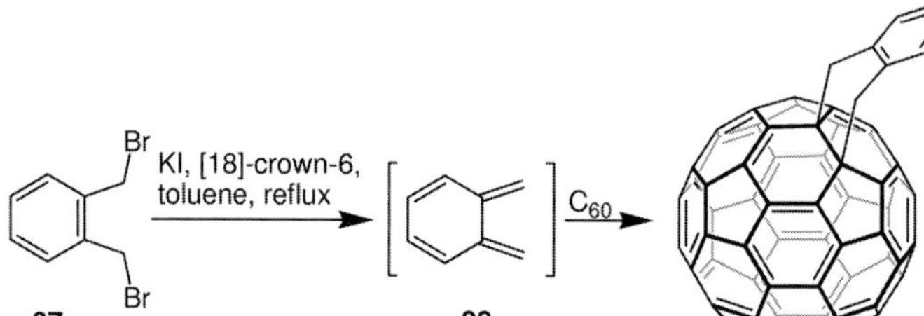

Scheme 4.5

The analogous reaction of C_{60} with bis-*o*-quinodimethane (Scheme 4.6) leads to a bridged system [41], which can be regarded as a part of a band polymer containing the C_{60} unit and exhibits a significantly decreased solubility. The bis-*o*-quinodimethane was prepared in situ from 1,2,4,5-tetrakis(bromomethyl)benzene **40**. Scheme 4.6 shows one of the possible conformers of the bisadduct **41**.

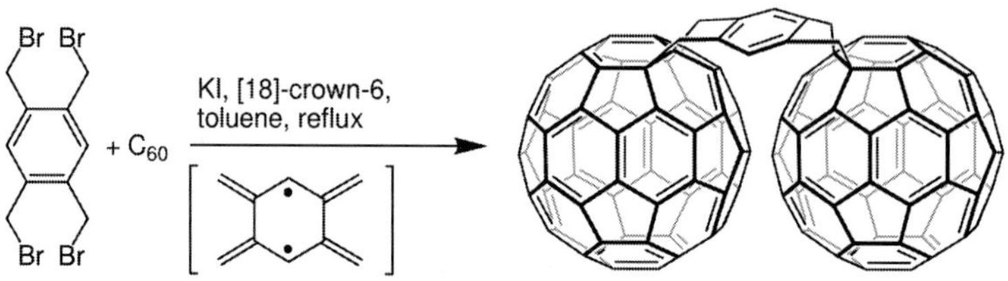

Scheme 4.6

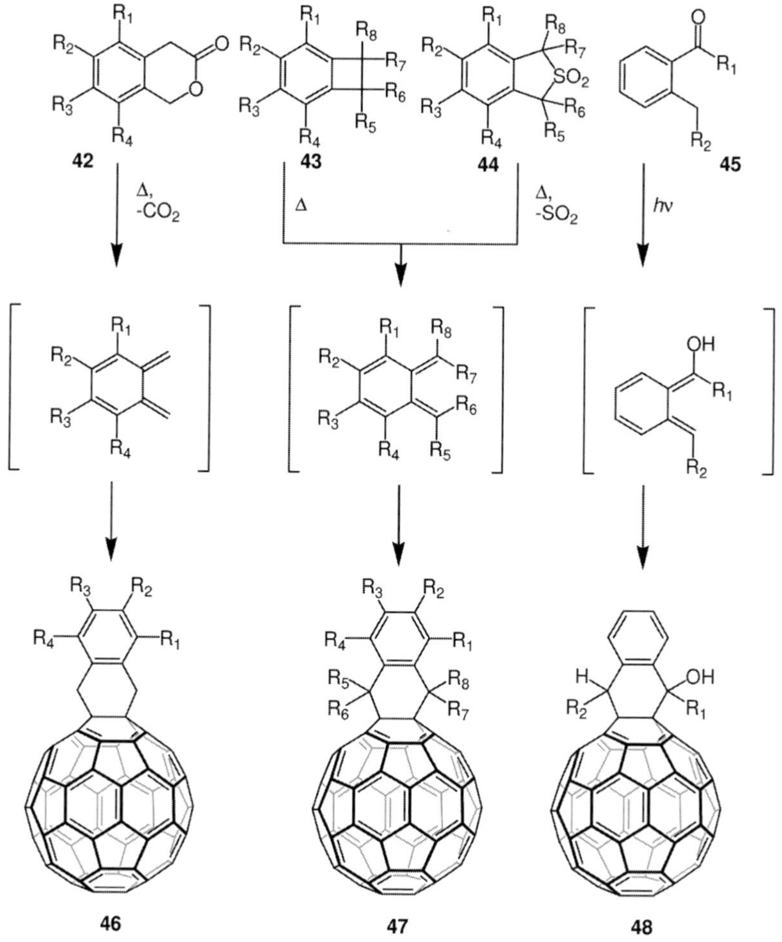

Scheme 4.7

These cycloadditions with *o*-quinodimethanes provide a broad variety of useful fullerene functionalizations, since *o*-quinodimethanes can be prepared using several routes and the resulting cycloadducts are thermally stable [42]. There exist several alternatives to the iodide-induced bromine 1,4-elimination of 1,2-bis(bromomethyl)-benzenes [44–47]. *o*-Quinodimethanes have been prepared by thermolysis of 3-isochromanone (**42**) [43], benzocyclobutenes (**43**) [48–50], isobenzothiophene 2,2-dioxides (**44**) [42] and sultines [51, 52] or by photolysis of *o*-alkylphenones such as **45** [53–55] and could be added to C_{60} in good yields (Scheme 4.7). Indene, thermally rearranged to isoindene, also adds to C_{60} in similar fashion to quinodimethanes [56].

Ring opening of the benzocyclobutenes or the extrusion of CO_2 and SO_2 requires high temperatures. Therefore, the cycloadditions are carried out in solvents such as toluene, chlorobenzene, *o*-dichlorobenzene or 1,2,4-trichlorobenzene at elevated temperatures. Some representative examples are shown in Table 4.2.

Table 4.2 [4+2]-Cycloaddition of C_{60} with different *o*-quinodimethane derivatives.

Substrate	*Conditions (% yield)*	*Ref.*
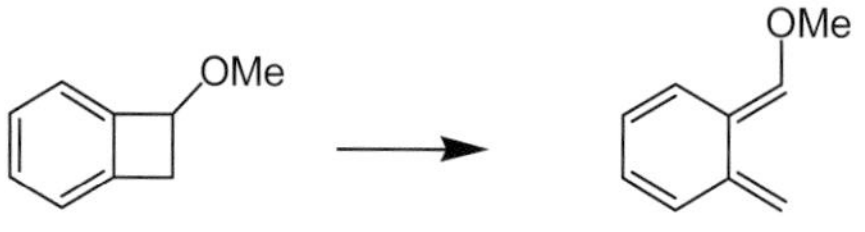 49 (OMe) → 50 (OMe)	Toluene, reflux (50%)	49
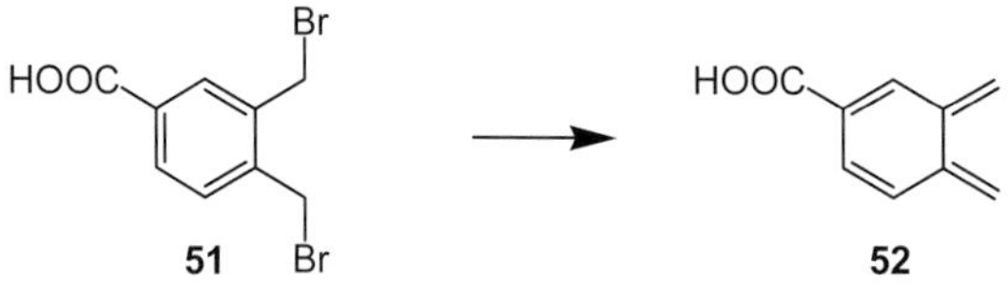 51 (HOOC, Br, Br) → 52 (HOOC)	Toluene, reflux, KI, 18-crown-6 (63%)	46
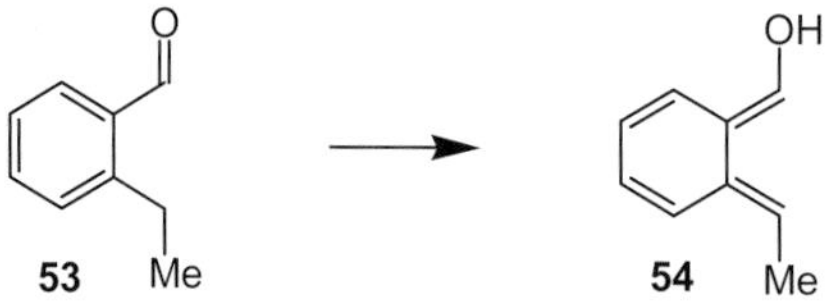 53 (O, Me) → 54 (OH, Me)	Benzene, *hν*, room temperature (17%)	53, 54
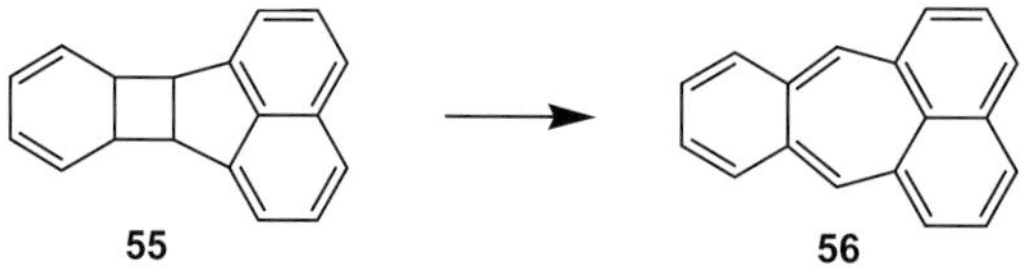 55 → 56	1,2,4-Trichloro-benzene, reflux, (57%)	50
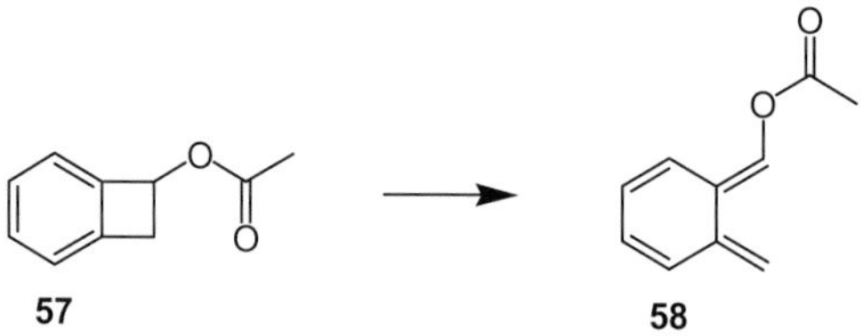 57 → 58	1,2,4-Trichloro-benzene, reflux, (17%)	50
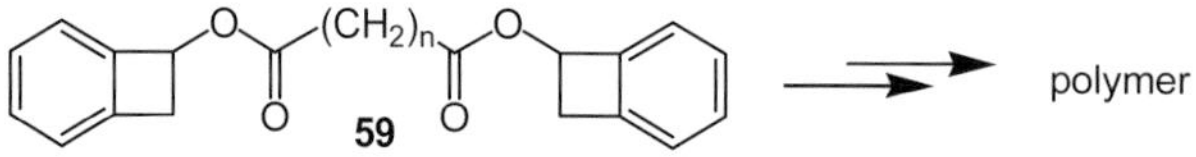 59 ($(CH_2)_n$) → polymer	Neat, 210 °C	50
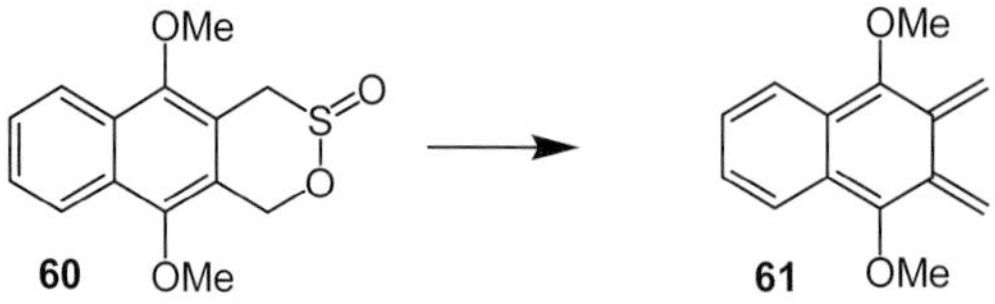 60 (OMe, S=O, O, OMe) → 61 (OMe, OMe)	Toluene, reflux (22%)	51, 52
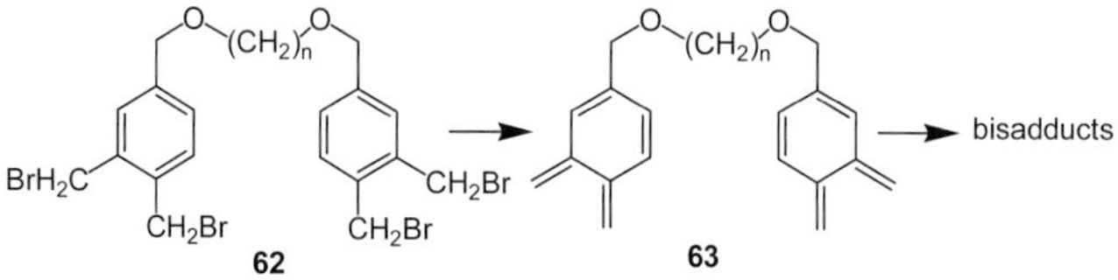 62 ($(CH_2)_n$, BrH_2C, CH_2Br, CH_2Br, CH_2Br) → 63 ($(CH_2)_n$) → bisadducts	Toluene, reflux, KI, 18-crown-6 (8–30%)	45, 47

Table 4.2 (continued)

Substrate	*Conditions (% yield)*	*Ref.*
 OH, Cl, OH, **64** → OH, Cl, OH, **65**	*o*-Dichloro-benzene, 240 °C (36%)	48, 57
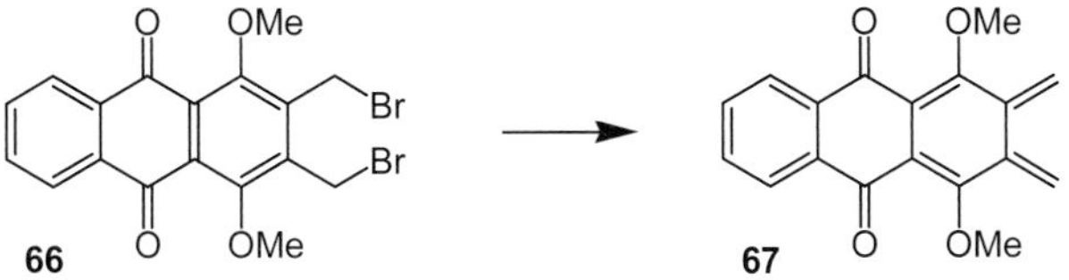 O, OMe, Br, Br, O, OMe, **66** → O, OMe, O, OMe, **67**	Toluene, reflux, KI, 18-crown-6 (63%)	44
 **68** → **69**	*o*-Dichloro-benzene, > 180 °C	56
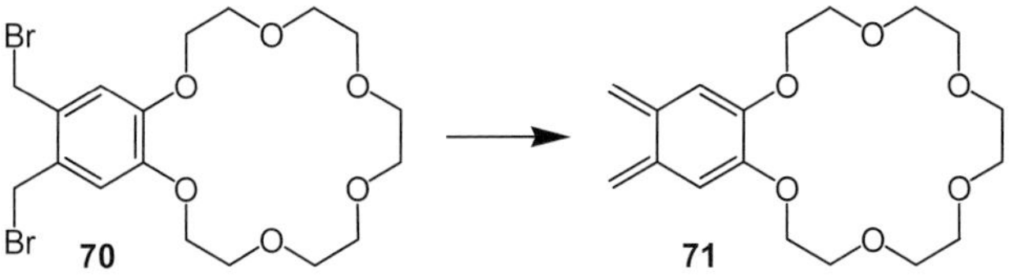 Br, Br, O, O, O, O, O, O, **70** → O, O, O, O, O, O, **71**		58

Information about the electronic structure of the 4,5-dimethoxy-*o*-quinodimethane adduct **46** in the solid state, prepared from 6,7-dimethoxy-3-isochromanone **42** ($R_1 = R_4 = H$; $R_2 = R_3 = OMe$), comes from ESCA measurements [43]. In accord with the crystal structure analysis [43, 58], this cycloadduct, which contains the electron-rich dimethoxyphenyl moiety, shows an intermolecular donor–acceptor interaction, in particular, of the methoxy oxygen to the C_{60} unit.

A benzo[18]crown-6 adduct (**72**) of C_{60} (not shown) has been synthesized by the addition of the corresponding *o*-quinodimethane **71** in toluene [58]. The solubility of **72** in protic solvents such as MeOH strongly increases after the complexation of K^+ ions, as shown by extraction experiments. The combination of the crown ether and the fullerene moiety in **72** provides a highly amphiphilic character. This behavior allowed the preparation of Langmuir–Blodgett films of monolayers on mica of **72** and its K^+ complex.

Reaction of carboxy- or hydroxy-substituted *o*-quinodimethanes such as **52**, **54** or deprotected **58** lead to C_{60} adducts with a functionalizable group. They can be derivatized for example by forming an ester [46, 49, 50] from the acid or by addition of the hydroxy functionality to double bonds [49]. Addition products of the para-dihydroxy derivative **65** can be easily oxidized to the corresponding benzoquinone-linked fullerene with DDQ [48]. The cycloadduct **73** of C_{60} with 4-(4-fluoro-3-

nitrobenzoyl)benzocyclobutene is very stable, losing the benzocyclobutene moiety above 400 °C (Scheme 4.8) [42]. The strong activation of the fluoro-substituted phenyl ring by the nitro- and keto-group enables nucleophilic additions to this adduct in very good to quantitative yields. With strong nucleophilic aliphatic amines, as for example 4,13-diaza-[18]-crown-6 or 1,6-bis(aminomethyl)hexane, substitution occurs within some minutes, whereas the use of less nucleophilic aromatic amines, for example 4-amino-azobenzene, requires longer reaction times. In the latter case, a chromophore is covalently attached to C_{60} (**74**, Scheme 4.8). All these cycloadducts are soluble in chloroform or toluene; therefore, their chromatographic separation can be achieved efficiently by using polystyrene gel as stationary phase [42, 43, 50, 59, 60]. The methods of further functionalization to provide new fullerene derivatives are limited due to the electronic properties of C_{60}. For example, treatment of **73** with 1,3-dimercaptobenzene in *N,N*-dimethylacetamide does not produce the expected substitution product but leads to oligophenylene disulfides via oxidative coupling. Here, the C_{60} derivative acts as a weak oxidizing reagent.

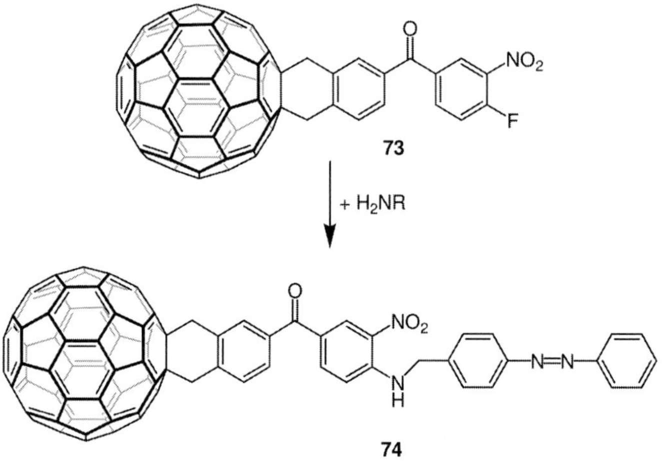

Scheme 4.8

Another method of stabilizing the obtained cycloadduct was employed by Rubin and co-workers [61], using 5,6-dimethylene-1,4-dimethy1-2,3-diphenylnorborn-2-en-7-one (**75**) as diene (Scheme 4.9). The stabilized adduct **77** is formed after loss of CO. Also in this case, the cyclohexene ring undergoes a conformational exchange between the two boat conformations of the molecule. Temperature-dependent ^{1}H NMR investigations revealed a coalescence temperature of 35 °C and an activation energy for the boat-to-boat barrier of inversion of $\Delta G^{\ddagger} = 14.6 \pm 0.1$ kcal mol^{-1}.

Since their discovery by Cava and Napier [62], *o*-quinodimethanes have been widely used in organic synthesis and also for the derivatization of C_{60}. Their heterocyclic analogues have been recognized only recently. Meanwhile, there are some examples of reactions with C_{60}. Preparation of the heterocyclic *o*-quinodimethanes proceeds analogously to the all-carbon *o*-quinodimethane via reactive intermediates that are formed from sultines [63], sulfolenes [63–66], α,α′-dihalides [64, 67] and cyclobutene-

Scheme 4.9

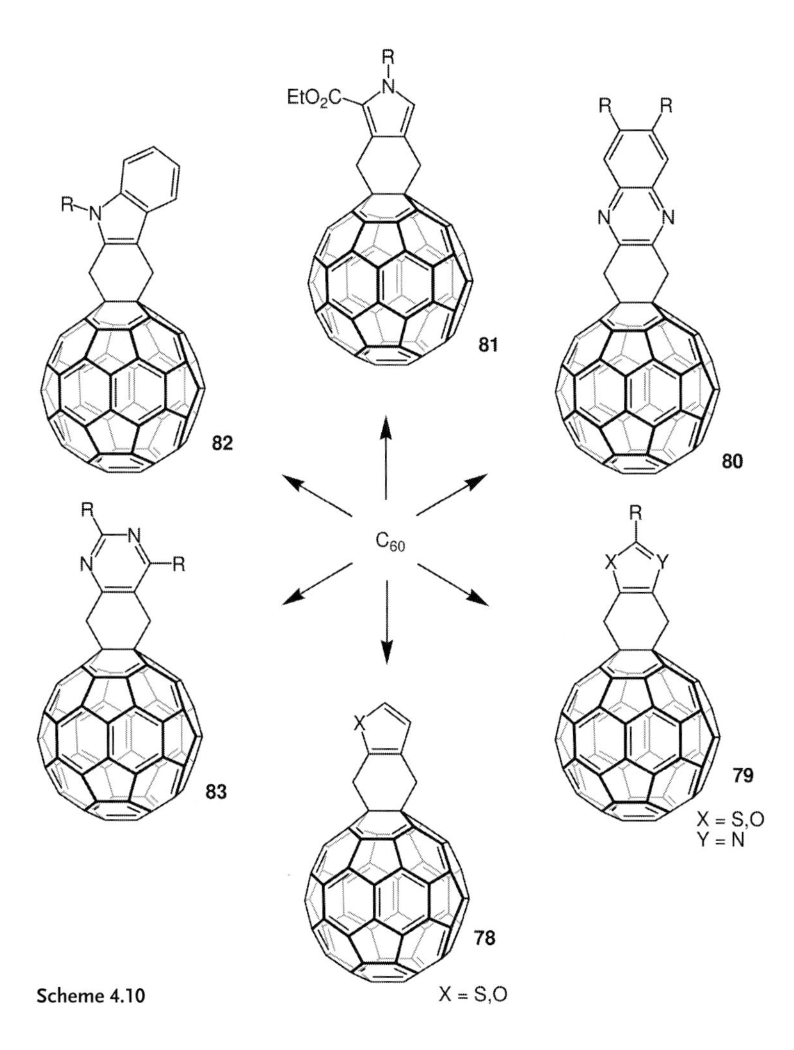

Scheme 4.10

fused [63, 68–70] heterocycles or via flash vacuum pyrolysis of *p*-chlorobenzoate ester [64]. Furan-, thiophen-, oxazol- and thiazol-linked C_{60} derivatives [64, 70] (**78**, **79**, Scheme 4.10) were among the first cycloaddition products between fullerene and heterocyclic *o*-quinodimethane. Aromatic systems such as indole [64] (**82**), quinoxaline [63, 64] (**80**), pyrrole [71] (**81**), pyrazine [63, 67] and pyrimidine [65, 66, 68, 69] (**83**) were also successfully attached to the fullerene core.

The pyrrole adduct **81** was obtained by reaction of 3,4-fused pyrrolo-3-sulfolenes, probably via a radical-mechanism [71]. Obviously, structure **81** can be regarded as part of a porphyrin-structure. Thus, fullereno-porphyrinates with up to four C_{60} units surrounding the porphyrin-core were synthesized [72–74].

A few examples of hetero-Diels–Alder adducts have been reported [75–81]. A thiochroman-fused fullerene adduct was synthesized by the reaction of *o*-thioquinone with C_{60} in *o*-dichlorobenzene at 180 °C [77]. The obtained cyclic sulfide **84** (Figure 4.4) can be oxidized to the corresponding sulfoxide and sulfone with *m*-chloroperoxybenzoic acid. Reaction of azadienes with C_{60} leads to hetero Diels–Alder adducts such as, for example, **85** (Figure 4.4) [79]. The tetrahydropyrido[60]-fullerene **85** is formed in refluxing *o*-dichlorobenzene.

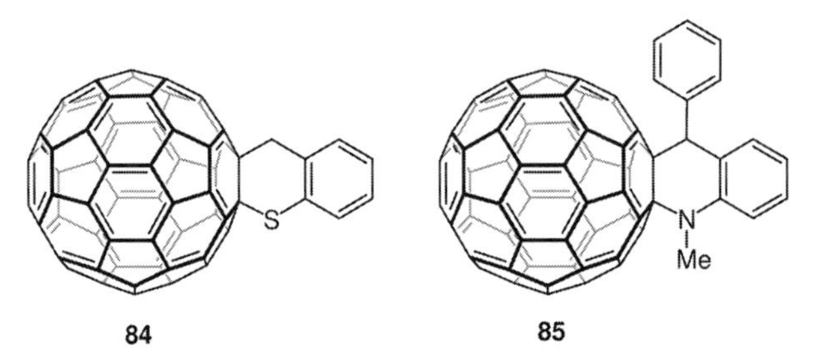

Figure 4.4 Two examples for hetero-Diels–Alder reactions to C_{60}.

Reaction of electron-deficient diaryl-1,2,4,5-tetrazine with C_{60} may also be considered as a hetero-Diels–Alder reaction (Scheme 4.11) [82–84]. The initially formed Diels–Alder product undergoes a rapid retro-Diels–Alder reaction under loss of nitrogen, which renders the cycloaddition irreversible. The obtained diaryl-dihydropyridazine monoadducts are isolable but unstable and they react readily with water under the influence of light to afford a 1,4-hydrogenated product [82].

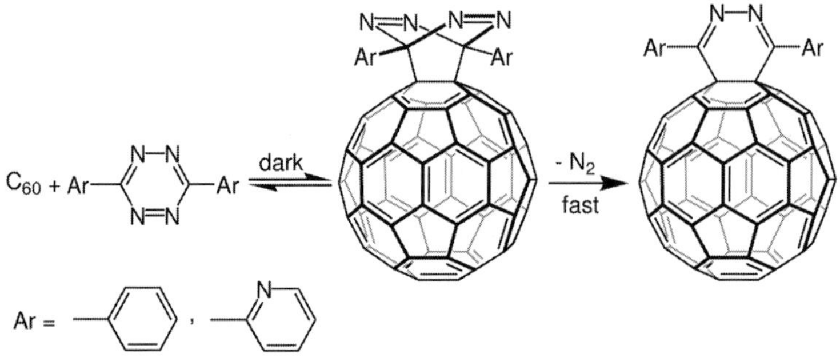

C_{60} + Ar–(tetrazine)–Ar ⇌ (dark) → ($-N_2$, fast)

Ar =

Scheme 4.11

Also, dimerization and rearrangements were observed under certain conditions [83, 84].

Tetrathiafulvalene is a promising candidate on the way to donor–acceptor-dyads of C_{60} [85]. TTF is a non-aromatic molecule that forms upon oxidation the 1,3-dithiolium cation, which possesses aromatic character [85]. This gain of aromaticity leads to the expectation of long-lived charge-transfer complexes of C_{60}-TTF dyads. Various TTF-linked fullerenes have been synthesized via [3+2]-cycloaddition or the Bingel-reaction [85]. The first C_{60}-TTF dyads prepared via [4+2] cycloaddition were synthesized simultaneously by Hudhomme [86, 87] (**89**) and Rovira (**87**, R = CO_2Me) (Scheme 4.12) [88]. The main difference between the Diels–Alder C_{60}-TTF derivatives and the other TTF-linked C_{60} dyads is in the cyclohexene connection between the two electroactive components. The fusion with a six-membered ring provides a short distance and a rigid bridge between donor and acceptor, giving rise to a well-defined spacing and orientation with respect to each other [89].

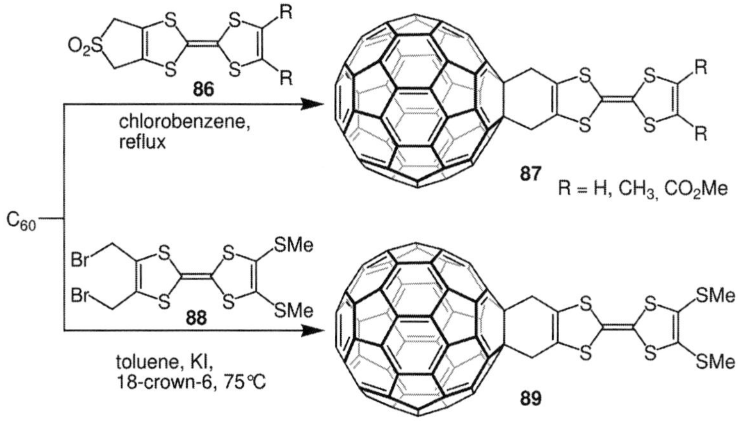

Scheme 4.12

Photoinduced intramolecular electron transfer in the donor–acceptor complex **87** (R = H) generates transient charge-separated open-shell species with the remarkably long lifetime of about 75 μs [89]. Dyads that contain π-extended tetrathiafulvalene units also form stable cationic species upon oxidation [90]. The dumbbell shaped triad **91** [91–93] (Scheme 4.13) was obtained by carrying out the reaction with the in situ generated bis-diene at room temperature, in the dark and in *o*-dichlorobenzene as a solvent in 50% yield. The product is thermally unstable and easily undergoes a retro-Diels–Alder reaction [91].

By allowing C_{60} to react with 2-trimethylsilyloxy-1,3-butadiene in toluene at reflux [38, 94], another stable Diels–Alder adduct was obtained (Scheme 4.14). The ketone **93** is formed after hydrolysis of the uncharacterized intermediate silyl enol ether under flash chromatographic conditions.

This ketone can be reduced with DIBAL-H in toluene at 20 °C to afford the corresponding racemic mixture of alcohols **94** in 93% yield. With **94**, coupling

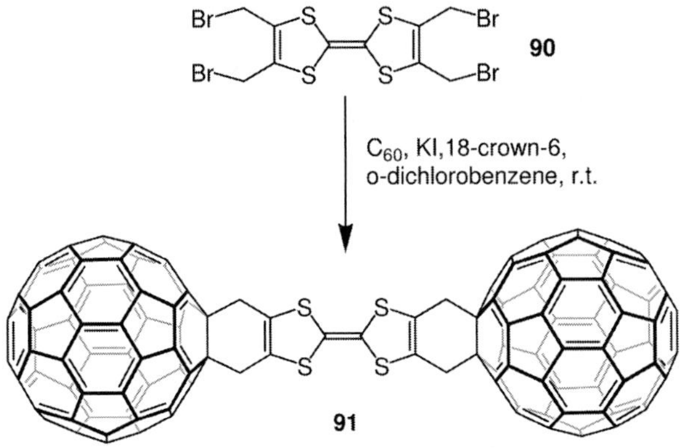

Br Br S S S S Br Br 90

C_{60}, KI,18-crown-6, o-dichlorobenzene, r.t.

91

Scheme 4.13

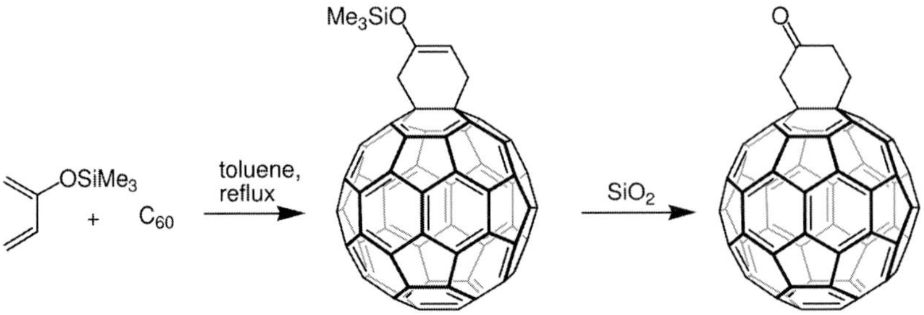

OSiMe₃ + C_{60} toluene, reflux Me₃SiO SiO₂

92 93

Scheme 4.14

reactions can be carried out, leading to a further functionalization of the fullerene sphere [94]. Esterification with protected amino acids, for example with *N*-(benzyl-oxycarbony1)-L-alanine or α-benzyl-*N*-(*tert*-butoxycarbony1)-L-glutamate in toluene in the presence of DCC and DMAP at room temperature, provides amino acid derivatives of C_{60} as a mixture of diastereomers in high yields (Scheme 4.15).

Similar conditions were employed to append the chromophore pyrene to C_{60} in 41% yield [95]. Reaction of **94** with an acid bromide yields almost quantitatively the bromo compound **96** [96]. Compound **96** can be further functionalized via nucleophilic substitution. This reaction was utilized to get access to a C_{60}-linked deoxyoligonucleotide. Upon exposure to light and oxygen a good selectivity of C_{60}-induced guanosine cleavage with a high activity could be observed [96].

Based on bis-silylated dienes another approach to quinoxaline derivatives such as **80** (Scheme 4.10) was found [97]. Fast [4+2] cycloaddition takes place by treatment of C_{60} with 2,3-bis(trimethylsilyloxy)butadiene **98**, yielding the acyloin-fused fullerene derivative **100** in good yields (Scheme 4.16). The silylated diene is formed in situ by treatment of **98** at 180 °C in *o*-dichlorobenzene. Controlled bromination of the intermediate **99** leads to the transient diketone **101**, which reacts readily in a one-pot reaction with various *o*-diaminoarenes to yield the quinoxaline-fused fullerenes **102**.

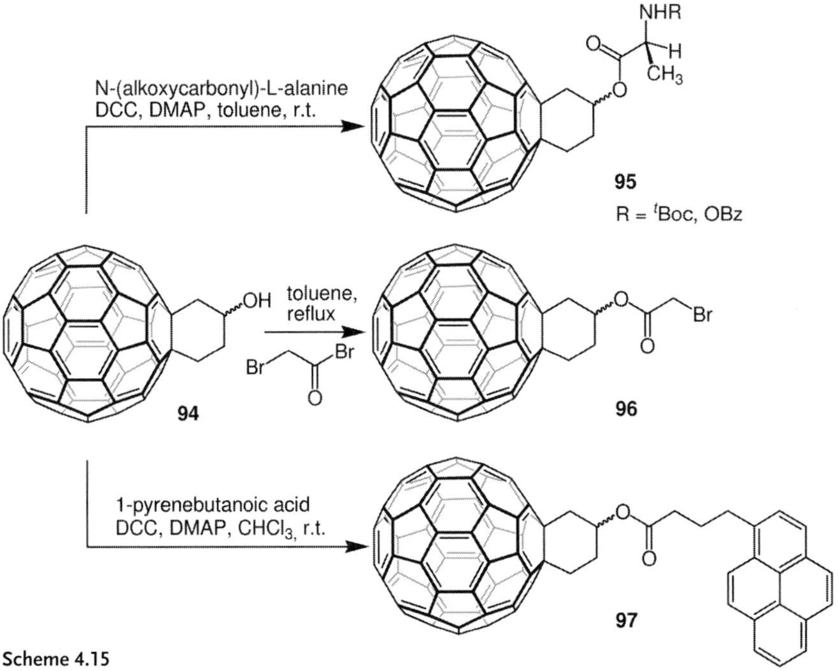

Scheme 4.15

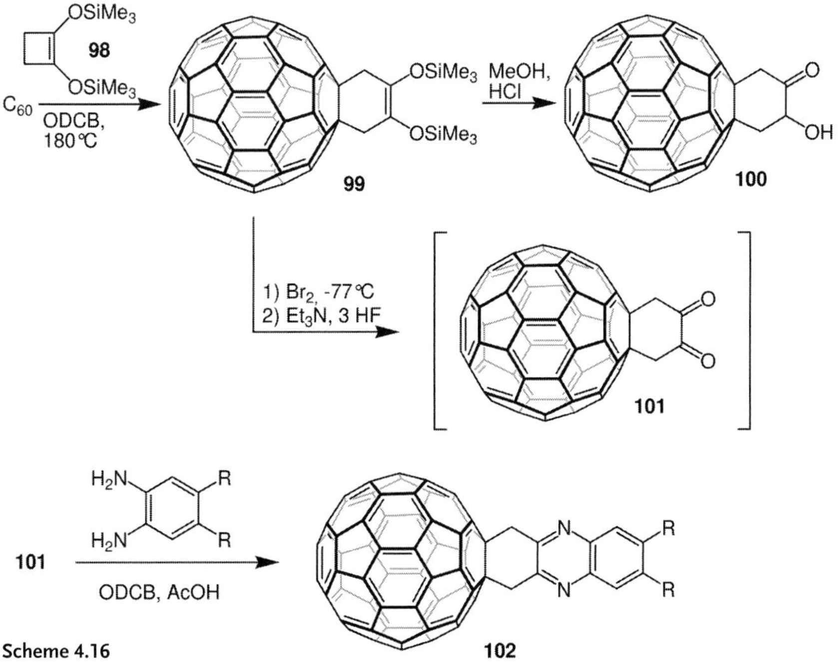

Scheme 4.16

Danishefsky dienes [98] cycloadd to C_{60} in refluxing toluene or benzene [5, 38, 99–101]. The diene **103** adds in 60% yield to C_{60} to give the desilylated ketone **104** [5, 101]. Acid-catalyzed methanol elimination then furnishes the enone **105** in 82% yield (Scheme 4.17). As already described, this enone can be reduced by DIBAL-H to the corresponding alcohol for further functionalization. The same α,β-unsaturated alcohol can also be obtained in better yield by Diels–Alder reaction of C_{60} with butadiene, followed by oxidation with singlet oxygen to the allylic hydroperoxide and PPh_3 reduction to the desired alcohol [101]. This sequence yields the allylic alcohol in 53%, starting from C_{60} without the need of isolating intermediates.

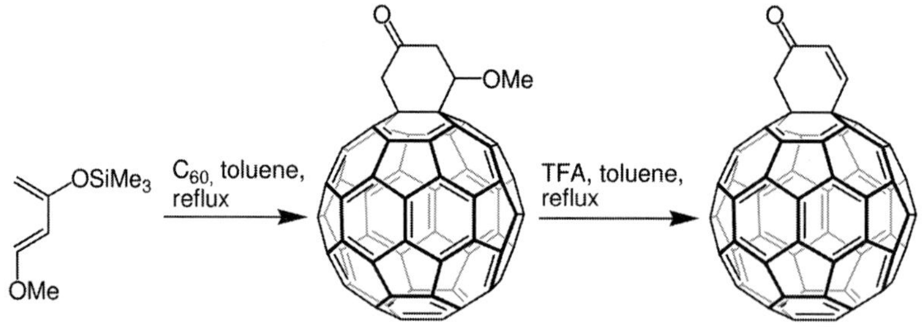

OSiMe3 OMe **103** — C_{60}, toluene, reflux → **104** (OMe) — TFA, toluene, reflux → **105**

Scheme 4.17

Treatment of the intermediate silylenolether, obtained by reaction of **106** with C_{60}, with SiO_2 and triethylamine leaves the methoxy group and yields a mixture of the two isomers **107** and **108** (Scheme 4.18) [99, 100]. This reaction can be carried out either photochemically or thermally. Because the *trans*-product **107** is the major product under both thermal and photochemical conditions, the mechanism of this addition is concluded to proceed stepwise. The first step is probably an electron transfer from the Danishefsky diene to C_{60}. At least for the photochemical pathway, this electron-transfer step could be proven [100].

Reaction of *trans*-2-stilbenecarboxaldehyde derivatives with *N*-methylglycine and C_{60} was performed to obtain a [3+2] cycloaddition product, but an unexpected [4+2]

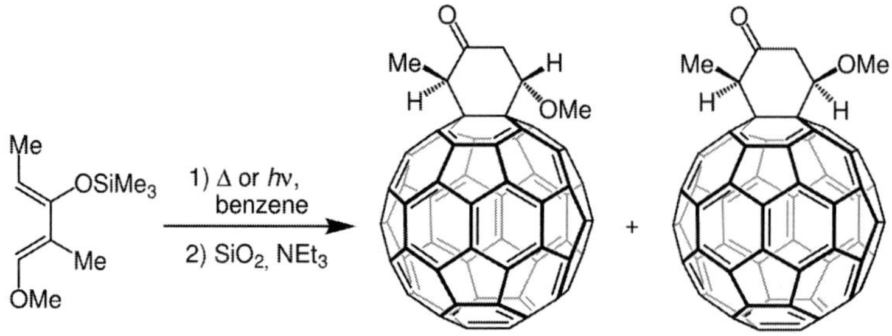

Me OSiMe3 Me OMe **106** — 1) Δ or *hν*, benzene 2) SiO_2, NEt_3 → **107** (Me, H, H, OMe) + **108** (Me, H, OMe, H)

Scheme 4.18

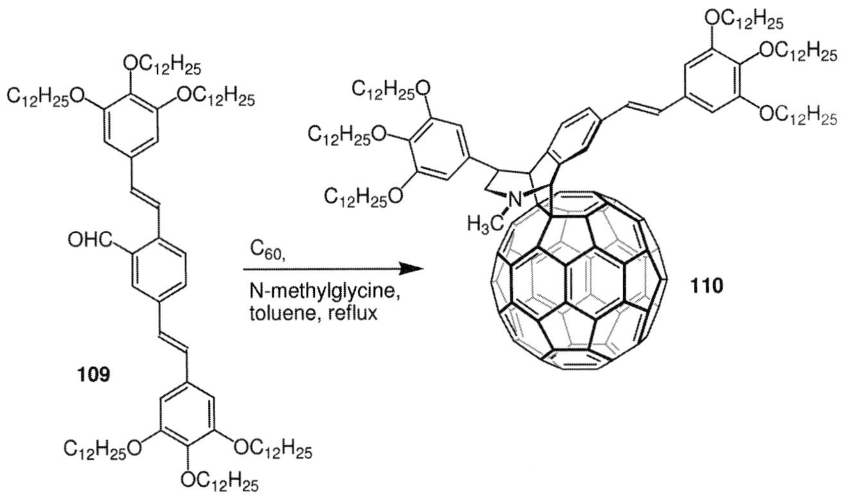

Scheme 4.19

cycloaddition took place (Scheme 4.19) [102]. This result could be generalized by using different stilbene derivatives, which underwent the same reaction in moderate yields.

4.3 [3+2] Cycloadditions

4.3.1 Addition of Diazomethanes, Diazoacetates and Diazoamides

A broad variety of methano-bridged fullerenes are accessible by the reaction of C_{60} with different diazomethanes [2, 3, 103–128], diazoacetates [104, 129, 130] and diazoamides [131]. Also, diazomethylphosphonate [132] and diazoketone [133] were used successfully. These chemical transformations of C_{60} with diazo-derivatives were discovered by Wudl and are based on the finding that C_{60} behaves as an 1,3-dipolarophile [1, 103, 104]. With the synthesis of the parent 1,2-methano[60]fullerene $C_{61}H_2$, by treating C_{60} with diazomethane in toluene, the pyrazoline intermediate **112** could be isolated and characterized (Scheme 4.20) [106, 108]. The first step of this reaction, the [3+2] cycloaddition of diazomethane, occurs at a [6,6] double bond. Extrusion of N_2 out of **112** can be achieved either photochemically or thermally by refluxing **112** in toluene. By these procedures, two different methano-bridged fullerenes are formed, namely the ring-closed system **113a**, where the bridging occurs at a 1,2-position [108], as well as the 1,6-bridged isomer with the ring-opened structure **113b** (fulleroid) (Scheme 4.20) [106]. Whereas the photolysis of **112** gives a mixture of **113a** and **113b**, thermolysis leads to the formation of **113b** with only traces of **113a** detectable by ^{1}H NMR spectroscopy. Importantly, individual photolysis

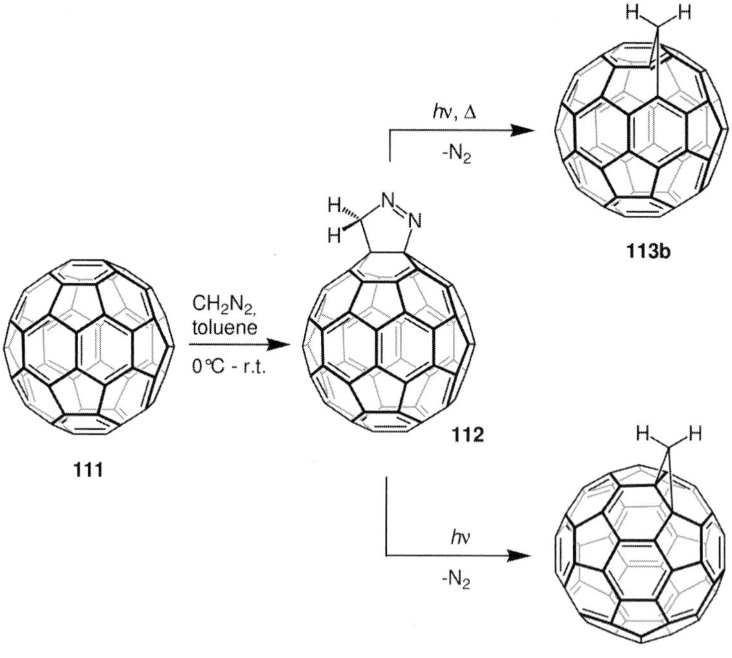

Scheme 4.20

of **113a** and **113b** did not lead to their interconversion and thermolysis of **113b** failed to generate **113a** [108]. Calculations at the semi-empirical MNDO and the HF (3-21G) level indicate that these experimentally isolated isomers, **113a** and **113b**, are indeed the two low-energy isomers of $C_{61}H_2$ [134]; the two isomers are quite close in energy. At the HF/3-21G level, **113a** is 6 kcal mol^{-1} more stable than the ring-opened form **113b**.

The addition of other diazo compounds, for example substituted diphenyldiazomethanes [110] or alkyl diazoacetates [129], also leads to mixtures of different isomers [2, 115]. Four different isomers **A–D** (R = R′) obtained from single additions of symmetric diazo compounds are principally possible (Figure 4.5). These are the corresponding ring-closed and ring-opened structures with the methylene bridges in 1,2- (**A**, **B**) and in 1,6-positions (**C**, **D**). For 1,6-bridges with asymmetrical methylene groups (R ≠ R′), two isomers of each **C** and **D** must be distinguished with one R laying above a pentagon and the other R above a hexagon and vice versa. Based on calculations on the MNDO level [110], with several R and R′ groups, the isomers **A** and **C** are the most stable. Indeed, the isomers found in the reaction mixtures are in all cases the closed 1,2-bridged structures **A** and the open 1,6-bridged structures **C**. These findings together with calculations of the stabilities of different regioisomeric addition products $C_{60}R_n$ [135–139], with R being a segregated group, lead to the following simple rule: for C_{60} as well as for C_{60} derivatives it is energetically unfavorable to introduce double bonds at [5,6] ring junctions or to reduce the greatest possible number of [5]radialene rings. As will be shown in Chapters 6, 8 and 9, this

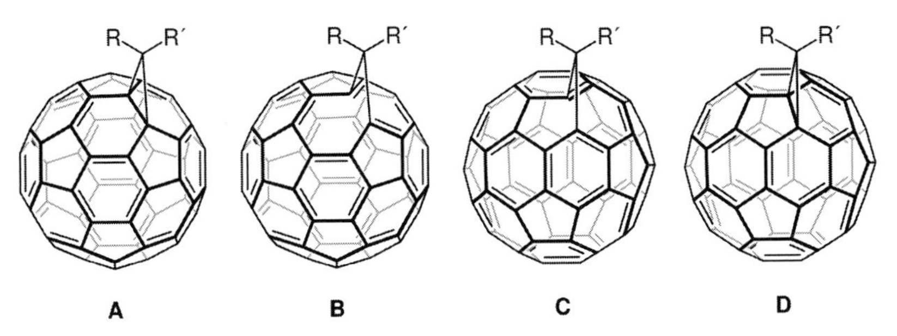

Figure 4.5 The four possible isomers from single addition of diazomethane to C_{60}.

is always valid as long as additional strain introduced by sterical demanding addends (eclipsing interactions) does not prevail. For methano bridges, however, those steric arguments do not have to be considered. Whereas in the structures **A** and **C** only [6,6] double bonds are present, structures **B** and **D** require the formal introduction of three and two [5,6] double bonds, respectively.

Structural assignment of the different methano-bridged isomers is based on NMR criteria (Figure 4.6). From the NMR spectra the symmetry of the isomers, for example C_s or C_{2v}, can be determined. More importantly, in analogy to the methanoannulenes [140–142], which can also exhibit ring-opened or ring-closed structures, the ^{13}C NMR chemical shifts of the bridgehead C atoms and the coupling constants $^{1}J_{(C,H)}$ of the methano bridge C atom, which strongly depend on the structure under consideration, can be used to distinguish between the different isomers [106, 108–110, 114–116, 118, 129, 143]. The bridgehead carbons of **A** resonate in the range $\delta = 70–90$, whereas the bridgehead carbons of **C** appear between $\delta = 130$ and 150. Typical coupling constants $^{1}J_{(C,H)}$ for the methano bridge C atoms of **A** lie in the range 165–170 Hz, whereas those of **C** lie between 140 and 145 Hz. In addition, the ^{1}H NMR chemical shift of a proton, which is in close proximity to the fullerene sphere, strongly depends on whether it is located above a five- or a six-membered ring. A proton located above a pentagon resonates up to 3.5 ppm more down-field than a proton above a hexagon. These results were

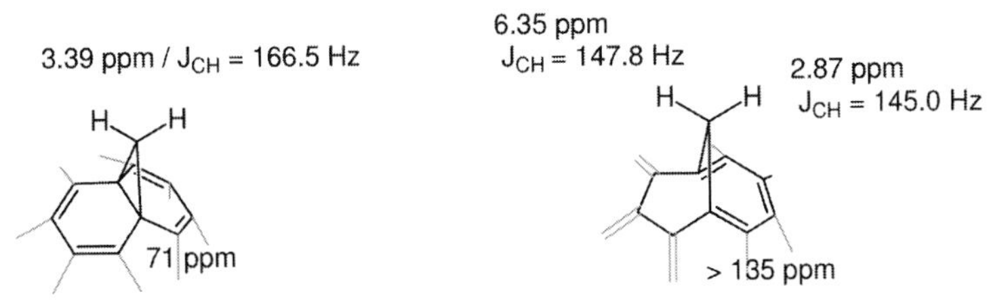

Figure 4.6 Chemical shifts for the bridgehead C atoms and the methylene H-atoms and coupling constants $^{1}J_{(C,H)}$ for the methano bridge C atoms in $C_{61}H_2$ for isomers **113a** and **113b**. The corresponding C atoms resonate in the region between 130 and 150 ppm together with all other sp^2-carbons of the fullerene sphere. For the 1,6-addition adduct of C_{60} with (*p*-methoxyphenyl)diazomethane, the peak position of the bridgehead C atoms was found by HETCOR analysis to be 138.65 ppm [110].

interpreted in terms of local and ring-current anisotropy effects [109, 144]. Calculations of local ring-currents and the NICS-values for the five- and six-membered rings in C_{60} further corroborate these findings [145, 146]. The existence of two types of ring currents is predicted. These are the strong paramagnetic ring currents in the pentagons and the mild diamagnetic currents supported by the hexagons. As a consequence, the strong paramagnetic ring currents are expected to deshield protons located above pentagons, and the mild diamagnetic ring currents should weakly shield the protons closely located above the center of the hexagons. Although this ring-current model was predicted for the parent C_{60}, it may be operative within the homo-rings of the ring-opened structures **C** since the π-system of C_{60} is preserved to a large extent in this family of fulleroids [144].

The addition of alkyl diazoacetates (Scheme 4.21) reproducibly yields the three isomers **114a-c** in the ratio of 1 : 1 : 3, as determined by ^{1}H NMR spectroscopy [129]. The reaction is carried out in refluxing toluene. Under these conditions the formation of the open structure **C** is kinetically favored. The predominance of isomer **114c** over **114b** by a ratio of 3 : 1 represents a remarkable diastereoselectivity. Since there is no obvious steric effect present, a stereoelectronic effect was taken into consideration. Although unambiguous evidence is given for [3+2] cycloadditions of diazomethanes [106] and diphenyl-diazomethanes [147] at room temperature, followed by the thermal or photochemical extrusion of N_2, a pyrazoline intermediate could not be detected in the synthesis of **114a-c** by the addition of alkyl diazoacetates in boiling toluene [129]. However, also in this case, it is very likely that the first step is a [3+2] cycloaddition since the [5,6] ring bridged as well as the [6,6] ring bridged isomers are formed simultaneously. A [2+1] cycloaddition of carbenes affords, exclusively, formation of the [6,6]-ring bridged isomers (Section 4.5.1). In a very mild rhodium-catalyzed reaction of diazoesters with fullerene the [6,6]-closed adduct was formed with a high selectivity. This reaction proceeds via a four-membered rhodium-based metallocycle and not via a [3+2]-cycloaddition [130].

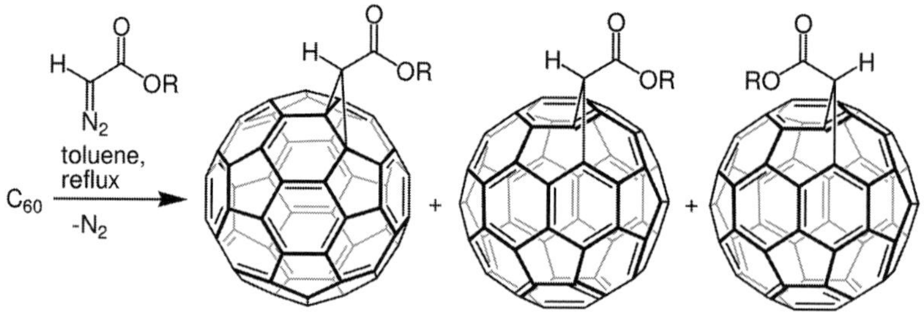

C_{60} —(H(N$_2$)C–COOR; toluene, reflux; $-N_2$)→ **114a** + **114b** + **114c**

R = Et, ^{t}Bu

Scheme 4.21

Significantly, purified mixtures of isomers of **A** and **C** equilibrate, in most cases, quantitatively to the ring-closed [6,6]-ring bridged cycloadducts **A** after refluxing in toluene for several hours. This implies that cycloadducts **A** are the thermodynamically most stable isomers [110, 118, 129]. Beside the thermal treatment [110,

116, 143, 148, 149], photochemical [115, 150] or electrochemical [150, 151] conditions or acid catalysis [152] can also lead to a rearrangement to the [6,6] isomer [115]. The ease of thermal isomerization from diaryl-, arylalkyl and dialkyl-fulleroids to the corresponding methanofullerene decreases in that order. As a mechanism for the isomerization to the thermodynamically stable isomers, a [1,5] shift of the methano bridge as well as retro-carbene additions coupled to additions leading to the stable isomer have been proposed [106, 110, 129]. Several studies report zero-order kinetics, which are inconsistent with this mechanism. Shevlin studied various products from alkyl-, aryl- and arylalkyl [6,5]-open methanofulleroids and found that they undergo both a thermal first-order rearrangement and a photochemical rearrangement with zero-order kinetics. It appears that these rearrangements involve a biradical intermediate that achieves a major part of its radical stabilization from the fullerene ring [115]. The thermal process proceeds by a higher energy pathway than the photochemical reaction, but radical-stabilizing substituents on the methano bridge may lower the thermal isomerization barrier.

The redox properties of C_{60} are largely retained in the methano-bridged fullerenes [103, 105, 106]. The reduction waves of both [5,6] and [6,6] ring-bridged cycloadducts undergo a small shift to more negative potentials in the range of about 100 mV [106]. Due to the higher similarity between the π-systems of [5,6]methanofullerene and C_{60} the potential shift in the ring-opened isomer is not as pronounced as in the ring-closed isomer [116]. The fact that the acceptor properties of fullerene derivatives are retained is an important requirement for further transformations (n-doping) into charge-transfer complexes with interesting properties as materials. Charge-transfer properties have been examined in various methanofullerenes. Some representative examples are derivatives functionalized with electroactive moieties such as π-extended TTF [124], quinone or anthraquinone [119, 120], acetylene dendrimers [125, 126], metal complexes [127, 128], porphyrin [153], aniline [121] or DMAP [123].

The electronic absorption spectra of C_{60} and [5,6] and [6,6] methano-bridged cycloadducts (monoadducts) exhibit almost the same features in the UV region, consisting of three dominant bands at 325, 259 and 219 nm [103, 106, 108, 126, 129, 154]. In the visible region, however, the spectra of the ring-closed 1,2- bridged adducts differ significantly from those of C_{60} and the open 1,6-bridged isomers (Figure 4.7) [129]. Three new bands, not present in C_{60}, are exhibited in the spectrum: a sharp band at 430, a broad band at 492 and a group of weak structures in the 650–700 nm region with the strongest absorption at about 695 nm [154]. Characteristic for the spectra of all the 1,2-bridged cycloadducts is the new sharp band with λ_{max} near 430 nm [154]. This band is also observed in the spectra of other adducts, such as the Diels–Alder adducts of C_{60} (Section 4.2) or the 1,2-dihydro[60]fullerenes (see Section 3.2), which, of course, also have closed transannular [6,6] bonds. In contrast, the spectra of the open 1,6-bridged isomers resemble much more the spectrum of C_{60} (Figure 4.7). These findings independently prove that in the open 1,6-bridged isomers the spherical chromophore of the pure fullerene is less perturbed than in the closed 1,2-bridged isomers. Whereas the purple color of solutions of 1,6-bridged isomers is almost the same as that of C_{60}, those of monoadducts of C_{60} with the addends at the closed [6,6] bonds are chestnut brown to wine red.

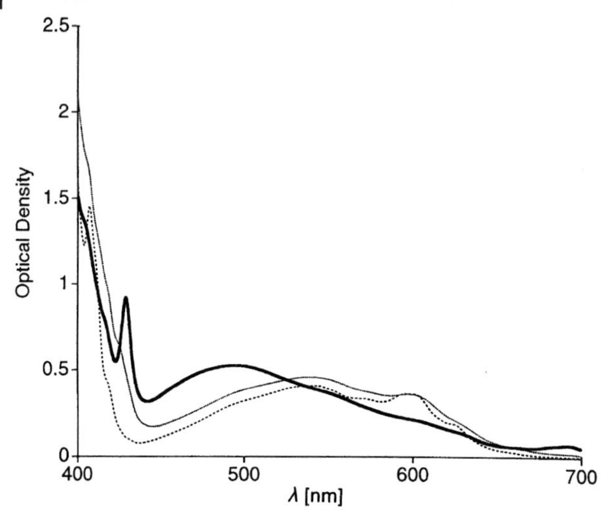

Figure 4.7 Electronic-absorption spectra recorded in toluene of a mixture of (ethoxycarbonyl)-methylene-bridged fullerenes **114** with the relative ratio (1 : 1 : 3) (—), of pure isomer **114a** (—), and of C_{60} (- - -) [129].

Cycloadditions of substituted phenyldiazomethanes provide access to a broad variety of fullerene compounds bearing functional groups that are important for further chemical modifications and applications of these materials. These reactions have been creatively exploited by Wudl [103–105, 107, 110–112] and others [155]. Substituents on the phenyl ring include amino, hydroxy, alkoxy, alkoxycarbonyl and nitro groups, as well as halogens and crown ethers [104]. In particular, *p,p′*-disubstituted diphenylmethano-bridged fullerenes [156] are easily prepared from readily available substituted benzophenones, which are converted into diazomethanes via the hydrazones [157]. Also, bis(diazo)compounds react specifically with C_{60}. The cycloaddition of phenylene-bis(diazomethane), for example (Scheme 4.22), leads to the phenyl-bridged fullerene dimer **116**, which can be regarded as a "two-pearl section" of the "pearl necklace polymer" with C_{60} as a part of the main chain [105].

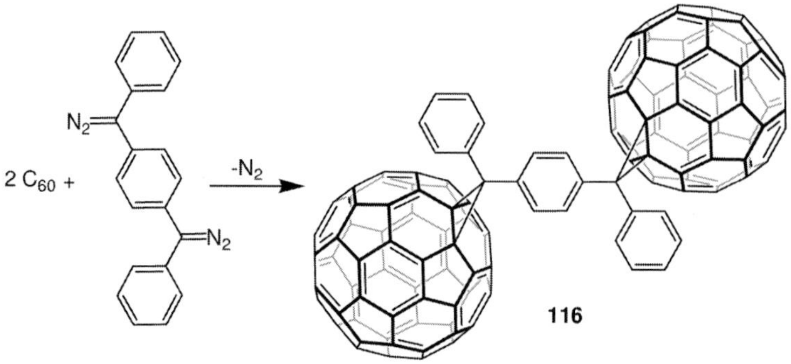

$2\ C_{60}$ + **115** $\xrightarrow{-N_2}$ **116**

Scheme 4.22

An enantiopure dimer **149** with a 1,1′-binaphthyl-bridge was prepared via the bis-tosylhydrazone (see Table 4.4, page 130/131) [122]. The electronic properties of these dimers, such as the electronic absorption spectra and cyclic voltammetry are indistinguishable from those of other methano-bridged fullere[illegible] data show clearly that the two C_{60}-units of the binaphthyl-dimer are reduced independently [122].

The synthesis of a diphenylmethano-bridged fullerene derivative with reactive functional groups on the phenyl rings is exemplified by the preparation of the diphenol derivative **118** (Scheme 4.23). It can be obtained from the corresponding methyl ether by treatment with BBr_3 in *o*-dichlorobenzene at 0 °C to room temperature in 94% yield. In contrast to the non-polar diphenyl-methano bridged fullerenes, **118** is soluble and stable in pyridine but sparingly soluble in benzene or toluene.

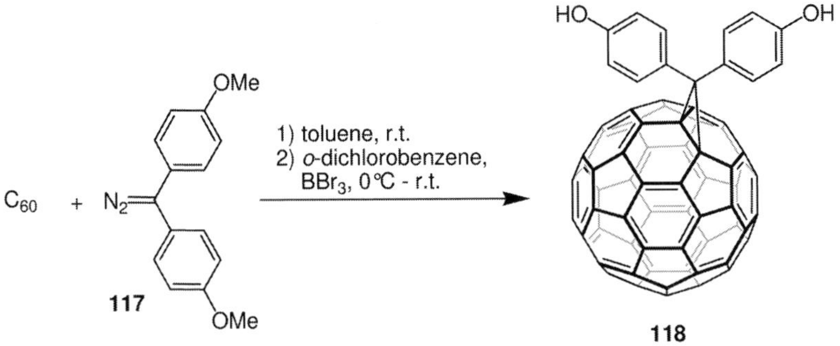

C_{60} + **117** → **118**
1) toluene, r.t.
2) *o*-dichlorobenzene, BBr_3, 0 °C - r.t.

Scheme 4.23

Copolymerization of **118** and sebacoyl chloride leads to a polymer with C_{60} pending on the side chain [107]. The corresponding poly(ester) is more soluble than the dimer **116** [105], for example in nitrobenzene or benzonitrile. Solution cyclic voltammetry of the polymer in THF shows that the electronic properties of the diphenylmethano-bridged fullerenes are retained. In a similar fashion, **118** can be copolymerized with hexamethylene diisocyanate to give the poly(urethane) as an insoluble brown powder [107]. The diphenol derivative **118** has also been used to synthesize fullerene-bound dendrimers by treatment with the fourth-generation dendrimer [G-4]-Br **119** in the presence of K_2CO_3 (Scheme 4. 24) [112]. The reaction was monitored by size-exclusion chromatography (SEC). Under optimized reaction conditions the dendritic system was obtained in 79% yield, after work up, as a light brown-colored glass. The dendritic macromolecules dramatically improve the solubility of the fullerene in organic solvents.

The synthesis of a water-soluble diphenylmethano-bridged fullerene **122** was achieved by hydrolyzing the bis(acetamide) **121** with acetic acid-aqueous hydrochloric acid and then converting it into the bis(succinamide) **122** by treatment with succinic anhydride (Scheme 4.25) [158]. Compound **122** is soluble in water at pH ≥ 7. This is an important requirement for the investigation of the biological activity of fullerenes. Remarkably, **122** is an inhibitor for the HIV enzymes protease (HIV-P) and reverse transcriptase (HIV-RT) [159]. As suggested by molecular modeling,

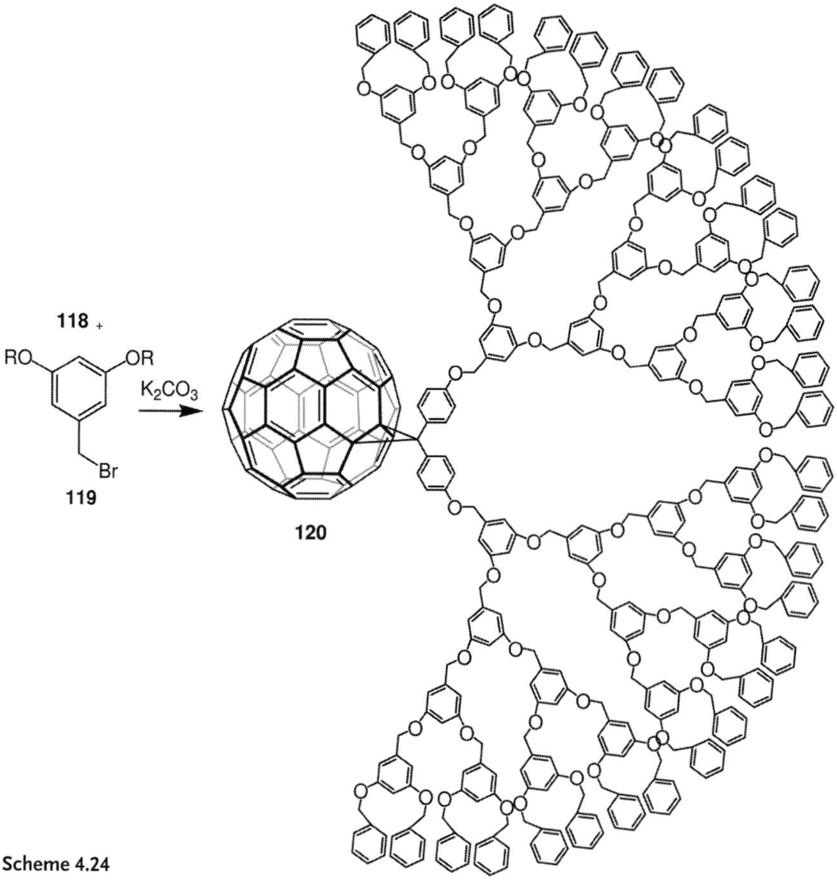

118 +

RO OR

K_2CO_3

Br

119

120

Scheme 4.24

specific C_{60} derivatives fit exactly in the active sites of the enzymes due to their steric and chemical complementarity. In particular, design considerations require that an "active" C_{60} derivative should have polar functional groups at one end of the carbon cluster, which is fulfilled for **122**. In addition, the compound is nontoxic to three different lines of test cells. This first example of a biologically active fullerene is already impressive and has initiated extensive research dealing with specific tailor-made derivatives [160–163].

Another versatile synthon for the preparation of fullerene derivatives with polar groups on the side chain is the 1,2-(carboxymethano)[60]fullerene **123**, which can be obtained either from the corresponding ethoxycarbonylmethyl carboxylate **124** or *tert*-butyl carboxylate **125**, which themselves are accessible by the reaction of the corresponding diazoacetate with C_{60} (Scheme 4.26) [113].

With the carboxylic acid **123**, dicyclohexylcarbodiimide (DCC) mediated esterification and amidation reactions can be carried out (Scheme 4. 27) [113]. For example, the reaction of **123** with EtOH in the presence of DCC and 1H-benzotriazol (BtOH) in bromobenzene with a catalytic amount of 4-(dimethylamino)pyridine (DMAP)

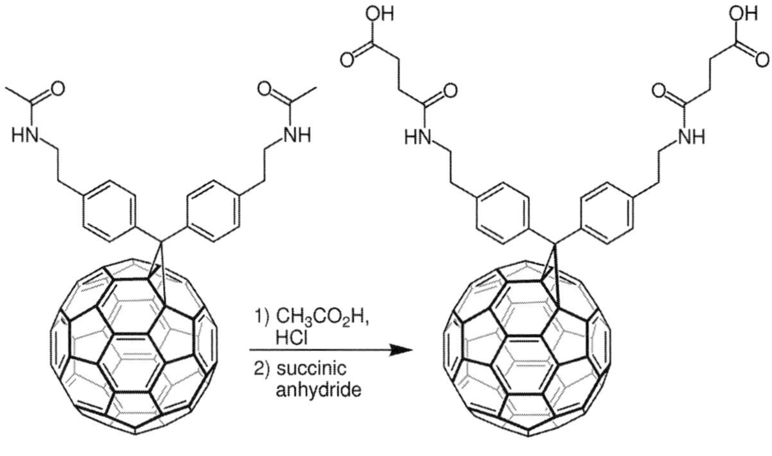

Scheme 4.25

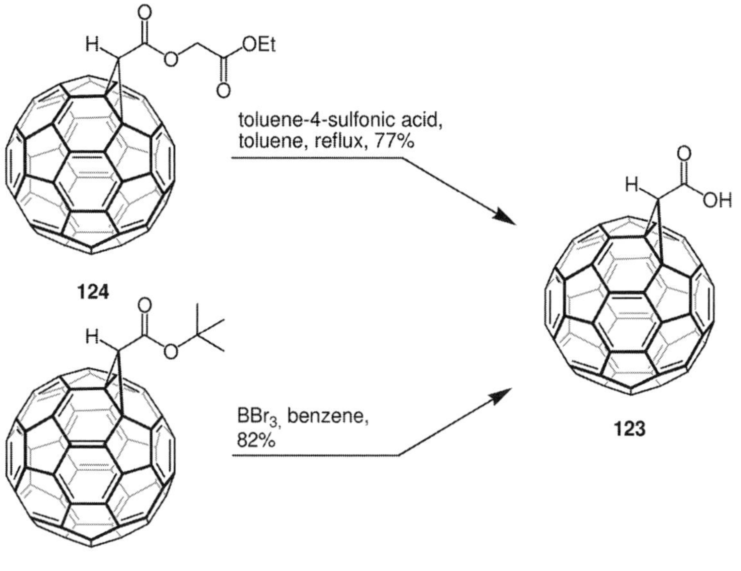

Scheme 4.26

affords the ethylester in 68% yield. Similar to such esterifications the *N*-[(methano-fullerene)carbonyl]-substituted amino acid esters, **127** and **128**, are formed in good yields. Whereas the glycine derivative **127** is poorly soluble in most organic solvents, the phenylalanine derivative **128** is sufficiently soluble in $CDCl_3$ to record its ^{13}C NMR spectra.

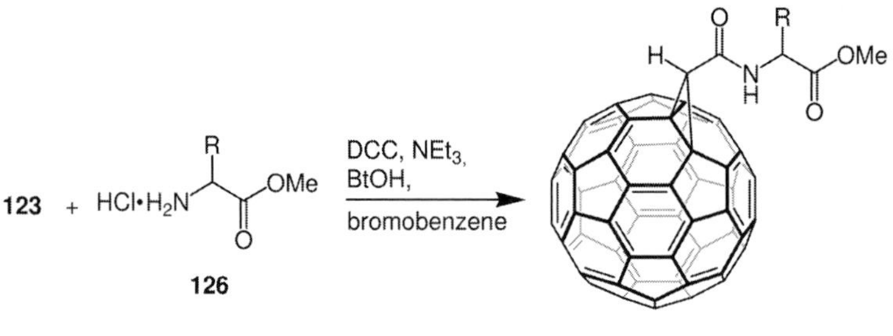

123 + HCl•H_2N ... OMe (126) → DCC, NEt_3, BtOH, bromobenzene →

127 R = H

128 R = $PhCH_2$

Scheme 4.27

Even an oligopeptide has been attached to C_{60} (Table 4.3, compound **130**) [111]. This was achieved by a coupling reaction of the carboxylic group in the side chain of the cyclopropane ring as well. First, the *tert*-butylcarboxylate **129** was synthesized by the reaction of the corresponding diazomethylbenzoate with C_{60}. After hydrolysis with trifluoromethanesulfonic acid, the acyl chloride was generated by treatment with oxalyl chloride. Finally, in a one-step procedure the fullerene peptide **130** was obtained by the reaction with the N-deprotected pentapeptide H-(L-Ala-Aib)$_2$-L-Ala-OMe.

Further examples of diazomethane substrates successfully attached to C_{60} are listed in Table 4.3. The isolated diazomethanes that are used in these examples are usually generated by oxidation from the corresponding hydrazones with silver oxide, nickel peroxide, mercuric oxide, manganese dioxide or lead dioxide [2, 116]. An interesting exception is given with the reaction of precursor **140** [114]. Reaction of 11-aminoundecanoic acid and subsequent nitrosylation yields **140**, which can be used directly without further purification. Formation of the corresponding diazo compound **141** can be controlled by the rate of the addition of KOH. Reaction of the in situ formed **141** with C_{60} gives the methanofullerene in good yields. Preparation of **141** with the usual methods leads only to 4% of the highly reactive aliphatic diazo compound.

In search of a convenient procedure for preparing diazo substrates for the cycloaddition to C_{60}, Wudl introduced the base-induced decomposition of tosylhydrazones [116]. This procedure allows the in situ generation of the diazo compound without the requirement of its purification prior to addition to C_{60}. Since they are rapidly trapped by the fullerene, even unstable diazo compounds can be successfully used in the 1,3-dipolar cycloaddition. In a one-pot reaction the tosylhydrazone is converted into its anion with bases such as sodium methoxide or butyllithium, which after decomposition readily adds to C_{60} (at about 70 °C). This method was first proven to be successful with substrate **142**. Some more reactions that indicate the versatility of this procedure are shown in Table 4.4. Reaction of **142** with C_{60} under the previously described conditions and subsequent deprotection of the *tert*-butyl ester leads to [6,6]-phenyl-C_{61}-butyric acid (PCBA) that can easily be functionalized by esterification or amide-formation [116]. PCBA was used to obtain the already described binaphthyl-dimer (obtained from **149** by twofold addition) in a DCC-coupling reaction [122].

Table 4.3 Diazomethane substrates attached to C_{60} in a [3+2] cycloaddition.

Substrate	*Conditions (% yield)*	*Ref.*
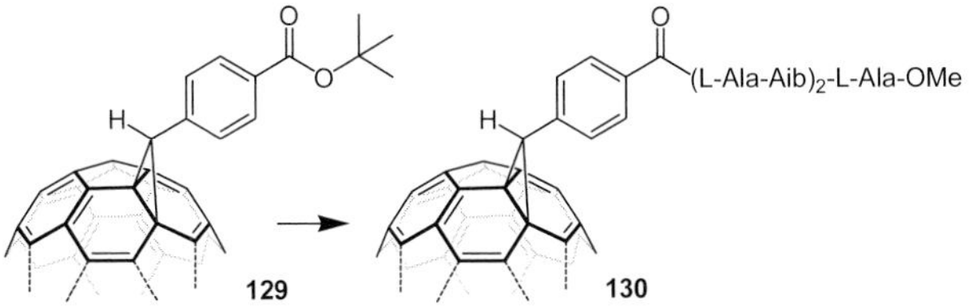 129 → 130 ((L-Ala-Aib)$_2$-L-Ala-OMe); Ala = alanine, Aib = α-aminoisobutyric acid	(1) Trifluoromethanesulfonic acid (2) oxalyl chloride (3) H-(L-Ala-Aib)$_2$-L-Ala-OMe	111
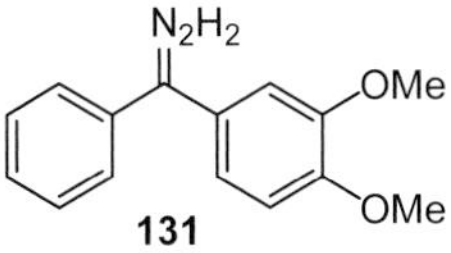 131 (N_2H_2, OMe, OMe)	$Pb(OAc)_4$, *n*-hexane, toluene, room temperature to reflux (18%)	155
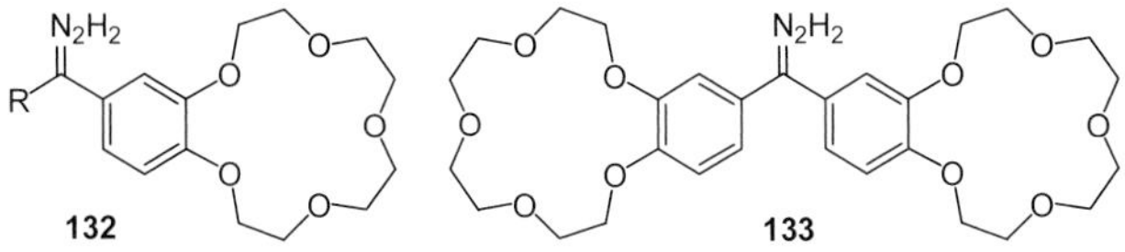 132 (R = H, Me, Ph), 133 (N_2H_2)	MnO_2–Ag_2O, toluene, room temperature (≈40%)	143, 155
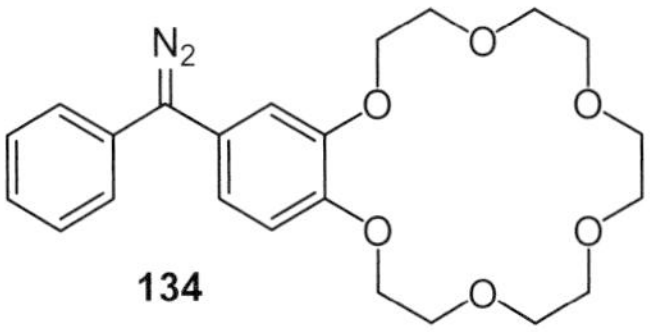 134 (N_2)	MnO_2, benzene, 10 °C (42%)	147
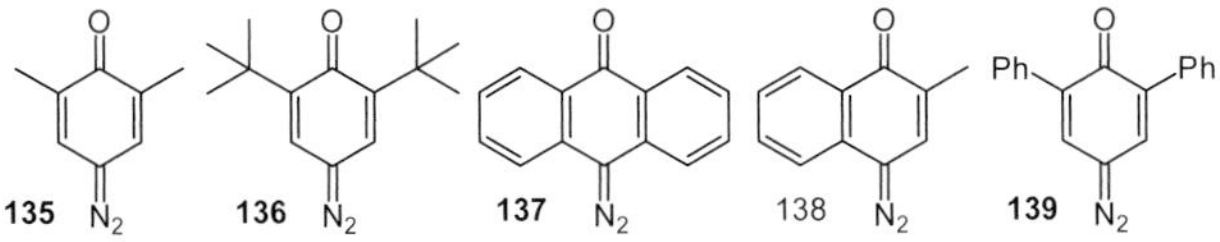 135, 136, 137, 138, 139 (N_2; Ph)	ODCB, *h*ν or Δ (or via the tosylhydrazone)	119, 120
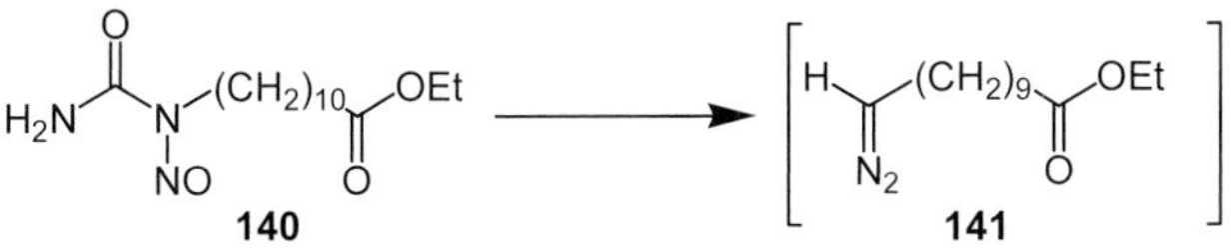 140 (H_2N, NO, $(CH_2)_{10}$, OEt) → [141] (H, $(CH_2)_9$, OEt, N_2)	aq. KOH, toluene, room temperature, (≈40%)	114

Table 4.4 Tosylhydrazone substrates for [3+2] cycloaddition.

Substrate	*Conditions (%yield)*	*Ref.*
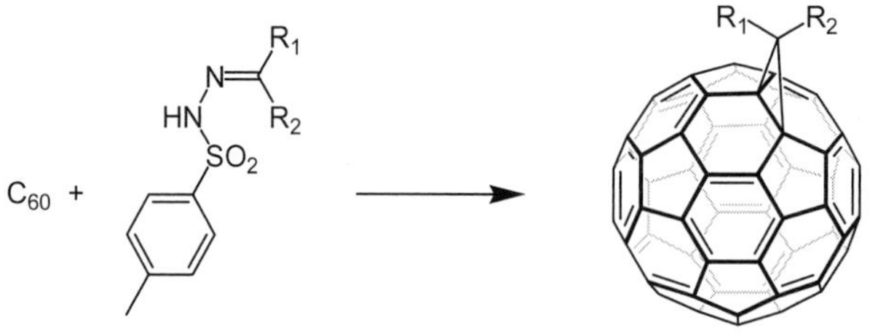 C_{60} + (N=C(R_1)(R_2), HN, SO_2) → (R_1, R_2)		
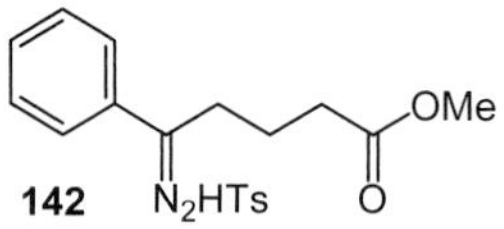 **142** (N_2HTs, OMe, O)	(1) Pyridine, NaOMe; C_{60}, ODCB, 70 °C (35%, [5,6] adduct); (2) ODCB, reflux (98%, [6,6] adduct)	116
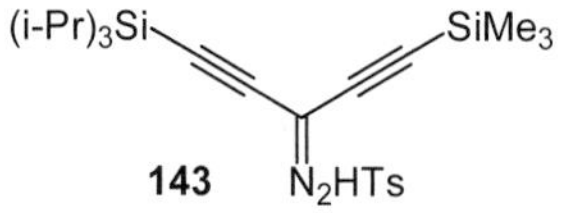 **143** ((i-Pr)$_3$Si, $SiMe_3$, N_2HTs)	*n*-BuLi, benzene, 80 °C (53%)	117, 164
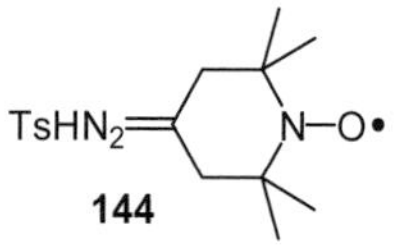 **144** ($TsHN_2$, N–O•)	(1) NaOMe (2) chlorobenzene, 125 °C (≈20%)	148
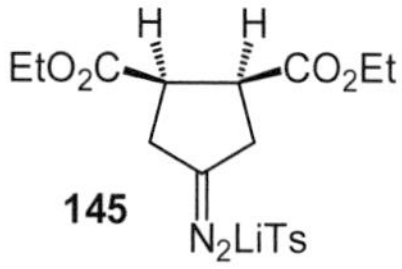 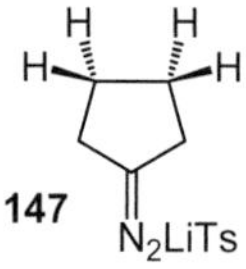 **145** 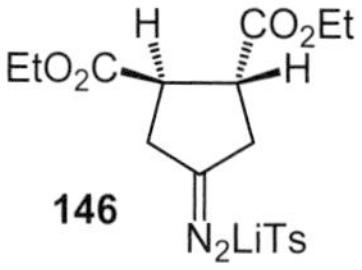 (H, H, EtO_2C, CO_2Et, N_2LiTs); **146** (H, CO_2Et, EtO_2C, H, N_2LiTs); **147** (H, H, H, H, N_2LiTs)	Toluene, reflux (20%-67%)	115, 118
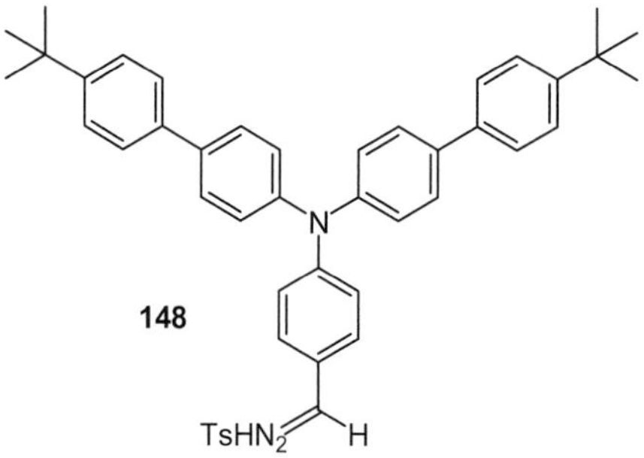 **148** (N, $TsHN_2$, H)	NaOMe, pyridine, ODCB, 50 °C (48%)	121

Table 4.4 (continued)

Substrate	*Conditions (%yield)*	*Ref.*
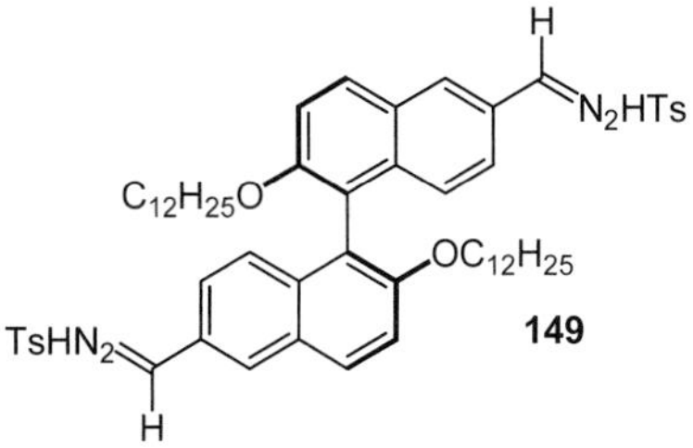 $C_{12}H_{25}O$, $OC_{12}H_{25}$, N_2HTs, $TsHN_2$, H, H **149**	NaOMe, pyridine, ODCB, 50 °C (9% bisadduct)	122
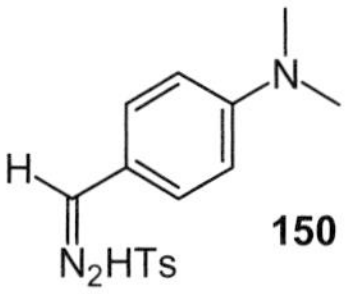 H, N_2HTs, N **150**	NaOMe, pyridine, ODCB, 50 °C (29%)	123
$R_1 = R_2 =$ 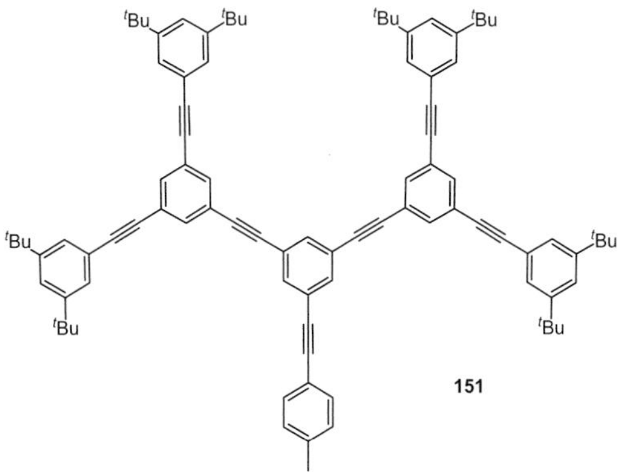 ^{t}Bu **151**	NaH, toluene, 70 °C (23%)	125, 126
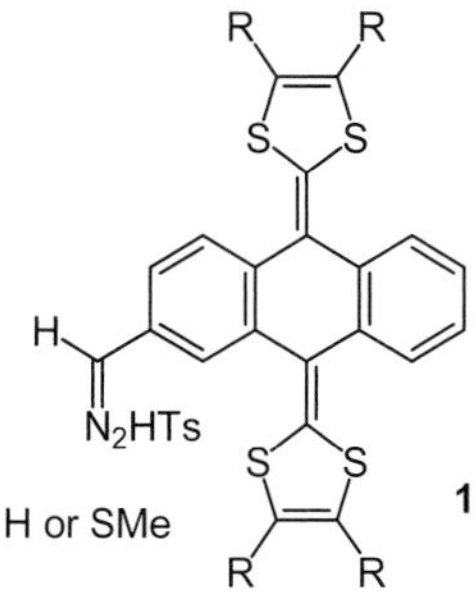 R, S, H, N_2HTs, R = H or SMe **152**	NaOMe, pyridine, toluene, reflux (> 30%)	124

Another example is a phthalocyanine–azacrown–fullerene system that can be synthesized with the same DCC supported esterification [153].

In most of the examples shown in Tables 4.3 and 4.4 the [5,6] adduct is the kinetically favored product whereas the [6,6] adduct is the thermodynamically more stable product. The ratio of the different isomers **A** or **C** (Figure 4.5) in the reaction mixtures after loss of N_2 depends on the diazomethane derivative used for the cycloaddition and strongly on the reaction temperature. Reaction of **142** at 70 °C leads to the formation of one kinetic product, which is the [5,6] adduct where the phenyl ring is over the former pentagon [116]. This adduct is converted into the more stable [6,6] adduct by prolonged heating at 180 °C. Only the [5,6] fulleroid could be obtained in the reaction of **140** [114] or **144** [148] with C_{60}. The reaction of the TTF derivative **152** [124] with C_{60} was carried out at 70 °C in toluene and leads exclusively to the [5,6] product whereas at reflux-temperature the [6,6] product is formed. The [6,6] product can also be obtained by conversion of the 5,6-product in toluene at reflux temperature. The same method was used for the synthesis of the addition product of **143** [117, 164] to C_{60} to get only one isomer, namely the [6,6] methanofullerene. Reaction of **148** [121], **150** [123] or **151** [125, 126] yields only the [6,6] methanofullerene, although relatively low reaction temperatures are applied. Control over the product distribution is not only possible via control of temperature but also by adjusting the reaction time. In refluxing toluene the tosylhydrazones **145–147** [115, 118] can be converted mainly into the [5,6] adducts with a high selectivity by applying short reaction times (about 15 to 25 min). Longer reaction times lead to an increased yield of the [6,6] adduct.

Besides diazomethanes and diazoacetates, cycloadditions of C_{60} with diazoamides or diazoketones have been successfully carried out [131, 133]. A large variety of diazoamides are easily accessible in high yields by direct diazoacetylation with succinimidyl diazoacetate [165] of primary and secondary amino groups in amino compounds, including protected amino acids. As with the addition of diazoacetates, the synthesis of the amido-, lysine- and phenylalanine derivatives **153–156** (Table 4.5) requires elevated temperatures and is carried out in refluxing toluene. The 1,2-methano-bridged and two 1,6-methano-bridged isomers of the cycloadduct, obtained by reaction of **153** with C_{60}, can be isolated by chromatography on silica gel with toluene as eluent. The 1,2-methano-bridged isomer is the most retentive. The different regioisomers of the cycloaddition products of **154–156** elute simultaneously and are less retentive on silica gel columns. Compared with the morpholenide moiety in **153**, the more extended substituents in **154–156** increase the solubility and at the same time reduce the difference in retention behavior of the regioisomers.

An efficient way to attach carboxylic acid derivatives directly to C_{60} was found with the reaction of the corresponding diazoketones **158–162** with C_{60} in toluene or methylnaphthalene at elevated temperatures [133]. The [6,6] closed cycloaddition products are formed in moderate yields although a significant amount of a side-product was produced. This side-product was found to be a dihydrofuran fused C_{60}-adduct. The direct reaction product of the malic acid derivative **162** without elimination of AcOH could not be isolated; instead the *trans*-alkene **163** was formed and isolated in 18% yield.

Table 4.5 Miscellaneous diazo-substrates for [3+2] cycloaddition to C_{60}.

Substrate	*Conditions (%yield)*	*Ref.*
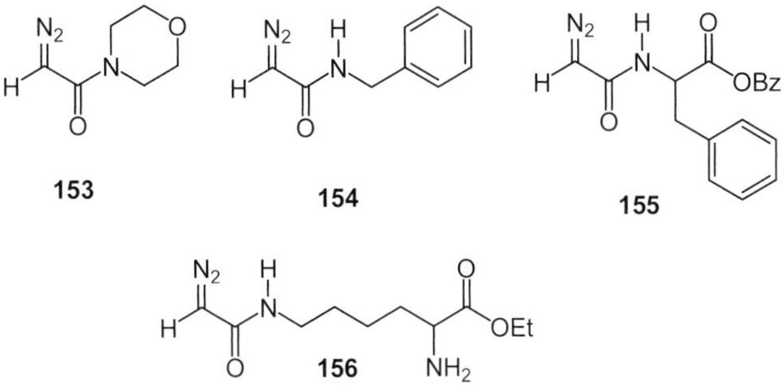 **153**, **154**, **155**, **156**	Toluene, reflux (20–30%)	131
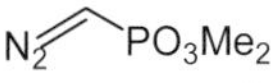 N_2=CH–PO_3Me_2 **157**	Toluene, reflux	132
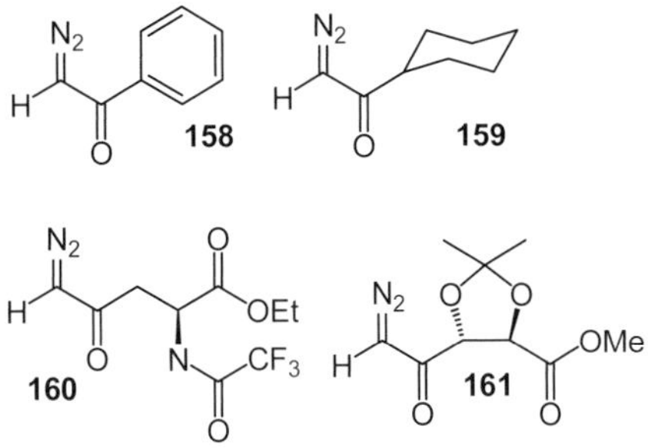 **158**, **159**, **160**, **161**	100–150 °C, toluene or methyl-naphthalene, (12–16%)	133
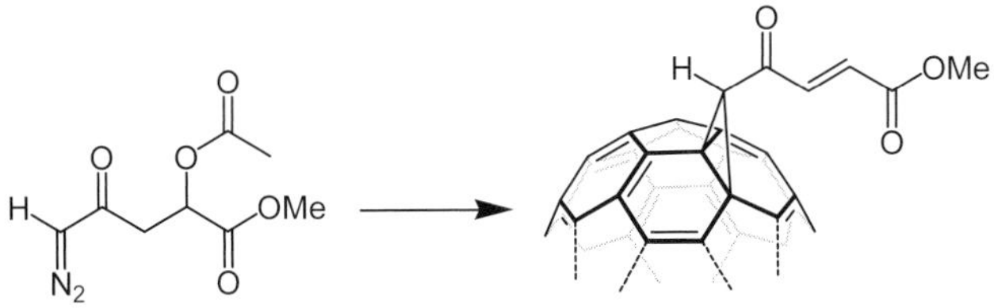 **162** → **163**	100–150 °C, toluene or methyl-naphthalene, (15–20%)	133

Dimethyl diazomethylphosphonate **157** reacts with C_{60} in refluxing toluene to afford a mixture of the [5,6]-closed and the [6,6]-open products in a ratio of 1 : 3 [132]. Only the [5,6] isomer with the phosphorous above the pentagon could be isolated. The obtained dimethyl methano fullerenephosphonate can be converted into the corresponding methano fullerenephosphonic acid in good yields by treatment with trifluoroacetic acid.

4.3.2
Addition of Azides

Organic azides can also act as 1,3-dipoles and undergo [3+2] cycloadditions to the [6,6] double bonds of C_{60}, yielding a [6,6] triazoline intermediate **164** (Scheme 4.28), which in some cases can be detected or even isolated [166–170].

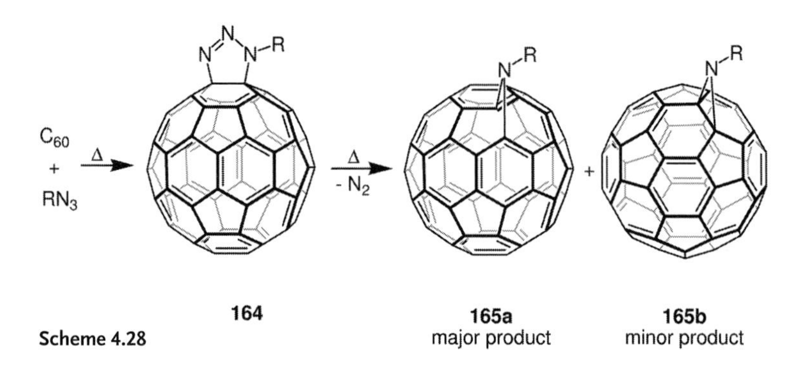

Scheme 4.28

Thermal nitrogen-extrusion leads to aza-bridged fullerenes. Theoretical investigations indicate a stepwise mechanism in which the cleavage of the N–N single bond precedes the breaking of the N–C bond. During extrusion of N_2 the steric effect of the leaving N_2 molecule prevents the addition of the nitrene substituent to the [6,6] bond and forces the addition to an adjacent [5,6] ring junction. This can explain the product distribution. In most of the alkyl azide additions the [5,6] product (**165a**) is the major product and the [6,6]-closed product (**165b**) is formed in only small amounts (up to 10%).

In analogy to the reaction with diazomethane (Section 4.3.1) only two out of the four principally possible addition products could be isolated. The [6,6]-open and [5,6]-closed isomers (analog to **B** and **D** in Figure 4.5) were not observed.

Beside the fact that azide addition to C_{60} is a convenient method for functionalizing fullerenes, the iminofullerenes also serve as precursors in the synthesis of nitrogen heterofullerenes [171]. Their chemistry is described in more detail in Chapter 12.

Reaction of nitrenes with C_{60} exclusively takes place at the [6,6] ring junctions. Similar behavior is observed in the reaction of azidoformiate and its derivatives. Heating the educts in tetrachloroethane or chloronaphthalene at 140–160 °C favors N_2 extrusion prior to a possible cycloaddition. Thus, instead of a [3+2] cycloaddition of an azide, a [2+1] cycloaddition of a nitrene intermediate leads to the expected attack at the [6,6] double bond under formation of the [6,6]-closed aziridines with a high selectivity. More examples of nitrene additions to C_{60} are given in Section 4.5.2. Allowing a concentrated solution of C_{60} in chloronaphthalene to react with N_3CO_2R at 60 °C enables a [3+2] cycloaddition [172]. After adding a 10-fold volume of toluene, the mixture is heated to reflux, causing the loss of N_2 and the formation of the expected [5,6] bridged isomer.

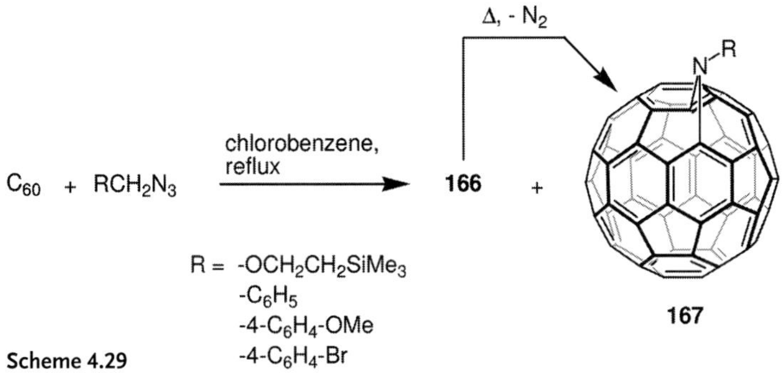

Δ, - N_2

C_{60} + RCH_2N_3 —chlorobenzene, reflux→ **166** + **167**

R = $-OCH_2CH_2SiMe_3$
$-C_6H_5$
$-4-C_6H_4-OMe$
$-4-C_6H_4-Br$

Scheme 4.29

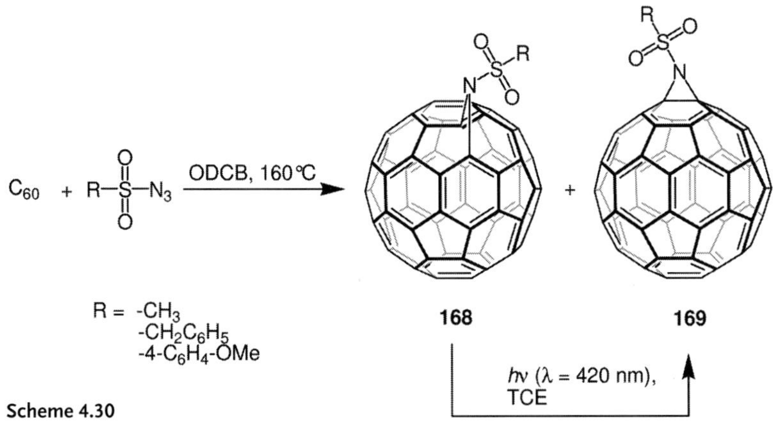

C_{60} + $R-SO_2-N_3$ —ODCB, 160 °C→ **168** + **169**

R = $-CH_3$
$-CH_2C_6H_5$
$-4-C_6H_4-OMe$

hν (λ = 420 nm), TCE

Scheme 4.30

Analogously to the methanofullerenes – obtained by the cycloaddition of diazomethane – the [5,6]-open azafulleroids **165a** can be thermally rearranged into the [6,6]-closed aziridines **165b**. In many cases the addition is carried out by refluxing equimolar amounts of C_{60} and the azide in chlorobenzene, toluene or 1-chloronaphthalene for several hours [167, 168, 170, 173–189]. The first examples of such reactions were the additions of [(trimethylsilyl)-ethoxy]methyl azide and some benzyl azides to C_{60} in refluxing chlorobenzene, yielding two major products, namely the triazoline **166** and the azafulleroid **167** (Scheme 4.29). The triazoline **166** was converted into the azafulleroid either by heating in refluxing chlorobenzene or in the solid state at 180 °C [167].

A mixture of sulfonyl-azafulleroid **168** and sulfonylaziridino-fullerene **169** was obtained by reaction of sulfonyl azide with C_{60} in *o*-dichlorobenzene at 160 °C (Scheme 4.30) [173]. The ratio of the two products depends on the substituent of the sulfonyl group. In all cases the aziridino fullerene can be obtained from the azafulleroids by irradiation.

Azides are easily prepared by nucleophilic substitution of alkylhalogenides with sodium azide and the yields of azide additions to C_{60} are generally sufficient. Thus the functionalization of C_{60} via this pathway leads to a broad range of products. Some examples are shown in Table 4.6.

Table 4.6 Examples of azides attached to C_{60} via [3+2] cycloaddition.

Substrate	*Conditions (%yield)*	*Ref.*
 **170**	Chlorobenzene, reflux (≈35%)	174
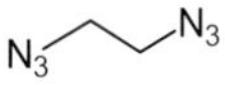 **171**	Chlorobenzene, reflux	190
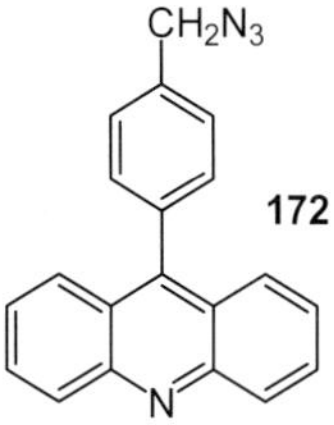 **172**	*o*-Dichlorobenzene, 130 °C (7%)	175
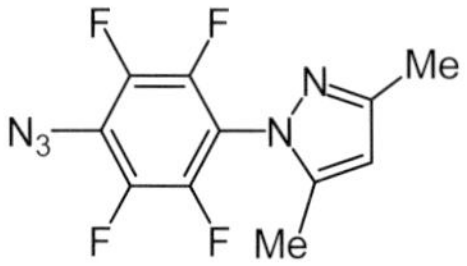 **173** 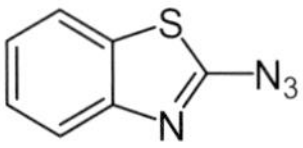 **174**	Toluene, reflux (7–13%)	176
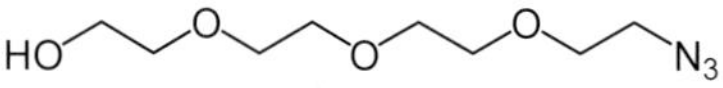 **175**	Chlorobenzene, reflux	177
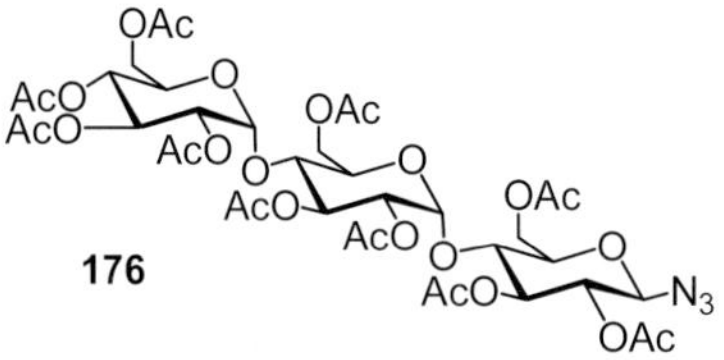 **176**	Chlorobenzene, reflux (16%)	178
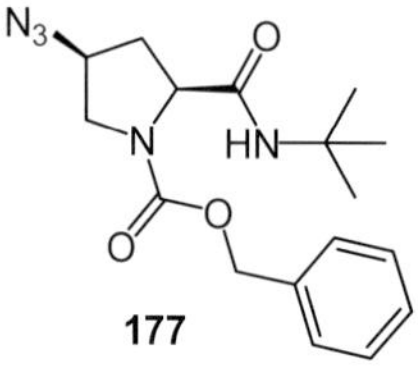 **177**	Chlorobenzene, 130 °C	191
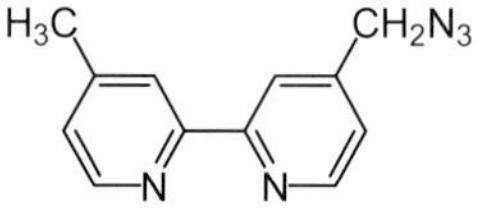 **178** 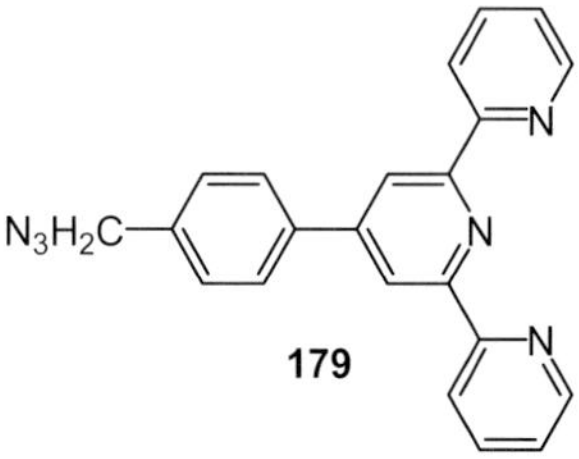 **179**	Chlorobenzene, reflux	180

Table 4.6 (continued)

Substrate	***Conditions (%yield)***	***Ref.***
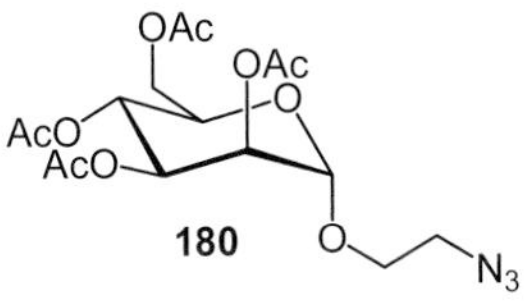 **180**	Chlorobenzene, reflux (19%)	181
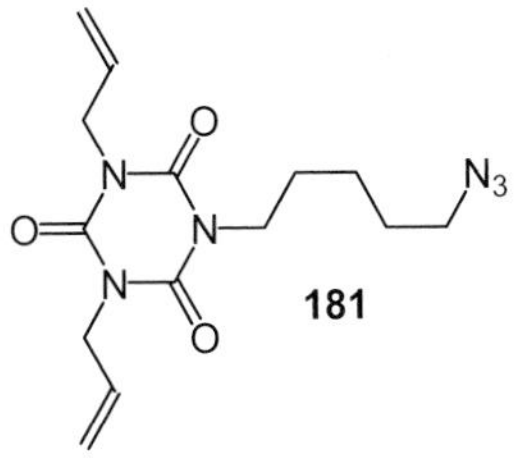 **181**	ODCB, 180 °C	182, 183
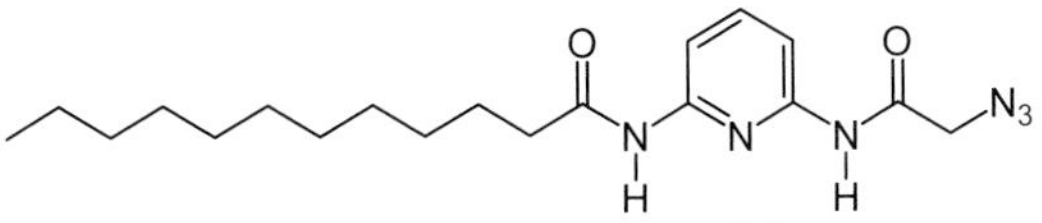 **182**	(1) Chlorobenzene, reflux, (2) *o*-dichlorobenzene, reflux (16%)	189, 192
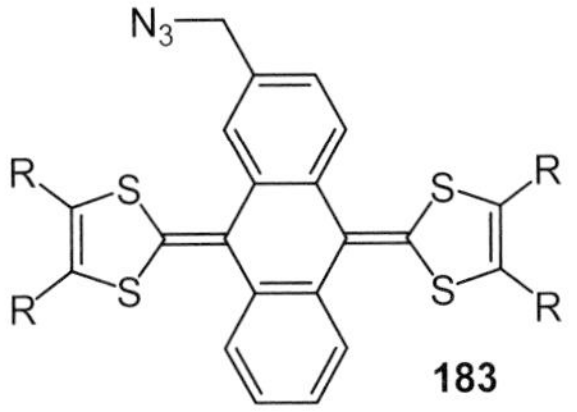 **183** R = H or SMe	ODCB, 120 °C	170
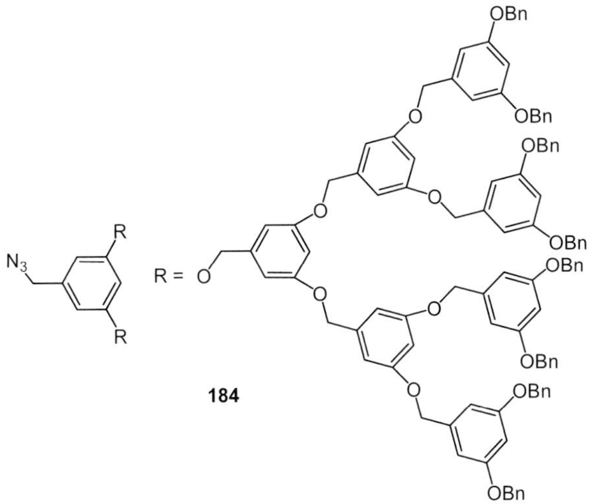 **184**	Chlorobenzene, reflux (68%)	193

Oligopyridine-fullerene derivatives derived from the reaction of C_{60} with **178** and **179** were used to either synthesize ruthenium complexes or to self-assemble on gold surfaces and gold nanoparticles [180, 184, 188]. Complexation of metals with pyridine derivatives is one possibility for forming supramolecular aggregates. Another way of building assemblies is based on hydrogen bonding. Iminofullerenes prepared by the addition of **182** to C_{60} can form dimers or bind to polymers via hydrogen-bonding interactions [189, 192, 194]. Polymers such as PPV can not only be functionalized via hydrogen bonding but also via chemical reaction with C_{60}. Azide functionalized PPV-polymer reacts with C_{60} under the usual conditions in good yields [179, 185–187].

Reaction of **175** with C_{60} yields a hydroxy-functionalized fullerene that can be further derivatized. This hydroxy-fullerene was coupled with a porphyrine unit via a polyethyleneglycol-linker. This linker can be arranged similarly to a crown-ether to complex metal cations. Complexation is used to tune the distance between the porphyrin unit and the C_{60}-moiety and thus tune the donor–acceptor properties of this porphyrin-fullerene hybrid [177].

4.3.3
Addition of Trimethylenemethanes

Stable five-membered ring adducts of C_{60} can be synthesized by [3+2] cycloadditions of trimethylenemethanes (TMM) [13, 195]. The TMM intermediates are prepared in situ by thermolysis of 7-alkylidene-2,3-diazabicycloheptene (non-polar TMM) or methylenecyclopropanone ketals (polar TMM). With the methylenecyclopropanone ketal addition (Scheme 4.31), **185** and **186** were isolated after chromatography on silica gel.

The structure of the α-methylenecyclopropanone ketal **185** is reminiscent of the addition mode of the corresponding TMM to C=O [196]. The ester **186** is probably the product of silica-gel-catalyzed hydrolysis of the ketene acetal **187** (Figure 4.8), which is the expected product in the reaction of TMM with electron-deficient olefins [197]. At higher temperatures **185** isomerizes into **187** [195]. NMR spectroscopic investigations of these adducts reveal that the cycloadditions occur at the [6,6] double bonds. Analogous products to **185–187** have been observed for the reaction of the dipolar TMM with C_{70} [198].

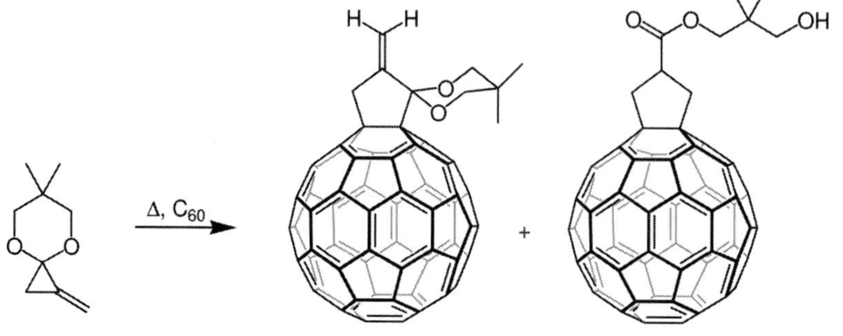

Δ, C_{60} → **185** + **186**

Scheme 4.31

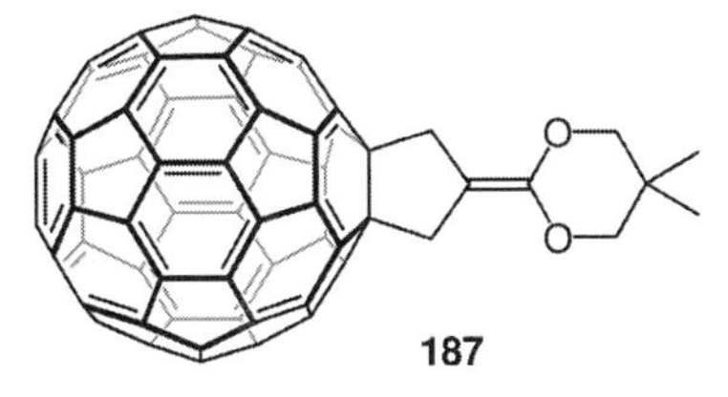

Figure 4.8 The ketene acetale **187** is the expected product of [3+2]-cycloaddition of methylene cyclopropanone ketal to C_{60}.

Various chemical transformations in the side chain have been carried out with the alcohol **186** as starting material [199]. The side-chain chemistry provides access to new fullerene derivatives, including those with sugar and amino acid moieties, which may be important for possible applications, for example investigations of their biological activity. In general, it was found that C_{60} survives acidic to weakly basic conditions as well as several oxidations and reductions required to carry out a range of C–O and C–C bond forming reactions (Table 4.7). In particular, the benzoate **189** can be obtained by treatment of **186** with benzoyl chloride and pyridine in toluene or alternatively by condensation with benzoic acid, dicyclohexylcarbodiimide (DCC) and 4-(dimethylamino)pyridine (DMAP) in CH_2Cl_2 (Table 4.7). Similar procedures lead to the methacrylic ester **192** in quantitative yield as well as to the amino acid derivative **194**. For the quantitative conversion of **186** into the corresponding derivatives, an excess of reagents is needed. C–O bond formation can also be achieved under acidic conditions. A tetrahydropyranyl ether has been synthesized by treatment of **186** with dihydropyran and pyridinium-*p*-toluenesulfonate in quantitative yields. Furthermore, the sugar derivative **196** was obtained by acidic etherification with tri-*O*-acetylglycal in the presence of *p*-toluenesulfonic acid.

Oxidation of the hydroxyl group in **186** with pyridinium chlorochromate (PCC) in CH_2Cl_2 affords the aldehyde **197**. The reduction of **197** back to **186** is possible in EtOH in the presence of $TiCl_4$, whereas upon treatment of **197** with diisobutylaluminium hydride a competitive reaction with the fullerene core was observed.

The aldehyde group in **197** bears a potential for C–C bond elongation reactions. For example, **198** was obtained in 89% yield by the reaction with a stabilized ylide as an E-isomer (> 95 : 5) (Figure 4.9). Additionaly Lewis-acid-mediated reactions are possible. The reaction of the aldehyde **197** with allyltributyltin and $TiCl_4$ gave the homoallylic alcohol **199** in 86% yield.

Some of these fullerene derivatives derived from **186** have been investigated with respect to their biological activity [200]. One requirement for the study of biological activity is water solubility. To obtain water-soluble fullerenes the alcohol **186** was transformed into the carboxylic acid **200** by reaction with succinic anhydride in the presence of DMAP (Figure 4.10). The succinate ester was found to cleave double-stranded DNA or to inhibit the growth of mammalian cells upon visible light irradiation. This activity was suggested to be due to the efficient conversion of triplet oxygen into highly reactive singlet oxygen by the photoexcited C_{60} moiety in its triplet state [201]. Inhibiton of enzyme activity was also noted, including that of

Table 4.7 Reactions of the alcohol **186**.

Reaction	*Conditions (%yield)*
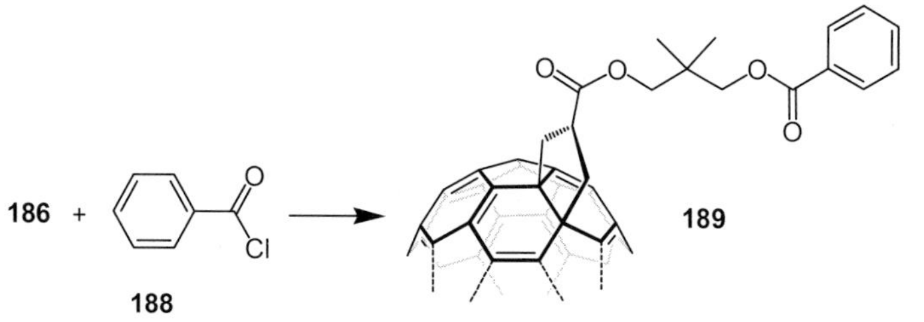 **186** + **188** (Cl) → **189**	Pyridine, toluene, 50 °C (62%)
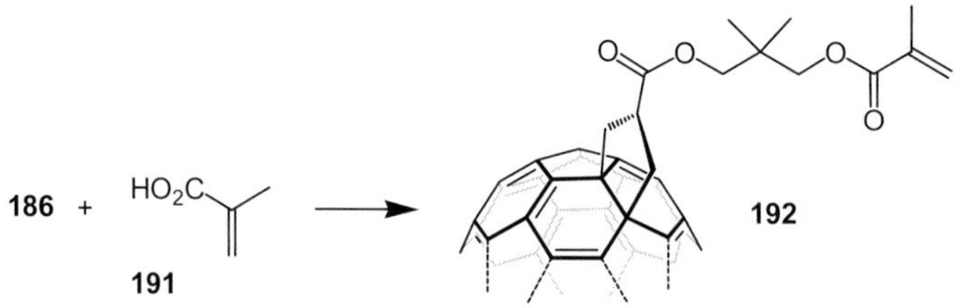 **186** + HO_2C **191** → **192**	DCC, DMAP, CH_2Cl_2, room temperature (100%)
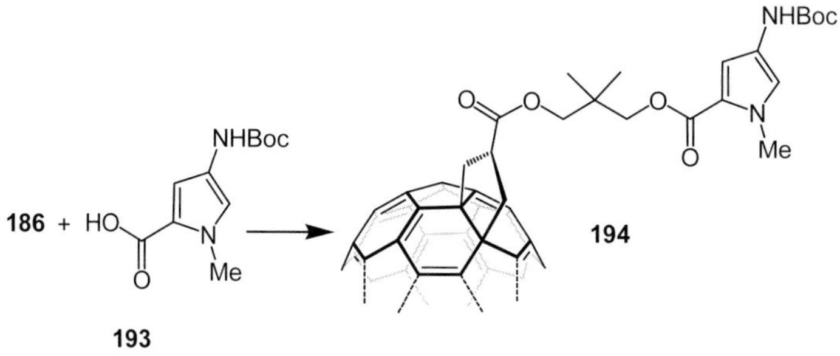 **186** + HO (NHBoc, N, Me) **193** → **194** (NHBoc, N, Me)	DCC, DMAP, CH_2Cl_2, room temperature (64%)
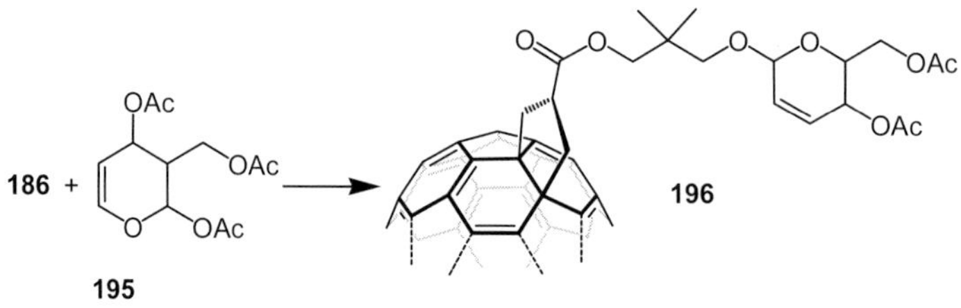 **186** + (OAc, OAc, OAc) **195** → **196** (OAc, OAc)	TsOH, CH_2Cl_2, room temperature (62%)
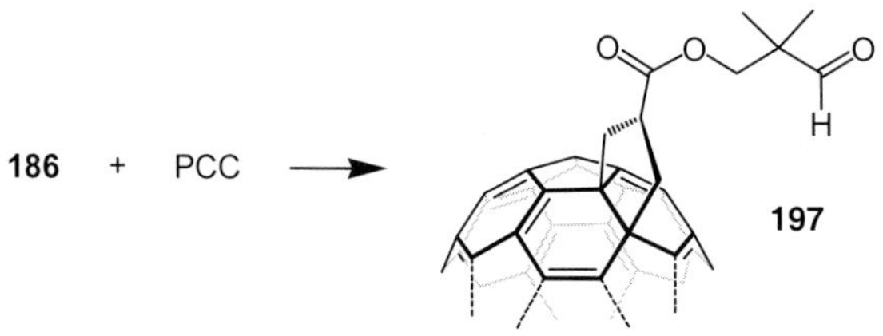 **186** + PCC → **197** (H)	AcONa, CH_2Cl_2, room temperature (54%)

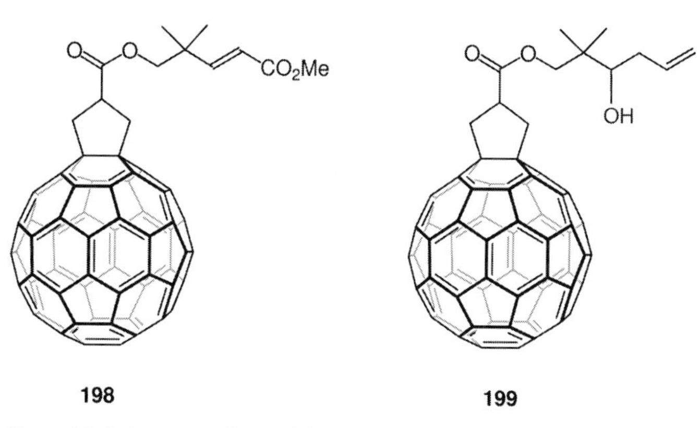

Figure 4.9 **197** reacts with a stabilized ylide to give **198** and with allyltributyltin to give **199**.

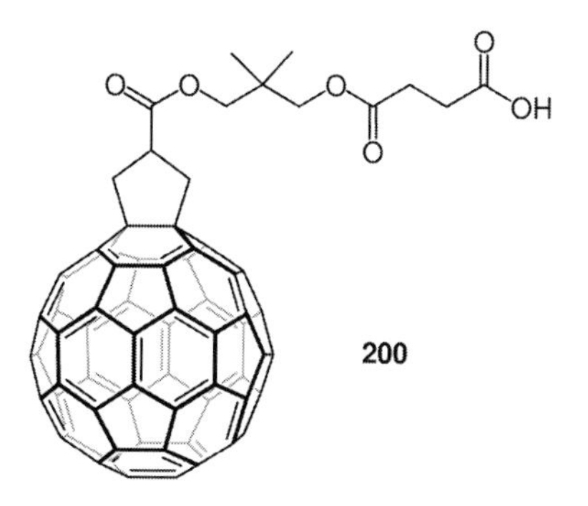

Figure 4.10 Esterification of **197** with succinic anhydride yields the water-soluble derivative **200**.

HIV-protease. Examination of radiolabeled succinate **200** [202] provided information on the pharmacokinetic behavior of this water-soluble fullerene. A behavior similar to hydrophobic steroids was found, accompanied by a very low level of acute toxicity and the ability of **200** to penetrate the blood–brain barrier [162, 195, 199, 200, 203].

4.3.4 Addition of Azomethine Ylides

Azomethine ylides, planar species of the general formula **201** (Figure 4.11), which exhibit 1,3-dipole character, react smoothly and in good yields with C_{60} to give fulleropyrrolidines [204–207]. This reaction was introduced by Prato and Maggini [204] and evolved into one of the most used methods of fullerene functionalization. The broad acceptance of this method is explained by the good selectivity (only the [6,6] bond is attacked) and a wide range of addends and functional groups that are tolerated. The nitrogen atom in the obtained fulleropyrrolidines is about six orders of magnitude less basic and three orders of magnitude less reactive than in the corresponding pyrrolidines without C_{60} [208, 209].

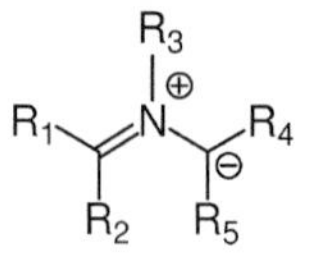

Figure 4.11 General formula of azomethine ylides.

Azomethine ylides can be generated in situ from various readily accessible starting materials. One of the easiest approaches to produce 1,3-dipoles involves the decarboxylation of immonium salts derived from condensation of α-amino acids with aldehydes or ketones [3, 204–206]. For example, the azomethine ylide **203**, obtained by decarboxylating the condensation product of *N*-methylglycine and paraformaldehyde in refluxing toluene, reacts with C_{60} to give the *N*-methyl-pyrrolidine derivative **204** in 41% yield (Scheme 4.32) [204].

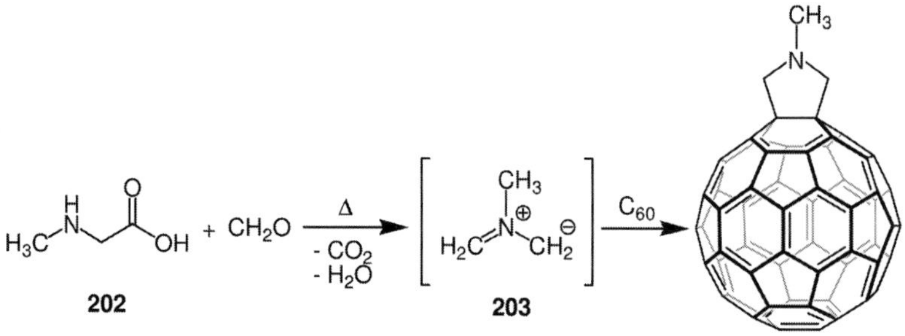

Scheme 4.32

Pyrrolidine derivatives such as **206** have been synthesized by the addition of the corresponding azomethine ylide generated by the thermal ring opening of the aziridine **205** (or similar aziridines) as shown in Scheme 4.33 [204, 210–212].

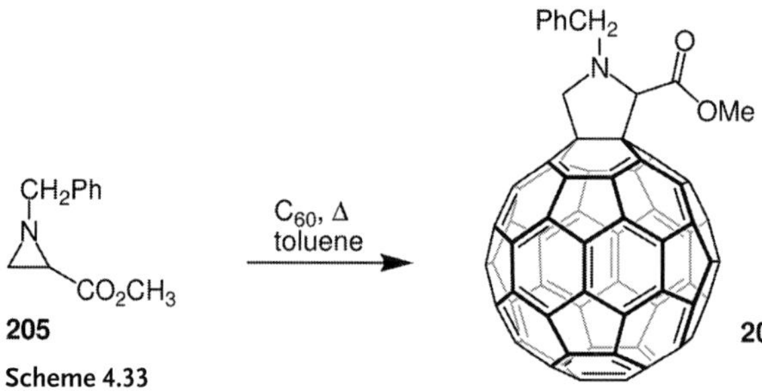

Scheme 4.33

Further methods [205] successfully employed to synthesize fulleropyrrolidines include acid-catalyzed [213] or thermal [214] desilylation of trimethylsilyl amino derivatives, tautomerization of α-aminoesters of immonium salts [215] and imines [216, 217], reaction with aldehydes in the presence of aqueous ammonia [218], reaction with oxazolidinone [204] or photochemical reaction with some amino derivatives [219–223]. The reaction with amino acids and aldehydes was also carried out without solvent either under microwave irradiation [224, 225] or by applying the high-speed vibration milling technique [226].

The vast majority of fulleropyrridolidines have been prepared via the Prato reaction, which is the already described reaction of C_{60} with any amino acid and an arbitrary aldehyde or ketone by simply heating the mixture in toluene at reflux [204–206]. The versatility of this reaction arises from the possibility of introducing different substituents into three different positions of the pyrrolidine ring depending on the used aldehyde/ketone and respective amino acid. Reaction with functionalized aldehydes furnishes 2-substituted fulleropyrrolidines, whereas reaction with N-substituted glycines leads to N-substituted fulleropyrrolidines. A theoretical maximum of five substituents can be introduced but this would lead to a sterically hindered system. Using ketone **207** and the tertiary amino acid **208**, for example, leads to the 2,2,5,5-tetrasubstituted pyrrolidine derivative **209** (Scheme 4.34) [227].

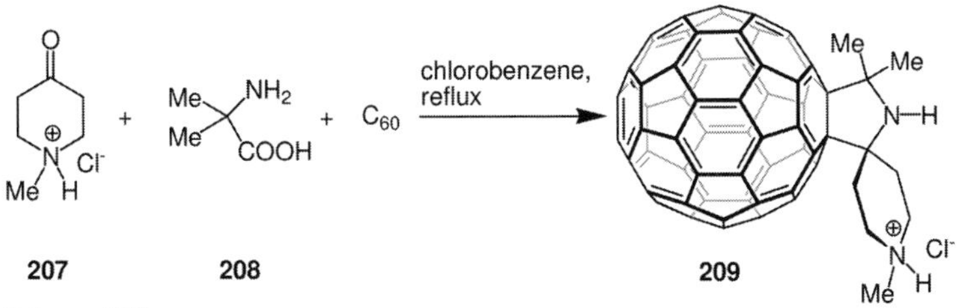

Scheme 4.34

Further side-chain chemistry can be carried out either at one of the side chains attached to the five-membered ring [228–232] or at the pyrrolidine nitrogen [215, 224, 231, 233–239]. For example, upon heating 3-triphenylmethyloxazolidin-5-one (**210**) at reflux for 16 h with C_{60} in equimolar amounts, **211** can be isolated in 39% yield [204]. Subsequent treatment of **211** with trifluoromethanesulfonic acid, then with pyridine and dansyl chloride (DnsCl) yields the dansyl pyrrolidine derivative **213** in 76% yield (Scheme 4.35). The amine **212** is an intermediate in this reaction.

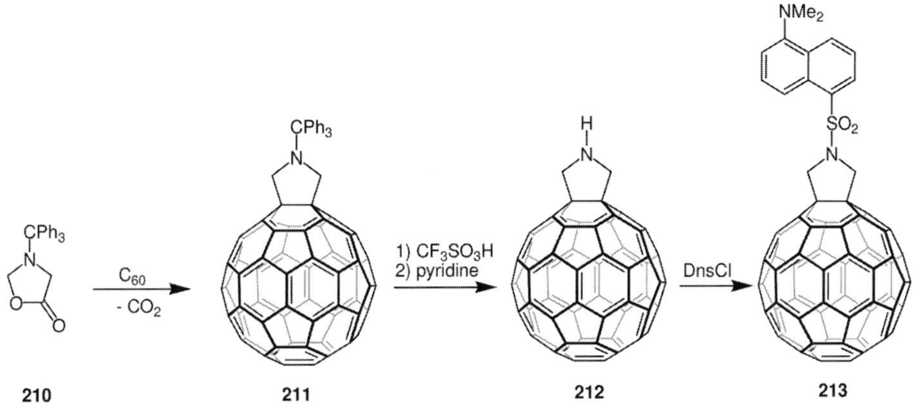

Scheme 4.35

Amine **212** was also coupled with peptides [233], acetic anhydride [215, 234], dinitrobenzoyl chloride [235] or a ferrocene carboxylic acid chloride [215, 236] in good yields. Reaction with dinitrochlorobenzene [237] or dinitrofluorobenzene [238] proceeds via nucleophilic aromatic substitution in the presence of K_2CO_3 or NaH. Without solvent and under microwave irradiation different alkyl bromides can be substituted with the pyrrolidine-amine within 10 min [224]. Preparation of a 2,5-disubstituted fulleropyrrolidine leads to a mixture of the *cis* and *trans* isomers. Resolution and determination of the absolute configuration of both isomers could be accomplished by reaction of the mixture with a chiral isocyanate [239].

An example of side-chain functionalization is the attachment of tetraphenyl-porphyrin carboxylic acid to a fullerene bound steroid (**214**) via standard EDCI coupling with the hydroxy group of the steroid (Scheme 4.36) [230].

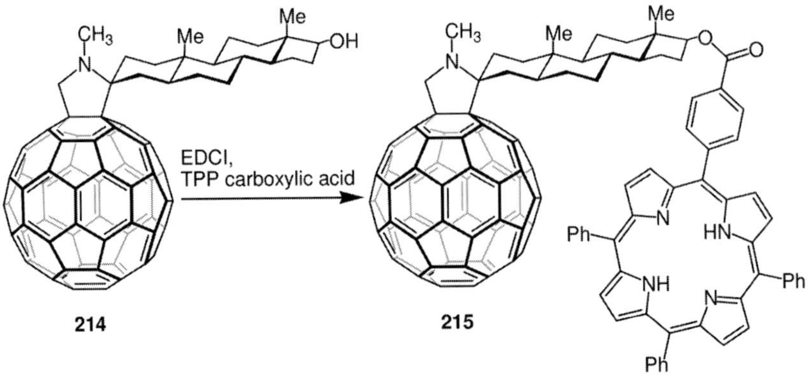

Scheme 4.36

An established procedure for attaching any functional molecule with aldehyde or ketone functionality to C_{60} is the original Prato reaction using sarcosine (*N*-methylglycine) as the amino acid component [204]. It has been used to append molecules such as porphyrins [229, 230, 240–248], ferrocene [215, 235, 243, 244, 249, 250] and other metal complexes [232, 233, 251], TTF [215, 252–257], steroids [230, 232] sugar [258], aniline and its derivatives [259–261], fluorescein [262], di- and tetracyanoquinonedimethane [263], crown ether [228], calixarenes [264, 265], dendrimers [266, 267], flavone [268], stilbene [269] or π-conjugated oligomers [270–276]. Most of the obtained fullerene derivatives were used to examine the interaction between donors (e.g. porphyrin, TTF, conducting polymers etc.) and C_{60}, which acts as an electron acceptor. Energy- and charge-transfer in these complexes emerged as a very important field of fullerene research. Factors such as structure and distance between donor and acceptor or solvent effects, have systematically been exploited to control charge-separation dynamics and efficiencies [3]. The most frequently attached donor molecules are probably porphyrins. Some examples of porphyrins carrying an aldehyde group suitable for the Prato reaction are given in Table 4.8 (see also Scheme 4.36).

Table 4.8 Porphyrin aldehydes subjected to the Prato reaction with sarcosin.

Porphyrin substrate	*Yield (%)*	*Ref.*
216	40	240, 241
217	62	229, 277
218 (R = H, $O(CH_2)_{11}SH$)	54	243–245, 247
219	79	246–248

Reaction of **217** with C_{60} leads to the amino-protected porphyrin-fulleropyrrolidine, which can easily be deprotected to the corresponding amine [229, 277]. By further functionalization via amide coupling an easy access to extended donor–acceptor systems is possible. A carotene–porphyrin–fullerene triad was prepared by reaction of the amine with the appropriate carotene acid chloride. The motivation for the synthesis of all these donor–acceptor systems is the attempt to understand and imitate the photosynthetic process. On that score, a model for an artificial photosynthetic antenna-reaction center complex has been achieved by attaching five porphyrin cores in a dendrimer-like fashion to the fullerene [242].

With the objective to enable long-lived charge-separated C_{60}-complexes (also Section 4.2) TTF and extended TTF were bound via [3+2] cycloaddition with different spacer-lengths (R_1) (Figure 4.12) [215, 252–257]. In these C_{60}-TTF dyads as well as in a bis-tetrathiafulvalene-C_{60} triad no significant interaction between the TTF units and the C_{60}-moiety in the ground state could be observed.

Further donors, mainly based on aniline or ferrocene, were subjected to the sarcosin-based Prato reaction. The bisaldehydes **224** [260] or **227** [250] react with two C_{60}-molecules, leading to dumbbell shaped dimers (Table 4.9). The C_{60} dimer obtained from **224** could be considered as an A-D-A triad but the CV-data indicate that the dimer behaves electrochemically as two independent donor–acceptor units.

Derivatizing C_{60} with a strong acceptor, such as the tetracyanoquinodimethane **231** yields organofullerenes that can accept up to eight electrons in solution [263, 278]. The acceptor ability of the quinodimethane moiety in the fullerene adduct is

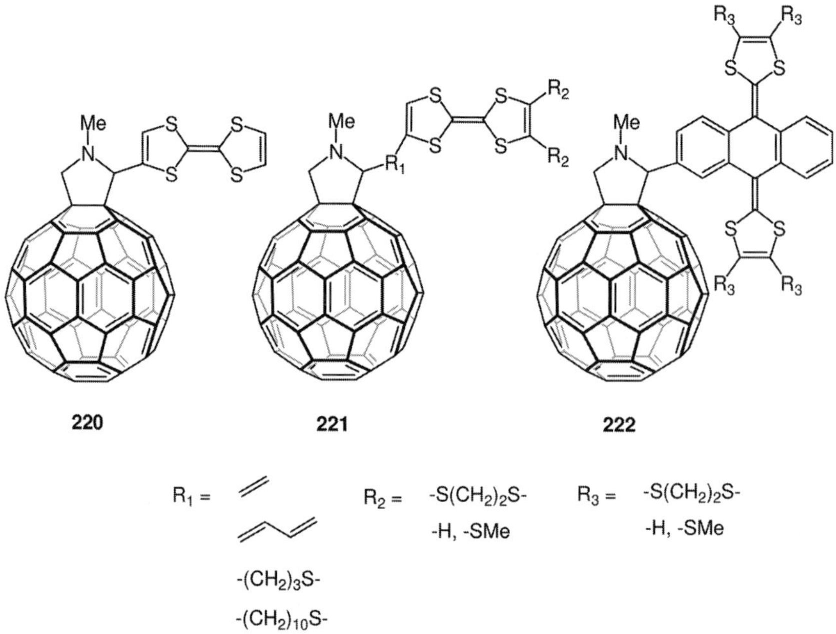

Figure 4.12 TTF and extended TTF was attached to C_{60} via the Prato-reaction in order to get dyads suitable for long-lived charge-separated states.

Table 4.9 Miscellaneous aldehydes subjected to the Prato reaction with sarcosin.

Substrate	*Yield (%)*	*Ref.*
223 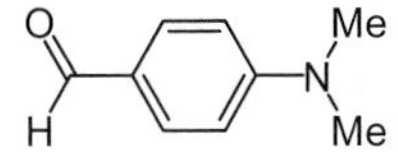	–	259
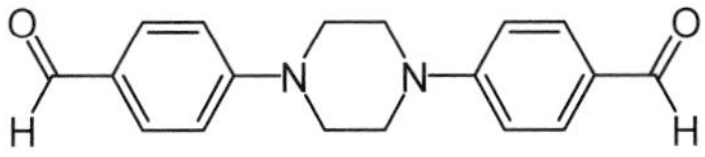 224	34 (bisadduct)	260
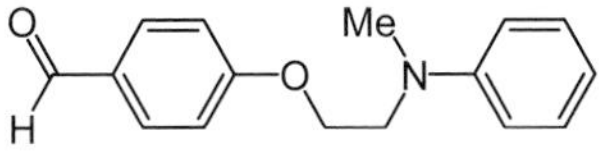 225	33	210, 261
226 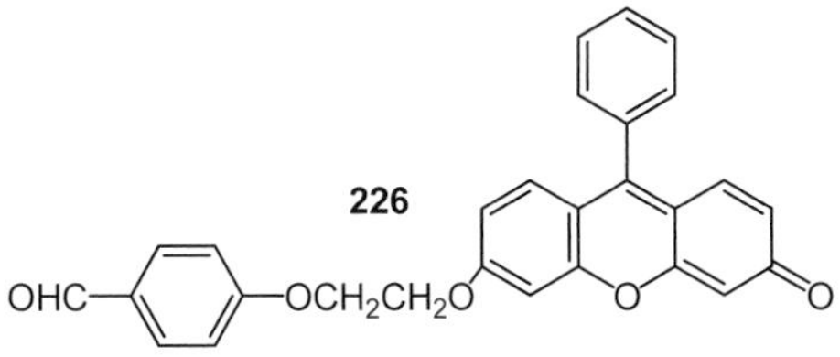	30	262
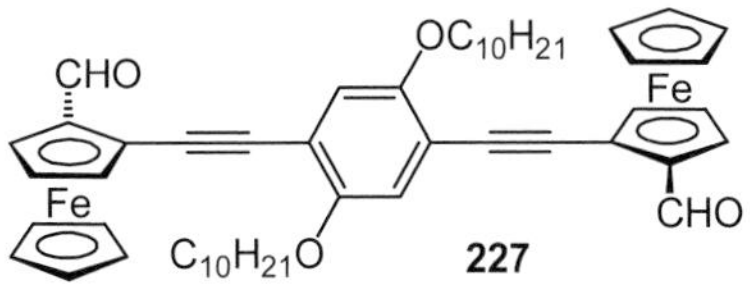 227	21	250
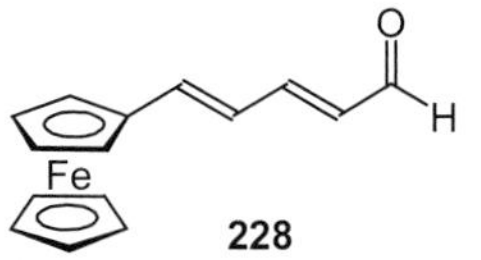 228, 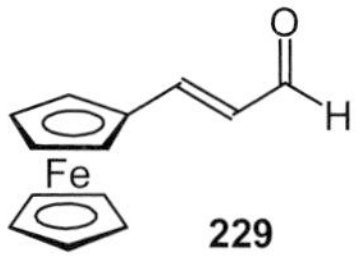 229	29, 31	249

enhanced compared with the parent quinodimethane. Owing to their remarkable electron-acceptor character these dyads are also referred to as electron sponges [278].

Supramolecular derivatives containing crown ether [228], calixarenes [264, 265] or dendrimers [267, 279] have been synthesized with the precursors shown in Table 4.10. Fullerene-flavonoid [268] or the fullerene glycoconjugate [258] derived from **237** and **235**, respectively, are expected to show biological activity.

The functionalization of the pyrrolidine nitrogen can be accomplished by substitution reactions, as described earlier in this chapter or by using other N-substituted amino acids instead of sarcosin. Glycine, which is N-functionalized with triethyleneglycol (**240**, Scheme 4.37), provides an access to water-soluble fullerene derivatives [281–284]. Condensation of the glycol-glycine **240** with several aldehydes [281–284] affords C_{60} derivatives **241**, which have moderate solubility in DMSO–water (9 : 1), in almost 40% yield [281]. The water solubility was drastically enhanced by introducing an amino group as the endgroup of the glycol chain and by addition of more than one of these solubilizing groups to C_{60} [282].

Table 4.10 Aldehydes subjected to the Prato-reaction with sarcosin.

Substrate	*Yield (%)*	*Ref.*
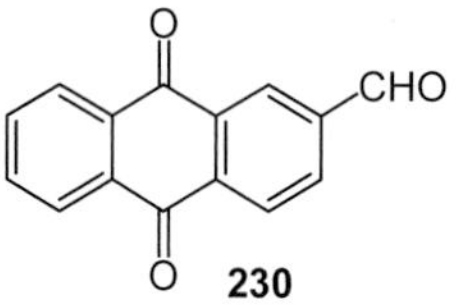 **230**, 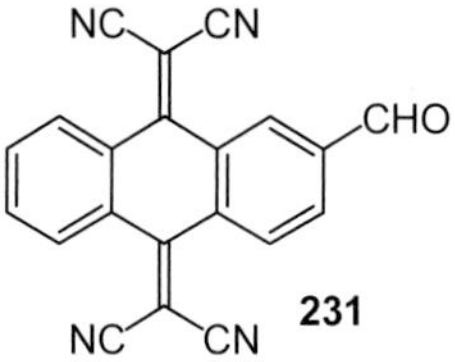 **231** (O, CHO; NC, CN, CHO)	20	263
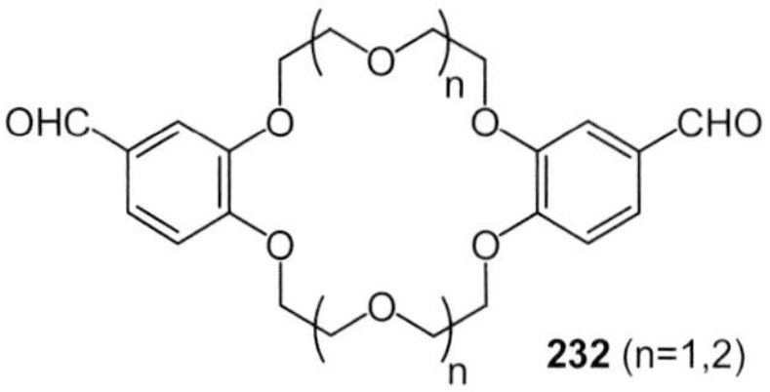 **232** (n=1,2) (OHC, CHO, O, n)	25 (only mono-adduct)	228
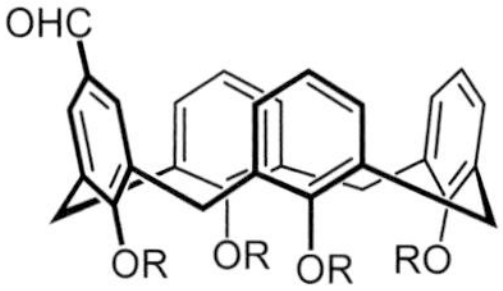 **233** R = H, Me, OBu (OHC, OR, RO)	64	264, 265
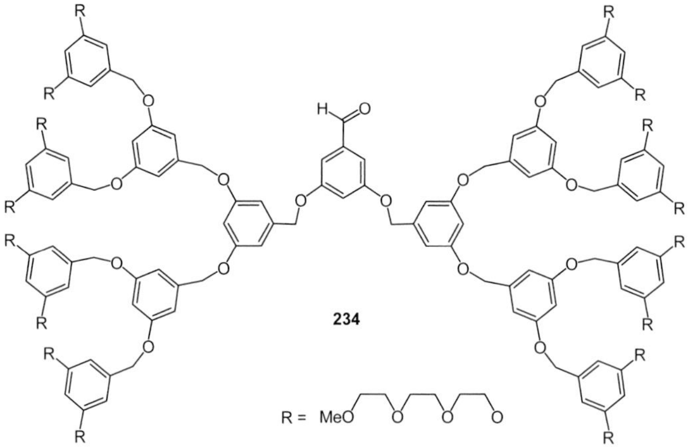 **234** R = MeO–O–O–O (H, O, R)	40	267
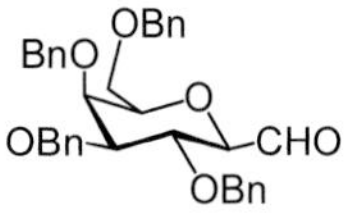 **235** (BnO, OBn, O, CHO)	10	258
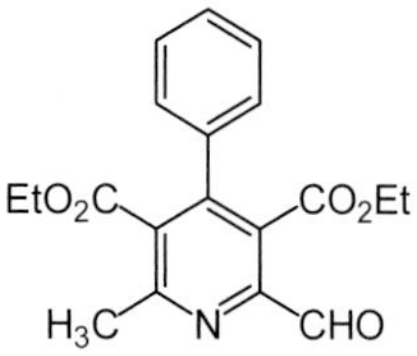 **236** (EtO_2C, CO_2Et, H_3C, N, CHO)	–	280

Table 4.10 (continued)

Substrate	*Yield (%)*	*Ref.*
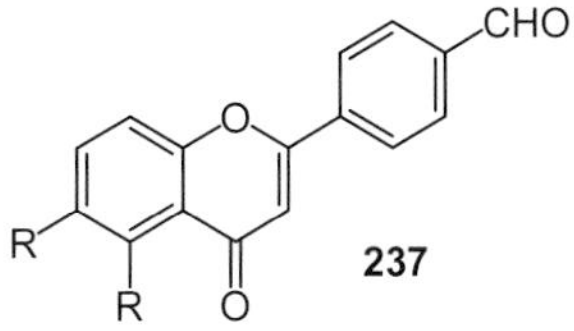 **237** R = H, OBn	30–50	268
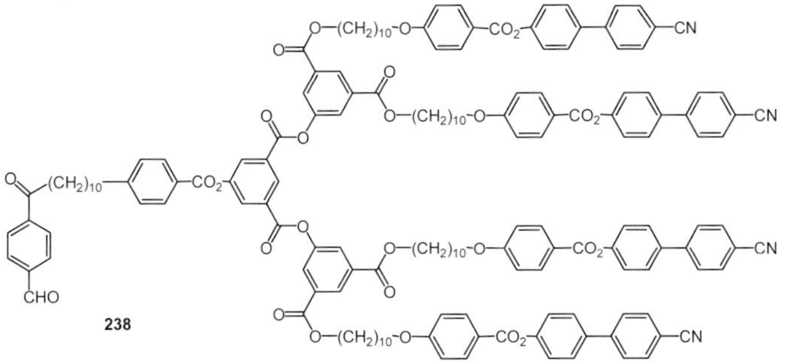 **238**		266
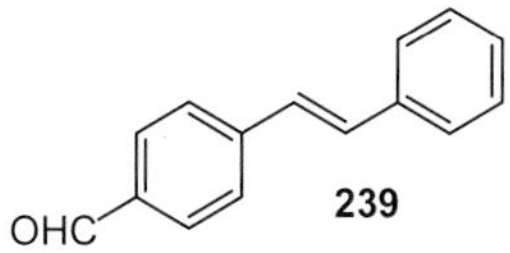 **239**		269

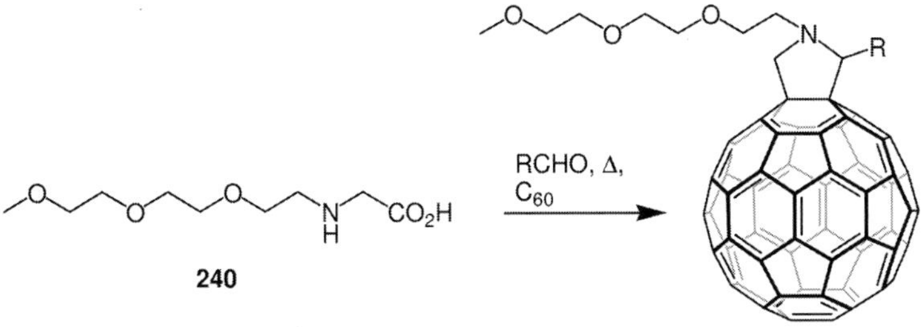

240 → RCHO, Δ, C_{60} → **241**

Scheme 4.37

Another simple method to increase water solubility of fulleropyrrolidines is quaternization of the nitrogen, e.g. with methyl iodide [285–287], to yield fulleropyrrolidinium salts.

Condensation of *N*-tritylglycin with donor or acceptor substituted aldehydes results in fulleropyrrolidines whose amino functionality can be easily deprotected with triflic acid and further functionalized with, e.g., another donor [236]. Thus, access to various triads is provided. An example for a donor–acceptor–donor triad is given in Scheme 4.38.

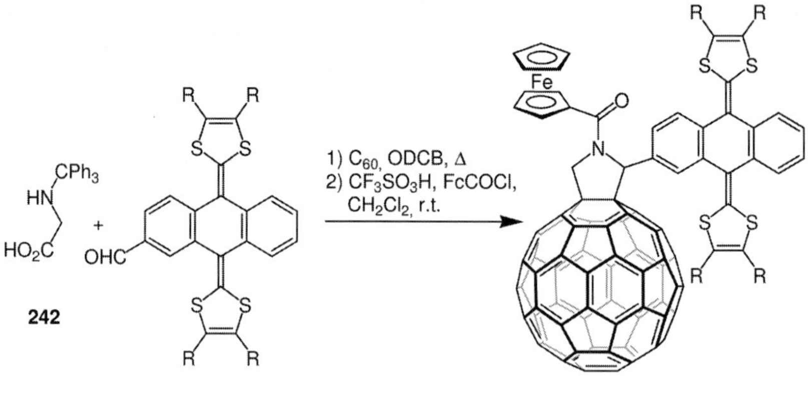

Scheme 4.38

Using derivatized glycine usually lowers the yields compared with sarcosin – nevertheless various functionalized glycine derivatives have been used. Amine **245**, obtained by reaction of a trimethoxy-indole functionalized amino acid with an aliphatic ω-amino-aldehyde, has been coupled with an oligonucleotide to increase the affinity of C_{60} conjugates to certain nucleic acids (Figure 4.13) [288].

Fulleropyrrolidines bearing one or two TEMPO-groups have been synthesized via the Prato reaction [289, 290]. The biradical adduct **248** was formed by condensation of the corresponding TEMPO aldehyde **246** and the TEMPO-functionalized amino acid **247** (Scheme 4.39) [289]. Probably due to steric hindrance, the yields of this reaction are modest.

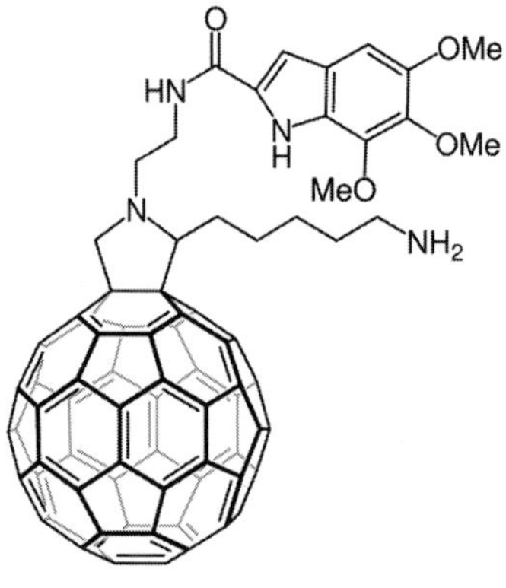

Figure 4.13 Amine **245** was obtained by reaction of a trimethoxy-indole functionalized amino-acid with an α-aminoaldehyde.

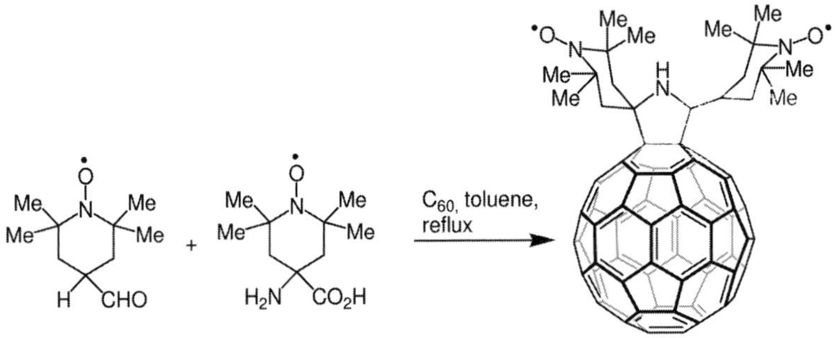

246 247 248

Scheme 4.39

4.3.5
Addition of Nitrile Oxides and Nitrile Imines

Isoxazoline derivatives of C_{60} such as **250** (Scheme 4.40) are accessible by 1,3-dipolar cycloadditions of nitrile oxides to [6,6] double bonds of the fullerene [2, 278, 291–305]. The nitrile oxides **249** with R = methyl, ethyl, ethoxycarbonyl and anthryl are generated in situ from the corresponding nitroalkane, phenyl isocyanate and triethylamine. The isoxazoline derivative of C_{60} **250** (with R = anthryl) crystallizes in black prisms out of a solvent mixture of CS_2 and acetone (3 : 2) [292]. X-ray crystal structure analysis shows that addition of the nitrile oxide occurs on a [6,6] double bond of the fullerene framework.

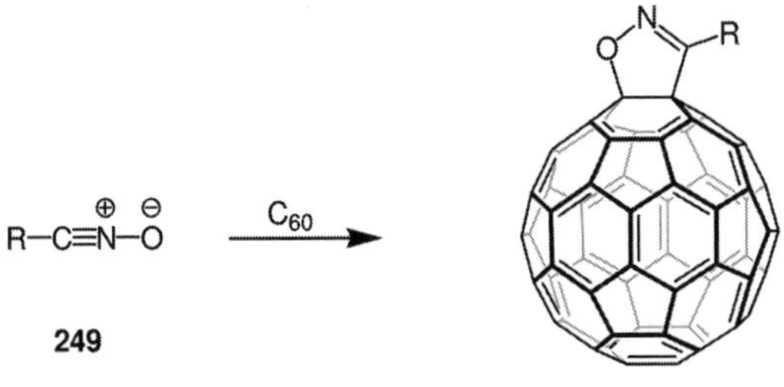

249 250

Scheme 4.40

Another methodology for the in situ preparation of nitrile oxide is the dehydrohalogenation of hydroxymoyl chlorides with triethylamine. Hydroxymoyl chlorides are accessible by the reaction of aldoximes with chlorinating agents such as NCS (*N*-chlorosuccinimide). Isoxazolines of C_{60} and C_{70} [293–295] with R = Ph, alkyl, 4-$C_6H_4OCH_3$, 4-C_6H_4CHO, amino acid [305], dialkoxyphosphoryl [296, 297] or ferrocene [298] have been synthesized in ca. 20–40% yields. The latter reaction is slower than the dehydration of nitroalkanes and requires one equivalent of hydroxymoyl chloride whereas excess nitroalkane is necessary for an optimum reaction [293].

Fullerene isoxazolines are stable compounds. Non-fulleroid alkyl isoxazolines easily undergo a reductive ring opening to yield various synthetically interesting

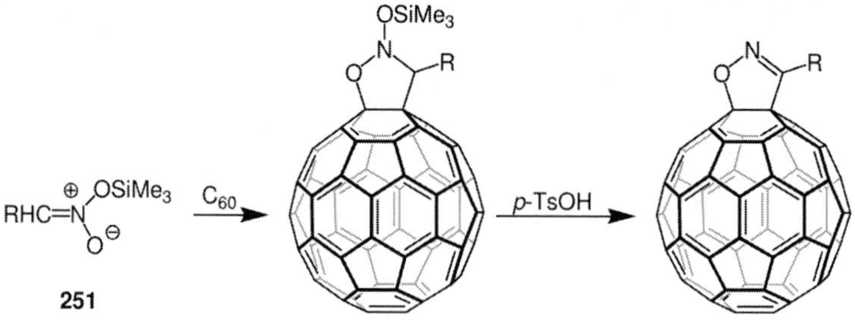

Scheme 4.41

products such as α-hydroxy-ketones or α-amino-alcohols. The corresponding fullerene derivatives tend to be less reactive and reduction usually does not lead to a ring opening. The cycloaddition of nitrile oxides is reversible at elevated temperatures higher than 250 °C [293]. C_{60} can also be recovered quantitatively by treatment of the isoxazolines with $Mo(CO)_6$ under conditions that usually favour N–O bond cleavage [304].

Cycloaddition of N-silyloxynitrones primarily leads to the isoxazolidines, which upon acid treatment undergo elimination to the corresponding fullero isoxazolines (Scheme 4.41) [299, 300]. N-Siloxynitrones (**251**) can be formed from nitroalkanes and Me_3SiCl in the presence of triethylamine. Addition of nitrones proceeds in good yields when R is not a sterically demanding group [R = H, CH_3, $(CH_2)_2CO_2Et$, CH_2OH].

A series of pyrazolino[60]fullerenes has been prepared in one-pot reactions by 1,3-dipolar cycloaddition of the corresponding nitrile imines [306–311]. In all cases

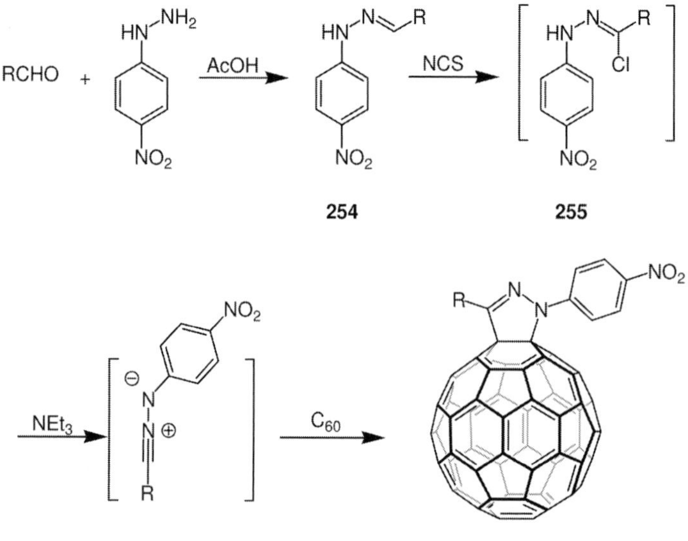

Scheme 4.42

di-aryl nitrile imines were used, whereas aryl can also be, for example, pyrazole or ferrocene. The reaction starts with the hydrazone **254**, which is chlorinated. The chlorinated intermediate **255** is eliminated and cycloadds to C_{60}, yielding pyrazolinofullerenes of the structure **257** (Scheme 4.42). The 4-nitrophenyl-group can be replaced by a 4-methoxyphenyl- or a phenyl substituent. In this reaction various aromatics and substituted aromatics are tolerated as residues R (e.g. furan, ferrocene, pyrazole or benzene and substituted benzene). The nitro group of the nitrophenyl residue can be reduced with Sn–HCl to the aniline derivative, which can be further functionalized by amide coupling with acid chlorides [311].

4.3.6
Addition of Sulfinimides and Thiocarbonyl Ylides

A sulfur-containing heterocyclic fullerene derivative **259** has been synthesized by the reaction of C_{60} with *N*-(1-adamantyl)bis(trifluoromethyl)sulfinimide (**258**, Scheme 4.43) [312]. The sulfinimide can be prepared from the corresponding sulfeneamide by 1,3-dehydrohalogenation. High yields of **259** with only a small amount of higher adducts can be obtained, if an excess of C_{60} is allowed to react with the sulfinimide. Compound **259** is stable up to 300 °C. The ^{19}F NMR spectrum of **259** exhibits one singlet, in contrast to that of the starting material. This provides evidence that the addition of the sulfinimide occurs at the [6,6] bond. Only this structure gives the magnetic equivalence of the two trifluoromethyl groups. C_{60} reacts with a large excess of the sulfinimide to afford the hexaadduct within one week.

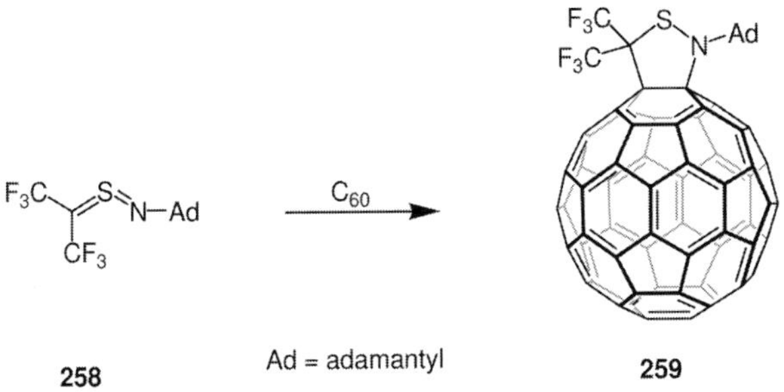

F_3C–C(CF_3)=S=N–Ad (**258**) $\xrightarrow{C_{60}}$ **259**

Ad = adamantyl

Scheme 4.43

5-Imino-1,2,4-thiadiazolidine-3-ones such as **260** (Scheme 4.44) react smoothly and in good yields with C_{60} in a [3+2]-cycloaddition mode to yield another sulfur- and nitrogen-containing C_{60}-fused heterocycle. The masked 1,3-dipole **260** adds to C_{60} under loss of phenyl isocyanate in refluxing toluene with yields of 30–70% [313].

Tetrahydrothiophene-fused C_{60} can be generated by its reaction with the thiocarbonyl ylide precursor bis(trimethylsilylmethyl) sulfoxide **262** [314, 315]. Thermal sila-Pummerer rearrangement leads in situ to the ylide **263**, which is readily added to C_{60} (Scheme 4.45).

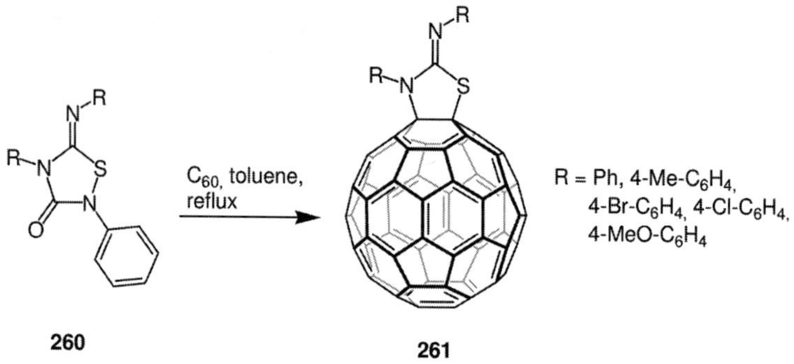

Scheme 4.44

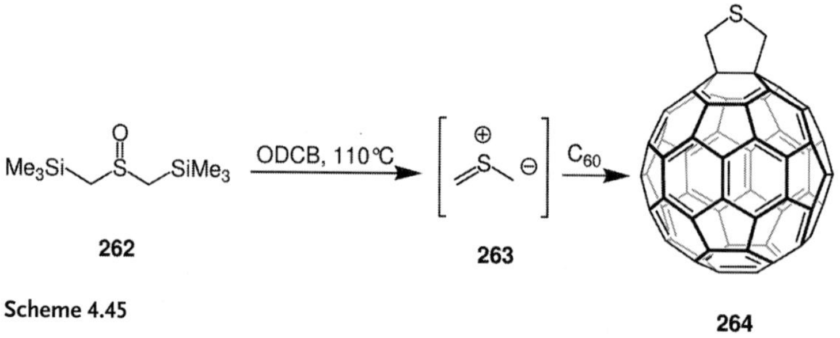

Scheme 4.45

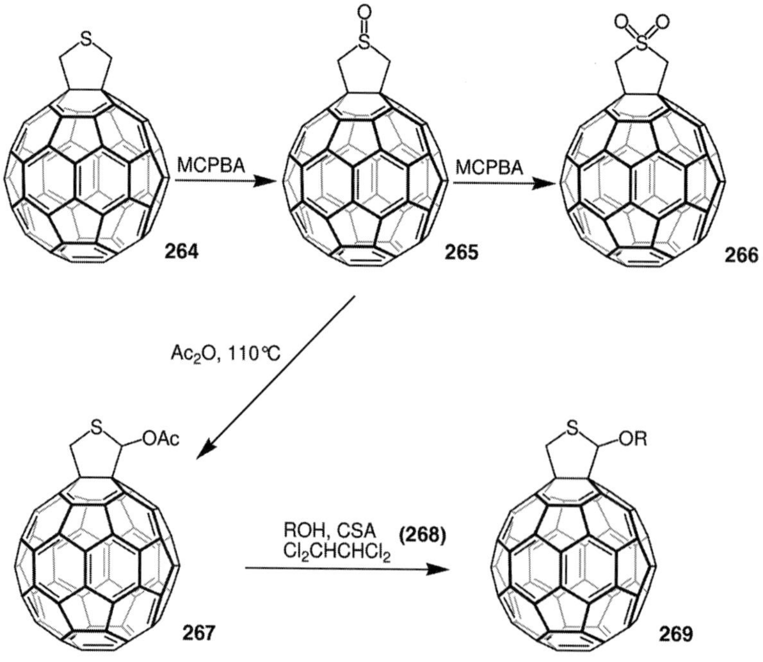

Scheme 4.46

benzyne monoadducts can be isolated, whereas the [6,6]-closed 1,2-isomer is the major product [333–335].

4.4.2 Addition of Enones

The photochemical [2+2] cycloaddition of enones to C_{60} is possible by irradiation of benzene solutions of the components with a high-pressure mercury lamp or with a XeCl excimer laser (Scheme 4.55) [336, 337]. Analysis of the reaction products by ^{1}H NMR, IR and HPLC [338] shows that the isolated monoadduct **293** is a mixture of two isomers, which can be assigned as the *cis* and *trans*-fused stereoisomers arising from the [2+2] cycloaddition to a [6,6] bond of C_{60}. Significantly, the IR carbonyl stretching frequencies of these two isomers differ remarkably, by about 20 cm^{-1}. The *cis* and *trans* isomers of **293** are chiral (C_1 symmetry). The corresponding enantiomers of each diastereomer for R = Me have been resolved and isolated by HPLC using chiral stationary phases [339, 340]. Chiroptical investigations indicate that the fullerene moiety in the strained *trans* cycloadducts has a skewed π-system with local C_2 symmetry. The enantiomers of the *trans* adducts show a much greater polarometric response than those of the *cis* adducts. These [2+2] cycloadditions cannot be achieved by irradiation at 532 nm, where C_{60} is the only light-absorbing component. This indicates that the fullerene triplet-state, being efficiently produced from singlet state [323], does not undergo addition to ground state enones. It was proposed that the photochemical [2+2] cycloadditions of enones to C_{60} proceeds by stepwise addition of the enone triplet excited state to the fullerene via an intermediate triplet 1,4-biradical, which is consistent with the observation that alkenes, such as cyclopentene or cyclohexene, do not photochemically cycloadd to C_{60} [336].

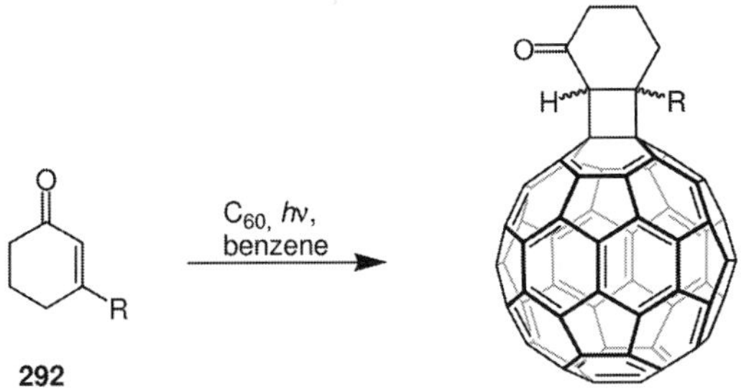

Scheme 4.55

A series of different cyclic enones has been subjected to photochemical [2+2] cycloaddition [337, 340]. Some examples are shown in Figure 4.14. As expected, enones substituted with bulky groups hinder addition; for example, the *tert*-butyl substituted enone **295** is unreactive toward C_{60}.

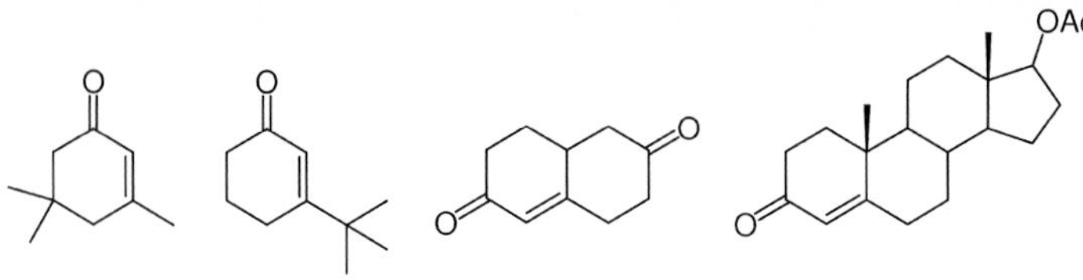

294 295 296 297

Figure 4.14 Some cyclic enones that were subjected to photochemical [2+2]-cycloaddition with C_{60}.

Cyclic diones or silylmethyl-protected diones react with olefines in a [2+2] fashion. Addition of these compounds to C_{60} [341] leads initially to the cyclobutane fused fullerenes, which are not stable and are readily oxidized and rearranged to the furanylfullerenes **300** and **301** (Scheme 4.56). The intermediate **299** probably reacts by either intermolecular oxidation with $^1O_2$ to yield **300** or by an intramolecular oxidation with the triplet fullerene moiety to yield **301**.

In respect of cycloreversion, cyclobutane-fused fullerenes derived from acyclic enones [342] are less stable than their bycyclic equivalents (e.g. **293**, Scheme 4.55). For the addition of mesityl oxide the equilibrium constant is so small that a 1000-fold excess of the enone is necessary to complete the reaction. The product of **302** is more stable and requires only a 100-fold excess of the enone. Reaction of **302**

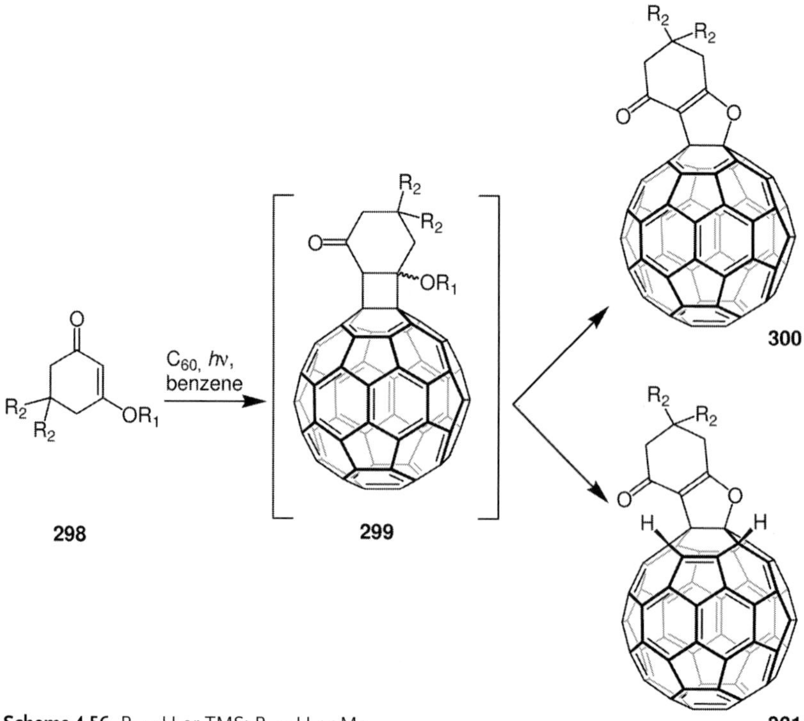

C_{60}, $h\nu$, benzene

298 299 300 301

Scheme 4.56 R_1 = H or TMS; R_2 = H or Me.

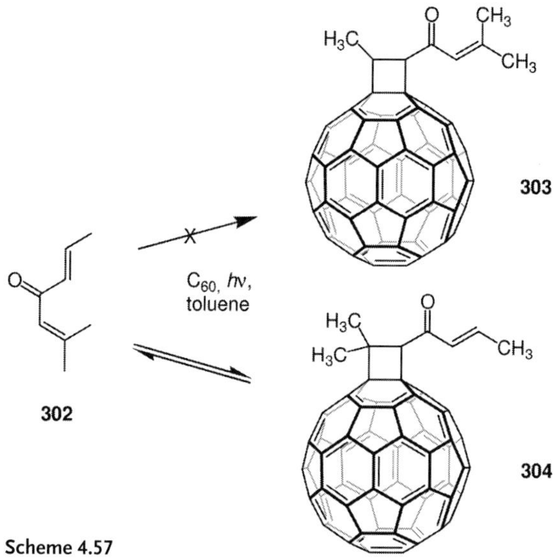

Scheme 4.57

with C_{60} under irradiation proceeds regiospecifically to afford exclusively one of two possible isomers (Scheme 4.57). The photocycloaddition occurs on the more substituted double bond, which can be explained by the higher stability of the intermediate biradicals.

4.4.3
Addition of Electron-rich Alkynes and Alkenes

The ease of the photoreaction of C_{60} with electron-rich organic molecules has been used to cycloadd *N,N*-diethylpropynylamine (**305**, R = Me) to C_{60} upon irradiation of oxygen-free toluene solutions of the components at room temperature for 20 min (Scheme 4.58) [343, 344].

The formation of the cycloadduct **306** [343] by thermal treatment of C_{60} with the alkyne **305** (R = Me) is very slow. The cycloadduct **306** is not stable, and upon exposure to air and room light for 2 h it cleanly produces the oxoamide **310** via a 1,2-dioxetane intermediate (Scheme 4.58). The SiO_2-catalyzed hydrolysis of **306** yields the amide **308** and not the cyclobutanone and diethylamine (Scheme 4.58). Apparently, the dihydrofullerene core is such a strong electron acceptor that it is a better leaving group than diethylamine. SiO_2-catalyzed hydrolysis yields **308** only in traces, but the reaction of **306** with *p*-toluenesulfonic acid hydrate in the dark affords **308** quantitatively [344]. Other alkynylamines have also been used (**305** with R = Me, CH=CMe_2, NEt_2, SEt or S^tBu) [345, 346]. The amines **305** with R = NEt_2, SEt or S^tBu were added to C_{70}, which reacts faster than C_{60} [346].

The oxoamide **310** with R = NEt_2 or SEt can be converted into the fullerene anhydride **311** by heating in toluene at 100 °C in the presence of *p*-TsOH (Scheme 4.59) [346].

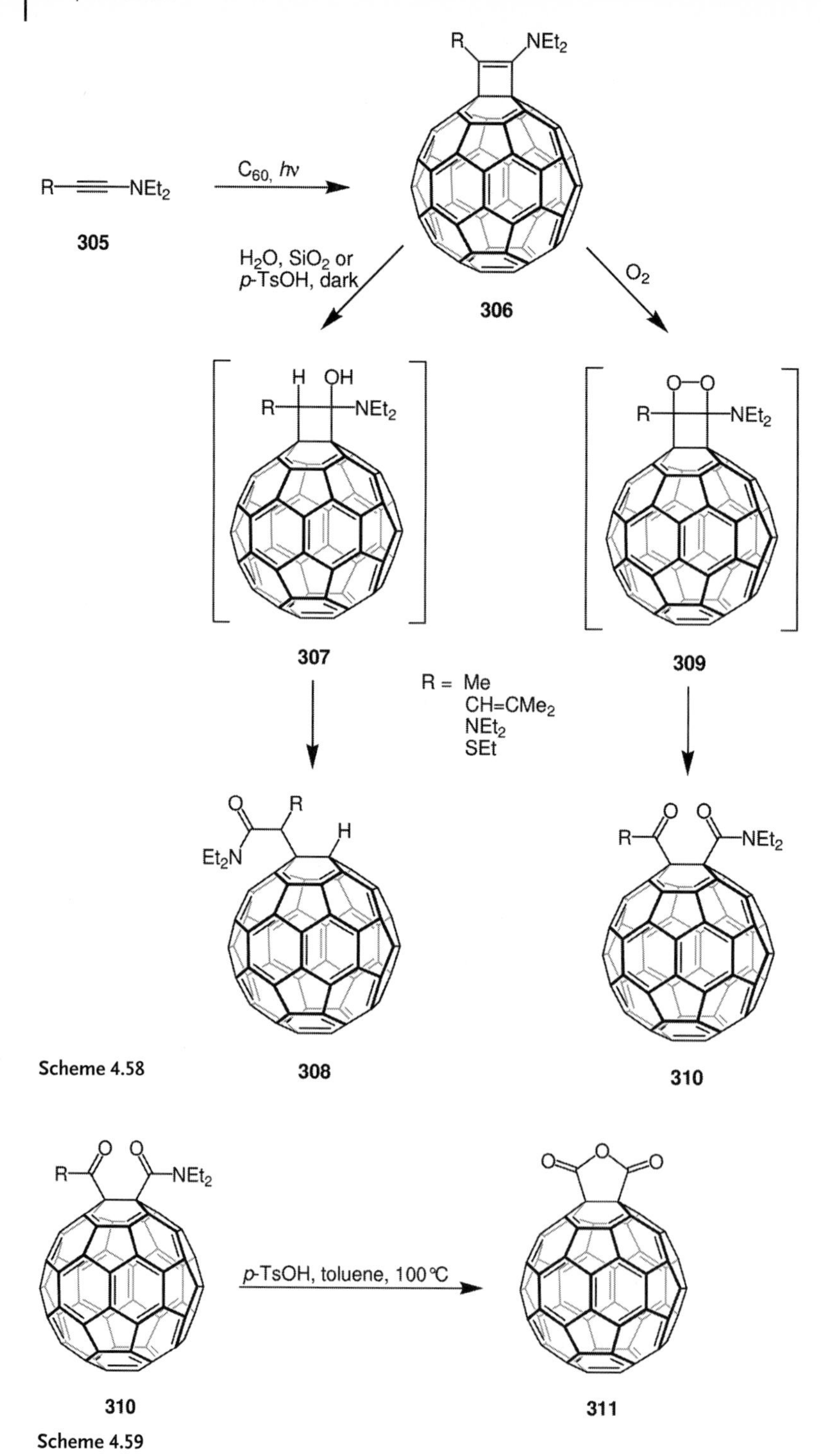

Scheme 4.58

Scheme 4.59

In the presence of oxygen and charcoal the amide **308** cyclizes to the lactone **312** (Scheme 4.60). The methyl lactone **312** is quite robust, surviving irradiation from a xenon-lamp and refluxing toluene [344].

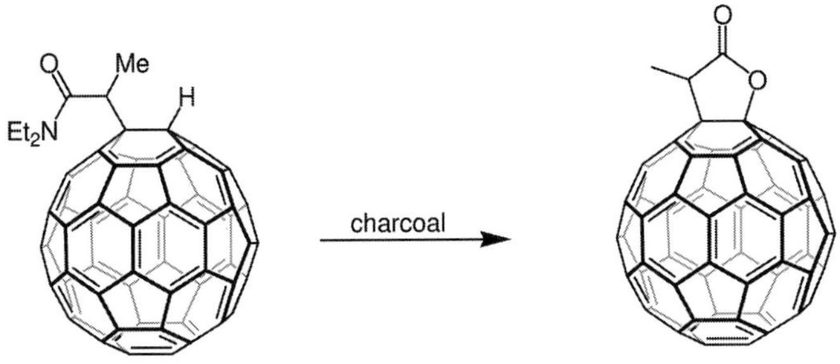

Me H Et_2N charcoal

308 **312**

Scheme 4.60

According to the proposed mechanism [343, 345, 347], the addition of electron-rich alkynes (**305**, R = Me, $CH{=}CMe_2$) to C_{60} involves the triplet excited state of C_{60} in the first step, generated by irradiation. The second step includes the addition of the alkyne to the triplet excited state of C_{60} via electron-, or at least charge-transfer from the electron-rich alkyne, followed by the rapid collapse of the initial ion pair or charge-transfer complex to the [2+2] cycloadducts. With the extremely electron-rich ynamines **305** with R = NEt_2 or SEt, the proposed charge transfer may occur without irradiation [346]. Cycloaddition with these alkynes can be carried out thermally at room temperature. Also, with the very electron-rich tetraethoxy ethylene a thermal [2+2] cycloaddition can be observed, but temperatures of 100 °C are required [346].

Only very few examples of alkene-[2+2] cycloadditions are known [345, 347, 348]. By using a large excess of the moderate electron-rich alkene *p*-propenyl-anisol [348] or even less electron-rich alkyl-substituted 1,3-butadienes [347] no thermal [2+2] cycloaddition occurs, but a photochemical cycloaddition can be enforced. The mechanism is proven to be stepwise via a biradical or dipolar intermediate [347–351], comparable to the addition of the alkynes. During the addition of *cis*- and *trans*-alkenes the existence of this relatively long lived intermediate leads to a loss of stereochemical integrity. Addition of *cis*-4-propenylanisol or *trans*-4-propenylanisol results in both cases exclusively in the *trans*-adduct (Scheme 4.61).

Thermal [2+2] cycloaddition with the electron-rich tetraferrocenyl[5]-cumulene **314** leads to a mixture of two products (Scheme 4.62), which are green and air stable [352, 353]. Shorter cumulenes bearing up to four double bonds do not react in a [2+2]-mode but form clathrates with C_{60}. Cyclovoltammetric experiments revealed complex interactions among the three redox-active components (ferrocenyl, cumulene and fullerene).

Contrary to the phosphine-catalyzed reaction of the allene buta-2,3-dienoate, which adds to C_{60} in a [3+2] cycloaddition [354], the allene amide **317** forms the cyclobutane annulated fullerene derivative in good yields (Scheme 4.63) [355]. The reaction was also performed with similar allene amides bearing a six- or seven-membered lactam.

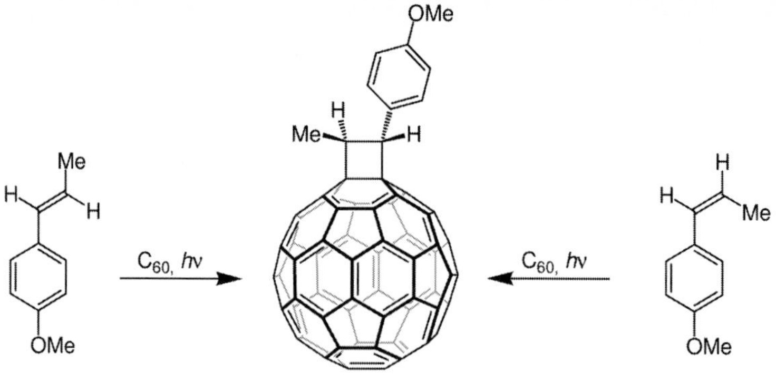

Scheme 4.61

trans-**313**

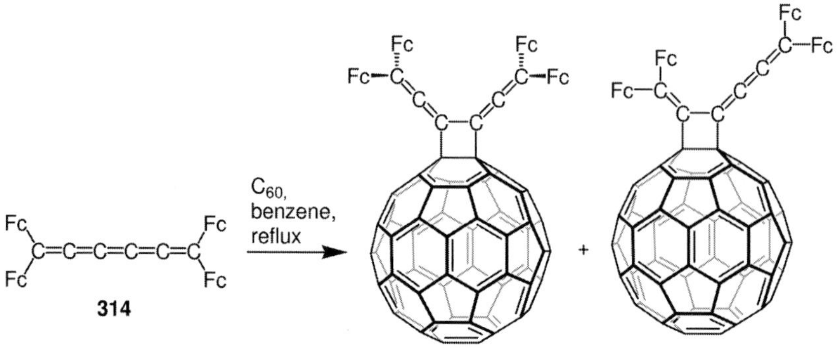

Scheme 4.62

314 **315** **316**

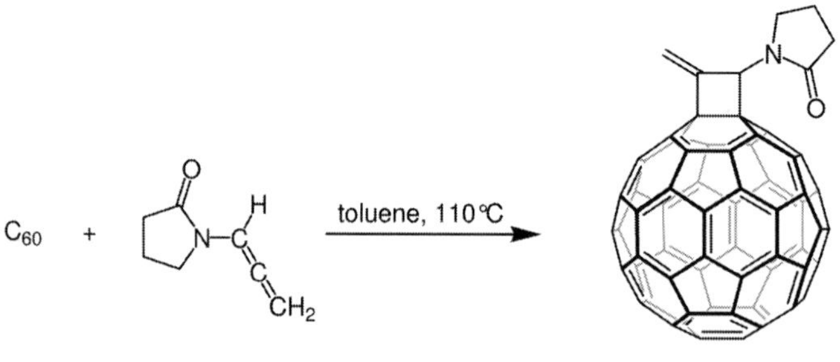

Scheme 4.63 **317** **318**

4.4.4 Addition of Ketenes and Ketene Acetals

Cycloaddition of ketenes to C_{60} is expected to be difficult due to their electrophilic character. Nevertheless, the reaction of C_{60} with some aryloxy- and alkoxy-ketenes led to the successful characterization of cycloaddition-products [356]. The reaction proceeds via a formal [2+2] cycloaddition, followed by enolization and acylation. Products **320** are stable and can be obtained in moderate yields (Scheme 4.64).

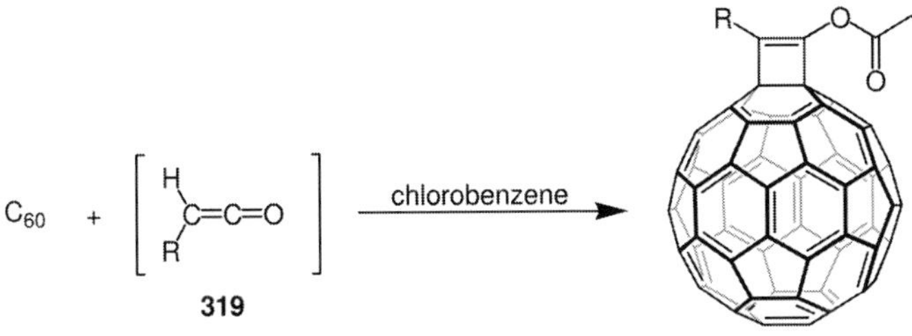

C_{60} + [319] —chlorobenzene→ 320

R = C_6H_5O, $C_6H_5CH_2O$, 4-Cl-C_6H_5, C_2H_5O, CH_3O

Scheme 4.64

Ketenes **319** were generated in situ by the reaction of the corresponding acid chlorides with triethylamine in chlorobenzene.

Prior to the addition of ketenes, attempts with ketene acetals were performed. With usual unstrained ketene acetals no cycladdition could be observed. However, the strained ketene acetal **322** reacts with remarkable ease (Scheme 4.65) – it is electron-rich and reacts readily with various electron-deficient olefines. Thermolysis of methylenecyclopropane **321** at 150 °C produces **322**, which reacts with C_{60} in toluene at room temperature. Cycloadduct **323** was obtained in 67% yield and

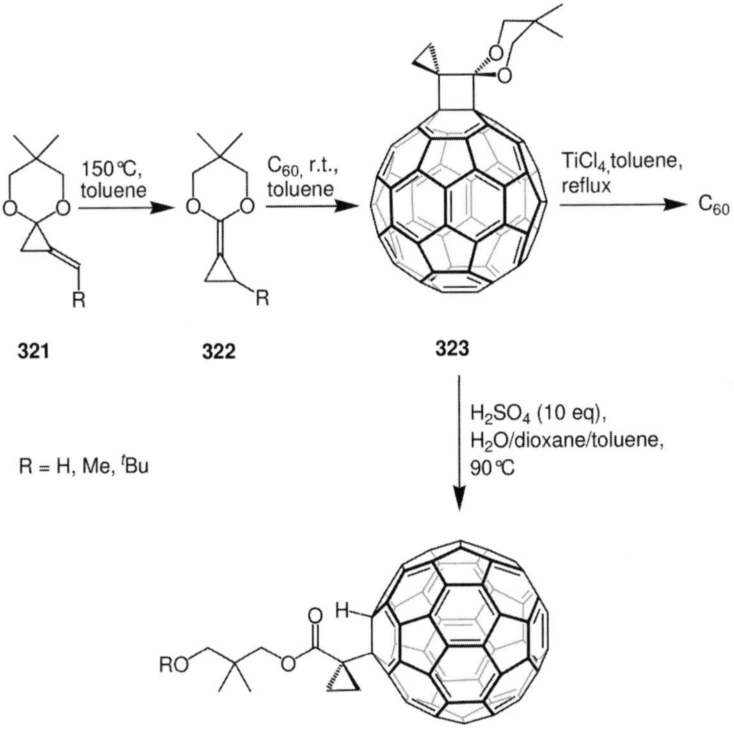

321 —150 °C, toluene→ **322** —C_{60}, r.t., toluene→ **323** —$TiCl_4$, toluene, reflux→ C_{60}

323 —H_2SO_4 (10 eq), H_2O/dioxane/toluene, 90 °C→ **324**

R = H, Me, ^{t}Bu

Scheme 4.65

showed no cycloreversion at room temperature but slowly cycloreverses upon heating at 100 °C or upon treatment with Lewis acids such as $TiCl_4$. Heating **323** in the presence of sulfuric acid in dioxane–toluene or TsOH in toluene yields either the fullerene carboxylic acid **324** (R = H) or its tosylate (R = Ts).

4.4.5
Addition of Quadricyclane

A thermal [2+2] cycloaddition of quadricyclane can be achieved by allowing a 10-fold excess to react with C_{60} in toluene at 80 °C (Scheme 4.66) [357].

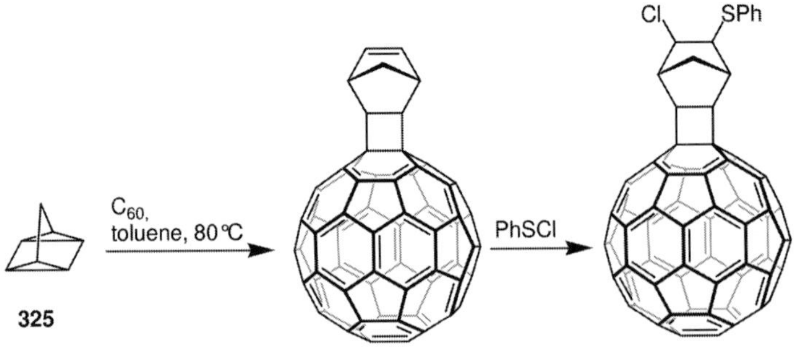

C_{60}, toluene, 80 °C; PhSCl

325 326 327 (Cl, SPh)

Scheme 4.66

The reaction proceeds smoothly and gives **326** in 43% yield. The stereochemistry of the addition is exo with respect to the norbornene moiety and in line with the usual cycloaddition behavior of quadricyclane [358]. The norbornene double bond in **326** is easily accessible by electrophiles, and, for example, the anti-addition of benzenesulfenyl chloride proceeds quantitatively at room temperature (Scheme 4.66).

4.4.6
Photodimerization of C_{60}

If oxygen-free C_{60} films are irradiated with visible or ultraviolet light, a phototransformation into a polymer occurs [359–363]. After phototransformation, the films are no longer soluble in toluene. Although photopolymerization of C_{60} films initiated a series of studies, resulting in the discovery of various fullerene polymers, the structure of phototransformed C_{60} is still unclear [364]. It was postulated that polymerization proceeds via a [2+2] cycloaddition mechanism between two [6,6] double bonds of two C_{60} units. Soon after the phototransformation experiments, further methods of polymerization were employed [365–367]. C_{60}-polymers were produced by applying high pressure or heat in the solid phase, but could also be obtained in solution. High-pressure polymerization of C_{60} leads to at least three different polymeric phases, i.e. rhombohedral and tetragonal two-dimensional phases or an orthorombic one-dimensional phase [366].

The smallest subunit of these fullerene polymers is the C_{120} dimer [368]. Synthesis and examination of the properties of this molecule are, therefore, of considerable interest. Beside the solid-state mechanochemical reaction of C_{60} with KCN – described in Section 3.2.3 – the dimer C_{120} can be synthesized at 5 GPa and 200 °C in good yields from a bis(ethylene-dithio)TTF-C_{60} complex [369]. C_{120} is a dark brown solid that is almost insoluble in most organic solvents. Its solubility decreases with increasing purity of the sample; this behavior is also known for pristine C_{60}. The dimer dissociates quantitatively to C_{60} upon heating to 175 °C or upon reduction and slowly even upon exposure to light [368].

Dimers with hetero-atoms bridging the two C_{60}-moieties have been prepared starting from the dimer $C_{120}O$, which can be synthesized by thermal reaction of C_{60} with the epoxide $C_{60}O$ (see Chapter 8) [370, 371]. The bisoxygen-bridged dimer **328** ($C_{120}O_2$) can be isolated after heating $C_{120}O$ to 400 °C at atmospheric pressure under argon for 1 h [372]. Its structure, which possesses a similar cyclobutan-fusion as C_{120}, is shown in Figure 4.15. Heating $C_{120}O$ in the presence of elemental sulfur under similar conditions, but at a lower temperature of 230 °C, yields the S-O-brigded dimer **329** ($C_{120}SO$) with an homologous structure to $C_{120}O_2$ [373].

Dimers with two methylene bridges and a cyclobutane fusion were synthesized by thermolysis of ethoxycarbonylmethano-1,2-dihydro[60]fullerene. They have a similar structure as **328** [374].

Photodimerization of dimers, where the two C_{60}-spheres are already linked via bisfunctional groups, can be carried out in very good yields [375, 376]. The dimer

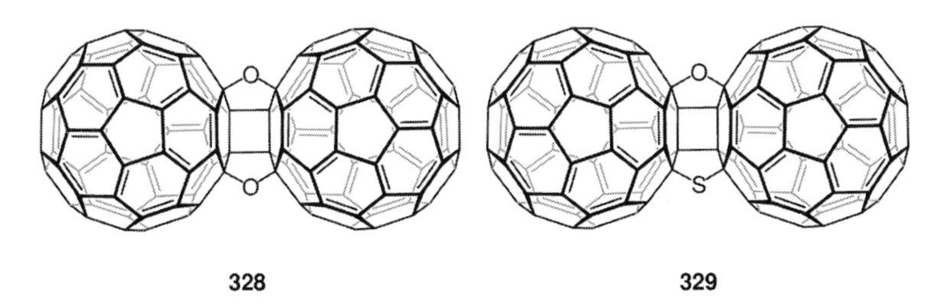

Figure 4.15 Hetero-atom bridged C_{60}-dimers.

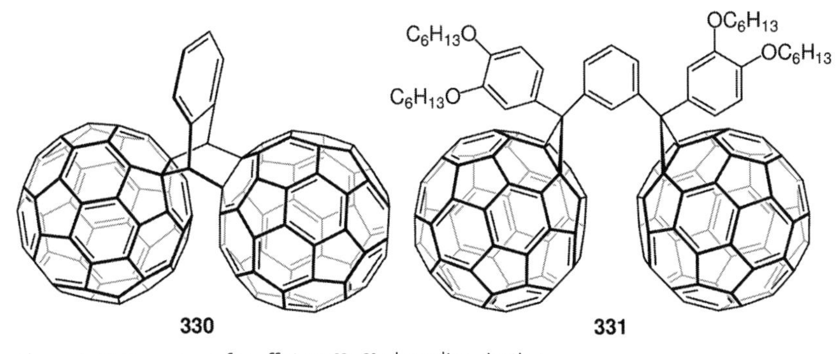

Figure 4.16 Precursors for efficient [2+2]-photodimerization.

330 (Figure 4.16), obtained by twofold Diels–Alder addition of two C_{60} to phthalazine, can be converted into the [2+2] cycloadduct in quantitative yield in chloroform solution [376]. The bis-linked product, which is derived from the bis-methanofullerene **331**, is one of the first examples of a controlled [2+2] cycloaddition process of fullerenes [375]. These examples make clear that in the case of fullerene oligomers with a close proximity of the fullerene fragments a [2+2] dimerization has always to be taken into account.

4.5
[2+1] Cycloadditions

4.5.1
Addition of Carbenes

Thermal extrusion of N_2 from O-benzyl- and O-pivaloyl-protected diazirine yields the corresponding carbenes, which react with C_{60} in toluene to give the fullerene sugars **333** (Scheme 4.67) [377, 378]. These [2+1] carbene additions cleanly lead to 1,2-methano-bridged sugar monoadducts.

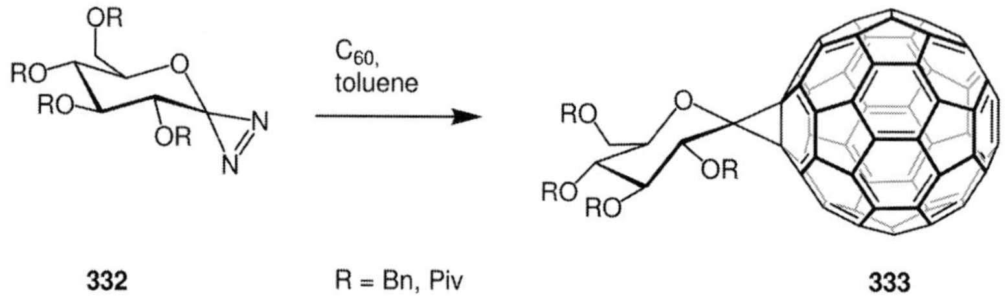

C_{60}, toluene

332 R = Bn, Piv **333**

Scheme 4.67

In contrast to the reaction with substituted or unsubstituted diazomethane, carbenes seem to add selectively to the [6,6] ring junction [378]. Support for this presumption was derived from the observation of a series of diazirine-additions [379–381]. Thermal reaction of chlorophenyldiazirine with C_{60} led to the same regioselectivity whereas this diazirine compound is known as a clean source of chlorophenylcarbene [379]. The photochemical reaction of diazirines such as adamantane diazirine [380] or chloro-isopropyldiazirine [381] yields not only a singlet carbene but also a diazo compound as an intramolecular rearrangement product. Since diazomethane derivatives can, finally, lead to [5,6]-bridged systems, this opens up the possibility of examining the regioselectivity of carbene and diazo compounds in one reaction. Irradiation of adamantane diazirine leads to adamantylidene and diazoadamantane in a ratio of 1 : 1. This ratio was determined in absence of any C_{60} by laser flash photolysis. Irradiation of the adamantanediazirine in the presence of C_{60} leads to a ratio of [6,6] to [5,6] adducts of 49/51. This result is a strong indication for the high [6,6] selectivity of carbenes and, furthermore, renders C_{60} a mechanistic probe for the formation of either carbene or diazo compound from diazirine.

The dimethoxymethanofullerene **335** has been synthesized via the corresponding carbene, which was prepared in situ upon thermolysis of the oxadiazole **334** in 32% yield (Scheme 4.68) [113].

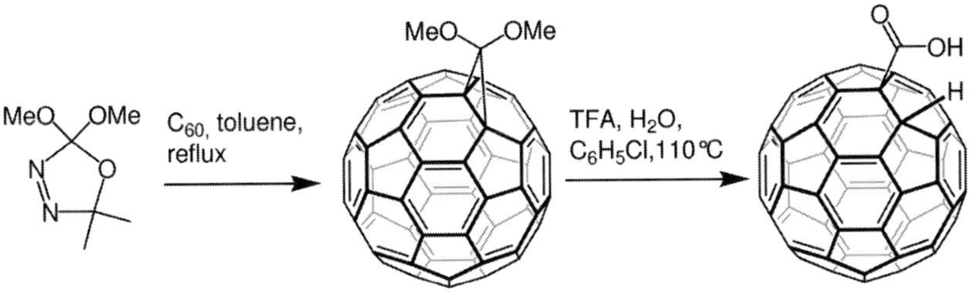

334 **335** **336**

Scheme 4.68

Attempts to synthesize water-soluble thioketals from **335** were not successful, but led to the isolation of the first fullerenecarboxylic acid derivative [382]; **336** is prepared in 71% yield from **335** by treatment with trifluoroacetic acid at 110 °C.

Methoxy[(trimethylsilyl)ethoxy]carbene – obtained by thermolysis of the corresponding oxadiazoline **337** – reacts with C_{60} in an unprecedented reaction pathway [383, 384]. The expected methanofullerene-[2+1]-cycloadduct could not be observed. Instead the two dihydrofullerene adducts **338** and **339** were isolated (Scheme 4.69). They are formed by an unusual addition–rearrangement mechanism that includes the migration of the trimethylsilyl group.

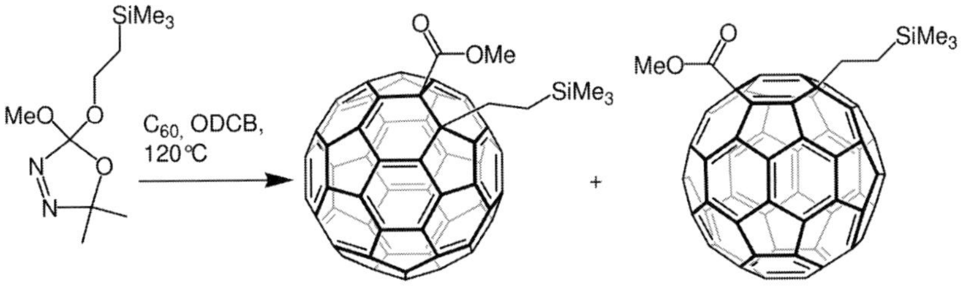

337 **338** **339**

Scheme 4.69

Further methods of generating carbenes used for the functionalization of C_{60} include thermolysis of Seyferth reagents $PhHgCBr_3$ [385] or $PhHgCCl_2Br$ [386], trichloroacetate [387] or cyclopropenone acetal [388]. With the mercury reagents or trichloroacetate the dihalogenemethanofullerenes were obtained. These halogenated compounds can be turned into methanofullerene carbene [385, 389, 390]. Dimerization of these carbenes can occur in a twofold manner to yield either C_{121} (**340**) or C_{122} (**341**) (Figure 4.17). C_{121} is the product of the reaction of one methanofullerene carbene with unreacted C_{60}, whereas C_{122} is formed upon dimerization of two carbenes. Also, the reaction of C_{60} with diazotetrazole, as a carbene precursor, yields

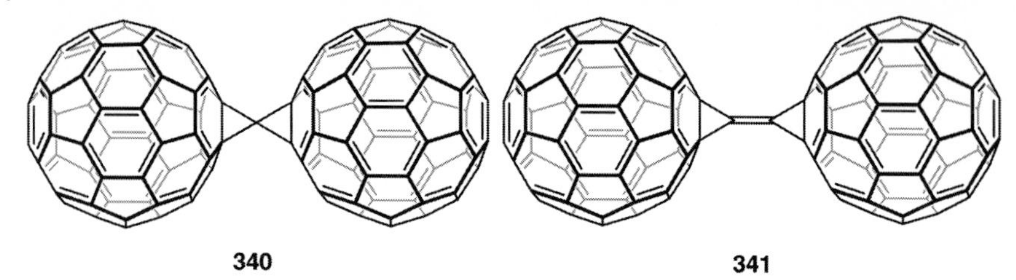

Figure 4.17 C_{60}-dimers C_{121} and C_{122} obtained by dimerization of the carbene C_{61} or by reaction of C_{61} with C_{60}.

the two dimers, but in this reaction dimer C_{122} can be obtained in isolable amounts [391].

Carbene addition to C_{70} results in the formation of different monoadduct isomers [386]. Treatment with $PhHgCCl_2Br$ in refluxing benzene produces a mixture of three isomers derived from addition to the C-1/C-2-bond, the C-5/C-6-bond and the C-7/C-8-bond in almost equimolar amounts. Contrary to this result, addition of trichloroacetate yields almost exclusively the C-5/C-6-product [386].

Reactions of C_{60} with metal carbene complexes also yield the [6,6] methanofullerenes [392]. These adducts are probably not formed via a carbene addition, but via a formal [2+2] cycloaddition under formation of a metalla cyclobutane intermediate. The Fischer carbene complex [methyl(methoxymethylene)]pentacarbonyl chromium can be utilized to prepare 1,2-methyl(methoxymethano)-fullerene in 20% yield [392]. A tungsten carbene complex was primarily used to initiate the formation of a polyacetylene polymer, but it was discovered that addition of C_{60} to the complex-polymer-mixture improves the polymerization and dramatically increases the catalytic activity of the carbene complex [393]. C_{60} can be integrated into the polymer via carbene addition.

4.5.2
Addition of Nitrenes

The addition of nitrenes leads predominantly to the closed [6,6] bridged isomers. The corresponding [5,6] bridged isomer is – if at all – formed only in small amounts, probably via a direct addition to the [5,6] bond [394]. Nitrenes have been generated by thermolysis of azido-formic esters [172, 395–400], photolysis of aroyl azides [401] or aryl azide [402], elimination of *O*-4-nitrophenylsulfonylalkylhydroxamic acid [403] or reaction of amines with $Pb(OAc)_4$ [404].

Azidoformic esters such as **342** react with C_{60} in a [2+1] addition (Scheme 4.70), if the temperature is high enough to induce the loss of nitrogen prior to addition, otherwise a [3+2] addition can be observed (Section 4.3.2) [172, 395, 397]. Typical conditions include heating of the mixture in solvents such as tetrachloroethane [395, 397, 398], chloronaphthalene [397] or toluene [396] at 110–160 °C. These conditions also afford multiple addition products [172]. To avoid potential hazard during purification of the azido formiates, they were also generated in situ in one pot by the reaction of chloro-formic ester with sodium azide [396].

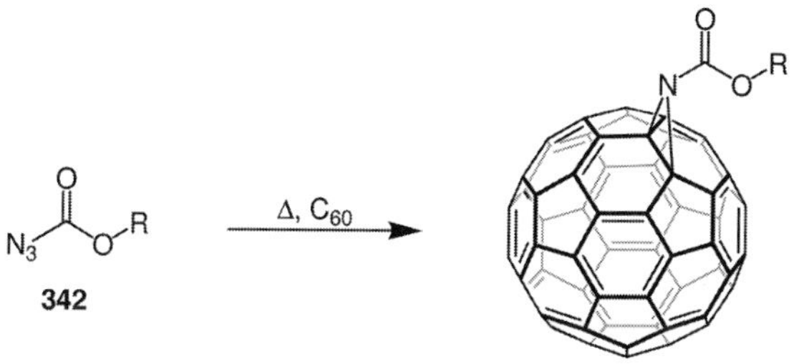

R = Me, Et, ^{t}Butyl, 2,4,6-tri-^{t}butylphenyl

Scheme 4.70

The *tert*-butyl derivative **343** quantitatively eliminates iso-butene and carbon dioxide on a neutral alumina column. The obtained aziridine can be esterified in good yields with acid chlorides in ODCB in the presence of pyridine [398, 399].

Photochemical reaction between aroylazide or arylazide in halogenated solvents proceeds, probably, also via the nitrene intermediate [401, 402]. This is supported by the fact that almost only the [6,6] adduct is formed. By photolysis of the aroylazides **344** in dichloromethane or tetrachloroethane the acylnitrenes are generated (Scheme 4.71) [401]. These are added in moderate yields to C_{60} to give the [6,6] adducts **345**, which can be thermally rearranged (tetrachloroethane, reflux) to the corresponding [6,6] fullerooxazole **346**.

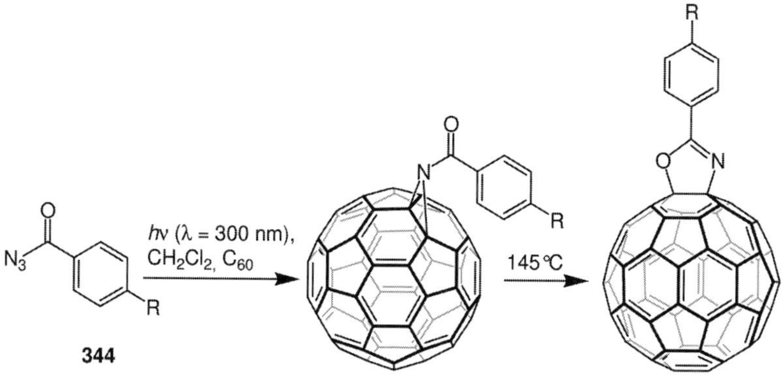

R = -OMe, -CN, -H, -Br

Scheme 4.71

Phthalimidonitrene **348** is derived by reaction of the corresponding amine **347** with lead tetraacetate in chlorobenzene (Scheme 4.72) [404]. Addition to C_{60} affords mono and multiple adducts with up to four phthalimido addends.

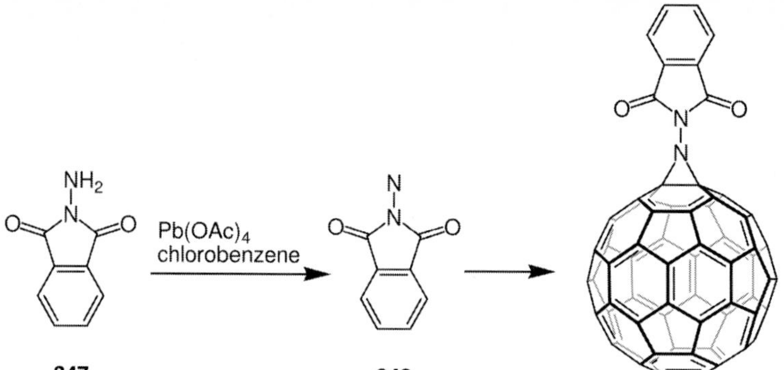

Scheme 4.72

4.5.3 Addition of Silylenes

The addition of bis(2,6-diisopropylphenyl)silylene, as a reactive divalent species, to C_{60} or C_{70} yields the [2+1] cycloadduct **351** as the ring-closed 1,2-bridged isomer (Scheme 4.73) [405–407]. The silylene was prepared in situ from the trisilane **350** by photolysis with a low-pressure lamp in toluene solution.

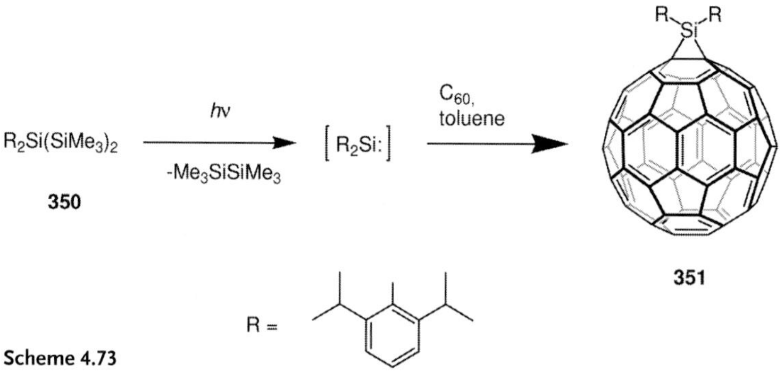

Scheme 4.73

References

1 F. Wudl, A. Hirsch, K. C. Khemani, T. Suzuki, P. M. Allemand, A. Koch, H. Eckert, G. Srdanov, H. M. Webb, *ACS Symp. Ser.* **1992**, *481*, 161.

2 M. A. Yurovskaya, I. V. Trushkov, *Russ. Chem. Bull.* **2002**, *51*, 367.

3 S. R. Wilson, D. I. Schuster, B. Nuber, M. S. Meier, M. Maggini, M. Prato, R. Taylor, *Full.: Chem., Phys. Technol.* **2000**, 91.

4 B. Kräutler, J. Maynollo, *Tetrahedron* **1996**, *52*, 5033.

5 S. R. Wilson, Q. Lu, *Tetrahedron Lett.* **1993**, *34*, 8043.

6 F. Langa, P. de la Cruz, E. Espildora, J. J. Garcia, M. C. Perez, A. de la Hoz, *Carbon* **2000**, *38*, 1641.

7 V. M. Rotello, J. B. Howard, T. Yadav, M. M. Conn, E. Viani, L. M. Giovane, A. L. Lafleur, *Tetrahedron Lett.* **1993**, *34*, 1561.
8 M. Tsuda, T. Ishida, T. Nogami, S. Kurono, M. Ohashi, *J. Chem. Soc., Chem. Commun.* **1993**, 1296.
9 J. A. Schlüter, J. M. Seaman, S. Taha, H. Cohen, K. R. Lykke, H. H. Wang, J. M. Williams, *J. Chem. Soc., Chem. Commun.* **1993**, 972.
10 B. Kräutler, T. Müller, A. Duarte-Ruiz, *Chem. Eur. J.* **2001**, *7*, 3223.
11 N. Chronakis, M. Orfanopoulos, *Tetrahedron Lett.* **2001**, *42*, 1201.
12 I. Lamparth, C. Maichle-Mössmer, A. Hirsch, *Angew. Chem.* **1995**, *107*, 1755; *Angew. Chem. Int. Ed. Engl.* **1995**, *34*, 1607.
13 M. Prato, T. Suzuki, H. Foroudian, Q. Li, K. Khemani, F. Wudl, J. Leonetti, R. D. Little, T. White, B. Rickborn, S. Yamago, E. Nakamura *J. Am. Chem. Soc.* **1993**, *115*, 1594.
14 Y. Takaguchi, T. Tajima, K. Ohta, J. Motoyoshiya, H. Aoyama, T. Wakahara, T. Akasaka, M. Fujitsuka, O. Ito, *Angew. Chem.* **2002**, *114*, 845; *Angew. Chem. Int. Ed. Engl.* **2002**, *41*, 817.
15 Y. Takaguchi, Y. Sako, Y. Yanagimoto, S. Tsuboi, J. Motoyoshiya, H. Aoyama, T. Wakahara, T. Akasaka, *Tetrahedron Lett.* **2003**, *44*, 5777.
16 M. A. Herranz, N. Martin, J. Ramey, D. M. Guldi, *Chem. Commun.* **2002**, 2968.
17 G. P. Miller, J. Mack, *Org. Lett.* **2000**, *2*, 3979.
18 G. P. Miller, J. Mack, J. Briggs, *Org. Lett.* **2000**, *2*, 3983.
19 Y. Murata, N. Kato, K. Fujiwara, K. Komatsu, *J. Org. Chem.* **1999**, *64*, 3483.
20 K. Mikami, S. Matsumoto, T. Tonoi, Y. Okubo, T. Suenobu, S. Fukuzumi, *Tetrahedron Lett.* **1998**, *39*, 3733.
21 M. F. Meidine, R. Roers, G. J. Langley, A. G. Avent, A. D. Darwish, S. Firth, H. W. Kroto, R. Taylor, D. R. M. Walton, *J. Chem. Soc., Chem. Commun.* **1993**, 1342.
22 B. Kräutler, M. Puchberger, *Helv. Chim. Acta* **1993**, *76*, 1626.
23 G. Torres-Garcia, J. Mattay, *Tetrahedron* **1996**, *52*, 5421.
24 S. I. Khan, A. M. Oliver, M. N. Paddon-Row, Y. Rubin, *J. Am. Chem. Soc.* **1993**, *115*, 4919.
25 Y.-Z. An, A. L. Viado, M.-J. Arce, Y. Rubin, *J. Org. Chem.* **1995**, *60*, 8330.
26 L. A. Paquette, R. J. Graham, *J. Org. Chem.* **1995**, *60*, 2958.
27 S. Cossu, G. Cuomo, O. De Lucchi, M. Maggini, G. Valle, *J. Org. Chem.* **1996**, *61*, 153.
28 H. P. Zeng, S. Eguchi, *Synlett* **1997**, 175.
29 M. Diekers, A. Hirsch, S. Pyo, J. Rivera, L. Echegoyen, *Eur. J. Org. Chem.* **1998**, 1111.
30 M. Ohno, Y. Shirakawa, S. Eguchi, *Synthesis* **1998**, 1812.
31 Y. Rubin, P. S. Ganapathi, A. Franz, Y.-Z. An, W. Qian, R. Neier, *Chem. Eur. J.* **1999**, *5*, 3162.
32 C. M. A. Alonso, M. G. P. M. S. Neves, A. C. Tome, A. M. S. Silva, J. A. S. Cavaleiro, *Tetrahedron Lett.* **2000**, *41*, 5679.
33 S. R. Wilson, M. E. Yurchenko, D. I. Schuster, A. Khong, M. Saunders, *J. Org. Chem.* **2000**, *65*, 2619.
34 N. Chronakis, M. Orfanopoulos, *Org. Lett.* **2001**, *3*, 545.
35 N. Chronakis, G. Froudakis, M. Orfanopoulos, *J. Org. Chem.* **2002**, *67*, 3284.
36 S. Ostrowski, A. Mikus, *Mol. Diversity* **2003**, *6*, 315.
37 G. Mehta, M. B. Viswanath, *Synlett* **1995**, 679.
38 M. Ohno, T. Azuma, S. Kojima, Y. Shirakawa, S. Eguchi, *Tetrahedron* **1996**, *52*, 4983.
39 W. Duczek, W. Radeck, H.-J. Niclas, M. Ramm, B. Costisella, *Tetrahedron Lett.* **1997**, *38*, 6651.
40 H. Ishida, K. Komori, K. Itoh, M. Ohno, *Tetrahedron Lett.* **2000**, *41*, 9839.
41 P. Belik, A. Gügel, J. Spickermann, K. Müllen, *Angew. Chem.* **1993**, *105*, 95; *Angew. Chem. Int. Ed. Engl.* **1993**, *32*, 78.
42 A. Gügel, A. Kraus, J. Spickermann, P. Belik, K. Müllen, *Angew. Chem.* **1994**, *106*, 601; *Angew. Chem. Int. Ed. Engl.* **1994**, *33*, 559.
43 P. Belik, A. Gügel, A. Kraus, J. Spickermann, V. Enkelmann, G. Frank, K. Müllen, *Adv. Mater.* **1993**, *5*, 854.
44 W. Bidell, R. G. Compton, J. C. Eklund, M. L. H. Green, T. O. Rebbitt, A. H. H. Stephens, *J. Organomet. Chem.* **1998**, *562*, 115.

45 M. Taki, S. Sugita, Y. Nakamura, E. Kasashima, E. Yashima, Y. Okamoto, J. Nishimura, *J. Am. Chem. Soc.* **1997**, *119*, 926.
46 P. Belik, A. Gügel, A. Kraus, M. Walter, K. Müllen, *J. Org. Chem.* **1995**, *60*, 3307.
47 Y. Nakamura, M. Taki, S. Tobita, H. Shizuka, H. Yokoi, K. Ishiguro, Y. Sawaki, J. Nishimura, *J. Chem. Soc., Perkin Trans. 2* **1999**, 127.
48 M. Iyoda, S. Sasaki, M. Yoshida, Y. Kuwatani, S. Nagase, *Tetrahedron Lett.* **1996**, *37*, 7987.
49 X. Zhang, C. S. Foote, *J. Org. Chem.* **1994**, *59*, 5235.
50 A. Kraus, A. Gügel, P. Belik, M. Walter, K. Müllen, *Tetrahedron* **1995**, *51*, 9927.
51 B. Illescas, N. Martin, C. Seoane, P. de la Cruz, F. Langa, F. Wudl, *Tetrahedron Lett.* **1995**, *36*, 8307.
52 B. M. Illescas, N. Martin, C. Seoane, E. Orti, P. M. Viruela, R. Viruela, A. de la Hoz, *J. Org. Chem.* **1997**, *62*, 7585.
53 H. Tomioka, M. Ichihashi, K. Yamamoto, *Tetrahedron Lett.* **1995**, *36*, 5371.
54 K. O-Kawa, Y. Nakamura, J. Nishimura, *Tetrahedron Lett.* **2000**, *41*, 3103.
55 Y. Nakamura, K. Okawa, S. Minami, T. Ogawa, S. Tobita, J. Nishimura, *J. Org. Chem.* **2002**, *67*, 1247.
56 A. Puplovskiss, J. Kacens, O. Neilands, *Tetrahedron Lett.* **1997**, *38*, 285.
57 M. Iyoda, F. Sultana, S. Sasaki, M. Yoshida, *J. Chem. Soc., Chem. Commun.* **1994**, 1929.
58 F. Diederich, U. Jonas, V. Gramlich, A. Herrmann, H. Ringsdorf, C. Thilgen, *Helv. Chim. Acta* **1993**, *76*, 2445.
59 A. Gügel, M. Becker, D. Hammel, L. Mindach, J. Räder, T. Simon, M. Wagner, K. Müllen, *Angew. Chem.* **1992**, *104*, 666; *Angew. Chem. Int. Ed. Engl.* **1992**, *31*, 644.
60 A. Gügel, K. Müllen, *J. Chromatogr.* **1993**, *628*, 23.
61 Y. Rubin, S. Khan, D. I. Freedberg, C. Yeretzian, *J. Am. Chem. Soc.* **1993**, *115*, 344.
62 M. P. Cava, D. R. Napier, *J. Am. Chem. Soc.* **1957**, *79*, 1701.
63 J.-H. Liu, A.-T. Wu, M.-H. Huang, C.-W. Wu, W.-S. Chung, *J. Org. Chem.* **2000**, *65*, 3395.
64 M. Ohno, N. Koide, H. Sato, S. Eguchi, *Tetrahedron* **1997**, *53*, 9075.
65 A. C. Tome, R. F. Enes, J. A. S. Cavaleiro, J. Elguero, *Tetrahedron Lett.* **1997**, *38*, 2557.
66 A. C. Tome, R. F. Enes, J. P. C. Tome, J. Rocha, M. G. P. M. S. Neves, J. A. S. Cavaleiro, J. Elguero, *Tetrahedron* **1998**, *54*, 11141.
67 U. M. Fernandez-Paniagua, B. Illescas, N. Martin, C. Seoane, P. de la Cruz, A. de la Hoz, F. Langa, *J. Org. Chem.* **1997**, *62*, 3705.
68 A. Herrera, R. Martinez, B. Gonzalez, B. Illescas, N. Martin, C. Seoane, *Tetrahedron Lett.* **1997**, *38*, 4873.
69 B. Gonzalez, A. Herrera, B. Illescas, N. Martin, R. Martinez, F. Moreno, L. Sanchez, A. Sanchez, *J. Org. Chem.* **1998**, *63*, 6807.
70 F. Effenberger, G. Grube, *Synthesis* **1998**, 1372.
71 H. Ishida, K. Itoh, S. Ito, N. Ono, M. Ohno, *Synlett* **2001**, 296.
72 F.-P. Montforts, O. Kutzki, *Angew. Chem.* **2000**, *112*, 612; *Angew. Chem. Int. Ed. Engl.* **2000**, *39*, 599.
73 A. Rieder, B. Kräutler, *J. Am. Chem. Soc.* **2000**, *122*, 9050.
74 V. B. Kräutler, C. S. Sheehan, A. Rieder, *Helv. Chim. Acta* **2000**, *83*, 583.
75 M. Ohno, S. Kojima, S. Eguchi, *J. Chem. Soc., Chem. Commun.* **1995**, 565.
76 M. Ohno, S. Kojima, Y. Shirakawa, S. Eguchi, *Heterocycles* **1997**, *46*, 49.
77 M. Ohno, S. Kojima, Y. Shirakawa, S. Eguchi, *Tetrahedron Lett.* **1995**, *36*, 6899.
78 M. Ohno, S. Kojima, Y. Shirakawa, S. Eguchi, *Tetrahedron Lett.* **1996**, *37*, 9211.
79 N. Martin, A. Martinez-Grau, L. Sanchez, C. Seoane, M. Torres, *J. Org. Chem.* **1998**, *63*, 8074.
80 M. Ohno, T. Azuma, S. Eguchi, *Chem. Lett.* **1993**, 1833.
81 M. Ohno, H. Sato, S. Eguchi, *Synlett* **1999**, 207.
82 G. P. Miller, M. C. Tetreau, *Org. Lett.* **2000**, *2*, 3091.
83 G. P. Miller, M. C. Tetreau, M. M. Olmstead, P. A. Lord, A. L. Balch, *Chem. Commun.* **2001**, 1758.
84 Y. Murata, M. Suzuki, K. Komatsu, *Chem. Commun.* **2001**, 2338.
85 J. L. Segura, N. Martin, *Angew. Chem.* **2001**, *113*, 1416; *Angew. Chem. Int. Ed. Engl.* **2001**, *40*, 1372.

86 C. Boulle, J. M. Rabreau, P. Hudhomme, M. Cariou, M. Jubault, A. Gorgues, J. Orduna, J. Garin, *Tetrahedron Lett.* **1997**, *38*, 3909.

87 C. Boulle, M. Cariou, M. Bainville, A. Gorgues, P. Hudhomme, J. Orduna, J. Garin, *Tetrahedron Lett.* **1997**, *38*, 81.

88 J. Llacay, M. Mas, E. Molins, J. Veciana, D. Powell, C. Rovira, *Chem. Commun.* **1997**, 659.

89 J. Llacay, J. Veciana, J. Vidal-Gancedo, J. L. Bourdelande, R. Gonzalez-Moreno, C. Rovira, *J. Org. Chem.* **1998**, *63*, 5201.

90 M. A. Herranz, N. Martin, *Org. Lett.* **1999**, *1*, 2005.

91 S. G. Liu, D. Kreher, P. Hudhomme, E. Levillain, M. Cariou, J. Delaunay, A. Gorgues, J. Vidal-Gancedo, J. Veciana, C. Rovira, *Tetrahedron Lett.* **2001**, *42*, 3717.

92 D. Kreher, M. Cariou, S.-G. Liu, E. Levillain, J. Veciana, C. Rovira, A. Gorgues, P. Hudhomme, *J. Mater. Chem.* **2002**, *12*, 2137.

93 P. Hudhomme, C. Boulle, J. M. Rabreau, M. Cariou, M. Jubault, A. Gorgues, *Synth. Methods* **1998**, *94*, 73.

94 Y. Z. An, J. L. Anderson, Y. Rubin, *J. Org. Chem.* **1993**, *58*, 4799.

95 T. Gareis, O. Köthe, J. Daub, *Eur. J. Org. Chem.* **1998**, 1549.

96 Y.-Z. An, C.-H. B. Chen, J. L. Anderson, D. S. Sigman, C. S. Foote, Y. Rubin, *Tetrahedron* **1996**, *52*, 5179.

97 G. Torres-Garcia, H. Luftmann, C. Wolff, J. Mattay, *J. Org. Chem.* **1997**, *62*, 2752.

98 S. J. Danishefsky, M. P. DeNinno, *Angew. Chem.* **1987**, *99*, 15; *Angew. Chem. Int. Ed. Engl.* **1987**, *26*, 15.

99 K. Mikami, S. Matsumoto, Y. Okubo, T. Suenobu, S. Fukuzumi, *Synlett* **1999**, 1130.

100 K. Mikami, S. Matsumoto, Y. Okubo, M. Fujitsuka, O. Ito, T. Suenobu, S. Fukuzumi, *J. Am. Chem. Soc.* **2000**, *122*, 2236.

101 Y.-Z. An, G. A. Ellis, A. L. Viado, Y. Rubin, *J. Org. Chem.* **1995**, *60*, 6353.

102 J.-F. Eckert, C. Bourgogne, J.-F. Nierengarten, *Chem. Commun.* **2002**, 712.

103 T. Suzuki, Q. Li, K. C. Khemani, F. Wudl, O. Almarsson, *Science* **1991**, *254*, 1186.

104 F. Wudl, *Acc. Chem. Res.* **1992**, *25*, 157.

105 T. Suzuki, Q. Li, K. C. Khemani, F. Wudl, O. Almarsson, *J. Am. Chem. Soc.* **1992**, *114*, 7300.

106 T. Suzuki, Q. Li, K. C. Khemani, F. Wudl, *J. Am. Chem. Soc.* **1992**, *114*, 7301.

107 S. Shi, K. C. Khemani, Q. Li, F. Wudl, *J. Am. Chem. Soc.* **1992**, *114*, 10656.

108 A. B. Smith, III, R. M. Strongin, L. Brard, G. T. Furst, W. J. Romanow, K. G. Owens, R. C. King, *J. Am. Chem. Soc.* **1993**, *115*, 5829.

109 M. Prato, T. Suzuki, F. Wudl, V. Lucchini, M. Maggini, *J. Am. Chem. Soc.* **1993**, *115*, 7876.

110 M. Prato, V. Lucchini, M. Maggini, E. Stimpfl, G. Scorrano, M. Eiermann, T. Suzuki, F. Wudl, *J. Am. Chem. Soc.* **1993**, *115*, 8479.

111 M. Prato, A. Bianco, M. Maggini, G. Scorrano, C. Toniolo, F. Wudl, *J. Org. Chem.* **1993**, *58*, 5578.

112 K. L. Wooley, C. J. Hawker, J. M. J. Frechet, F. Wudl, G. Srdanov, S. Shi, C. Li, M. Kao, *J. Am. Chem. Soc.* **1993**, *115*, 9836.

113 L. Isaacs, F. Diederich, *Helv. Chim. Acta* **1993**, *76*, 2454.

114 C. Zhu, Y. Xu, Y. Liu, D. Zhu, *J. Org. Chem.* **1997**, *62*, 1996.

115 M. H. Hall, H. Lu, P. B. Shevlin, *J. Am. Chem. Soc.* **2001**, *123*, 1349.

116 J. C. Hummelen, B. W. Knight, F. LePeq, F. Wudl, J. Yao, C. L. Wilkins, *J. Org. Chem.* **1995**, *60*, 532.

117 P. Timmerman, H. L. Anderson, R. Faust, J.-F. Nierengarten, T. Habicher, P. Seiler, F. Diederich, *Tetrahedron* **1996**, *52*, 4925.

118 Z. Li, K. H. Bouhadir, P. B. Shevlin, *Tetrahedron Lett.* **1996**, *37*, 4651.

119 T. Ohno, N. Martin, B. Knight, F. Wudl, T. Suzuki, H. Yu, *J. Org. Chem.* **1996**, *61*, 1306.

120 M. W. J. Beulen, J. A. Rivera, M. A. Herranz, B. Illescas, N. Martin, L. Echegoyen, *J. Org. Chem.* **2001**, *66*, 4393.

121 T. Ohno, K. Moriwaki, T. Miyata, *J. Org. Chem.* **2001**, *66*, 3397.

122 F. Giacalone, J. L. Segura, N. Martin, *J. Org. Chem.* **2002**, *67*, 3529.

123 D. Liu, J. Li, H. Pan, Y. Li, Z.-X. Guo, D. Zhu, *Synth. Methods* **2003**, *135–136*, 851.

124 N. Martin, L. Sanchez, D. M. Guldi, *Chem. Commun.* **2000**, 113.
125 A. G. Avent, P. R. Birkett, F. Paolucci, S. Roffia, R. Taylor, N. K. Wachter, *J. Chem. Soc., Perkin Trans. 2* **2000**, 1409.
126 M. Schwell, N. K. Wachter, J. H. Rice, J. P. Galaup, S. Leach, R. Taylor, R. V. Bensasson, *Chem. Phys. Lett.* **2001**, *339*, 29.
127 K.-Y. Kay, L. H. Kim, I. C. Oh, *Tetrahedron Lett.* **2000**, *41*, 1397.
128 M. D. Meijer, M. Rump, R. A. Gossage, J. H. T. B. Jastrzebski, G. van Koten, *Tetrahedron Lett.* **1998**, *39*, 6773.
129 L. Isaacs, A. Wehrsig, F. Diederich, *Helv. Chim. Acta* **1993**, *76*, 1231.
130 R. Pellicciari, D. Annibali, G. Costantino, M. Marinozzi, B. Natalini, *Synlett* **1997**, 1196.
131 A. Skiebe, A. Hirsch, *J. Chem. Soc., Chem. Commun.* **1994**, 335.
132 R. Pellicciari, B. Natalini, L. Amori, M. Marinozzi, R. Seraglia, *Synlett* **2000**, 1816.
133 H. J. Bestmann, C. Moll, C. Bingel, *Synlett* **1996**, 729.
134 K. Raghavachari, C. Sosa, *Chem. Phys. Lett.* **1993**, *209*, 223.
135 D. A. Dixon, N. Matsuzawa, T. Fukunaga, F. N. Tebbe, *J. Phys. Chem.* **1992**, *96*, 6107.
136 N. Matsuzawa, D. A. Dixon, T. Fukunaga, *J. Phys. Chem.* **1992**, *96*, 7594.
137 N. Matsuzawa, T. Fukunaga, D. A. Dixon, *J. Phys. Chem.* **1992**, *96*, 10747.
138 A. Hirsch, A. Soi, H. R. Karfunkel, *Angew. Chem.* **1992**, *104*, 808.
139 C. C. Henderson, C. M. Rohlfing, P. A. Cahill, *Chem. Phys. Lett.* **1993**, *213*, 383.
140 E. Vogel, *Pure Appl. Chem.* **1969**, *20*, 237.
141 E. Vogel, *Pure Appl. Chem.* **1982**, *54*, 1015.
142 E. Vogel, *Pure Appl. Chem.* **1993**, *65*, 143.
143 J. Osterodt, A. Zett, F. Vögtle, *Tetrahedron* **1996**, *52*, 4949.
144 M. Bühl, A. Hirsch, *Chem. Rev.* **2001**, *101*, 1153.
145 A. Pasquarello, M. Schlüter, R. C. Haddon, *Science* **1992**, *257*, 1660.
146 P. v. R. Schleyer, C. Märker, A. Dransfeld, H. Jiao, N. J. R. van Eikema Hommes, *J. Am. Chem. Soc.* **1996**, *118*, 6317.
147 S. R. Wilson, Y. Wu, *J. Chem. Soc., Chem. Commun.* **1993**, 784.
148 T. Ishida, K. Shinozuka, T. Nogami, M. Kubota, M. Ohashi, *Tetrahedron* **1996**, *52*, 5103.
149 A. B. Smith, III, R. M. Strongin, L. Brard, G. T. Furst, W. J. Romanow, K. G. Owens, R. J. Goldschmidt, R. C. King, *J. Am. Chem. Soc.* **1995**, *117*, 5492.
150 P. Ceroni, F. Conti, C. Corvaja, M. Maggini, F. Paolucci, S. Roffia, G. Scorrano, A. Toffoletti, *J. Phys. Chem. A* **2000**, *104*, 156.
151 M. Eiermann, F. Wudl, M. Prato, M. Maggini, *J. Am. Chem. Soc.* **1994**, *116*, 8364.
152 R. Gonzalez, J. C. Hummelen, F. Wudl, *J. Org. Chem.* **1995**, *60*, 2618.
153 A. Sastre, A. Gouloumis, P. Vazquez, T. Torres, V. Doan, B. J. Schwartz, F. Wudl, L. Echegoyen, J. Rivera, *Org. Lett.* **1999**, *1*, 1807.
154 R. V. Bensasson, E. Bienvenue, C. Fabre, J.-M. Janot, E. J. Land, S. Leach, V. Leboulaire, A. Rassat, S. Roux, P. Seta, *Chem. Eur. J.* **1998**, *4*, 270.
155 J. Osterodt, M. Nieger, P. M. Windscheif, F. Vögtle, *Chem. Ber.* **1993**, *126*, 2331.
156 F. Wudl, *Buckminsterfullerenes* **1993**, 317.
157 J. March, *Advanced Organic Chemistry*, 3rd Ed.; Wiley: New York, 1985; p. 1062.
158 R. Sijbesma, G. Srdanov, F. Wudl, J. A. Castoro, C. Wilkins, S. H. Friedman, D. L. DeCamp, G. L. Kenyon, *J. Am. Chem. Soc.* **1993**, *115*, 6510.
159 S. H. Friedman, D. L. DeCamp, R. P. Sijbesma, G. Srdanov, F. Wudl, G. L. Kenyon, *J. Am. Chem. Soc.* **1993**, *115*, 6506.
160 A. W. Jensen, S. R. Wilson, D. I. Schuster, *Bioorg. Med. Chem.* **1996**, *4*, 767.
161 T. Da Ros, M. Prato, *Chem. Commun.* **1999**, 663.
162 E. Nakamura, H. Tokuyama, S. Yamago, T. Shiraki, Y. Sugiura, *Bull. Chem. Soc. Jpn.* **1996**, *69*, 2143.
163 E. Nakamura, H. Isobe, *Acc. Chem. Res.* **2003**, *36*, 807.
164 Y.-Z. An, Y. Rubin, C. Schaller, S. W. McElvany, *J. Org. Chem.* **1994**, *59*, 2927.
165 A. Ouihia, L. René, J. Guilhem, C. Pascard, B. Badet, *J. Org. Chem.* **1993**, *58*, 1641.

166 M. Cases, M. Duran, J. Mestres, N. Martin, M. Sola, *J. Org. Chem.* **2001**, *66*, 433.
167 M. Prato, Q. C. Li, F. Wudl, V. Lucchini, *J. Am. Chem. Soc.* **1993**, *115*, 1148.
168 T. Grösser, M. Prato, V. Lucchini, A. Hirsch, F. Wudl, *Angew. Chem.* **1995**, *107*, 1462; *Angew. Chem. Int. Ed. Engl.* **1995**, *34*, 1343.
169 B. Nuber, F. Hampel, A. Hirsch, *Chem. Commun.* **1996**, 1799.
170 S. Gonzalez, N. Martin, A. Swartz, D. M. Guldi, *Org. Lett.* **2003**, *5*, 557.
171 J. C. Hummelen, C. Bellavia-Lund, F. Wudl, *Top. Curr. Chem.* **1999**, *199*, 93.
172 G. Schick, A. Hirsch, H. Mauser, T. Clark, *Chem. Eur. J.* **1996**, *2*, 935.
173 L. Ulmer, J. Mattay, *Eur. J. Org. Chem.* **2003**, 2933.
174 C. J. Hawker, P. M. Saville, J. W. White, *J. Org. Chem.* **1994**, *59*, 3503.
175 Y. N. Yamakoshi, T. Yagami, S. Sueyoshi, N. Miyata, *J. Org. Chem.* **1996**, *61*, 7236.
176 N. Jagerovic, J. Elguero, J.-L. Aubagnac, *Tetrahedron* **1996**, *52*, 6733.
177 P. S. Baran, R. R. Monaco, A. U. Khan, D. I. Schuster, S. R. Wilson, *J. Am. Chem. Soc.* **1997**, *119*, 8363.
178 A. Yashiro, Y. Nishida, M. Ohno, S. Eguchi, K. Kobayashi, *Tetrahedron Lett.* **1998**, *39*, 9031.
179 H. Okamura, K. Miyazono, M. Minoda, T. Miyamoto, *Macromol. Rapid Commun.* **1999**, *20*, 41.
180 C. Du, Y. Li, S. Wang, Z. Shi, S. Xiao, D. Zhu, *Synth. Methods* **2001**, *124*, 287.
181 H. Kato, A. Yashiro, A. Mizuno, Y. Nishida, K. Kobayashi, H. Shinohara, *Bioorg. Med. Chem. Lett.* **2001**, *11*, 2935.
182 I. P. Romanova, G. G. Yusupova, S. G. Fattakhov, A. A. Nafikova, V. I. Kovalenko, V. V. Yanilkin, V. E. Kataev, N. M. Azancheev, V. S. Reznik, O. G. Sinyashin, *Russ. Chem. Bull.* **2001**, *50*, 445.
183 O. G. Sinyashin, I. P. Romanova, G. G. Yusupova, A. A. Nafikova, V. I. Kovalenko, N. M. Azancheev, S. G. Fattakhov, V. S. Reznik, *Russ. Chem. Bull.* **2001**, *50*, 2162.
184 C. Du, B. Xu, Y. Li, C. Wang, S. Wang, Z. Shi, H. Fang, S. Xiao, D. Zhu, *New J. Chem.* **2001**, *25*, 1191.
185 S. Xiao, S. Wang, H. Fang, Y. Li, Z. Shi, C. Du, D. Zhu, *Macromol. Rapid Commun.* **2001**, *22*, 1313.
186 S. Wang, S. Xiao, Y. Li, Z. Shi, C. Du, H. Fang, D. Zhu, *Polymer* **2002**, *43*, 2049.
187 H. Fang, S. Wang, S. Xiao, Y. Li, Y. Liu, L. Fan, Z. Shi, C. Du, D. Zhu, *Synth. Methods* **2002**, *128*, 253.
188 H. Fang, C. Du, S. Qu, Y. Li, Y. Song, H. Li, H. Liu, D. Zhu, *Chem. Phys. Lett.* **2002**, *364*, 290.
189 S. Xiao, Y. Li, H. Fang, H. Li, H. Liu, Z. Shi, L. Jiang, D. Zhu, *Org. Lett.* **2002**, *4*, 3063.
190 L.-L. Shiu, K.-M. Chien, T.-Y. Liu, T.-I. Lin, G.-R. Her, T.-Y. Luh, *J. Chem. Soc., Chem. Commun.* **1995**, 1159.
191 M. Iglesias, B. Gomez-Lor, A. Santos, *J. Organomet. Chem.* **2000**, *599*, 8.
192 S. Xiao, Y. Li, H. Fang, H. Li, H. Liu, L. Jiang, D. Zhu, *Synth. Methods* **2003**, *135–136*, 839.
193 C. J. Hawker, K. L. Wooley, J. M. J. Frechet, *J. Chem. Soc., Chem. Commun.* **1994**, 925.
194 H. Fang, Z. Shi, Y. Li, S. Xiao, H. Li, H. Liu, D. Zhu, *Synth. Methods* **2003**, *135–136*, 843.
195 E. Nakamura, S. Yamago, *Acc. Chem. Res.* **2002**, *35*, 867.
196 S. Yamago, E. Nakamura, *J. Org. Chem.* **1990**, *55*, 5553.
197 S. Yamago, E. Nakamura, *J. Am. Chem. Soc.* **1989**, *111*, 7285.
198 S. Yamago, E. Nakamura, *Chem. Lett.* **1996**, 395.
199 S. Yamago, H. Tokuyama, E. Nakamura, M. Prato, F. Wudl, *J. Org. Chem.* **1993**, *58*, 4796.
200 H. Tokuyama, S. Yamago, E. Nakamura, T. Shiraki, Y. Sugiura, *J. Am. Chem. Soc.* **1993**, *115*, 7918.
201 J. W. Arbogast, C. S. Foote, *J. Am. Chem. Soc.* **1991**, *113*, 8886.
202 S. Yamago, H. Tokuyama, E. Nakamura, K. Kikuchi, S. Kananishi, K. Sueki, H. Nakahara, S. Enomoto, F. Ambe, *Chem. Biol.* **1995**, *2*, 385.
203 K. Irie, Y. Nakamura, H. Ohigashi, H. Tokuyama, S. Yamago, E. Nakamura, *Biosci., Biotechnol., Biochem.* **1996**, *60*, 1359.
204 M. Maggini, G. Scorrano, M. Prato, *J. Am. Chem. Soc.* **1993**, *115*, 9798.

205 N. Tagmatarchis, M. Prato, *Synlett* **2003**, 768.
206 M. Prato, M. Maggini, *Acc. Chem. Res.* **1998**, *31*, 519.
207 A. Bianco, T. Da Ros, M. Prato, C. Toniolo, *J. Pept. Sci.* **2001**, *7*, 208.
208 A. Bagno, S. Claeson, M. Maggini, M. L. Martini, M. Prato, G. Scorrano, *Chem. Eur. J.* **2002**, *8*, 1015.
209 F. D'Souza, M. E. Zandler, G. R. Deviprasad, W. Kutner, *J. Phys. Chem. A* **2000**, *104*, 6887.
210 K. G. Thomas, V. Biju, M. V. George, D. M. Guldi, P. V. Kamat, *J. Phys. Chem. A* **1998**, *102*, 5341.
211 A. Bianco, M. Maggini, G. Scorrano, C. Toniolo, G. Marconi, C. Villani, M. Prato, *J. Am. Chem. Soc.* **1996**, *118*, 4072.
212 A. Bianco, F. Gasparrini, M. Maggini, D. Misiti, A. Polese, M. Prato, G. Scorrano, C. Toniolo, C. Villani, *J. Am. Chem. Soc.* **1997**, *119*, 7550.
213 X. Zhang, M. Willems, C. S. Foote, *Tetrahedron Lett.* **1993**, *34*, 8187.
214 M. Iyoda, F. Sultana, M. Komatsu, *Chem. Lett.* **1995**, 1133.
215 M. Prato, M. Maggini, C. Giacometti, G. Scorrano, G. Sandona, G. Farnia, *Tetrahedron* **1996**, *52*, 5221.
216 L.-H. Shu, G.-W. Wang, S.-H. Wu, *Tetrahedron Lett.* **1995**, *36*, 3871.
217 S.-H. Wu, W.-Q. Sun, D.-W. Zhang, L.-H. Shu, H.-M. Wu, J.-F. Xu, X.-F. Lao, *J. Chem. Soc., Perkin Trans. 1* **1998**, 1733.
218 A. Komori, M. Kubota, T. Ishida, H. Niwa, T. Nogami, *Tetrahedron Lett.* **1996**, *37*, 4031.
219 G. E. Lawson, A. Kitaygorodskiy, B. Ma, C. E. Bunker, Y.-P. Sun, *J. Chem. Soc., Chem. Commun.* **1995**, 2225.
220 K.-F. Liou, C.-H. Cheng, *Chem. Commun.* **1996**, 1423.
221 S.-H. Wu, D.-W. Zhang, G.-W. Wang, L.-H. Shu, H.-M. Wu, J.-F. Xu, X.-F. Lao, *Synth. Commun.* **1997**, *27*, 2289.
222 D. Zhou, H. Tan, C. Luo, L. Gan, C. Huang, J. Pan, M. Lu, Y. Wu, *Tetrahedron Lett.* **1995**, *36*, 9169.
223 L. Gan, D. Zhou, C. Luo, H. Tan, C. Huang, M. Lue, J. Pan, Y. Wu, *J. Org. Chem.* **1996**, *61*, 1954.
224 P. De La Cruz, A. De La Hoz, L. M. Font, F. Langa, M. C. Perez-Rodriguez, *Tetrahedron Lett.* **1998**, *39*, 6053.
225 F. Langa, P. De la Cruz, A. De la Hoz, E. Espildora, F. P. Cossio, B. Lecea, *J. Org. Chem.* **2000**, *65*, 2499.
226 G.-W. Wang, T.-H. Zhang, E.-H. Hao, L.-J. Jiao, Y. Murata, K. Komatsu, *Tetrahedron* **2002**, *59*, 55.
227 Z. Shi, J. Jin, Y. Li, Z. Guo, S. Wang, L. Jiang, D. Zhu, *New J. Chem.* **2001**, *25*, 670.
228 Z. Ge, Y. Li, Z. Guo, Z. Shi, D. Zhu, *Tetrahedron Lett.* **1999**, *40*, 5759.
229 P. A. Liddell, D. Kuciauskas, J. P. Sumida, B. Nash, D. Nguyen, A. L. Moore, T. A. Moore, D. Gust, *J. Am. Chem. Soc.* **1997**, *119*, 1400.
230 R. Fong, II, D. I. Schuster, S. R. Wilson, *Org. Lett.* **1999**, *1*, 729.
231 D. Pantarotto, A. Bianco, F. Pellarini, A. Tossi, A. Giangaspero, I. Zelezetsky, J.-P. Briand, M. Prato, *J. Am. Chem. Soc.* **2002**, *124*, 12543.
232 D. M. Guldi, M. Maggini, E. Menna, G. Scorrano, P. Ceroni, M. Marcaccio, F. Paolucci, S. Roffia, *Chem. Eur. J.* **2001**, *7*, 1597.
233 A. Polese, S. Mondini, A. Bianco, C. Toniolo, G. Scorrano, D. M. Guldi, M. Maggini, *J. Am. Chem. Soc.* **1999**, *121*, 3446.
234 M. Maggini, A. Karlsson, L. Pasimeni, G. Scorrano, M. Prato, L. Valli, *Tetrahedron Lett.* **1994**, *35*, 2985.
235 F. D'Souza, M. E. Zandler, P. M. Smith, G. R. Deviprasad, K. Arkady, M. Fujitsuka, O. Ito, *J. Phys. Chem. A* **2002**, *106*, 649.
236 M. A. Herranz, B. Illescas, N. Martin, C. Luo, D. M. Guldi, *J. Org. Chem.* **2000**, *65*, 5728.
237 P. De la Cruz, A. De la Hoz, F. Langa, N. Martin, M. C. Perez, L. Sanchez, *Eur. J. Org. Chem.* **1999**, 3433.
238 G. R. Deviprasad, M. S. Rahman, F. D'Souza, *Chem. Commun.* **1999**, 849.
239 X. Tan, D. I. Schuster, S. R. Wilson, *Tetrahedron Lett.* **1998**, *39*, 4187.
240 T. Drovetskaya, C. A. Reed, P. Boyd, *Tetrahedron Lett.* **1995**, *36*, 7971.
241 D. Kuciauskas, S. Lin, G. R. Seely, A. L. Moore, T. A. Moore, D. Gust, T. Drovetskaya, C. A. Reed, P. D. W. Boyd, *J. Phys. Chem.* **1996**, *100*, 15926.
242 D. Kuciauskas, P. A. Liddell, S. Lin, T. E. Johnson, S. J. Weghorn,

J. S. Lindsey, A. L. Moore, T. A. Moore, D. Gust, *J. Am. Chem. Soc.* **1999**, *121*, 8604.
243 H. Imahori, H. Yamada, Y. Nishimura, I. Yamazaki, Y. Sakata, *J. Phys. Chem. B* **2000**, *104*, 2099.
244 H. Imahori, H. Yamada, S. Ozawa, Y. Sakata, K. Ushida, *Chem. Commun.* **1999**, 1165.
245 H. Imahori, H. Norieda, H. Yamada, Y. Nishimura, I. Yamazaki, Y. Sakata, S. Fukuzumi, *J. Am. Chem. Soc.* **2001**, *123*, 100.
246 K. Tamaki, H. Imahori, Y. Sakata, Y. Nishimura, I. Yamazaki, *Chem. Commun.* **1999**, 625.
247 H. Imahori, K. Tamaki, D. M. Guldi, C. Luo, M. Fujitsuka, O. Ito, Y. Sakata, S. Fukuzumi, *J. Am. Chem. Soc.* **2001**, *123*, 2607.
248 C. Luo, D. M. Guldi, H. Imahori, K. Tamaki, Y. Sakata, *J. Am. Chem. Soc.* **2000**, *122*, 6535.
249 D. M. Guldi, M. Maggini, G. Scorrano, M. Prato, *J. Am. Chem. Soc.* **1997**, *119*, 974.
250 V. Mamane, O. Riant, *Tetrahedron* **2001**, *57*, 2555.
251 T. Da Ros, M. Prato, M. Carano, P. Ceroni, F. Paolucci, S. Roffia, L. Valli, D. M. Guldi, *J. Organomet. Chem.* **2000**, *599*, 62.
252 N. Martin, L. Sanchez, C. Seoane, R. Andreu, J. Garin, J. Orduna, *Tetrahedron Lett.* **1996**, *37*, 5979.
253 K. B. Simonsen, V. V. Konovalov, T. A. Konovalova, T. Kawai, M. P. Cava, L. D. Kispert, R. M. Metzger, J. Becher, *J. Chem. Soc., Perkin Trans. 2* **1999**, 657.
254 M. A. Herranz, N. Martin, L. Sanchez, C. Seoane, D. M. Guldi, *J. Organomet. Chem.* **2000**, *599*, 2.
255 J. L. Segura, E. M. Priego, N. Martin, *Tetrahedron Lett.* **2000**, *41*, 7737.
256 N. Martin, L. Sanchez, M. A. Herranz, D. M. Guldi, *J. Phys. Chem. A* **2000**, *104*, 4648.
257 M. C. Diaz, M. A. Herranz, B. M. Illescas, N. Martin, N. Godbert, M. R. Bryce, C. Luo, A. Swartz, G. Anderson, D. M. Guldi, *J. Org. Chem.* **2003**, *68*, 7711.
258 A. Dondoni, A. Marra, *Tetrahedron Lett.* **2002**, *43*, 1649.
259 R. M. Williams, J. M. Zwier, J. W. Verhoeven, *J. Am. Chem. Soc.* **1995**, *117*, 4093.
260 A. I. de Lucas, N. Martin, L. Sanchez, C. Seoane, *Tetrahedron Lett.* **1996**, *37*, 9391.
261 K. G. Thomas, V. Biju, D. M. Guldi, P. V. Kamat, M. V. George, *J. Phys. Chem. A* **1999**, *103*, 10755.
262 B. Jing, D. Zhang, D. Zhu, *Tetrahedron Lett.* **2000**, *41*, 8559.
263 B. Illescas, N. Martin, C. Seoane, *Tetrahedron Lett.* **1997**, *38*, 2015.
264 J. Wang, C. D. Gutsche, *J. Org. Chem.* **2000**, *65*, 6273.
265 T. Gu, C. Bourgogne, J. F. Nierengarten, *Tetrahedron Lett.* **2001**, *42*, 7249.
266 T. Chuard, R. Deschenaux, *J. Mater. Chem.* **2002**, *12*, 1944.
267 Y. Rio, J. F. Nicoud, J. L. Rehspringer, J. F. Nierengarten, *Tetrahedron Lett.* **2000**, *41*, 10207.
268 M. D. L. de la Torre, G. L. Marcorin, G. Pirri, A. C. Tome, A. M. S. Silva, J. A. S. Cavaleiro, *Tetrahedron Lett.* **2002**, *43*, 1689.
269 D. I. Schuster, B. Nuber, S. A. Vail, S. MacMahon, C. Lin, S. R. Wilson, A. Khong, *Photochem. Photobiol. Sci.* **2003**, *2*, 315.
270 T. Gu, J. F. Nierengarten, *Tetrahedron Lett.* **2001**, *42*, 3175.
271 J.-F. Eckert, J.-F. Nicoud, J.-F. Nierengarten, S.-G. Liu, L. Echegoyen, F. Barigelletti, N. Armaroli, L. Ouali, V. Krasnikov, G. Hadziioannou, *J. Am. Chem. Soc.* **2000**, *122*, 7467.
272 N. Armaroli, F. Barigelletti, P. Ceroni, J.-F. Eckert, J.-F. Nierengarten, *Int. J. Photoenergy* **2001**, *3*, 33.
273 J. L. Segura, N. Martin, *Tetrahedron Lett.* **1999**, *40*, 3239.
274 P. A. Van Hal, J. Knol, B. M. W. Langeveld-Voss, S. C. J. Meskers, J. C. Hummelen, R. A. J. Janssen, *J. Phys. Chem. A* **2000**, *104*, 5974.
275 E. Peeters, P. A. van Hal, J. Knol, C. J. Brabec, N. S. Sariciftci, J. C. Hummelen, R. A. J. Janssen, *J. Phys. Chem. B* **2000**, *104*, 10174.
276 C. Martineau, P. Blanchard, D. Rondeau, J. Delaunay, J. Roncali, *Adv. Mater.* **2002**, *14*, 283.
277 D. Kuciauskas, P. A. Liddell, A. L. Moore, T. A. Moore, D. Gust, *J. Am. Chem. Soc.* **1998**, *120*, 10880.
278 B. M. Illescas, N. Martin, *J. Org. Chem.* **2000**, *65*, 5986.

279 G. Accorsi, N. Armaroli, J.-F. Eckert, J.-F. Nierengarten, *Tetrahedron Lett.* **2002**, *43*, 65.

280 B. Illescas, M. A. Martinez-Grau, M. L. Torres, J. Fernandez-Gadea, N. Martin, *Tetrahedron Lett.* **2002**, *43*, 4133.

281 T. Da Ros, M. Prato, F. Novello, M. Maggini, E. Banfi, *J. Org. Chem.* **1996**, *61*, 9070.

282 S. Bosi, L. Feruglio, D. Milic, M. Prato, *Eur. J. Org. Chem.* **2003**, 4741.

283 B. M. Illescas, R. Martinez-Alvarez, J. Fernandez-Gadea, N. Martin, *Tetrahedron* **2003**, *59*, 6569.

284 J. L. Segura, E. M. Priego, N. Martin, C. Luo, D. M. Guldi, *Org. Lett.* **2000**, *2*, 4021.

285 T. Mashino, D. Nishikawa, K. Takahashi, N. Usui, T. Yamori, M. Seki, T. Endo, M. Mochizuki, *Bioorg. Med. Chem. Lett.* **2003**, *13*, 4395.

286 D. M. Guldi, H. Hungerbühler, K.-D. Asmus, *J. Phys. Chem. A* **1997**, *101*, 1783.

287 T. Da Ros, M. Prato, M. Carano, P. Ceroni, F. Paolucci, S. Roffia, *J. Am. Chem. Soc.* **1998**, *120*, 11645.

288 M. Bergamin, T. Da Ros, G. Spalluto, M. Prato, A. Boutorine, *Chem. Commun.* **2001**, 17.

289 F. Conti, C. Corvaja, M. Maggini, G. Scorrano, P. Ceroni, F. Paolucci, S. Roffia, *Phys. Chem. Chem. Phys.* **2001**, *3*, 3518.

290 F. Arena, F. Bullo, F. Conti, C. Corvaja, M. Maggini, M. Prato, G. Scorrano, *J. Am. Chem. Soc.* **1997**, *119*, 789.

291 M. S. Meier, M. Poplawska, *J. Org. Chem.* **1993**, *58*, 4524.

292 H. Irngartinger, C. M. Köhler, U. Huber-Patz, W. Krätschmer, *Chem. Ber.* **1994**, *127*, 581.

293 M. S. Meier, M. Poplawska, *Tetrahedron* **1996**, *52*, 5043.

294 M. S. Meier, M. Poplawska, A. L. Compton, J. P. Shaw, J. P. Selegue, T. F. Guarr, *J. Am. Chem. Soc.* **1994**, *116*, 7044.

295 H. Irngartinger, A. Weber, T. Escher, *Eur. J. Org. Chem.* **2000**, 1647.

296 O. G. Sinyashin, I. P. Romanova, F. R. Sagitova, V. A. Pavlov, V. I. Kovalenko, Y. V. Badeev, N. M. Azancheev, A. V. Il'yasov, A. V. Chernova, I. I. Vandyukova, *Mendeleev Commun.* **1998**, 79.

297 L. V. Ermolaeva, V. E. Kataev, S. I. Strobykin, A. P. Timosheva, V. I. Kovalenko, I. P. Romanova, O. G. Sinyashin, *Russ. Chem. Bull.* **2002**, *51*, 593.

298 E. Kowalska, P. Byszewski, M. Poplawska, L. Gladczuk, J. Suwalski, R. Diduszko, J. Radomska, *J. Therm. Anal. Cal.* **2001**, *65*, 647.

299 A. Yashiro, Y. Nishida, K. Kobayashi, M. Ohno, *Synlett* **2000**, 361.

300 M. Ohno, A. Yashiro, S. Eguchi, *Synlett* **1996**, 815.

301 H. Irngartinger, P. W. Fettel, *Tetrahedron* **1999**, *55*, 10735.

302 H. Irngartinger, A. Weber, *Tetrahedron Lett.* **1996**, *37*, 4137.

303 P. De la Cruz, E. Espildora, J. J. Garcia, A. De la Hoz, F. Langa, N. Martin, L. Sanchez, *Tetrahedron Lett.* **1999**, *40*, 4889.

304 T. Da Ros, M. Prato, F. Novello, M. Maggini, M. De Amici, C. De Micheli, *Chem. Commun.* **1997**, 59.

305 B. Illescas, J. Rife, R. M. Ortuno, N. Martin, *J. Org. Chem.* **2000**, *65*, 6246.

306 S. Muthu, P. Maruthamuthu, R. Ragunathan, P. R. V. Rao, C. K. Mathews, *Tetrahedron Lett.* **1994**, *35*, 1763.

307 P. de la Cruz, A. Diaz-Ortiz, J. J. Garcia, M. J. Gomez-Escalonilla, A. de la Hoz, F. Langa, *Tetrahedron Lett.* **1999**, *40*, 1587.

308 F. Langa, P. de la Cruz, E. Espildora, A. de la Hoz, J. L. Bourdelande, L. Sanchez, N. Martin, *J. Org. Chem.* **2001**, *66*, 5033.

309 F. Langa, P. de la Cruz, J. L. Delgado, M. J. Gomez-Escalonilla, A. Gonzalez-Cortes, A. de la Hoz, V. Lopez-Arza, *New J. Chem.* **2002**, *26*, 76.

310 E. Espildora, J. L. Delgado, P. de la Cruz, A. de la Hoz, V. Lopez-Arza, F. Langa, *Tetrahedron* **2002**, *58*, 5821.

311 J. L. Delgado, P. de la Cruz, V. Lopez-Arza, F. Langa, *Tetrahedron Lett.* **2004**, *45*, 1651.

312 D. Brizzolara, J. T. Ahlemann, H. W. Roesky, K. Keller, *Bull. Soc. Chim. Fr.* **1993**, *130*, 745.

313 W. Duczek, F. Tittelbach, B. Costisella, H.-J. Niclas, *Tetrahedron* **1996**, *52*, 8733.

314 H. Ishida, M. Ohno, *Tetrahedron Lett.* **1999**, *40*, 1543.

315 H. Ishida, K. Itoh, M. Ohno, *Tetrahedron* **2001**, *57*, 1737.
316 N. Jagerovic, J. Elguero, J.-L. Aubagnac, *J. Chem. Soc., Perkin Trans. 1* **1996**, 499.
317 V. Nair, D. Sethumadhavan, K. C. Sheela, G. K. Eigendorf, *Tetrahedron Lett.* **1999**, *40*, 5087.
318 V. Nair, D. Sethumadhavan, K. C. Sheela, S. M. Nair, G. K. Eigendorf, *Tetrahedron* **2002**, *58*, 3009.
319 J. Averdung, E. Albrecht, J. Lauterwein, H. Luftmann, J. Mattay, H. Mohn, W. H. Müller, H. U. ter Meer, *Chem. Ber.* **1994**, *127*, 787.
320 J. Averdung, J. Mattay, *Tetrahedron* **1996**, *52*, 5407.
321 A. A. Ovcharenko, V. A. Chertkov, A. V. Karchava, M. A. Yurovskaya, *Tetrahedron Lett.* **1997**, *38*, 6933.
322 Y. Tsunenishi, H. Ishida, K. Itoh, M. Ohno, *Synlett* **2000**, 1318.
323 J. W. Arbogast, C. S. Foote, M. Kao, *J. Am. Chem. Soc.* **1992**, *114*, 2277.
324 J. W. Arbogast, A. P. Darmanyan, C. S. Foote, F. N. Diederich, R. L. Whetten, Y. Rubin, M. M. Alvarez, S. J. Anz, *J. Phys. Chem.* **1991**, *95*, 11.
325 R. J. Sension, A. Z. Szarka, G. R. Smith, R. M. Hochstrasser, *Chem. Phys. Lett.* **1991**, *185*, 179.
326 T. Akasaka, W. Ando, K. Kobayashi, S. Nagase, *J. Am. Chem. Soc.* **1993**, *115*, 10366.
327 T. Akasaka, E. Mitsuhida, W. Ando, K. Kobayashi, S. Nagase, *J. Am. Chem. Soc.* **1994**, *116*, 2627.
328 A. Han, T. Wakahara, Y. Maeda, Y. Niino, T. Akasaka, K. Yamamoto, M. Kako, Y. Nakadaira, K. Kobayashi, S. Nagase, *Chem. Lett.* **2001**, 974.
329 T. Wakahara, A. Han, Y. Niino, Y. Maeda, T. Akasaka, T. Suzuki, K. Yamamoto, M. Kako, Y. Nakadaira, K. Kobayashi, S. Nagase, *J. Mater. Chem.* **2002**, *12*, 2061.
330 M. Tsuda, T. Ishida, T. Nogami, S. Kurono, M. Ohashi, *Chem. Lett.* **1992**, 2333.
331 S. H. Hoke II, J. Molstad, D. Dilettato, M. J. Jay, D. Carlson, B. Kahr, R. G. Cooks, *J. Org. Chem.* **1992**, *57*, 5069.
332 Y. Nakamura, N. Takano, T. Nishimura, E. Yashima, M. Sato, T. Kudo, J. Nishimura, *Org. Lett.* **2001**, *3*, 1193.
333 A. D. Darwish, A. K. Abdul-Sada, G. J. Langley, H. W. Kroto, R. Taylor, D. R. M. Walton, *J. Chem. Soc., Chem. Commun.* **1994**, 2133.
334 A. D. Darwish, A. G. Avent, R. Raylor, D. R. M. Walton, *J. Chem. Soc., Perkin Trans. 2* **1996**, 2079.
335 M. S. Meier, G.-W. Wang, R. C. Haddon, C. P. Brock, M. A. Lloyd, J. P. Selegue, *J. Am. Chem. Soc.* **1998**, *120*, 2337.
336 S. R. Wilson, N. Kaprinidis, Y. Wu, D. I. Schuster, *J. Am. Chem. Soc.* **1993**, *115*, 8495.
337 D. I. Schuster, J. Cao, N. Kaprinidis, Y. Wu, A. W. Jensen, Q. Lu, S. R. Wilson, *J. Am. Chem. Soc.* **1996**, *118*, 5639.
338 C. J. Welch, W. H. Pirkle, *J. Chromatogr.* **1992**, *609*, 89.
339 S. R. Wilson, Y. Wu, N. A. Kaprinidis, D. I. Schuster, C. J. Welch, *J. Org. Chem.* **1993**, *58*, 6548.
340 S. R. Wilson, Q. Lu, J. Cao, Y. Wu, C. J. Welch, D. I. Schuster, *Tetrahedron* **1996**, *52*, 5131.
341 A. W. Jensen, A. Khong, M. Saunders, S. R. Wilson, D. I. Schuster, *J. Am. Chem. Soc.* **1997**, *119*, 7303.
342 G. Vassilikogiannakis, M. Orfanopoulos, *J. Org. Chem.* **1999**, *64*, 3392.
343 X. Zhang, A. Romero, C. S. Foote, *J. Am. Chem. Soc.* **1993**, *115*, 11024.
344 R. Bernstein, C. S. Foote, *Tetrahedron Lett.* **1998**, *39*, 7051.
345 X. Zhang, A. Fan, C. S. Foote, *J. Org. Chem.* **1996**, *61*, 5456.
346 X. Zhang, C. S. Foote, *J. Am. Chem. Soc.* **1995**, *117*, 4271.
347 G. Vassilikogiannakis, N. Chronakis, M. Orfanopoulos, *J. Am. Chem. Soc.* **1998**, *120*, 9911.
348 G. Vassilikogiannakis, M. Orfanopoulos, *Tetrahedron Lett.* **1997**, *38*, 4323.
349 G. Vassilikogiannakis, M. Orfanopoulos, *J. Am. Chem. Soc.* **1997**, *119*, 7394.
350 G. Vassilikogiannakis, M. Hatzimarinaki, M. Orfanopoulos, *J. Org. Chem.* **2000**, *65*, 8180.
351 M. Hatzimarinaki, G. Vassilikogiannakis, M. Orfanopoulos, *Tetrahedron Lett.* **2000**, *41*, 4667.
352 B. Bildstein, M. Schweiger, H. Angleitner, H. Kopacka, K. Wurst, K.-H. Ongania, M. Fontani, P. Zanello, *Organometallics* **1999**, *18*, 4286.

353 B. Bildstein, *Coord. Chem. Rev.* **2000**, *206–207*, 369.
354 L.-H. Shu, W.-Q. Sun, D.-W. Zhang, S.-H. Wu, H.-M. Wu, J.-F. Xu, X.-F. Lao, *Chem. Commun.* **1997**, 79.
355 V. Nair, D. Sethumadhavan, S. M. Nair, P. Shanmugam, P. M. Treesa, G. K. Eigendorf, *Synthesis* **2002**, 1655.
356 S. Matsui, K. Kinbara, K. Saigo, *Tetrahedron Lett.* **1999**, *40*, 899.
357 M. Prato, M. Maggini, G. Scorrano, V. Lucchini, *J. Org. Chem.* **1993**, *58*, 3613.
358 T. L. Gilchrist, R. C. Storr, *Organic Reactions and Orbital Symmetry*, 2nd Ed.; Cambridge University Press: Cambridge, **1979**.
359 A. M. Rao, P. Zhou, K. A. Wang, G. T. Hager, J. M. Holden, Y. Wang, W. T. Lee, X. X. Bi, P. C. Elkund, D. S. Cornett, M. A. Duncan, I. J. Amster, *Science* **1993**, *259*, 955.
360 D. S. Cornett, I. J. Amster, M. A. Duncan, A. M. Rao, P. C. Eklund, *J. Phys. Chem.* **1993**, *97*, 5036.
361 P. Zhou, Z.-H. Dong, A. M. Rao, P. C. Eklund, *Chem. Phys. Lett.* **1993**, *211*, 337.
362 Y. Wang, J. M. Holden, Z.-H. Dong, X.-X. Bi, P. C. Eklund, *Chem. Phys. Lett.* **1993**, *211*, 341.
363 A. Ito, T. Morikawa, T. Takahashi, *Chem. Phys. Lett.* **1993**, *211*, 333.
364 S. Pekker, E. Kovats, K. Kamaras, T. Pusztai, G. Oszlanyi, *Synth. Methods* **2003**, *133–134*, 685.
365 Y. Iwasa, T. Arima, R. M. Fleming, T. Siegrist, O. Zhou, R. C. Haddon, L. J. Rothberg, K. B. Lyons, H. L. Carter, Jr, A.F. Hebard, R. Tycko, G. Dabbagh, J.J Krajewski, G.A. Thomas, T. Yagi, *Science* **1994**, *264*, 1570.
366 B. Sundqvist, *Fullerene: Chem., Phys. Technol.* **2000**, 611.
367 B. Sundqvist, *Phys. Stat. Sol. B* **2001**, *223*, 469.
368 J. L. Segura, N. Martin, *Chem. Soc. Rev.* **2000**, *29*, 13.
369 Y. Iwasa, K. Tanoue, T. Mitani, A. Izuoka, T. Sugawara, T. Yagi, *Chem. Commun.* **1998**, 1411.
370 S. Lebedkin, S. Ballenweg, J. Gross, R. Taylor, W. Krätschmer, *Tetrahedron Lett.* **1995**, *36*, 4971.
371 A. B. Smith III, H. Tokuyama, R. M. Strongin, G. T. Furst, W. J. Romanow, B. T. Chait, U. A. Mirza, I. Haller, *J. Am. Chem. Soc.* **1995**, *117*, 9359.
372 A. Gromov, S. Lebedkin, S. Ballenweg, A. G. Avent, R. Taylor, W. Krätschmer, *Chem. Commun.* **1997**, 209.
373 S. Giesa, J. H. Gross, R. Gleiter, A. Gromov, W. Krätschmer, W. E. Hull, S. Lebedkin, *Chem. Commun.* **1999**, 465.
374 N. Dragoe, H. Shimotani, M. Hayashi, K. Saigo, A. De Bettencourt-Dias, A. L. Balch, Y. Miyake, Y. Achiba, K. Kitazawa, *J. Org. Chem.* **2000**, *65*, 3269.
375 J. Knol, J. C. Hummelen, *J. Am. Chem. Soc.* **2000**, *122*, 3226.
376 Y. Murata, N. Kato, K. Komatsu, *J. Org. Chem.* **2001**, *66*, 7235.
377 A. Vasella, P. Uhlmann, C. A. A. Waldraff, F. Diederich, C. Thilgen, *Angew. Chem.* **1992**, *104*, 1383; *Angew. Chem. Int. Ed. Engl.* **1992**, *31*, 1388.
378 F. Diederich, L. Isaacs, D. Philp, *Chem. Soc. Rev.* **1994**, *23*, 243.
379 K. Komatsu, A. Kagayama, Y. Murata, N. Sugita, K. Kobayashi, S. Nagase, T. S. M. Wan, *Chem. Lett.* **1993**, 2163.
380 T. Akasaka, M. T. H. Liu, Y. Niino, Y. Maeda, T. Wakahara, M. Okamura, K. Kobayashi, S. Nagase, *J. Am. Chem. Soc.* **2000**, *122*, 7134.
381 T. Wakahara, Y. Niino, T. Kato, Y. Maeda, T. Akasaka, M. T. H. Liu, K. Kobayashi, S. Nagase, *J. Am. Chem. Soc.* **2002**, *124*, 9465.
382 W. W. Win, M. Kao, M. Eiermann, J. J. McNamara, F. Wudl, D. L. Pole, K. Kassam, J. Warkentin, *J. Org. Chem.* **1994**, *59*, 5871.
383 R. Gonzalez, F. Wudl, D. L. Pole, P. K. Sharma, J. Warkentin, *J. Org. Chem.* **1996**, *61*, 5837.
384 P. K. Sharma, M.-l. Dawid, J. Warkentin, R. M. Vestal, F. Wudl, *J. Org. Chem.* **2001**, *66*, 7496.
385 J. Osterodt, F. Vögtle, *Chem. Commun.* **1996**, 547.
386 A. F. Kiely, R. C. Haddon, M. S. Meier, J. P. Selegue, C. P. Brock, B. O. Patrick, G.-W. Wang, Y. Chen, *J. Am. Chem. Soc.* **1999**, *121*, 7971.
387 M. Tsuda, T. Ishida, T. Nogami, S. Kurono, M. Ohashi, *Tetrahedron Lett.* **1993**, *34*, 6911.

388 H. Tokuyama, M. Nakamura, E. Nakamura, *Tetrahedron Lett.* **1993**, *34*, 7429.

389 N. Dragoe, S. Tanibayashi, K. Nakahara, S. Nakao, H. Shimotani, L. Xiao, K. Kitazawa, Y. Achiba, K. Kikuchi, K. Nojima, *Chem. Commun.* **1999**, 85.

390 N. Dragoe, H. Shimotani, J. Wang, M. Iwaya, A. de Bettencourt-Dias, A. L. Balch, K. Kitazawa, *J. Am. Chem. Soc.* **2001**, *123*, 1294.

391 T. S. Fabre, W. D. Treleaven, T. D. McCarley, C. L. Newton, R. M. Landry, M. C. Saraiva, R. M. Strongin, *J. Org. Chem.* **1998**, *63*, 3522.

392 C. A. Merlic, H. D. Bendorf, *Tetrahedron Lett.* **1994**, *35*, 9529.

393 B. Z. Tang, H. Xu, J. W. Y. Lam, P. P. S. Lee, K. Xu, Q. Sun, K. K. L. Cheuk, *Chem. Mater.* **2000**, *12*, 1446.

394 M. Cases, M. Duran, M. Sola, *J. Mol. Mod.* **2000**, *6*, 205.

395 A. B. Smith III, H. Tokuyama, *Tetrahedron* **1996**, *52*, 5257.

396 T. Ishida, K. Tanaka, T. Nogami, *Chem. Lett.* **1994**, 561.

397 G. Schick, T. Grösser, A. Hirsch, *J. Chem. Soc., Chem. Commun.* **1995**, 2289.

398 J. Averdung, H. Luftmann, J. Mattay, K.-U. Calus, W. Abraham, *Tetrahedron Lett.* **1995**, *36*, 2957.

399 J. Averdung, C. Wolff, J. Mattay, *Tetrahedron Lett.* **1996**, *37*, 4683.

400 F. Djojo, A. Herzog, I. Lamparth, F. Hampel, A. Hirsch, *Chem. Eur. J.* **1996**, *2*, 1537.

401 J. Averdung, J. Mattay, D. Jacobi, W. Abraham, *Tetrahedron* **1995**, *51*, 2543.

402 M. Yan, S. X. Cai, J. F. W. Keana, *J. Org. Chem.* **1994**, *59*, 5951.

403 M. R. Banks, J. I. G. Cadogan, I. Gosney, P. K. G. Hodgson, P. R. R. Langridge-Smith, J. R. A. Millar, A. T. Taylor, *Tetrahedron Lett.* **1994**, *35*, 9067.

404 S.-y. Kuwashima, M. Kubota, K. Kushida, T. Ishida, M. Ohashi, T. Nogami, *Tetrahedron Lett.* **1994**, *35*, 4371.

405 T. Akasaka, W. Ando, K. Kobayashi, S. Nagase, *J. Am. Chem. Soc.* **1993**, *115*, 1605.

406 T. Akasaka, E. Mitsuhida, W. Ando, K. Kobayashi, S. Nagase, *J. Chem. Soc., Chem. Commun.* **1995**, 1529.

407 T. Wakahara, Y. Maeda, M. Kako, T. Akasaka, K. Kobayashi, S. Nagase, *J. Organomet. Chem.* **2003**, *685*, 177.

5 Hydrogenation

5.1 Introduction

From the beginning of the "fullerene era" an obvious challenge to chemists was to hydrogenate these fascinating new unsaturated molecules to "fulleranes". Indeed, hydrogenations of C_{60} were among the first reported attempts to chemically modify fullerenes. However, this task proved very difficult and it took some time to achieve structural characterization of some hydrofullerenes.

Hydrides of fullerenes with a low degree of hydrogenation, such as $C_{60}H_2$, $C_{60}H_4$, $C_{60}H_6$ or $C_{70}H_2$, have been specifically synthesized and completely characterized. Also, the structure of some isomers of the polyhydrofullerenes $C_{60}H_{18}$ and $C_{60}H_{36}$ could be proven.

Polyhydrofullerenes were observed up to a degree of hydrogenation of about 44 H-atoms per C_{60} or even higher. But at this high hydrogen-content, bond breaking in the C_{60}-cage usually takes place and the fullerene loses its structural integrity. Completely hydrogenated $C_{60}H_{60}$ has not yet been synthesized because polyhydrofullerenes are instable. Why are the polyhydrofullerenes, even those not completely saturated such as $C_{60}H_{36}$, unstable? This question has been answered, at least partly, on the basis of theoretical investigations, systematically carried out on several $C_{60}H_{2n}$s. An increasing degree of hydrogenation introduces considerable strain into the framework of the spheres. This instability is one of the characteristic and unique properties of fullerene chemistry.

Hydrogenated fullerenes may be interesting for applications such as hydrogen storage; the storage capacity for example of $C_{60}H_{36}$ is 4.8%. Another possible application is the use of $C_{60}H_n$ as an additive for lithium ion cells. The lifetime of these cells could significantly be prolonged by addition of a few % of hydrofullerenes.

Fullerenes: Chemistry and Reactions. Andreas Hirsch and Michael Brettreich

ISBN: 3-527-30820-2

5.2
Oligohydrofullerenes $C_{60}H_n$ and $C_{70}H_n$ ($n = 2$–12)

5.2.1
Hydrogenation via Hydroboration and Hydrozirconation

One approach to the synthesis of hydrofullerenes is the defined stoichiometrically controlled addition to one or a few reactive [6,6] double bonds of C_{60} and C_{70}. Presuming that the addition patterns are similar to those of nucleophilic additions and cycloadditions (see Chapters 3 and 4), a limited number of energetically favored isomers can be expected in the reaction mixtures. This would allow the chromatographic isolation and also the systematic characterization of isomerically pure hydrofullerenes with a low degree of hydrogenation. Especially with the simplest hydrocarbon derivatives $C_{60}H_2$ and $C_{70}H_2$, the number of thermodynamically and kinetically favored isomers is very low. The high regioselectivity of nucleophilic additions to C_{60} and C_{70} in which two-step mechanisms are involved imply that similar two-step procedures, for example, a hydrometalation followed by hydrolysis, should be promising methods to synthesize oligohydrofullerenes. This has been demonstrated by the synthesis of $C_{60}H_2$ via hydroboration and hydrozirconation (Scheme 5.1) [1–3]. This simplest of the hydrocarbon derivatives of C_{60} has also been prepared by a defined zinc/acid reduction [4], by reduction with Zn/Cu [5] or hydrazine [3, 6] and with various other reduction methods [7]. In each case only a single regioisomer of dihydro[60]fullerene ($C_{60}H_2$) is formed, which can be isolated from the reaction mixture by HPLC. Analysis of the ^{1}H NMR spectra of $C_{60}H_2$ and the deuterated $C_{60}HD$ prepared by hydrolysis of the reaction intermediate $C_{60}HBH_2$ with D_2O demonstrates complete accordance with the results of the hydroalkylation reactions (see Chapter 3) [8, 9]. This regioisomer is the 1,2-addition product. The 1,4-addition product has not been observed neither by reduction via hydroboration nor with other reductants [7]. As with the 1-organo-1,2-dihydrofullerenes, the fullerenyl protons in the parent 1,2-dihydrofullerene $C_{60}H_2$ resonate at low field, ca. $\delta = 5.9$ ppm in toluene-d_8 [1]. The acidity of this fullerenyl proton has been estimated by voltammetric techniques in dimethyl sulfoxide as solvent. The first and second dissociation constants are 4.7 and 16, respectively [10]. Another fingerprint, the UV/Vis spectra [1, 2] of the chestnut-brown solutions of $C_{60}H_2$, especially the sharp band at 430 nm, independently reveal a characteristic property of a 1,2-monoadduct of C_{60} [8].

The electrochemical behavior of C_{60} determined by cyclic voltammetry [4, 10] is very similar to that of $C_{60}HR$ (R = alkyl) [11]. The reduction waves are slightly shifted to more negative potentials than for C_{60}. Interestingly, the reduction of $C_{60}H_2$ at slower scan rates is accompanied by a new set of reoxidation waves at potentials that closely match the anodic peak potentials of the corresponding parent C_{60} species. The formation of C_{60} from $C_{60}H_2$ during cyclic voltammetry was confirmed by HPLC analysis. The experiments indicate the instability of anionic hydrofullerenes. The conversion into C_{60} is solvent and temperature dependent [12].

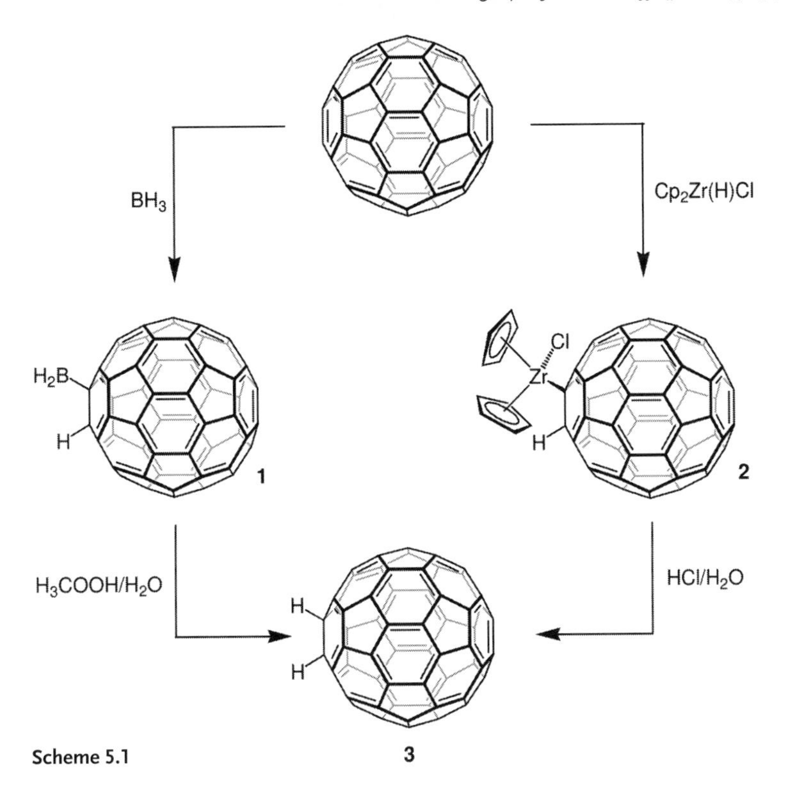

Scheme 5.1

The very soluble intermediate $(\eta^5\text{-}C_5H_5)_2ZrClC_{60}H$ (Scheme 5.1) is accessible to further hydrozirconations [2]. Therefore, by using a two-fold excess of $(\eta^5\text{-}C_5H_5)_2Zr(H)Cl$, the higher adducts $[(\eta^5\text{-}C_5H_5)_2ZrCl]_nC_{60}H_n$ (n = 2,3) are formed as by-products. Their hydrolysis leads to $C_{60}H_4$ and $C_{60}H_6$ as a mixture of different regioisomers. The defined regioisomer 1,2,3,4-$C_{60}H_4$ of tetrahydro[60]fullerene was obtained in 10–15% yield as the major product by the hydroboration and subsequent hydrolysis of $C_{60}H_2$ [13]. This most polar cis-1 regioisomer makes up approximately 50% of the total amount of $C_{60}H_4$. Attempts to form $C_{60}H_6$ in a similar procedure by hydroboration of $C_{60}H_4$ failed [14]. A mixture of $C_{60}H_4$ isomers partially isomerizes on a Pt-contaminated Buckyclutcher I column to 1,2,3,4-$C_{60}H_4$, which also indicates that this regioisomer is the major kinetic and apparent thermodynamic product.

The 1,2,3,4-isomer is also the major product if other reduction reagents [7] such as anhydrous hydrazine [6], diimide [3] or palladium hydride wrapped in gold foil [15] are used. Contrary to this result, reduction with wet Zn/Cu couple (Section 5.2.2) does not lead to the *cis*-1-adduct. Instead the *e*-isomer and the *trans*-3-isomer are formed as major products [5].

Upon hydroboration of C_{70} followed by hydrolysis of the presumed intermediates $C_{70}HBH_2$, two isomers of $C_{70}H_2$ are obtained. These are the 1,2-dihydro[70]fullerene (**4**) as the major and the 5,6-dihydro[70]fullerene (**5**) as the minor reaction product (Figure 5.1) [16, 17].

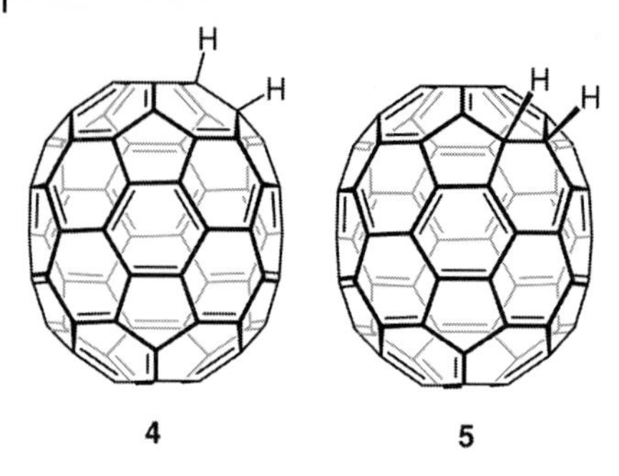

Figure 5.1 1,2-$C_{70}H_2$ and 5,6-$C_{70}H_2$.

In hydroalkylation or -arylation reactions of C_{70}, only the thermodynamically most stable 1,2-isomers of alkylated or arylated $C_{70}HR$ were isolated from the reaction mixture [8]. Both isomers **4** (1,2-adduct) and **5** (5,6-adduct) of $C_{70}H_2$ are kinetic products of the two-step hydrogenation reaction. Isomerization was not observed in either pure solution at room temperature for several weeks or at 100 °C for 1 h [16, 17]. The 5,6-product is less stable toward decomposition than the 1,2-isomer, but both isomers are indefinitely stable in toluene–hexane solution at −20 °C. A conversion of pure isomer **4** into a mixture of **4** and **5** (isomerization), and C_{70} (decomposition) over platinum on silica catalyst, was observed at room temperature. The energy difference ΔG_{295} of these isomers was experimentally determined to be 1.4 ± 0.2 kcal mol^{-1} with 1,2-$C_{70}H_2$ being lower in energy.

5.2.2
Reduction with Reducing Metals (Zn/Cu)

Hydrogenation of C_{60} or C_{70} has been successfully carried out with reducing metals such as Mg, Ti, Al or Zn in the presence of a proton source [5]. Treatment of fullerenes with wet Zn/Cu couple turned out to be the most efficient and selective method [5, 18–21]. Reductions with Mg, Ti or Al are inefficient and the resulting hydrofullerene mixtures are very difficult, if not impossible, to separate [5, 22]. Reduction with a Zn/Cu couple is usually performed in toluene with a small amount of water as the proton donor. In this reaction water was the most suitable proton source. The hydrofullerenes $C_{60}H_2$, $C_{60}H_4$ and $C_{60}H_6$ can be synthesized with this method in good yields. The product distribution and the number of formed isomers can be controlled via reaction time, efficiency of stirring and the ratio of metal to C_{60} [5]. The smallest hydrofullerene $C_{60}H_2$, for example, can be obtained with 1 h reaction time in 66% yield after purification with GPC [5, 18]. After 2 h reaction time the major product is $C_{60}H_4$ and after 4 h it is $C_{60}H_6$.

Three isomers of $C_{60}H_4$, namely the *e*-isomer 1,2,18,36-$C_{60}H_4$ (**8**), the *trans*-3-isomer 1,2,33,50-$C_{60}H_4$ (**7**) and an unidentified isomer, are formed as major products [5] in a ratio of 1 : 1 : 0.3 (Figure 5.2). After 4 h reaction time a further reduction of the two major $C_{60}H_4$ isomers to $C_{60}H_6$ obviously took place, forming a mixture of two $C_{60}H_6$ isomers in the ratio 6 : 1 with some $C_{60}H_6O$ side products [5, 20]. $C_{60}H_6$

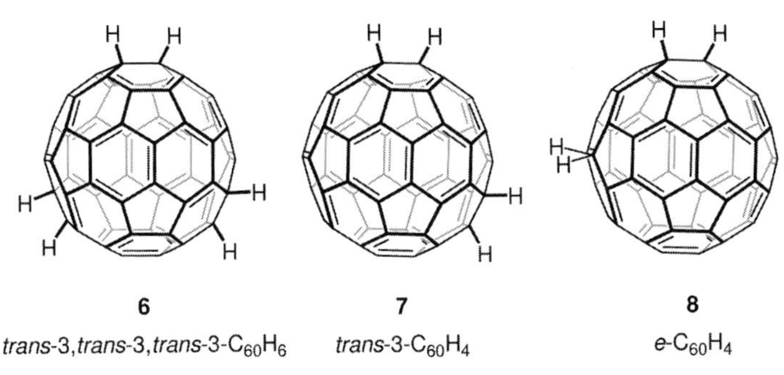

Figure 5.2 Major isomers of $C_{60}H_4$ and $C_{60}H_6$ formed in the reduction of C_{60} with Zn/Cu couple.

can be obtained in about 35% yield. Through NMR spectroscopy the major isomer was proven to be the *trans*-3,*trans*-3,*trans*-3-isomer 1,2,33,41,42,50-$C_{60}H_6$ (**6**), whose precursor must be *trans*-3-$C_{60}H_4$. The structure of the other isomer could not be proven but a 1,2,18,22,23,36 addition pattern was suggested. Due to the high symmetry (D_3) compound **6** (*trans*-3,*trans*-3,*trans*-3) shows only 10 signals in the ^{13}C NMR spectrum and one singlet in the ^1H NMR spectrum at 5.18 ppm. In the ^{13}C NMR spectrum the six sp^3-carbons show only one signal at 52.3 ppm, and the sp^2-carbons show nine resonances with equal intensity [20]. All signals were completely assigned with the help of 2D INADEQUATE NMR spectroscopy by using the ^{13}C-enriched form of the $C_{60}H_6$-isomer [23].

A powerful tool to examine the product distribution of hydrogenation reactions is ^3He NMR spectroscopy of endohedral He-fullerene complexes [24]. As each different endohedral fullerene derivative gives a single, distinct and sharp peak in the ^3He NMR spectrum, the number of generated isomers is simply correlated with the number of signals. With this method a couple of hydrogenation reactions have been examined [7, 24, 25]. Hydrogenation of $^3He@C_{60}$ with the Zn/Cu couple under conditions that should lead to the hexakisadduct $^3He@C_{60}H_6$ was carried out [25]. Instead of two different signals for the expected two major isomers, which should be formed as described above, one more signal was found. Assuming each derivative should give a single signal, a new isomer of $C_{60}H_6$ was found. These three isomers show resonances at −16.35 ppm for the minor isomer, −15.31 ppm for the trans-3,trans-3,trans-3-isomer and −14.24 ppm for the new, unidentified isomer.

Applying the Zn/Cu reduction to C_{70} the reduction proceeds to a greater extent than the reaction with C_{60} did [5, 7, 21, 25, 26]. Some distinct isomers of $C_{70}H_n$ with $n = 2$, 4, 6, 8 and 10 can be isolated. $C_{70}H_{12}$ is formed, but only in small amounts and was not yet separated. HPLC, ^1H, ^{13}C and ^3He NMR spectroscopy together with calculations helped to resolve the structure of the obtained hydro[70]fullerenes. The reduction proceeds in different possible pathways [21]. One of these reduction manifolds leads – besides some minor isomers – to the two major isomers of $C_{70}H_n$ with $n = 2$ (**9**) and $n = 4$ (**10–12**), where the hydrogens are – as expected – located at the poles of C_{70} (Table 5.1). The other manifold leads to the adduct $C_{70}H_8$ (**13**) with a completely different addition pattern, where the hydrogens are added to the equatorial belt of C_{70}. Further reduction of $C_{70}H_8$ leads to $C_{70}H_{10}$ (**14**) with the two additional hydrogens also in equatorial positions. The unique structure of this oligoadduct was also confirmed by ^{3}He NMR spectroscopy [25]. The addition near the equator leads to a different magnetic environment inside the cage compared with the "pole"-adducts. This is reflected in the spectrum by observation of the most down-field shifted ^{3}He NMR signals for $C_{70}H_8$ and $C_{70}H_{10}$ among the neutral C_{70}-derivatives (Table 5.1).

Table 5.1 Major isomers of $C_{70}H_n$.

	Major isomers	*Proven structures*	*^{3}He shift (ppm)*
$C_{70}H_2$	1,2	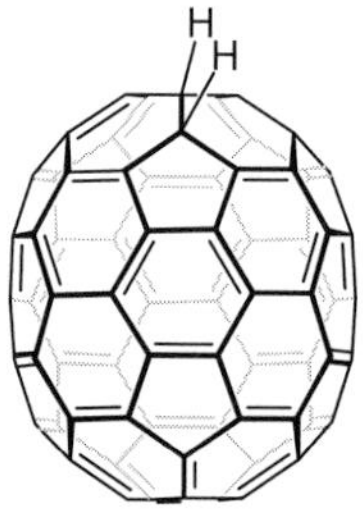 **9**	−27.18
$C_{70}H_4$	1,2,56,57; 1,2,41,58; 1,2,67,68	**10**, 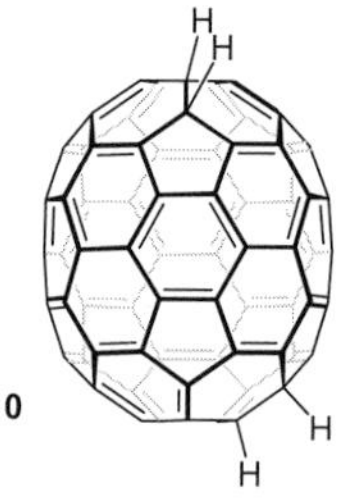 **11**, 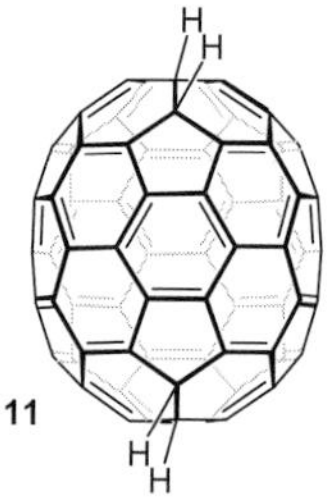 **12** 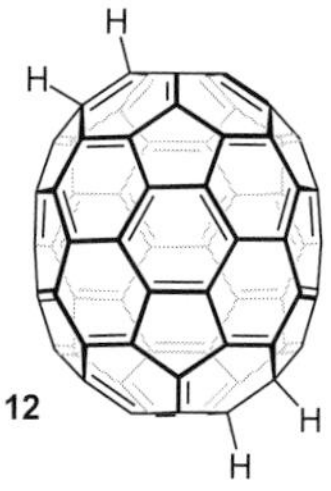	−25.33 −24.77 −23.76
$C_{70}H_8$	7,19,23,27, 33,37,44,53	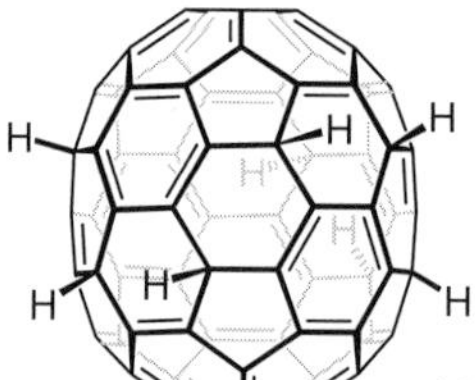 **13**	−17.84
$C_{70}H_{10}$	7,8,19,26, 33,37,45,49, 53,63	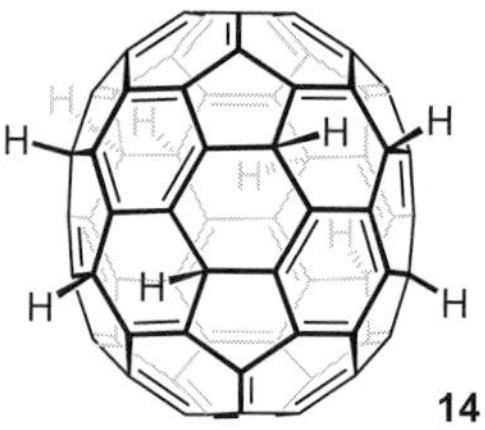 **14**	−17.17

5.2.3
Hydrogenation with Hydrazine and with Organic Reducing Agents

Photoinduced electron transfer [22] from reductants such as 1-benzyl-1,4-dihydronicontinamide [27], the Hantzsch-ester [22] (diethyl-2,6-dimethyl-1,4-dihydropyridine-3,5-dicarboxylate) or 10-methyl-9,10-dihydroacridine [27, 28] to the fullerene and successive proton transfer leads selectively to 1,2-dihydro[60]fullerene. These reductions usually proceed under mild conditions.

The major products of the hydrogenation of C_{60} with diimide are $C_{60}H_2$ and different isomers of $C_{60}H_4$ [3, 6, 24]. In smaller amounts, $C_{60}H_6$ and $C_{60}H_8$ are formed and with a large excess of the reductant $C_{60}H_{18}$ and $C_{60}H_{36}$ can be produced [6]. Diimide was formed in situ via reaction with anhydrous hydrazine in benzene [6], hydrazine hydrate with copper(II)sulfate [3], or by thermolysis of toluenesulfonehydrazide [24, 29]. The assumption that these reactions proceed via a similar mechanism, i.e. via the diimide, is supported by the similar yields and the relative ratio of the products. This reaction has been primarily employed to synthesize $C_{60}H_4$ and to examine the relative ratio of the eight different possible bisadduct isomers [3]. Based on the preference of the 1,2-addition pattern, only the eight isomers, which are also known from the cycloaddition reaction with C_{60} (Chapter 10), can be observed. At least seven out of eight isomers can be detected via ^{13}C NMR spectroscopy and by analyzing the number and pattern of the signals in the ^{1}H NMR spectrum. At present, it is not possible to separate and assign all isomers. Through HPLC only the most abundant isomer can be separated [6]. The structure of this *cis*-1-isomer – 1,2,3,4-tetrahydro[60]fullerene – was positively assigned from its ^{1}H NMR spectrum. 1,2,3,4-Tetrahydrofullerene is the only isomer that shows a AA′BB′-type spectrum. The other isomers should show AB-type quartets (*cis*-2-, *cis*-3-, *trans*-2-, *trans*-3- and *trans*-4-isomer), an AB-quartet with a singlet for the *e*-isomer or a singlet for the *trans*-1-isomer. ^{3}He NMR spectroscopy is not capable of assigning the structure of specific isomers, but the number of obtained compounds is accessible [24]. The ^{3}He NMR spectrum of a mixture of tetrahydro[60]fullerenes obtained by diimide reduction shows six signals, the two missing signals were assumed to be too small to be seen.

The products of the copper-supported hydrogenation of C_{70} with hydrazine were not separated but analyzed by ^{1}H NMR spectroscopy of the reaction mixture. Beside the 1,2-dihydro[70]fullerene and the 5,6-dihydro[70]fullerene, six tetrahydro[70]-fullerenes were observed in the ^{1}H NMR spectrum [3].

5.2.4
Theoretical Investigations

In principle, 23 regioisomers of the dihydrofullerene $C_{60}H_2$ are possible. The formal addition of an A-B molecule, for example H_2, to the externally C_{60} sphere could proceed in three ways [30]: (1) Addition to one double bond of the low-energy Kekulé structure (Figure 5.3), which would leave all the other bonds unchanged ([6,6] double bonds and [5,6] single bonds); (2) conjugate addition of two atoms, which requires the formal introduction of [5,6] double bonds to retain a closed-shell Kekulé structure (Figure 5.3); and (3) hydrogenolysis that accompanies cleavage of a bond in C_{60}. Various calculations of different isomers of $C_{60}H_2$ and $C_{70}H_2$ have been carried out at the MNDO level [8, 16, 30, 31] and at the AM1 level [32]. The results of the MNDO calculations [30] and the AM1 calculations [32] are almost identical for all 23 regioisomers of $C_{60}H_2$ (Table 5.2). The most stable isomer of $C_{60}H_2$ is indeed the 1,2-addition product, which was exclusively found experimentally. The next three most stable isomers are the 1,4-, the 1,16- and the 1,6-adducts respectively. In the former, and in the latter, two [5,6] double bonds have to be introduced into the canonical Kekulé structure. For the closed-shell isomers an additional introduction of a [5,6] double bond costs about 8–9 kcal mol^{-1} (averaged value) (Table 5.2 and Figure 5.3).

Table 5.2 Calculated MNDO and AM1 heats of formation (ΔH_f°) and number of [5,6] double bonds for $C_{60}H_2$ [30, 32].

Bond alterations [a]	*Isomer*	*ΔH_f° MNDO (kcal mol^{-1})*	*Δ_{MNDO} [$\Delta H_f^\circ(1,2) - \Delta H_f^\circ$]*	*ΔH_f° AM1 (kcal mol^{-1})*	*Δ_{AM1} [$\Delta H_f^\circ(1,2) - \Delta H_f^\circ$]*
	C_{60}	811.7			
0	1,2	776.1	0	931.2	0
1	1,4	780.0	3.9	935.6	4.4
2	1,6	794.5	18.4	950.0	18.8
2	1,16	791.6	15.5	947.3	16.1
3 [b]	1,21; 1,35	800.8	24.7	956.9	25.7
4 [b]	1,19; 1,15; 1,22; 1,51	809.3	33.2	965.8	34.6
5 [b]	1,7; 1,20; 1,36; 1,40; 1,55	819.0	42.9	975.6	44.4
6 [b]	1,10; 1,37; 1,39; 1,49; 1,53	824.9	48.8	981.8	50.6
6 [c]	1,38; 1,52; 1,56	825.2		985.98	
[c]	1,52	831.8		989.05	
[c]	1,56	835.9		990.51	

a) Values are averaged over all isomers with this number of bond alterations.
b) Number of newly formed double bonds in five-membered rings.
c) For (1,52)-$C_{60}H_2$ and (1,56)-$C_{60}H_2$ an alternating structure of the C–C single bonds and double bonds is not obtained. In this case the number of bond alterations is not defined.

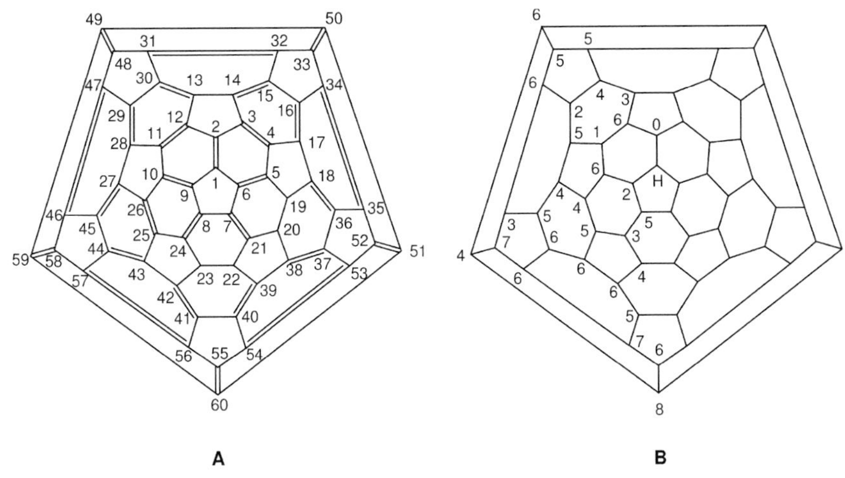

Figure 5.3 (A) Numbering system and lowest energy canonical Kekulé structure of C_{60}. (B) Topologically counted minimum number of [5,6] double bonds that have to be introduced by the addition of two hydrogens [30].

The 1,2- and the 1,4-isomers differ only by about 4 kcal mol^{-1}. In the 1,2-isomer, other than removing one π-bond, the bonding of the framework does not change. However, there is an eclipsing interaction present of the two hydrogen atoms in the neighboring 1 and 2 positions, which was estimated to be 3–5 kcal mol^{-1}. In the 1,4-isomer one [5,6] double bond is introduced, which costs 8–9 kcal mol^{-1} but there are no eclipsing interactions present. Thus, the 1,2-isomer should be about 4–6 kcal mol^{-1} more stable than the 1,4-isomer, which is in agreement with the calculated energy difference (Table 5.2 and Figure 5.4).

Enhancing the eclipsing interaction in the 1,2-position by introducing more steric requiring addents will, therefore, lead to a further destabilization and the 1,4-isomer will eventually become the most stable structure. An eclipsing interaction is also present in the 1,6-isomer, which has two [5,6] double bonds and should, therefore, be about $2 \cdot 8.5$ kcal mol^{-1} less stable than the 1,2-isomer, which again is in good agreement with the calculated 18.4 kcal mol^{-1} (Table 5.2). Conversely, the 1,16-isomer having no eclipsing interactions and also two [5,6] double bonds is about 3 kcal mol^{-1} more stable than the 1,6-isomer.

These four most stable isomers of $C_{60}H_2$ have also been calculated using ab initio methods at the HF/3-21G and HF/6-31G* levels [16, 33]. The investigations show that the energy ordering obtained by semiempirical calculations is preserved (Table 5.3). The energy difference between the isomers, however, becomes more pronounced using ab initio methods. A simpler method for determining the stability of hydrofullerene isomers has been developed by using a generalized Pauling bond order method [34]. The Pauling bond order describes the ratio between the number of Kekulé structures in an isomer of $C_{60}H_2$ or $C_{60}H_n$ to that in C_{60} itself. The prediction of reactivity of a specific carbon site in C_{60} to hydrogenation via the Pauling bond order roughly corresponds with the reactivity values derived from

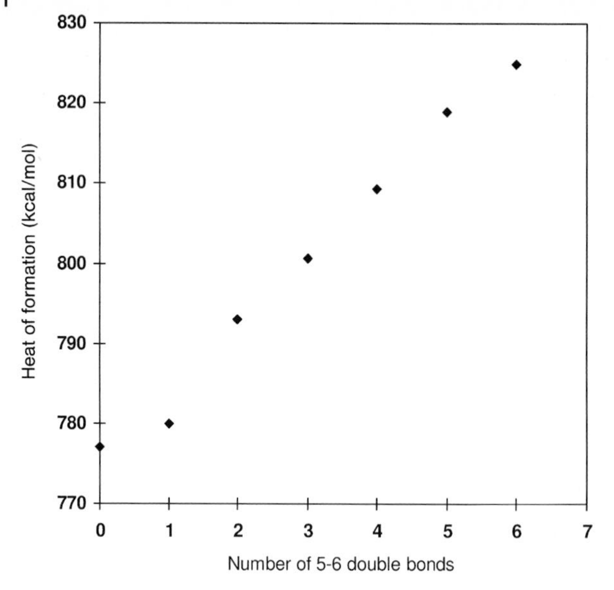

Figure 5.4 Dependence of the MNDO heats of formations (ΔH_f°) of the $C_{60}H_2$ isomers on the number of [5,6] double bonds introduced by the hydrogen addition [30]. The heats of formation are average values of the different structures from Table 5.2 with the same number of [5,6] double bonds.

MNDO or ab initio calculations (Table 5.3). Further confirmation for the preference of 1,2-addition was established by ab initio calculation of the C–H bond energy in hydrogenated fullerenes [35]. Hybrid density functional theory using the B3LYP functional with the 6-31 G(d,p) basis set leads to the bond energies shown in Table 5.3. The most stable bond is found in 1,2 adducts with a bond energy of 2.86 eV, followed by a bond energy of 2.69 eV in 1,4-adducts. All the other addition patterns such as 1,3 addition or addition to a [5,6] bond lead to less stable C–H bonds (Table 5.3).

Table 5.3 Relative energies of the most stable isomers of $C_{60}H_2$ and $C_{70}H_2$. Isomers **A–D** see Figure 5.5.

	MNDO/PM-3 (kcal mol^{-1}) [16, 30]	*HF/3-21G (kcal mol^{-1})* [16]	*HF/6-31G* (kcal mol^{-1})* [16]	*Pauling bond order P* [34]	*C–H bond energy (ev)* [35]
1,2-$C_{60}H_2$	0	0	0	0.440	2.86
1,4-$C_{60}H_2$	3.9	7.8	7.6	0.300	2.69
1,16-$C_{60}H_2$	15.5	23.1	20.9	0.238	2.43
1,6-$C_{60}H_2$	18.4	26.4	24.0	0.280	–
1,2-$C_{70}H_2$ (**A**)	0	0	0	–	–
5,6-$C_{70}H_2$ (**B**)	–1.1	0.2	1.3	–	–
8,22-$C_{70}H_2$ (**C**)	0.3	2.1	4.5	–	–
2,5-$C_{70}H_2$ (**D**)	1.4	5.8	6.4	–	–

The C–H bonds in hydrofullerenes are weak. To determine the thermal stability of $C_{60}H_2$ the thermally induced dehydrogenation was examined theoretically as well as experimentally [36]. Density functional theory calculations at a B3LYP/6-311G** level were carried out and showed that the thermal dehydrogenation in the gas phase is probably a multistep radical reaction and requires an activation energy of 61 kcal mol^{-1}. A concerted H_2-elimination via a single transition step would require a significantly higher energy of 92 kcal mol^{-1}. Thermolysis by heating $C_{60}H_2$ in dichlorobenzene gave C_{60} and H_2 in a pseudo-first-order reaction with an activation barrier of 61.4 kcal mol^{-1}.

For $C_{70}H_2$, 143 regioisomers are, in principle, possible. The four most stable isomers calculated by the semiempirical AM1 and MNDO methods [16, 31] are represented in Figure 5.5. Additions to the "C_{60}-like" double bonds in C_{70} at the pole in the 1,2-position and in the 5,6-position are the most favorable. Whereas at the AM1 and MNDO level the 5,6-isomer is slightly favored over the 1,2-isomer, this order is reversed at both ab initio levels (Table 5.3) [16, 17]. Predictions from the ab initio calculations are consistent with the experimental results. In the synthesis of $C_{70}H_2$ from C_{70} and BH_3 [16] as well as in the synthesis from C_{70} and Zn/Cu couple [21], the 1,2-isomer is the most abundant in the reaction mixture, which contains the 5,6-isomer as the minor product. The calculated energy difference of 1,2-$C_{70}H_2$ (**A**) and 5,6-$C_{70}H_2$ (**B**) at the HF/6-31G* level is in excellent agreement with the experimentally observed $\Delta G_{295} = 1.4 \pm 0.2$ kcal mol^{-1} [17].

If tetrahydro[60]fullerene ($C_{60}H_4$) is formed by additions to two [6,6] double bonds, which are two 1,2-additions with respect to the cyclohexatriene units in C_{60}, then eight regioisomers are possible [37, 38] (see Chapter 10). This very plausible assumption is corroborated by the theoretical investigations of multiple additions to C_{60} in a 1,2- and a 1,4-mode. These investigations predict 1,2-additions to be favorable over 1,4-additions up to the formation of $C_{60}H_{12}$ (see also Table 5.5 below) [38].

The eight different isomers of $C_{60}H_4$ exhibit a similar AM1 heat of formation, with the *cis* isomers being slightly energetically disfavored (Table 5.4) [32, 34, 38]. However, according to ab initio calculations [33, 39] and to calculations of the Pauling bond order [34] the *cis*-1-isomer exhibits a significantly lower energy than the other seven isomers [13]. In addition, the energy spread is more pronounced. Indeed, the *cis*-1-isomer (1,2,3,4-$C_{60}H_4$) of the tetrahydro[60]fullerenes is the major product

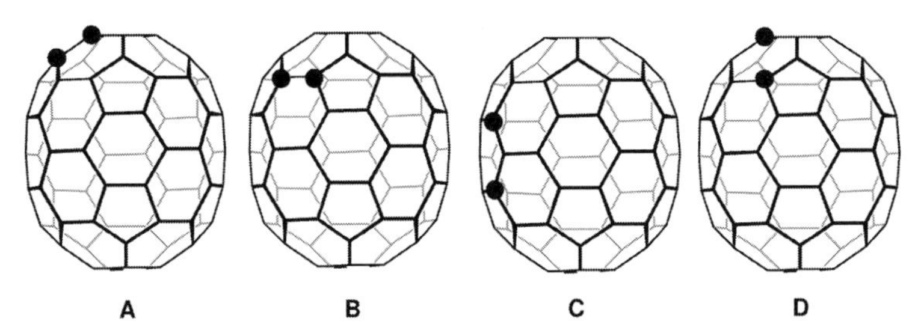

Figure 5.5 Four most stable regioisomers of $C_{70}H_2$ [16, 31]. Dots represent C–H units.

found in the reaction mixture [17]. Conversely, upon cyclopropanation reactions of C_{60} (see Chapter 3) the corresponding *cis*-1-isomer of $C_{62}(COOEt)_4$ does not form at all, which is mainly due to the much higher steric requirement of the bulky bis(diethoxycarbonyl)methylene groups [40]. Interestingly the e-isomer is the second most stable isomer (with only 1,2-additions) of $C_{60}H_4$ (Table 5.3). The *e*-isomer of $C_{62}(COOEt)_4$ is the major product formed by the biscyclopropanation of C_{60}.

One isomer that does not derive exclusively from 1,2-additions – the isomer 1,2,4,15 – is calculated to be more stable than all of the other mentioned isomers except the *cis*-1-isomer (Table 5.4) [33]. Nevertheless, the three isomers, which are calculated as the most stable 1,2-addition-products, are the major isomers in the reductions with either borane or hydrazine (*cis*-1-isomer, see Sections 5.2.1 and 5.2.3) or with the Zn/Cu-couple (*trans*-3 and *e*-isomer, see Section 5.2.2), whereas the 1,2-1,4-addition product (1,2,4,15-isomer) was not observed as a main product in a reduction.

Table 5.4 Relative energies of the eight regioisomers of $C_{60}H_4$ formed by two 1,2-additions to [6,6] double bonds of C_{60}.

	Carbon sites	*MNDO/AM1 (kcal mol^{-1})* [34, 38]	*HF/3-21G (kcal mol^{-1})* [33, 39]	*HF/6-31G* (kcal mol^{-1})* [33]	*Pauling bond order P* [34]
cis-1	1,2,3,4	3.3	0	0	0.258
cis-2	1,2,7,21	3.5	6.1	6.6	0.174
cis-3	1,2,16,17	2.4	8.1	8.5	0.190
e	1,2,18,36	0	3.2	4.0	0.199
trans-4	1,2,34,35	0.2	4.1	4.9	0.210
trans-3	1,2,33,50	0.4	3.5	4.2	0.194
trans-2	1,2,51,52	0.6	3.9	4.8	0.198
trans-1	1,2,55,60	0.5	3.9	4.8	0.200

For $C_{60}H_6$, 46 regioisomers are possible, assuming that only 1,2-additions in C_{60} take place [37]. Significantly, in the most stable regioisomer the hydrogens are bound to [6,6] double bonds, which are all in *e*-positions to each other. This *e,e,e*-isomer was found by reduction of C_{60} with Zn/Cu couple as the minor isomer of the two major products. The major isomer in this reduction is the *trans*-3,*trans*-3,*trans*-3-isomer. In nucleophilic cyclopropanation of C_{60} (see Chapter 3) the corresponding *e,e,e*-isomer was formed out of the e-isomeric bisadduct as the major product [40]. One of the least stable isomers of $C_{60}H_6$ is that with the hydrogens bound in *cis*-1 positions [37]. Also, in ab initio calculations at the Hartree–Fock/3-21G [33, 41] level the *cis*-1,*cis*-1,*cis*-1 isomer is less stable then the other calculated isomers. The other 16 isomers that were calculated in this study were isomers with a 1,2,3,4-addition pattern (*cis*-1) in addition to one other hydrogenated double bond and also one isomer with a different addition pattern consisting of 1,4- and 1,2-additions (1,2,4,11,15,30). The latter is more stable than all the other *cis*-1 isomers.

5.3
Polyhydrofullerenes $C_{60}H_n$ and $C_{70}H_n$ ($n = 14–60$)

5.3.1
Birch–Hückel Reduction

The first attempt to hydrogenate C_{60} was performed by using the Birch–Hückel reduction with Li in liquid NH_3 in the presence of ^{t}BuOH [42]. Polyhydrofullerenes could be so-obtained and this method emerged as one of the standard procedures for the synthesis of highly hydrogenated fullerenes. Thereby the purple C_{60} is converted into a light cream to off-white substance. The major products of this reduction are the isomers of $C_{60}H_{18}$ and $C_{60}H_{36}$. These polyhydrofullerenes were among the first synthesized C_{60}-derivatives.

The kinetic instability of the polyhydrofullerenes is shown by the reaction with 2,3-dichloro-5,6-dicyanobenzoquinone (DDQ) in refluxing toluene. Thereby, the Birch–Hückel products completely convert into C_{60}, which shows that the hydrogenation is completely reversible (Scheme 5.2) [42].

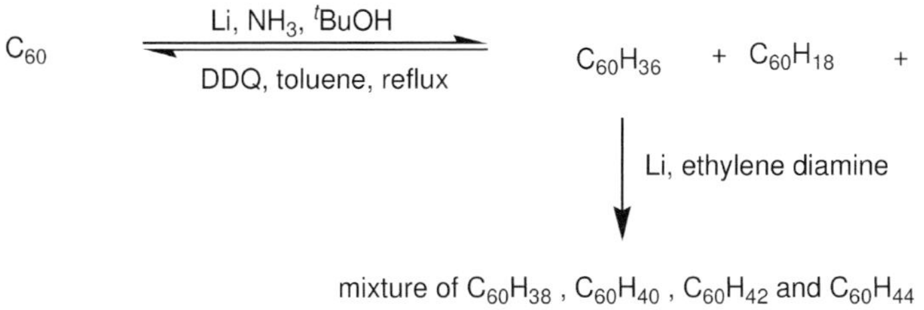

C_{60} ⇄ (Li, NH_3, ^{t}BuOH / DDQ, toluene, reflux) $C_{60}H_{36}$ + $C_{60}H_{18}$ + $C_{60}H_n$

↓ Li, ethylene diamine

mixture of $C_{60}H_{38}$, $C_{60}H_{40}$, $C_{60}H_{42}$ and $C_{60}H_{44}$

Scheme 5.2 Reduction of C_{60} under Birch–Hückel conditions and further reduction under Benkeser conditions.

Polyfullerenes $C_{60}H_n$ with n reaching from 18 to 44 were observed under Birch conditions (for $n > 44$ see Section 5.3.4). Isomers with more than 36 hydrogens could not be obtained with the usual Birch procedure. Much milder conditions were necessary and were found with the Benkeser reduction [43]. $C_{60}H_{36}$, obtained with Birch reduction, was subjected to a reduction with Li in refluxing ethylenediamine and yielded four new polyhydrofullerenes, $C_{60}H_n$ with $n = 38$, 40, 42 and 44 [43]. These derivatives could be separated by preparative HPLC and characterized by mass spectrometry.

It is not yet established if $C_{60}H_{18}$ is one of the products of the Birch reaction or a pyrolysis product of $C_{60}H_{36}$ [7, 42]. Characterization of highly reduced fullerenes turned out to be rather difficult. Polyfullerenes are sensitive to air and light, especially in solution [7]. Thus characterization has to be carried out immediately after workup. As mass spectrometry is a fundamental method for analyzing hydrofullerenes it is important to control the fragmentation of $C_{60}H_n$ during the ionization process. Exclusion or at least minimization of fragmentation has been successfully established with ionization methods such as field desorption (FD) mass spectrometry [44–46], matrix-assisted laser desorption ionization (MALDI) [47], atmospheric pressure chemical ionization (APCI), chemical ionization (CI) or electron impact (EI) [48]. After some debate [49], the early assumption, that $C_{60}H_{36}$ is the major product of the Birch–Hückel reduction could be confirmed by methods mentioned above [47, 48]. For the outstanding stability of $C_{60}H_{36}$ over all the other possible polyfullerenes two main explanations were given [7]. Firstly, 36 is the number of hydrogens required to leave a single unconjugated double bond on each pentagon of C_{60} and Birch–Hückel reduction should not be able to reduce single double bonds. Secondly, the increasing number of hydrogens attached to C_{60} leads to decreasing bond angle strain, resulting from sp^2 to sp^3 hybridization, and at the same time increasing strain due to hydrogen–hydrogen repulsion. Combined strain reaches a minimum when $n = 36$. The first explanation proved not to be very important, since most experimentally and theoretically found stable isomers do not show isolated double bonds and, for example, the T_h isomer with only isolated double bonds is a high-energy-isomer relative to other possible isomers (Figure 5.10 below, Section 5.3.5). Instead a criterion of stability seems to be the formation of benzene-like hexagons with conjugated double bonds (Section 5.3.5).

About $6 \cdot 10^{14}$ different possible isomers of $C_{60}H_{36}$ were estimated [50]. This number, the similarity between this isomers and the instability of hydrofullerenes make it almost impossible to get the exact structure of the isomers that are included in the complex mixture of $C_{60}H_{36}$ isomers obtained by Birch reduction [51]. Based on ^{1}H and ^{3}He NMR spectroscopy and on calculations (Section 5.3.5) C_1-, C_3-, D_{3d}-, S_6-isomers and a T-isomer are supposed to be among the preferred isomers [7, 48, 51–53]. Some examples of probable structures are given in Figure 5.10 below (Section 5.3.5).

5.3.2
Reduction with Zn/HCl

Contrary to the reduction of C_{60} with Zn/Cu couple (Section 5.2.2) or with Zn and half-concentrated HCl [54] that leads only to oligohydrofullerenes, the reaction with Zn and concentrated HCl in toluene or benzene solution leads in a smooth and fast reaction mainly to $C_{60}H_{36}$ [7, 55, 56]. The reaction takes place within 1 h and at room temperature and only a small amount of side products were found. EI-mass spectroscopy, run immediately after the reaction, carried out under a nitrogen atmosphere gave very good spectra of the product almost without fragmentation [55, 56]. These spectra revealed that 75% of the product was $C_{60}H_{36}$. The remaining 25% were assigned to $C_{60}H_{38}$ and $C_{60}H_{40}$. The major product $C_{60}H_{36}$ was first proposed to be the T-isomer (Figure 5.10, Section 5.3.5) [56]. Later, based on vibrational spectroscopy, a S_6-symmetry was suggested [57]. Proof of the structure is not yet given but different isomers are formed than with the Birch–Hückel reduction [58].

Under exclusion of light and air $C_{60}H_{36}$ has good stability towards high temperature [55]. Nevertheless, extended heat treatment of the polyhydrofullerene mixture leads to an increasing peak of $C_{60}H_{18}$. Indeed, $C_{60}H_{18}$ formation can predominate if the reaction is carried out at high temperature and pressure [58].

Since deuterochloric acid DCl is easily available, the reduction with Zn and conc. DCl provides an easy access to deuterated fullerenes [55]. Reducing C_{60} under the same, already described conditions yields polydeuterofullerenes in good yields and in short time. Interestingly, the product distribution is different from the HCl reduction. The C–D bond is more stable than the C–H bond. This is probably why deuteration yields not only $C_{60}D_{36}$ but also $C_{60}D_{38}$ and, in slightly smaller percentage, also $C_{60}D_{40-44}$ as major products.

C_{70} is less reactive in terms of hydrogenation than C_{60}. Zn–HCl reduction was complete after 1.5 h and gave a pale yellow product, consisting of $C_{70}H_{36}$ and $C_{70}H_{38}$ as major products and $C_{70}H_{40-44}$ as further products [55]. Deuteration of C_{70} shows the same shift of the degree of hydrogenation to higher numbers. In this case $C_{70}D_{42}$ is the dominant component.

5.3.3
Transfer Hydrogenation of C_{60} and C_{70}

Hydrogenation to polyhydrofullerenes is also possible by transfer hydrogenation [59] using 9,10-dihydroanthracene as a hydrogen source [44, 51, 60, 61]. The treatment of C_{60} in a sealed glass tube in a melt of 9,10-dihydroanthracene at 350 °C under N_2 leads to a color change of the dissolved fullerene in 30 min from brown via rubin-red, orange and yellow to colorless (Scheme 5.3). The resulting product also shows a base peak in the mass spectra (EI, FAB, FD) centered on 756 u, corresponding to $C_{60}H_{36}$ [44]. Extending the reaction time to 24 h recolorizes the reaction mixture and the isolated reaction product shows a mass distribution, determined by the same techniques as above, centered at 738 u, corresponding to $C_{60}H_{18}$. Based on this observation, conditions were established that allow either synthesis of nearly pure $C_{60}H_{36}$ or of nearly pure $C_{60}H_{18}$ [7, 44]. Impurities are small amounts of other hydrofullerenes and anthracene. The anthracene impurities can be removed by sublimation. Polydeuterated fullerenes $C_{60}D_n$ can be obtained analogously using 9,10-dideuteroanthracene as a deuterium source. The reaction temperature of the transfer hydrogenation can be lowered to 250 °C by addition of 7*H*-benzanthrene or 7,7′-dideutero-7*H*-benzanthrene as a catalyst [60]. In this way, higher degrees of hydrogenation (n up to 44) are obtained for C_{60}.

The predominance $C_{60}H_{36}$ and $C_{60}H_{18}$ formation under the applied conditions has been conclusively proven via ^{3}He NMR [48, 52], ^{1}H NMR and ^{13}C NMR [51, 61,

$$C_{60} \underset{550\ ^\circ C}{\overset{250\text{-}350\ ^\circ C;\ R = H \text{ or } D}{\rightleftharpoons}} C_{60}H_{36} / C_{60}D_{36} \longrightarrow C_{60}H_{18} / C_{60}D_{18}$$

(Reagents shown above arrow: 9,10-R,R-dihydroanthracene or 7,7-R,R-benzanthrene structures labelled "R R", "or", "R R")

Scheme 5.3

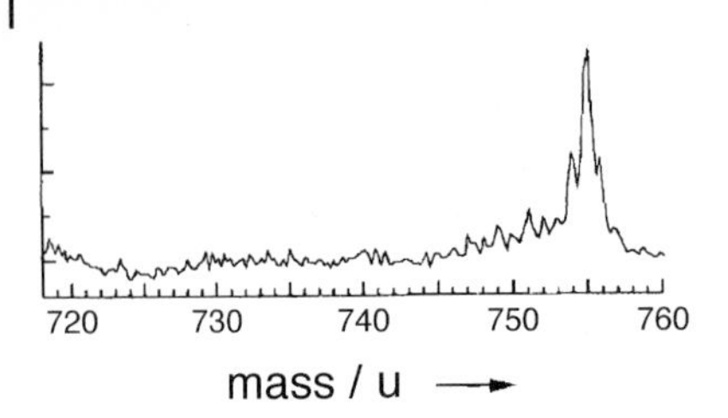

Figure 5.6 MALDI-RETOF mass spectrum of $C_{60}H_{36}$ with matrix 5-methoxysalicylic acid and comatrix $NaBF_4$ (1 : 10 : 10). $C_{60}H_{36}$ can be detected as the $C_{60}H_{35}^+$ ion at 755 u [63].

62] spectroscopy and with different mass spectrometric ionization techniques such as EI [44], FD [44, 51] and MALDI [47, 63]. Especially with MALDI, by using a combination of two matrix molecules [63], it was possible to obtain a mass spectrum consisting almost entirely of $C_{60}H_{35}^+$, which can be taken as direct evidence for the exclusive production of $C_{60}H_{36}$ (Figure 5.6).

The clear solutions of the polyhydrofullerenes in various organic solvents become inhomogeneous upon the formation of a precipitate. This, together with the broad peaks in the ^{1}H NMR spectra, shows the instability of these $C_{60}H_n$s. Thermal treatment of $C_{60}H_n$ in the solid state at 550 °C leads to a complete reversion to C_{60} (Scheme 5.3). Sublimation of $C_{60}H_{18}$ and $C_{60}H_{36}$ at lower temperatures (273–412 °C) was accompanied by partial loss of hydrogen. Decomposition of $C_{60}H_{36}$ was confirmed to be a stepwise process with formation of $C_{60}H_{18}$ as an intermediate product [64].

Compared with the Birch–Hückel reduction of C_{60}, transfer hydrogenation produces less isomers. As well as mainly $C_{60}H_{36}$ and $C_{60}H_{18}$, in good yields, a smaller number of different structural isomers are also formed. Nevertheless, due to the huge number of possible isomers, thermal decomposition and, probably, isomerization and the instability against air and light, identification turned out to be almost as complicated as with the Birch reduction product mixtures. For $C_{60}H_{36}$, structures with the symmetries T [52, 61, 65, 66], D_{3d} [57, 66], S_6 [57, 66], C_3 [51, 52, 61] and C_1 [51, 61] were claimed. Another reason for such various possible structures for $C_{60}H_{36}$ may be the dependence of the product distribution on the reaction conditions [51]. Surprisingly, and almost independent of all these factors, only one major isomer of $C_{60}H_{18}$ was found. The structure was elucidated by Taylor and co-workers [62], who showed with ^{1}H NMR spectroscopy that $C_{60}H_{18}$ is a crown-shaped molecule with C_{3v} symmetry that may be considered as a substructure of T-$C_{60}H_{36}$ (Figure 5.7). All hydrogens are located on one side of the C_{60}-ball, surrounding an isolated benzene ring and leaving an extended conjugated system on the opposite side.

In the ^{3}He NMR spectra of a mixture of $^3He@C_{60}H_{36}$ and $^3He@C_{60}H_{18}$ and also of a nearly pure $^3He@C_{60}H_{18}$ sample prepared via the "Rüchardt"-procedure only one isomer of $^3He@C_{60}H_{18}$ with a typical shift of –16.45 ppm was found [48].

After 10 years of attempts to elucidate the structure of the $C_{60}H_{36}$ isomers formed during hydrogenation with dihydroanthracene the work of Billups and co-workers

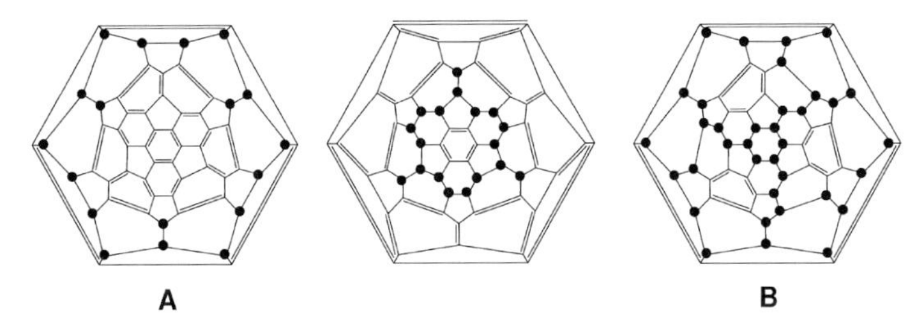

Figure 5.7 (A) Schlegel diagram of C_{3v}-$C_{60}H_{18}$ (two different views), which is a substructure of (B) T-$C_{60}H_{36}$.

lead to a breakthrough [51]. Using HPLC for purification instead of sublimation and combined ^1H, ^{13}C and ^3He NMR spectroscopy showed clearly that only two major isomers are formed and these isomers must have C_3 and C_1 symmetry (Figure 5.8). The two major isomers are formed in a ratio of C_1 to C_3 of about 3 : 1. This structure elucidation was made on the basis of 80 signals (32 sp^2 and 48 sp^3 signals) in the ^{13}C NMR spectrum and of only two distinct but very close ^3He signals. The exact structure of the C_1- and C_3 isomers could not be proven but it was suggested that they have the structures shown in Figure 5.8. Based on this work, Gakh and co-workers, using 2D ^1H NMR spectroscopy recorded at 800 MHz, demonstrated that these suggestions were right [61]. Moreover they found a further minor isomer with T-symmetry (Figure 5.8), which was already calculated to be the most stable isomer (Section 5.3.5).

Evidently the two major isomers have very similar structures, as shown in the almost identical ^3He NMR shifts of –8.014 and –8.139 ppm. This also shows impressively how effectively ^3He NMR spectroscopy can be utilized to distinguish different C_{60} isomers, even if their structures are very similar.

Also, the less reactive C_{70} can be hydrogenated and deuterated in this way [60]. Interestingly, the base peak in the EI mass spectra of $C_{70}H_n$ at 876 u shows that, in this case, a polyhydrofullerene with 36 hydrogens also exhibits an enhanced stability. Beside the predominant $C_{70}H_{36}$ substantial amounts of $C_{70}H_{38}$, $C_{70}H_{40}$, $C_{70}H_{42}$, $C_{70}H_{44}$ and $C_{70}H_{46}$ are formed and could be detected by field desorption (FD) mass spectrometry [45].

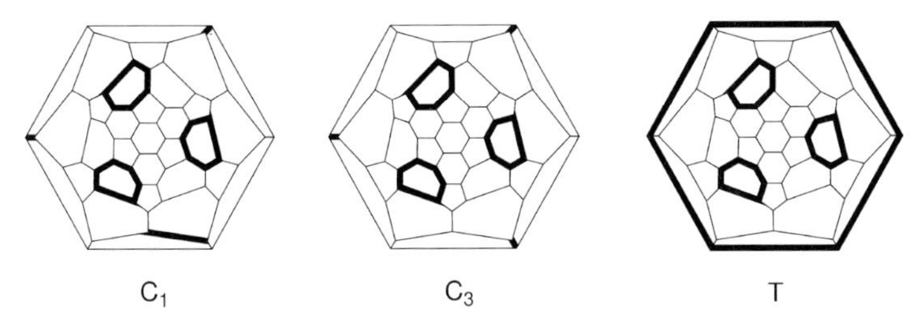

Figure 5.8 Structures of the major $C_{60}H_{36}$ isomers formed during transfer hydrogenation of C_{60}. Bold lines represent double bonds; bold hexagons represent a hexagon with three conjugated double bonds.

5.3.4 Reduction with Molecular Hydrogen

A radical-induced hydrogenation of C_{60} and C_{70} can be carried out with iodoethane as the hydrogen radical promoter [67–69]. In this method the fullerenes are placed in a glass vessel inside an autoclave with an excess of iodoethane and are pressurized with hydrogen to 6.9 MPa. Hydrogenation is carried out at 400 °C for 1 h. The polyhydrofullerenes are obtained as a light brown solid. In the absence of iodoethane, no hydrogenation of the fullerenes takes place. In contrast to the material obtained by the Birch–Hückel reduction (Section 5.3.1) and transfer hydrogenation (Section 5.3.3) these polyhydrofullerenes are insoluble in many organic solvents and are only slightly soluble in nitrobenzene. Analysis by fast atom bombardment (FAB) mass spectrometry revealed a mixture of hydrofullerenes consisting mainly of $C_{60}H_{36}$ and $C_{70}H_{36}$. Reactions at higher temperatures and pressures result in a lower degree of hydrogenation.

Catalytic hydrogenation of C_{60} is also possible on activated carbon with Ru as catalyst in refluxing toluene [70, 71]. Thereby, comparatively high degrees of hydrogenation (up to $C_{60}H_{50}$) are obtained. The degree of hydrogenation of C_{60} increases with increasing hydrogen gas pressure and by elevating the reaction temperature. A complete reversion to C_{60} takes place upon the treatment of $C_{60}H_n$ with DDQ in refluxing toluene. C_{70} has also been catalytically hydrogenated in this way, leading to mixtures of C_{70}-polyhydrofullerenes that consist mainly of $C_{70}H_{36}$. Dehydrogenation of $C_{70}H_n$ to C_{70} with DDQ proceeds quantitatively.

Hydrogenation of C_{60} on alumina-supported nickel leads selectively to $C_{60}H_{36}$ [72]. Degradation products or other hydrofullerenes such as $C_{60}H_{18}$ or $C_{60}H_{44}$ were not observed. The reduction was carried out in toluene in an autoclave at 50–250 °C with the hydrogen pressure ranging from 2.5 to 7.5 MPa for 1 to 24 h.

More insight into the activity and selectivity of different catalytically active metals was gained by a systematic investigation of this reaction [73]. Various metal and noble metal catalysts are usable for this hydrogenation. All metals were used on alumina support. Reactions with Ru, Rh and Ir as catalysts give mainly $C_{60}H_{18}$, while Pd, Pt, Co and Ni lead predominantly to $C_{60}H_{36}$, and Au and Fe have very little activity for C_{60} hydrogenation. The %-d character of the transitions metals is, seemingly, responsible for the selectivity, with Ru, Rh and Ir having a larger %-d character than Pd, Pt, Co and Ni. The best selectivity for $C_{60}H_{36}$ is with the Ni/Al_2O_3-catalyst [73].

The advantage of the above-described catalytic procedures is the possibility of preparing large amounts of hydrofullerenes that can be synthesized selectively (depending on the metal). Compared with other hydrogenations, the products can easily be separated by filtration of the catalyst and evaporation of the solvent [73].

In the absence of a catalyst, hydrogenation can also take place but either very high pressure or temperature is required to yield any products. High pressure hydrogenation has been performed at hydrogen or deuterium pressures of 3.0 GPa and 650–700 K [74]. The major product under these conditions is $C_{60}H_{36}$. High temperature but low pressure was used in a novel method, the chemical vapor modification (CVM) technique [75]. C_{60} is sublimed by exposure to a tungsten filament, which has a temperature of 1900 K. The substrate temperature at these conditions is 960 K and a reaction time of 30 min is used. At a hydrogen pressure of 20 Torr mainly $C_{60}H_{18}$ is formed.

5.3.5
Theoretical Investigations

Several systematic calculations on various structures of the polyhydrofullerenes $C_{60}H_n$ have been carried out at different theoretical levels [8, 16, 30–32, 35, 38, 51, 53, 76–88]. Since the number of the theoretically possible regioisomers of the several $C_{60}H_n$ adducts is very high, not every single structure has been calculated. The situation becomes even more complicated if not only isomers with externally but also with hydrogens added internally to the C_{60} cage are considered [77, 79, 80, 89, 90]. However, based on the results available, some trends of the stability of different polyhydrofullerenes, depending on the addition mode and the degree of hydrogenation of C_{60} and of C_{70}, can be recognized.

As pointed out in Section 5.2.4, the addition of hydrogen to C_{60} leading to hydrofullerenes $C_{60}H_n$ with a low degree of hydrogenation ($n \leq 12$) is energetically favored if it proceeds in a 1,2-mode to the [6,6] bonds of the cyclohexatriene units of the C_{60}-framework [35, 38]. This mode avoids the energetically unfavorable introduction of [5,6] double bonds. Conversely, the eclipsing interaction of the hydrogens resulting from one 1,2-addition costs about 3–5 kcal mol^{-1}. Thus, the strain energy, due to eclipsing interactions of the hydrogens, is expected to become ever more important upon increasing the degree of hydrogenation. Therefore, the polyhydrofullerenes will eventually be unstable. Indeed, calculations of several $C_{60}H_n$ isomers formed by a 1,2- or a 1,4-addition mode with the semiempirical MNDO method would predict 1,4-additions to be more favorable than 1,2-additions starting for $n > 12$ (Table 5.5) [38]. But, only up to $n = 24$, exclusive 1,4-additions are possible.

Table 5.5 Calculated MNDO heats of formation of $C_{60}H_n$ formed by 1,2- and 1,4-addition [38].

n	ΔH_f° 1,2-addition (kcal mol^{-1})	ΔH_f° 1,4-addition (kcal mol^{-1})
0 [a)]	973.3	973.3
2	931.2	935.6
4	889.4	899.0
8	805.3	814.0
12	720.5	720.9
16	653.9	652.2
20	587.2	576.9
24	518.1	504.3
36	354.2	–
48	242.2	–
60	335.0	–

a) $n = 0$ corresponds to C_{60}.

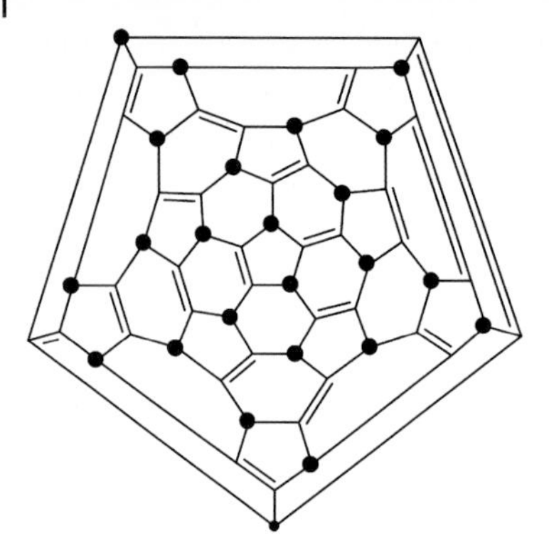

Figure 5.9 Schlegel diagram of T_h-$C_{60}H_{24}$ (dots represent C–H units).

Higher degrees of hydrogenation require formal 1,2-additions. The polyhydrofullerene formed by the complete and exclusive 1,4-addition mode is T_h-$C_{60}H_{24}$ (Figure 5.9).

To obtain a measure for the dependence of n (number of hydrogens in $C_{60}H_n$) on the stability of $C_{60}H_n$, the normalized [38] calculated heats of formation ΔH_f° were introduced:

$$\Delta(\ \Delta H_f^\circ)_n = [\Delta H_f^\circ(C_{60}H_n)\ -\ \Delta H_f^\circ(C_{60}H_2)]/(n/2)$$

Whereas, for 1,2-additions, a destabilization begins for $n > 12$ with $\Delta(\ \Delta H_f^\circ)$ becoming endothermic, the opposite is observed for 1,4-additions, where the values become more exothermic. Even for larger values (n up to 24) the normalized values are more exothermic than for $n = 2$ in this addition mode [38]. The large destabilization for $C_{60}H_{60}$ ($n = 60$) of $\Delta(\ \Delta H_f^\circ) = 20.8\ \text{kcal mol}^{-1}$ shows, significantly, the accumulating influence on continuing 1,2-additions to C_{60}.

The polyhydrofullerenes $C_{60}H_{36}$ and $C_{60}H_{48}$ are the most stable molecules in the $C_{60}H_{12n}$ ($n = 1$–5) series of exo-hydrogenated C_{60}, according to ab initio calculations at the Hartree–Fock level [80]. For some specific isomers of this series the C–H bond energies were calculated with an ab initio molecular orbital theory [35]. Although the employed hydrogenated isomers may not be the most stable isomers, the averaged C–H bond energy seems to be maximized in $C_{60}H_{36}$ (Table 5.6).

Table 5.6 C–H bond energies for some isomers of the series $C_{60}H_{12n}$ with $n = 1$–5 [35].

Isomer	*Symmetry*	*C–H bond energy (eV)*
$C_{60}H_{12}$	S_6	2.51
$C_{60}H_{24}$	T	2.61
$C_{60}H_{36}$	T	2.94
$C_{60}H_{36}$	D_{3d}	2.90
$C_{60}H_{36}$	C_3	2.88
$C_{60}H_{48}$	S_6	2.82
$C_{60}H_{60}$	I_h	2.61

Extensive AM1 calculations and density functional calculations were performed for numerous isomers of $C_{60}H_n$, with n varying from 2 to 60, by Clare and Kepert [32, 53, 82–85]. They used AM1 for the geometry optimization and either AM1 or the density functional method B3LYP for single point calculations of the energies. Among numerous calculated structures, which compared with the existing number of isomers is still minute, a couple of isomers turn out to have salient stability. Among other isomers, these are the hydrofullerenes $C_{60}H_{18}$, $C_{60}H_{36}$ and $C_{60}H_{48}$. Comparing calculations of ΔH_f° carried out either with the AM1 method or density functional methods reveals divergent results in some cases [53]. In AM1 calculations structures with the highest number of isolated double bonds were calculated as the most stable structures. In contrast, calculations with density functional methods result in low energies for structures with a high number of isolated benzenes such as hexagons.

The theoretical investigations support the experimental findings [42, 44, 49, 67, 70, 71] (see previous chapters) that (1) complete hydrogenation of C_{60} is difficult, and has not yet been observed, and (2) even polyhydrogenated fullerenes with a lower degree of hydrogenation, such as $C_{60}H_{36}$, are not stable and will slowly decompose. The reasons for the instabilities of polyhydrofullerenes in terms of simple topological arguments are inter alia (1) enhanced eclipsing interactions of the H atoms as in the case of *cis*-1-additions in the 1,2-mode; (2) strain within the C-network of polyhydrofullerenes, especially due to the deviation from the tetrahedral angle of sp^3 C-atoms; and (3) introduction of [5,6] double bonds. Initially generated reaction mixtures may rearrange to form more stable compounds. This can cause a predominant occurrence of the species $C_{60}H_{36}$, $C_{70}H_{36}$ and $C_{60}H_{18}$ after thermal treatment of polyhydrofullerenes, which were obtained, for example, from Birch–Hückel reductions or transfer-hydrogenations. The applied temperature is very important for this annealing process. During the synthesis of $C_{60}H_{36}$ via transferhydrogenation, an equilibrium between different isomers probably exists. At 340 °C the mixture anneals to a mixture containing only the most stable isomers, in this case a C_1-, C_3- and a T-isomer (Section 5.3.3). Higher temperatures (> 340 °C) lead to the dehydrogenation product $C_{60}H_{18}$, which appears to be the final hydrogenation product at elevated temperatures [61].

To date only one isomer of $C_{60}H_{18}$ with C_{3v}-symmetry has been experimentally proven (Section 5.3.3). Also in calculations this isomer appears as a particularly stable compound [83, 84]. From calculations of the bond length of this crown shaped isomer the central, isolated benzenoic ring can be considered essentially planar aromatic [91]. For $C_{60}H_{36}$, various structures have been suggested and extensively treated theoretically [42, 51, 83, 85–88, 91, 92]. Some of these structures are shown in Figure 5.10.

In structures such as **29** most of the double bonds are located in pentagons, which may be unfavorable, whereas in structures such as **15** only [6,6] double bonds and pentagons constructed with single bond edges are present. In addition, similar to T_h-$C_{60}H_{12}$, four benzenoid rings are formed, which may provide further stabilization. Considering all the energy values of the different isomers obtained by different methods (Table 5.7) reveals the benzenoid ring as a very important stabilizing

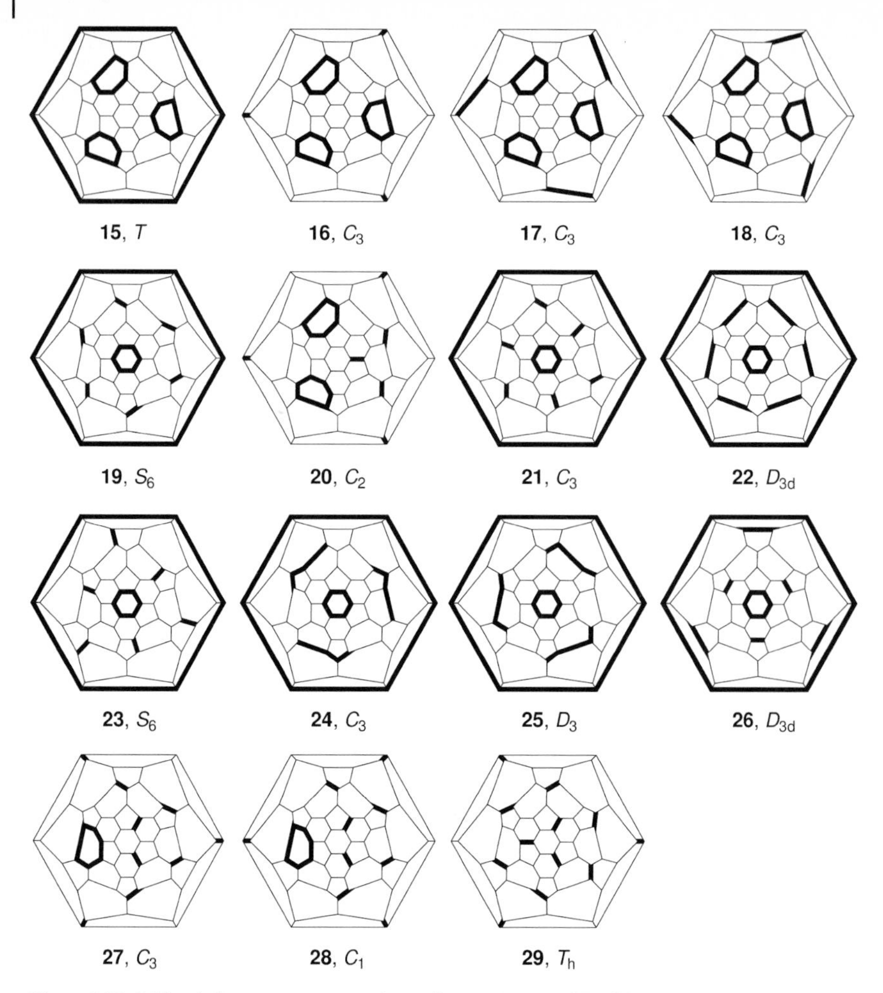

Figure 5.10 Schlegel diagram representations of some isomers [7] of $C_{60}H_{36}$ (out of 10^{14} theoretically possible isomers). Bold lines represent double bonds; bold hexagons represent a hexagon with three conjugated double bonds.

element. The more isolated benzenoid 6-rings the isomer holds the more stable is this isomer. In most calculations the *T*-isomer was found to be the most stable isomer. This derivative has four benzenoid rings, the maximum number. The three product isomers of the reduction with transfer hydrogenation (Section 5.3.3), whose structures are proven, have at least three benzenoid rings. Remarkably, instead of the two major isomers (Figure 5.8) actually formed, the minor isomer with *T*-symmetry (Figure 5.8) was predicted to be most stable. The S_6 isomer with only two benzenoid rings, which was often claimed to be among the most stable isomers, does not follow this trend.

Putting any of the 36 hydrogens inside the fullerene leads to conformers with higher energies [76].

Table 5.7 Heats of formation, normalized relative to isomer **15** at various levels of theory.

Isomer	*Symmetry*	*C_6 rings*	*ΔH_f° AM1* [a] *(kcal mol^{-1})*	*ΔH_f° DF* [b] *(kcal mol^{-1})*	*ΔH_f° D* [c] *(kcal mol^{-1})*	*ΔH_f°* [d] *(kcal mol^{-1})*
15	T	4	0.00	0.00	0.0	0.0
16	C_3	3	–7.78	1.02	5.2	–
17	C_3	3	2.33	16.58	20.4	–
18	C_3	3	26.59	44.83	45.6	–
19	S_6	2	–12.42	6.10	14.9	–
20	C_2	2	–7.03	13.16	21.0	–
21	C_3	2	–4.13	21.12	29.3	–
22	D_{3d}	2	–0.78	25.56	31.6	21.9
23	S_6	2	0.79	26.32	34.5	29.5
24	C_3	2	8.25	24.27	–	–
25	D_3	2	11.09	19.25	–	–
26	D_{3d}	2	–	–	–	77.8
27	C_3	1	1.13	34.87	45.1	–
28	C_1	1	2.30	38.97	49.0	–
29	T_h	0	15.64	64.70	–	58.6

a) AM1 with MOPAC 6.0 [53].
b), c) Density functional technique B3LYP/6-31G* with Gaussian 98, (b) [53], (c) [51].
d) Optimized with MNDO, reoptimized at the SCF/3-21G level [87].

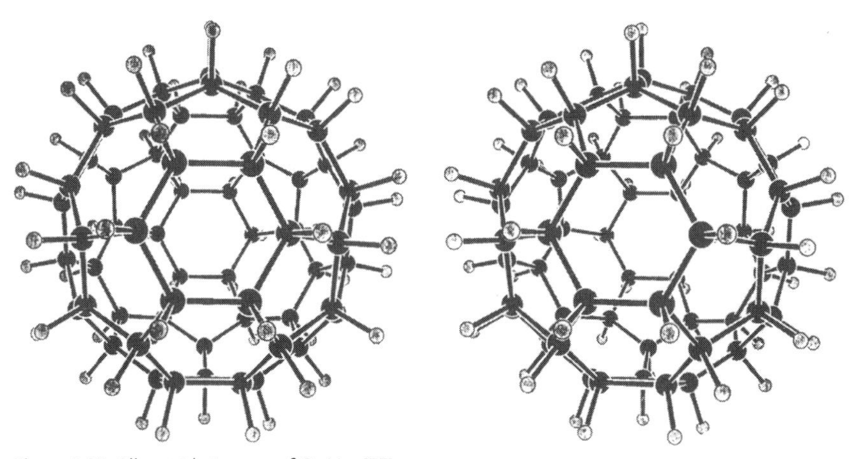

Figure 5.11 All-outside isomer of $C_{60}H_{60}$ [77].

The icosahedral $C_{60}H_{60}$ (Figure 5.11) is neither the only nor the most stable conformer of completely hydrogenated C_{60}. Molecular mechanics calculations (MM3) on $C_{60}H_{60}$ show that moving just one hydrogen inside the cage and forming a new conformer is an exothermic process and decreases the energy by 53 kcal mol^{-1} [77].

A conformer of $C_{60}H_{60}$ with 10 hydrogens added inside the cage is the lowest energy isomer (Figure 5.12). This minimum energy isomer with C_1-symmetry is predicted to have a heat of formation 400 kcal mol^{-1} lower than that for the all-

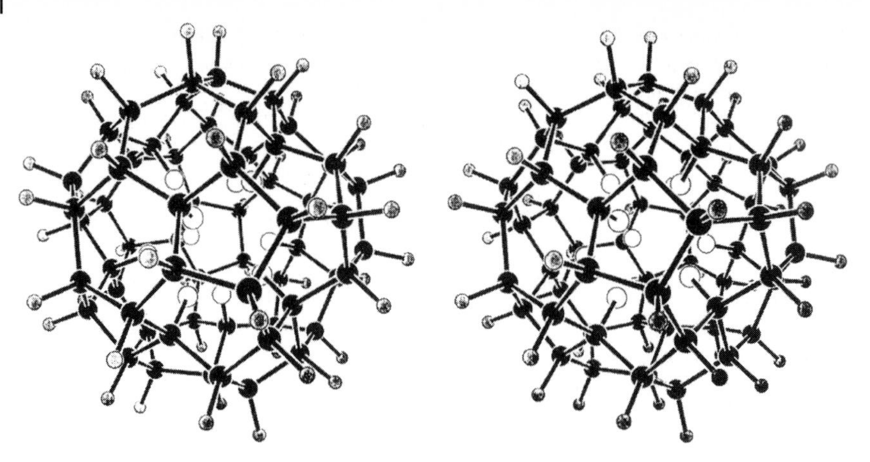

Figure 5.12 Most stable isomer of $C_{60}H_{60}$ with ten hydrogens inside [77].

outside isomer [77]. In another attempt, an isomer with D_5-symmetry but also with 10 hydrogens inside the cage was found as the most stable isomer [89]. The all-outside isomer is highly strained due to eclipsing hydrogen–hydrogen interactions and due to the 120° angles of the sp^3 carbons (20 planar cyclohexane rings). Putting hydrogens inside leads to fewer eclipsing interactions and also to a decrease of many C–C–C bond angles. In another study, comparison on the basis of ab initio calculations of the stabilities of totally exo-hydrogenated $C_{60}H_{60}$ and T_h-$C_{60}H_{60}$, with 12 hydrogens added inside the cage, predicts the latter conformer to be the more stable [80]. The calculated energy barrier for the hydrogen penetration from outside to inside the C_{60} cage, however, is very high (at least 2.7 eV $atom^{-1}$). To date no $C_{60}H_{60}$ could be isolated, regardless of the hydrogens pointing in or out of the cage.

References

1 C. C. Henderson, P. A. Cahill, *Science* **1993**, *259*, 1885.

2 S. Ballenweg, R. Gleiter, W. Krätschmer, *Tetrahedron Lett.* **1993**, *34*, 3737.

3 A. G. Avent, A. D. Darwish, D. K. Heimbach, H. W. Kroto, M. F. Meidine, J. P. Parsons, C. Remars, R. Roers, O. Ohashi, R. Taylor, D. R. M. Walton, *J. Chem. Soc., Perkin Trans. 2* **1994**, 15.

4 T. F. Guarr, M. S. Meier, V. K. Vance, M. Clayton, *J. Am. Chem. Soc.* **1993**, *115*, 9862.

5 R. G. Bergosh, M. S. Meier, J. A. L. Cooke, H. P. Spielmann, B. R. Weedon, *J. Org. Chem.* **1997**, *62*, 7667.

6 W. E. Billups, W. Luo, A. Gonzalez, D. Arguello, L. B. Alemany, T. Marriott, M. Saunders, H. A. Jimenez-Vazquez, A. Khong, *Tetrahedron Lett.* **1997**, *38*, 171.

7 J. Nossal, R. K. Saini, L. B. Alemany, M. Meier, W. E. Billups, *Eur. J. Org. Chem.* **2001**, 4167.

8 A. Hirsch, A. Soi, H. R. Karfunkel, *Angew. Chem.* **1992**, *104*, 808.

9 A. Hirsch, T. Grösser, A. Skiebe, A. Soi, *Chem. Ber.* **1993**, *126*, 1061.

10 M. E. Niyazymbetov, D. H. Evans, S. A. Lerke, P. A. Cahill, C. C. Henderson, *J. Phys. Chem.* **1994**, *98*, 13093.

11 P. J. Fagan, P. J. Krusic, D. H. Evans, S. A. Lerke, E. Johnston, *J. Am. Chem. Soc.* **1992**, *114*, 9697.

12 L. Echegoyen, L. E. Echegoyen, *Acc. Chem. Res.* **1998**, *31*, 593.

13 C. C. Henderson, C. M. Rohlfing, R. A. Assink, P. A. Cahill, *Angew. Chem.* **1994**, *106*, 803; *Angew. Chem. Int. Ed. Engl.* **1994**, *33*, 786.

14 R. G. Bergosh, M. S. Meier, H. P. Spielmann, G.-W. Wang, B. R. Weedon, *Proc. – Electrochem. Soc.* **1997**, *97-14*, 240.

15 B. Morosin, C. Henderson, J. E. Schirber, *Appl. Phys. A: Solids Surf.* **1994**, *A59*, 179.

16 C. C. Henderson, C. M. Rohlfing, P. A. Cahill, *Chem. Phys. Lett.* **1993**, *213*, 383.

17 C. C. Henderson, C. M. Rohlfing, K. T. Gillen, P. A. Cahill, *Science* **1994**, *264*, 397.

18 M. S. Meier, H. P. Spielmann, R. C. Haddon, R. G. Bergosh, M. E. Gallagher, M. A. Hamon, B. R. Weedon, *Carbon* **2000**, *38*, 1535.

19 M. S. Meier, J. A. L. Cooke, B. R. Weedon, H. P. Spielmann, *Proc. – Electrochem. Soc.* **1996**, *96-10*, 1193.

20 M. S. Meier, B. R. Weedon, H. P. Spielmann, *J. Am. Chem. Soc.* **1996**, *118*, 11682.

21 H. P. Spielmann, G.-W. Wang, M. S. Meier, B. R. Weedon, *J. Org. Chem.* **1998**, *63*, 9865.

22 B. P. Tarasov, N. F. Goldshleger, A. P. Moravsky, *Russ. Chem. Rev.* **2001**, *70*, 131.

23 M. S. Meier, H. P. Spielmann, R. G. Bergosh, R. C. Haddon, *J. Am. Chem. Soc.* **2002**, *124*, 8090.

24 R. J. Cross, H. A. Jimenez-Vazquez, Q. Lu, M. Saunders, D. I. Schuster, S. R. Wilson, H. Zhao, *J. Am. Chem. Soc.* **1996**, *118*, 11454.

25 G.-W. Wang, B. R. Weedon, M. S. Meier, M. Saunders, R. J. Cross, *Org. Lett.* **2000**, *2*, 2241.

26 H. P. Spielmann, B. R. Weedon, M. S. Meier, *J. Org. Chem.* **2000**, *65*, 2755.

27 S. Fukuzumi, T. Suenobu, M. Patz, T. Hirasaka, S. Itoh, M. Fujitsuka, O. Ito, *J. Am. Chem. Soc.* **1998**, *120*, 8060.

28 S. Fukuzumi, T. Suenobu, S. Kawamura, A. Ishida, K. Mikami, *Chem. Commun.* **1997**, 291.

29 R. V. Bensasson, E. Bienvenue, J. M. Janot, S. Leach, P. Seta, D. I. Schuster, S. R. Wilson, H. Zhao, *Chem. Phys. Lett.* **1995**, *245*, 566.

30 N. Matsuzawa, D. A. Dixon, T. Fukunaga, *J. Phys. Chem.* **1992**, *96*, 7594.

31 H. R. Karfunkel, A. Hirsch, *Angew. Chem.* **1992**, *104*, 1529; *Angew. Chem. Int. Ed. Engl.* **1992**, *31*, 1468.

32 B. W. Clare, D. L. Kepert, *Theochem* **2003**, *621*, 211.

33 P. A. Cahill, C. M. Rohlfing, *Tetrahedron* **1996**, *52*, 5247.

34 S. Narita, T. Morikawa, T. I. Shibuya, *Theochem* **2000**, *528*, 263.

35 Y. Okamoto, *J. Phys. Chem. A* **2001**, *105*, 7634.

36 H. F. Bettinger, A. D. Rabuck, G. E. Scuseria, N.-X. Wang, V. A. Litosh, R. K. Saini, W. E. Billups, *Chem. Phys. Lett.* **2002**, *360*, 509.

37 H. R. Karfunkel, A. Hirsch, unpublished results.

38 N. Matsuzawa, T. Fukunaga, D. A. Dixon, *J. Phys. Chem.* **1992**, *96*, 10747.

39 C. C. Henderson, C. M. Rohlfing, R. A. Assink, P. A. Cahill, *Angew. Chem.* **1994**, *106*, 803; *Angew. Chem. Int. Ed. Engl.* **1994**, *33*, 786.

40 A. Hirsch, I. Lamparth, H. R. Karfunkel, *Angew. Chem.* **1994**, *106*, 453; *Angew. Chem. Int. Ed. Engl.* **1994**, *33*, 437.

41 P. A. Cahill, *Chem. Phys. Lett.* **1996**, *254*, 257.

42 R. E. Haufler, J. Conceicao, L. P. F. Chibante, Y. Chai, N. E. Byrne, S. Flanagan, M. M. Haley, S. C. O'Brien, C. Pan, Z. Xiao, W. E. Billups, M. A. Ciufolini, R. H. Hauge, J. L. Margrave, L. J. Wilson, R. F. Curl, R. E. Smalley, *J. Phys. Chem.* **1990**, *94*, 8634.

43 A. Peera, R. K. Saini, L. B. Alemany, W. E. Billups, M. Saunders, A. Khong, M. S. Syamala, R. J. Cross, *Eur. J. Org. Chem.* **2003**, 4140.

44 C. Rüchard, M. Gerst, J. Ebenhoch, H. D. Beckhaus, E. E. B. Campbell, R. Tellgmann, H. Schwarz, T. Weiske, S. Pitter, *Angew. Chem.* **1993**, *105*, 609; *Angew. Chem. Int. Ed. Engl.* **1993**, *32*, 584.

45 A. S. Lobach, A. A. Perov, A. I. Rebrov, O. S. Roschupkina, V. A. Tkacheva, A. N. Stepanov, *Russ. Chem. Bull.* **1997**, *46*, 641.
46 A. S. Lobach, Y. M. Shul'ga, O. S. Roshchupkina, A. I. Rebrov, A. A. Perov, Y. G. Morozov, V. N. Spector, A. A. Ovchinnikov, *Fullerene Sci. Technol.* **1998**, *6*, 375.
47 Y. Vasil'ev, D. Wallis, T. Drewello, M. Nuchter, B. Ondruschka, A. Lobach, *Chem. Commun.* **2000**, 1233.
48 W. E. Billups, A. Gonzalez, C. Gesenberg, W. Lui, T. Marriott, L. B. Alemany, M. Saunders, H. A. Jimenez-Vazquez, A. Khong, *Tetrahedron Lett.* **1997**, *38*, 175.
49 M. R. Banks, M. J. Dale, I. Gosney, P. K. G. Hodgson, R. C. K. Jennings, A. C. Jones, J. Lecoultre, P. R. R. Langridge-Smith, J. P. Maier, *J. Chem. Soc., Chem. Commun.* **1993**, 1149.
50 K. Balasubramanian, *Chem. Phys. Lett.* **1991**, *182*, 257.
51 J. Nossal, R. K. Saini, A. K. Sadana, H. F. Bettinger, L. B. Alemany, G. E. Scuseria, W. E. Billups, M. Saunders, A. Khong, R. Weisemann, *J. Am. Chem. Soc.* **2001**, *123*, 8482.
52 O. V. Boltalina, M. Buhl, A. Khong, M. Saunders, J. M. Street, R. Taylor, *J. Chem. Soc., Perkin Trans. 2.* **1999**, 1475.
53 B. W. Clare, D. L. Kepert, *Theochem* **2002**, *589–590*, 195.
54 M. S. Meier, P. S. Corbin, V. K. Vance, M. Clayton, M. Mollman, M. Poplawska, *Tetrahedron Lett.* **1994**, *35*, 5789.
55 A. D. Darwish, A. a. K. Abdul-Sada, G. J. Langley, H. W. Kroto, R. Taylor, D. R. M. Walton, *J. Chem. Soc., Perkin Trans. 2* **1995**, 2359.
56 A. D. Darwish, A. K. Abdul-Sada, G. J. Langley, H. W. Kroto, R. Taylor, D. R. M. Walton, *Synth. Methods* **1996**, *77*, 303.
57 R. Bini, J. Ebenhoch, M. Fanti, P. W. Fowler, S. Leach, G. Orlandi, C. Rüchardt, J. P. B. Sandall, F. Zerbetto, *Chem. Phys.* **1998**, *232*, 75.
58 D. K. Palit, H. Mohan, J. P. Mittal, *J. Phys. Chem. A* **1998**, *102*, 4456.
59 C. Rüchardt, M. Gerst, M. Nölke, *Angew. Chem.* **1992**, *104*, 1516; *Angew. Chem. Int. Ed. Engl.* **1992**, *31*, 1523.
60 M. Gerst, H. D. Beckhaus, C. Rüchardt, E. E. B. Campbell, R. Tellgmann, *Tetrahedron Lett.* **1993**, *34*, 7729.
61 A. A. Gakh, A. Y. Romanovich, A. Bax, *J. Am. Chem. Soc.* **2003**, *125*, 7902.
62 A. D. Darwish, A. G. Avent, R. Taylor, D. R. M. Walton, *J. Chem. Soc., Perkin Trans. 2* **1996**, 2051.
63 I. Rogner, P. Birkett, E. E. B. Campbell, *Int. J. Mass Spectrom. Ion Processes* **1996**, *156*, 103.
64 P. A. Dorozhko, A. S. Lobach, A. A. Popov, V. M. Senyavin, M. V. Korobov, *Chem. Phys. Lett.* **2001**, *336*, 39.
65 A. V. Okotrub, L. G. Bulusheva, I. P. Asanov, A. S. Lobach, Y. M. Shulga, *J. Phys. Chem. A* **1999**, *103*, 716.
66 R. V. Bensasson, T. J. Hill, E. J. Land, S. Leach, D. J. McGarvey, T. G. Truscott, J. Ebenhoch, M. Gerst, C. Rüchardt, *Chem. Phys.* **1997**, *215*, 111.
67 A. M. Vassallo, M. A. Wilson, M. I. Attalla, *Energy & Fuels* **1988**, *2*, 539.
68 M. I. Attalla, M. A. Wilson, R. A. Quezada, A. M. Vassallo, *Energy & Fuels* **1989**, *3*, 59.
69 M. I. Attalla, A. M. Vassallo, B. N. Tattam, J. V. Hanna, *J. Phys. Chem.* **1993**, *97*, 6329.
70 K. Shigematsu, K. Abe, M. Mitani, K. Tanaka, *Chem. Express* **1993**, *8*, 37.
71 K. Shigematsu, K. Abe, M. Mitani, K. Tanaka, *Chem. Express* **1992**, *7*, 957.
72 T. Osaki, T. Tanaka, Y. Tai, *Phys. Chem. Chem. Phys.* **1999**, *1*, 2361.
73 T. Osaki, T. Hamada, Y. Tai, *React. Kinet. Catal. Lett.* **2003**, *78*, 217.
74 K. P. Meletov, S. Assimopoulos, I. Tsilika, I. O. Bashkin, V. I. Kulakov, S. S. Khasanov, G. A. Kourouklis, *Chem. Phys.* **2001**, *263*, 379.
75 S. M. C. Vieira, W. Ahmed, P. R. Birkett, C. A. Rego, *Chem. Phys. Lett.* **2001**, *347*, 355.
76 B. I. Dunlap, D. W. Brenner, J. W. Mintmire, R. C. Mowrey, C. T. White, *J. Phys. Chem.* **1991**, *95*, 5763.
77 M. Saunders, *Science* **1991**, *253*, 330.
78 D. A. Dixon, N. Matsuzawa, T. Fukunaga, F. N. Tebbe, *J. Phys. Chem.* **1992**, *96*, 6107.
79 D. Bakowies, W. Thiel, *Chem. Phys. Lett.* **1992**, *192*, 236.
80 T. Guo, G. E. Scuseria, *Chem. Phys. Lett.* **1992**, *191*, 527.

81 J. Cioslowski, *Chem. Phys. Lett.* **1991**, *181*, 68.
82 B. W. Clare, D. L. Kepert, *Theochem* **2002**, *589–590*, 209.
83 B. W. Clare, D. L. Kepert, *Theochem* **2003**, *622*, 185.
84 B. W. Clare, D. L. Kepert, *Theochem* **1996**, *363*, 179.
85 B. W. Clare, D. L. Kepert, *Theochem* **1999**, *466*, 177.
86 D. I. Dunlap, D. W. Brenner, G. W. Schriver, *J. Phys. Chem.* **1994**, *98*, 1756.
87 M. Bühl, W. Thiel, U. Schneider, *J. Am. Chem. Soc.* **1995**, *117*, 4623.
88 L. D. Book, G. E. Scuseria, *J. Phys. Chem.* **1994**, *98*, 4283.
89 H. Dodziuk, K. Nowinski, *Chem. Phys. Lett.* **1996**, *249*, 406.
90 H. Dodziuk, O. Lukin, K. S. Nowinski, *Pol. J. Chem.* **1999**, *73*, 299.
91 S. Jenkins, M. I. Heggie, R. Taylor, *J. Chem. Soc., Perkin Trans. 2* **2000**, 2415.
92 S. J. Austin, R. C. Batten, P. W. Fowler, D. B. Redmond, R. Taylor, *J. Chem. Soc., Perkin Trans. 2* **1993**, 1383.

6 Radical Additions

6.1 Introduction

Similar to nucleophiles (see Chapter 3), a large variety of radicals easily add to C_{60} and C_{70}, forming diamagnetic or paramagnetic adducts. Fullerenes behave like a radical sponge. Defined radical species, such as $R_nC_{60}^{\bullet}$ ($n = 1, 3, 5$), have been intensively investigated by ESR spectroscopy, which revealed various interesting phenomena of these adducts. One of them is the formation of fullerene dimers. In addition to such spectroscopic in situ investigations, the addition of radicals to C_{60} has been used to synthesize new materials, including polymers and perfluoroalkylated fullerenes, which may have importance for technological applications. The high affinity of C_{60} and C_{60}-derivatives to radicals makes them a potential radical scavenger. Some C_{60}-derivatives have already demonstrated efficiency in trapping radicals such as superoxide radical or hydroxyl radical, which play a role in neurodegenerative diseases.

6.2 ESR Investigations of Radical Additions

6.2.1 Addition of Single Radicals

ESR studies of a large variety of monoradical adducts, $RC_{60}^{\bullet}$ and $RC_{70}^{\bullet}$, have provided a valuable insight into the electronic situation on the fullerene surface and other physical phenomena of these species [1–19]. The organic radical species $R^{\bullet}$, which is to be added to the fullerene, can be generated in situ either photochemically or thermally by established free radical reactions from suitable precursor molecules. The photochemical generation of the radicals $R^{\bullet}$ can be achieved by the direct UV irradiation in the ESR cavity of saturated benzene or *tert*-butylbenzene solutions of C_{60} containing a small molar excess of a radical precursor, such as alkyl bromides, carbon tetrachloride, dialkylmercury compounds, hydrocarbons (RH), and di-*tert*-butylperoxide. Thereby, the radicals $R^{\bullet}$ are either generated directly from the halides RX, for example with *tert*-butylbromide, or indirectly using *tert*-butoxy radicals as intermediates (Scheme 6.1).

Fullerenes: Chemistry and Reactions. Andreas Hirsch and Michael Brettreich

ISBN: 3-527-30820-2

Scheme 6.1 X = Hal, HgR, H

Alkyl radicals have also been prepared by reaction of alkylbromides with photolytically generated $^{\bullet}Re(CO)_5$ (from $Re_2(CO)_{10}$) [17], photolysis of cobalt-alkyl complexes [20], photolysis of AIBN [17, 21, 22] or thermolysis of TEMPO adducts [23].

The stability of the radical species $RC_{60}^{\bullet}$ strongly depends on the nature of R. Sterically demanding groups, such as the *tert*-butyl group, give rise to rather persistent radicals $RC_{60}^{\bullet}$ [1–3, 12], but the less stable $HC_{60}^{\bullet}$ can also be studied in situ [5, 12]. The ESR spectrum of $^{t}BuC_{60}^{\bullet}$ (Figure 6.1) consists of ten narrow lines appropriate for a hyperfine interaction of the unpaired electron with nine equivalent protons of the *tert*-butyl group [1, 12].

Besides $^{t}BuC_{60}^{\bullet}$, a series of other monoalkyl- and also monoaryl-radical adducts have been similarly investigated [1–4, 12, 15–17, 20–23]. The observed hyperfine structures are those expected for the corresponding group R. An unusual long-

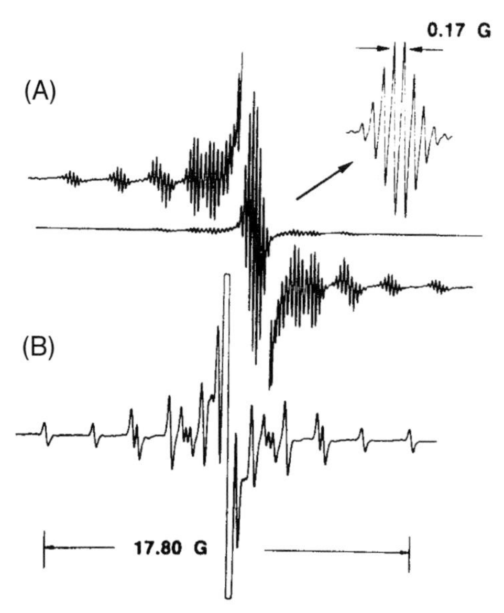

Figure 6.1 ESR spectra of $^{t}BuC_{60}^{\bullet}$ in benzene at 80 °C: (A) $(CH_3)_3CC_{60}^{\bullet}$ shows ^{13}C hyperfine satellites and the 0.17 G hyperfine splitting due to the 9 methyl protons. (B) $(CD_3)_3CC_{60}^{\bullet}$ shows ^{13}C satellites [1].

range proton hyperfine interaction was observed in adamantyl- and bicyclooctyl-C_{60}-radical adducts [1, 3, 10]. In adamantyl-$C_{60}^{\bullet}$ the six protons nearest to the C_{60} surface have smaller hyperfine interactions than the three more remote tertiary protons. Another unusual hyperfine splitting occurs for the C_6H_5-radical [15]. The phenyl monoadduct radical shows a triplet with the intensity 1 : 2 : 1. Only the pair of protons in the meta-position interacts with the unpaired electron. There is no hyperfine interaction with the ortho and para protons. This assignment was established by the spectral changes observed when these protons are replaced by various substituents.

Fluorinated alkyl groups (R_F) can be radically attached to C_{60}, for example by reaction of mercury-alkyl compounds under irradiation [10, 24] or by reaction with fluorinated alkyl iodides or bromides [25, 26]. Photochemical reaction with $(R_3Sn)_2$ or thermal reaction with R_3SnH generates the R_F radical from the corresponding iodides or bromides.

From the ^{13}C satellites in the ESR spectra, the symmetry of the radicals as well as the information about the localization of the unpaired electron in $RC_{60}^{\bullet}$ can be obtained (Figure 6.2). In $^tBuC_{60}^{\bullet}$, for example, ten pairs of ^{13}C satellites are clearly resolved (Figure 6.1). A C_s-symmetry can be deduced from their relative intensities and splittings [1, 12]. The unpaired electron is mostly located on two fused six-membered rings of the C_{60} surface, which rules out an extensive delocalization on the C_{60} framework.

Initially, the highest spin densities were assigned to the carbons C-1, C-3/C-3′ and C-5/C-5′, which lead to the formulation of the major canonical resonance forms represented in Figure 6.2 [1]. More measurements and quantum chemical calculations led to a reassignment of the density of the unpaired spin [12, 27]. The largest hyper fine interaction (Figure 6.1) belongs to C-1, which places the highest spin density on this carbon. The second largest interaction was assigned to the tertiary

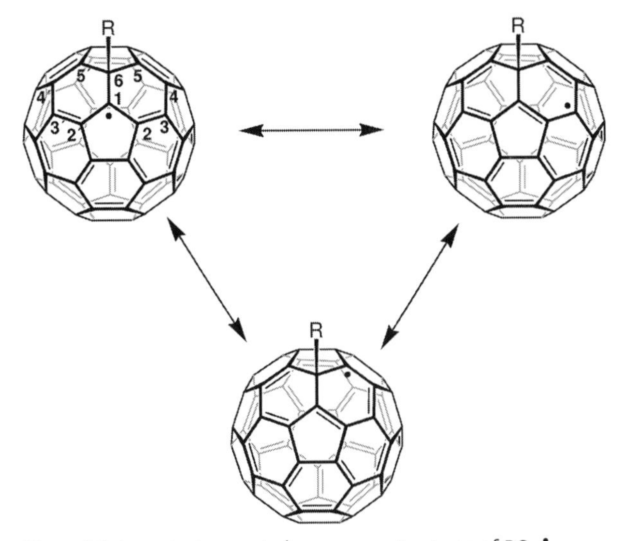

Figure 6.2 Important canonical resonance structures of $RC_{60}^{\bullet}$.

carbon of the *tert*-butyl-group and the two next-largest hyperfine interactions were assigned to C-2/C-2′ and C-3/C-3′. Contrary to earlier findings, almost no spin density is found on C-5/C-5′, which rules out the third canonical resonance structure in Figure 6.2. C-1 and C-3/C-3′ show large positive hyperfine interactions whereas C-2 and C-2′ have large, negative hyperfine interactions. For the latter a spin-polarization effect is probably important. The analogous investigations with other radicals $RC_{60}^{\bullet}$, for example with $CCl_3C_{60}^{\bullet}$ or $C_{60}H^{\bullet}$, the simplest hydrogenated fullerene radical, lead qualitatively to the same results [3, 5].

$C_{60}H^{\bullet}$ was generated in situ by irradiating saturated degassed benzene solutions of C_{60} in the presence of tri-*n*-butyltin hydride, 1,4-cyclohexadiene or thiophenol with the focused light of a high-pressure Hg/Xe arc (Scheme 6.2). Alternatively, ESR spectra were also recorded in solid neon [5]. The proton hyperfine splitting of $C_{60}H^{\bullet}$ was determined to be 33.07 G [5, 12].

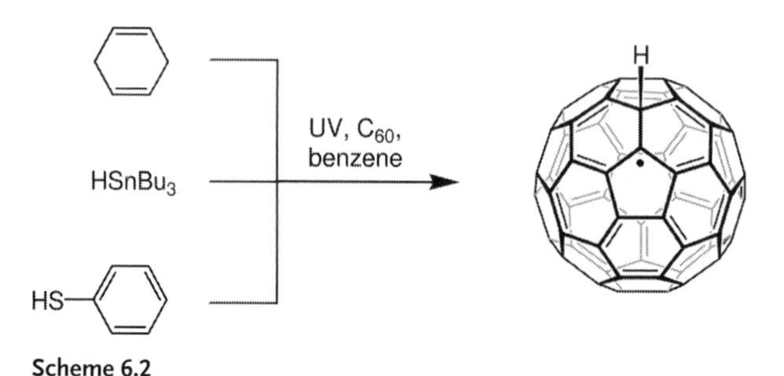

Scheme 6.2

In the monoalkyl radical adducts, $RC_{60}^{\bullet}$ exhibits a substantial hindered rotation around the C_{60}–R bond [10, 12]. This was concluded inter alia from the temperature-dependent line-shape changes in the ESR spectra and from the observation of distinct rotamers for $(^{13}CH_3)(CH_3)_2CC_{60}^{\bullet}$ with staggered conformations at low temperatures (Figure 6.3) [4]. Simulations [28–30] of the temperature-dependent spectral line shapes led to an activation energy for the internal rotation in $^tBuC_{60}^{\bullet}$ of 8.2 kcal mol^{-1}. A similar value (9.3 kcal mol^{-1}) was found in the anion $^tBuC_{60}^{-}$ [31]. Whereas for $RC_{60}^{\bullet}$ (R = ethyl, isopropyl) the internal rotations are also hindered with similar values for the activation energy barrier and the alkyl groups adopt a preferred conformation on the C_{60} surface even at elevated temperatures, the methyl group in $CH_3C_{60}^{\bullet}$ above 270 K undergoes a fast rotation on the ESR time scale [4].

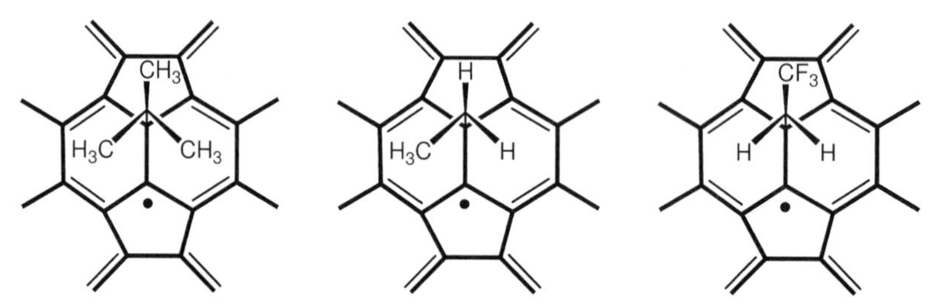

Figure 6.3 Preferred conformations in the radicals $^tBuC_{60}^{\bullet}$, $CH_3CH_2C_{60}^{\bullet}$ and $CF_3CH_2C_{60}^{\bullet}$.

In $CH_3CH_2C_{60}^{\bullet}$ and $CF_3CH_2C_{60}^{\bullet}$, the preferred equilibrium conformations are asymmetrical. The terminal methyl group in $CH_3CH_2C_{60}^{\bullet}$ prefers the position directly above a hexagon (Figure 6.3) [4], whereas the terminal group in $CF_3CH_2C_{60}^{\bullet}$ prefers the position above a pentagon [32–34]. These results can be generalized. It appears that, in radicals of the type $XYZCC_{60}^{\bullet}$, where X,Y and Z are CF_3, F, H or CH_3, a CH_3 group will never gain the pentagon position and a CF_3 group will never gain a hexagon position, except when there is no alternative [12].

Alkylthio and alkoxy radicals can also add to C_{60} (Schemes 6.3 and 6.4) [7, 10, 12, 35, 36]. With the following radical precursors the radicals $RS^{\bullet}$ and $RO^{\bullet}$ are generated in situ upon irradiation: alkyl-disulfides (RSSR) or bis(alkylthio)mercury compounds (RSHgSR) for the formation of $RS^{\bullet}$ and dialkyl peroxides (ROOR) or dialkoxy-disulfides (ROSSOR) for the formation of $RO^{\bullet}$. The radicals $RSC_{60}^{\bullet}$ are not persistent. Their ESR spectra are only observable as long as light is shining through the sample. The intensities of the signals in the spectra of $RSC_{60}^{\bullet}$ also decrease above room temperature. This behavior points to a reversal of the alkylthio radical addition due to the weak fullerene–sulfur bond [7]. This became especially evident upon the gradual change from amber for phenylthiyl radical adducts to the original purple of C_{60} [37]. In some cases, the cleavage of the carbon–sulfur bond of RSSR (R = benzyl, *tert*-butyl) was observed, which leads exclusively or partly to the radicals $RC_{60}^{\bullet}$. This is particularly the case when the light is not filtered through a Ni/Co sulfate solution [7].

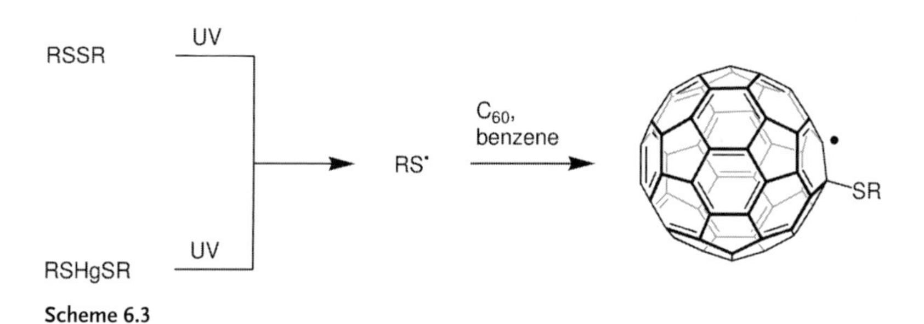

Scheme 6.3

Since dialkylperoxides are dangerous to handle, only three different $ROC_{60}^{\bullet}$ radicals have prepared by reaction with these reagents (Scheme 6.4). Dialkoxy-disulfides can easily be synthesized with any kind of R and are, therefore, a more convenient source of $RO^{\bullet}$ radicals [36].

For $CH_3SC_{60}^{\bullet}$ four local conformational minima are obtained by molecular mechanics calculations using the MMX force field (Figure 6.4) [7]. Thereby, as shown in Figure 6.4, the asymmetric conformation for this radical is preferred.

Phosphoryl radicals [10, 18, 38–42] tend to add to double bonds. Owing to the exceptionally high constants of hyperfine coupling of the unpaired electron with the phosphorus nucleus, phosphoryl radicals can be utilized as "paramagnetic reporters" [10]. Phosphoryl radicals have been prepared by photolysis of diphosphoryl mercury compounds (Scheme 6.5).

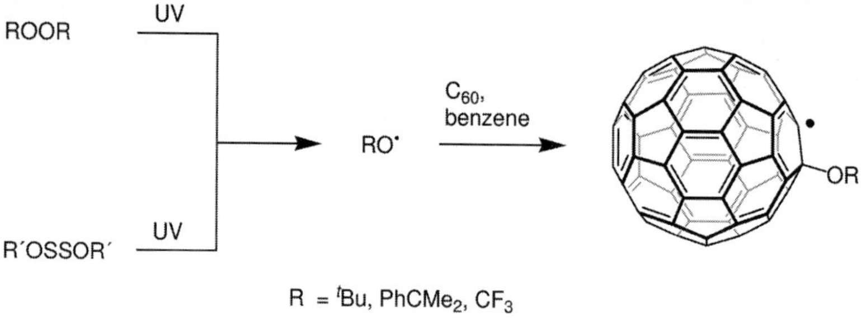

ROOR $\xrightarrow{UV}$ / R′OSSOR′ $\xrightarrow{UV}$ → RO• $\xrightarrow{C_{60},\ benzene}$ •C_{60}–OR

R = ^{t}Bu, $PhCMe_2$, CF_3

R′ = Me, Et, ^{i}Pr, ^{t}Bu, $^i PrCH_2$

Scheme 6.4

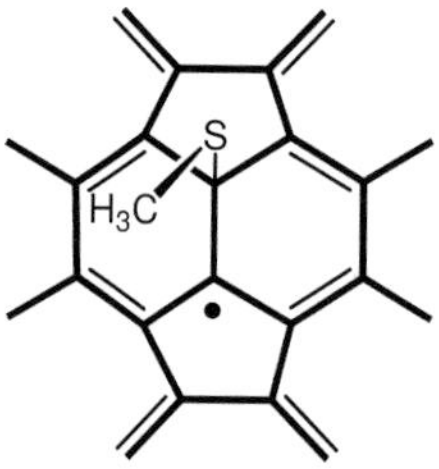

H_3C S

Figure 6.4 Structures of the preferred local conformational minima of the radical $CH_3SC_{60}^{\bullet}$ obtained by molecular mechanics calculations [7].

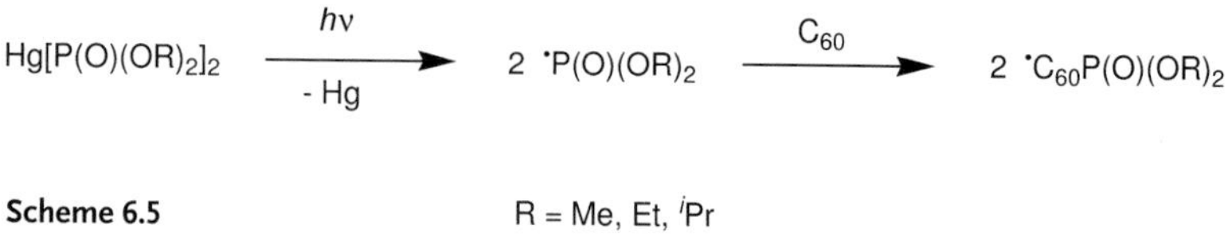

$Hg[P(O)(OR)_2]_2 \xrightarrow[-\ Hg]{h\nu} 2\ {}^{\bullet}P(O)(OR)_2 \xrightarrow{C_{60}} 2\ {}^{\bullet}C_{60}P(O)(OR)_2$

R = Me, Et, ^{i}Pr

Scheme 6.5

In the phosphoryl fullerenyl radical the unpaired electron is – similar to the alkylfullerenyl radical – delocalized over two six-membered rings adjacent to the C–P bond [10]. The rotation barrier for the radical ${}^{\bullet}C_{60}P(O)(O^iPr)_2$ was determined to be 4.8 kcal mol^{-1}. Another phosphorus-containing radical fullerenyl adduct, determined via ESR spectroscopy, is $PF_3(O^tBu)C_{60}^{\bullet}$. It can be obtained by photolytic reaction of C_{60} with HPF_4 and *tert*-butylperoxide in *tert*-butylbenzene [43].

ESR spectra of alkyl radical adducts $RC_{60}^{\bullet}$ (R = alkyl) show a characteristic temperature dependence of the signal intensities. For $^tBuC_{60}^{\bullet}$, for example, the intensity significantly increases as the temperature is raised from 30 to 130 °C and decreases again upon cooling to room temperature and disappears at ca. 10 °C [2, 3]. Such a cycle can be repeated several times without significant radical decay. This behavior, similar to that of the classical Gomberg radical $Ph_3C^{\bullet}$ [44], was interpreted in terms of an equilibrium of the radicals $RC_{60}^{\bullet}$ with their diamagnetic dimers $RC_{60}C_{60}R$ (Scheme 6.6) [2, 3].

A comparison of the temperature behavior of various $RC_{60}^{\bullet}$ radicals shows that the dimer bond strength depends on the size of the entering radical (Table 6.1).

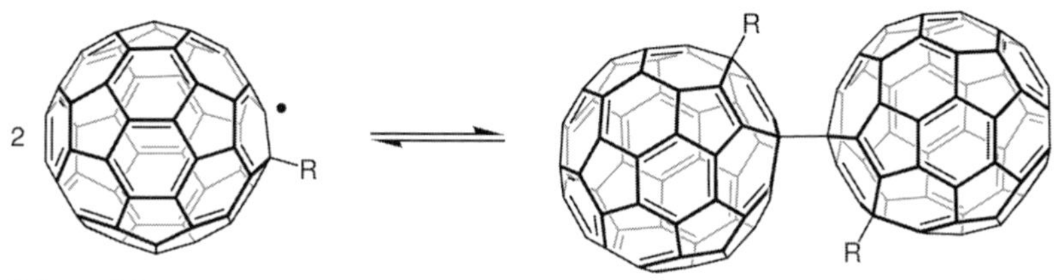

Scheme 6.6

Table 6.1 Enthalpies of dimerization for $RC_{60}^{\bullet}$ radicals [3].

R	ΔH (kcal mol^{-1})	*T* (K)
iso-Propyl	35.5	420–450
CCl_3	17.1	250–310
CBr_3	17.0	300–380
tert-Butyl	22.0	300–400
Adamantyl	21.5	300–400

A measure for the dimer bond strengths are the enthalpies of the dimerization (Scheme 6.6), obtained from plots of ln($T \times$ intensity) against $1000/T$ [2, 3].

The dependence of the dimer bond strength on the size of R suggests that bond formation occurs in close proximity to the C-atom carrying R. This accords with the observed predominant localization of the unpaired electron at C-1, C-3 and C-3′ (Figure 6.2). Dimerization at C-3 or C-3′ seems to be most likely for steric reasons [2, 3]. Due to the weakness of the carbon–carbon bond between the C_{60}-molecules, the dimers can also be cleaved with visible light. Excitation of the alkylfullerene dimers at 532 nm leads quantitatively to the alkylfullerene radicals [16]. Fullerene dimers that contain fluorinated alkyl groups (R_F) have been efficiently synthesized by reacting fluorinated alkyl iodides in the presence of $(R_3Sn)_2$. Dimers with various R_F groups can be obtained in about 50% yield [26].

The addition of a bulky radical to C_{70} gives rise to three different regioisomeric radical adducts of $RC_{70}^{\bullet}$ [6]. Radicals such as $(CH_3)C^{\bullet}$, $CCl_3^{\bullet}$ or $(MeO)_2PO^{\bullet}$ have been used. The latter can be generated in situ upon photolysis of di-*tert*-butyl peroxide in benzene containing dimethyl phosphite. The more reactive aryl and fluoroalkyl radicals give four isomers, except for the trifluoromethyl radical, which yielded, for the first time, five isomers [15]. The very reactive methoxy radical also yields five ESR detectable isomers of $MeOC_{70}^{\bullet}$ [35]. Since C_{70} contains five types of C atoms, five different radical mono-adducts $RC_{70}^{\bullet}$ are possible (Figure 6.5) [10, 15, 45]. The ESR signals of one radical species of $RC_{70}^{\bullet}$ can be attributed to structure **A** in Figure 6.5. This is the most "C_{60}-like" structure and produces a similar g value (2.0021) and comparable hyperfine interactions as observed in $RC_{60}^{\bullet}$ (R = *tert*-butyl). The temperature dependence of the ESR signal intensity of this radical adduct $RC_{70}^{\bullet}$ (structure **A**) is also characteristic for the dissociation of a dimer (ΔH = ca. 11 kcal mol^{-1}) [6]. The regiochemistry of radical addition to C_{70} is not yet well understood. In various additions the relative yields follow the order D > C > B > A > E, which is unusual when compared with other addition reactions to C_{70} [10].

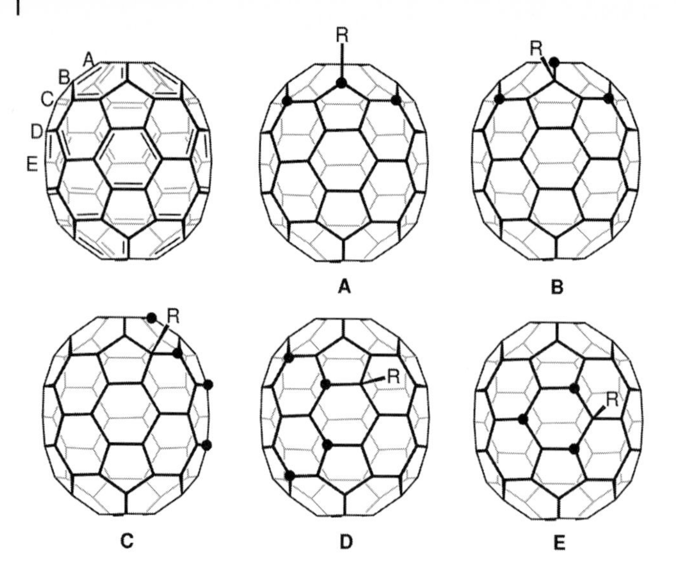

Figure 6.5 The five possible structures of $RC_{70}^{\bullet}$. Dots represent the carbons with the highest spin density, as calculated for the $HC_{70}^{\bullet}$ isomers [45].

6.2.2
Multiple Radical Additions

Prolonged irradiation of C_{60} solutions in the presence of excess radical precursors leads to multiple radical additions [10, 12, 37, 46], giving rise to broad, featureless ESR spectra. Some of these multiple radical species are rather persistent. Furthermore, radicals with higher spin multiplicities have been observed [37]. Valuable insight of such multiple radical additions came from the ESR spectroscopic investigations of benzyl radicals, ^{13}C labeled at the benzylic positions [46]. These radicals can be prepared in situ by photolysis of saturated solutions of C_{60} in labeled toluene containing about 5% di-*tert*-butyl peroxide [47]. Thereby, the photochemically generated *tert*-butoxy radicals readily abstract a benzylic hydrogen atom from the toluene. Two radical species with a different microwave power saturation behavior can be observed [46]. The ESR absorptions of these species do not decay when the light is extinguished. The spectrum of each species can be recorded without being superimposed by the others, simply by changing the microwave power level. One radical species can be attributed to an allylic radical (**3**) and the other to a cyclopentadienyl radical (**5**) formed by the addition to three and five adjacent [5]radialene double bonds, respectively (Scheme 6.7).

In **3** a coupling occurs to three nuclei (two identical and one different) and in **5** to five equivalent ^{13}C nuclei (Figure 6.6). In these experiments no evidence for the radical **1** is found, which is very likely a short-lived species [46].

The ESR spectra of **3** and **5** do not provide information on whether the corresponding radical species carry an even number of benzyl groups attached elsewhere on

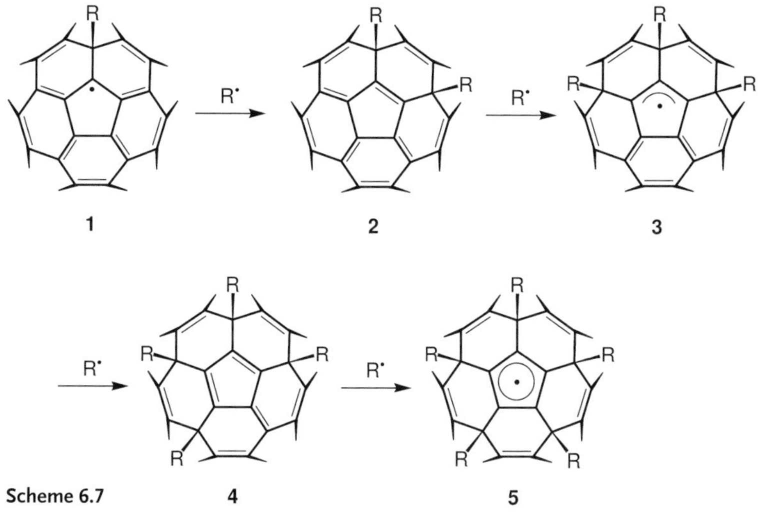

Scheme 6.7

the surface of the C_{60} molecule. Looking at the high reactivity of radicals towards C_{60}, it is very unlikely that only single radical adducts, such as $R_3C_{60}^{\bullet}$ or $R_5C_{60}^{\bullet}$, are responsible for the observed ESR spectra. However, very important information about the electronic and chemical properties of a certain area of the fullerene surface, namely a corannulene substructure, is extractable from these experiments. The orbitals with a spin density in radical species containing a moiety **5** are highly degenerated due to the local five-fold symmetry. Such radicals have an efficient spin-relaxation behavior and do not power saturate very readily [8]. The formation of species **3** and **5** from C_{60} proceeds by initial addition of one benzyl radical, leading

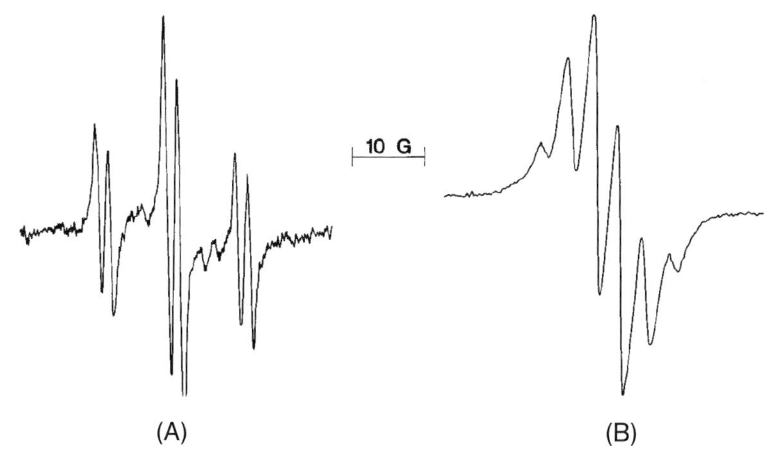

Figure 6.6 ESR spectra of (A) **3** at 0.2 mW (low microwave power), $g = 2.00250$, hyperfine interaction for two equivalent carbons of 9.70 G and for one carbon of 1.75 G (doublet of triplet); and of (B) **5** at 200 mW (high microwave power), showing hyperfine interaction for five equivalent carbons of 3.56 G (sextet) [12].

to the monoadduct **1**. The unpaired spin in this radical is mostly localized on C-1, C-3 and C-3′ (Figure 6.2). This electronic localization as well as the steric requirement of the benzyl group will direct a second attack accompanied with a radical recombination to C-3 or C-3′. A third attack, on the diamagnetic **2**, can occur anywhere on the fullerene surface. However, upon formation of **3** the unfavorable [5,6] double bond [48] in **2** disappears and a resonance-stabilized allyl radical is formed. Indeed, semiempirical calculations on two different adducts of $H_3C_{60}^{\bullet}$, **6** and **7**, show that **6**, in which the three hydrogens are all added in adjacent 1,4-positions, is about 8.5 kcal mol^{-1} lower in energy than **7**, where the hydrogens are added in non-adjacent positions (Figure 6.7) [8].

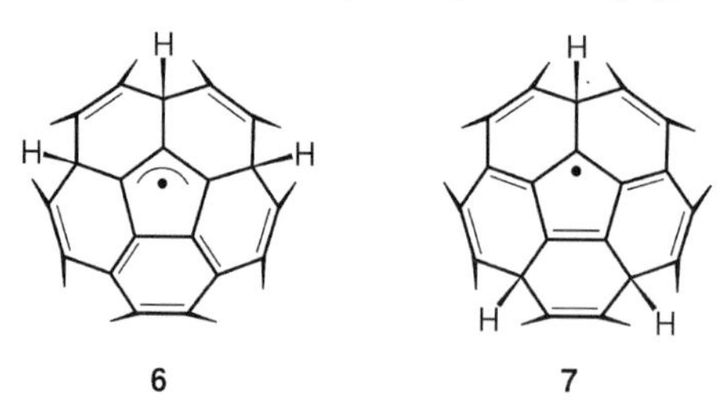

Figure 6.7 Two possible structures of $H_3C_{60}^{\bullet}$.

A radical $R_3C_{60}^{\bullet}$, in which a third attack occurs far from the groups already attached in $C_{60}R_2$ (**2**), would be very unstable due to a facile formation of a diamagnetic $C_{60}R_4$, which is not detectable in the ESR spectra. Another reason for the stability of **3** is the steric protection provided by the three attached benzyl groups. Radical recombination of **3** with a benzyl radical leads to **4** for the same reasons for which **2** is formed from **1**. The stability of **4** is restricted due to the two unfavorable [5,6] double bonds, which directs another attack to form the resonance stabilized **5**. In addition, steric arguments are responsible for the remarkable stability of **5**. For methyl radical additions, this steric hindrance is by far less important. This may be why the corresponding species **3** and **5** of methyl adducts of C_{60} are not observable by ESR spectroscopy [46].

A stepwise addition of benzyl radicals could be observed by using the dimer $[F(C_6H_4)C(CF_3)_2]_2$ as a radical precursor [9, 49]. Every 10 to 20 min, one more radical is added to C_{60} and these subsequent additions can be studied in the ESR spectrum. For the bulkier 3,5-dimethylphenylmethyl radical only four additions and no formation of the pentaadduct radical could be detected [49].

Mass spectrometric investigations of reaction products obtained by multiple radical additions show that up to eight benzyl groups are added to C_{60}. If dibenzyl ketone is used as a radical source, up to 15 benzyl groups are added. The addition of methyl radicals leads to products with up to 34 methyl groups attached to C_{60} [46].

Additions of perfluoroalkyl groups generated from perfluoroalkyl iodides (R_FI) and perfluoroacyl peroxides [$R_FC(O)OO(O)CR_F$] have also been carried out [50]. Another usable source of CF_3 radicals is the stable $[(CF_3)_2CF]_2{}^{\bullet}CCF_2CF_3$ radical (Scheme 6.8) [51]. As with alkyl radical additions, ESR spectroscopic investigations

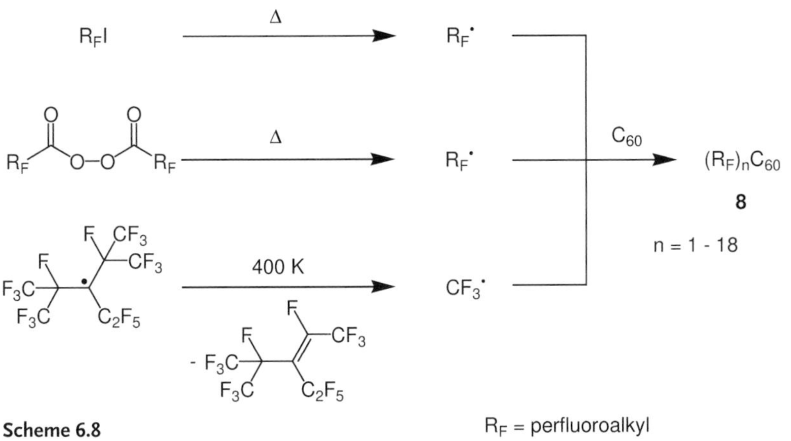

n = 1 - 18

Scheme 6.8

R_F = perfluoroalkyl

of fluoroalkyl monoadducts show a hindered rotation about the R_F–C_{60} bond and the tendency to dimerization. Perfluoralkylated fullerenes **8** (Scheme 6.8) can be produced by treating C_{60} with the radical precursors R_FI or $R_fC(O)OO(O)CR_F$ in a sealed glass tube at 200 °C for 24 h, with fluorocarbons or halofluorocarbons as solvents [50]. This leads to a substantial coverage of the C_{60} surface with up to 16 [50] or 18 [51] fluoroalkyl groups covalently bound. Owing to steric crowding, coverage of the C_{60} surface with more than 24 perfluoroalkyl groups seems unrealistic [50]. Whereas C_{60} is rather insoluble in fluorocarbons or halofluorocarbons, **8** shows a remarkable solubility in these solvents. Using other solvents to synthesize perfluoroalkylated fullerenes, such as 1,2,4-trichlorobenzene, leads to polyadducts also having hydrogen bound to C_{60}. The solid perfluoroalkylated fullerenes **8** are amorphous glassy materials, being more volatile than the parent C_{60} [50]. Under high vacuum, the perfluorohexylated **8** can be sublimed quantitatively to deposit a thin film on a glass substrate. These materials are also remarkably stable – they do not decompose until 270 °C and resist treatment with aqueous sulfuric acid and sodium hydroxide.

More details about multiple radical adducts of C_{60} can be found in Chapter 10.

6.3 Addition of Tertiary Amines

Primary and secondary amines can add to C_{60} as nucleophiles (Section 3.3). Tertiary amines can not form similar addition products, rather an electron transfer under formation of zwitterions is often observed (Section 3.3). However, a photochemical reaction of tertiary amines with C_{60} is possible and leads to complex mixtures of addition products [52–62]. The product distribution strongly depends on conditions such as temperature and the presence of either light or oxygen. If oxygen is thoroughly excluded, **9** is the major product (Figure 6.8) in the photoaddition of triethylamine [56, 59]. It can be isolated in low yields.

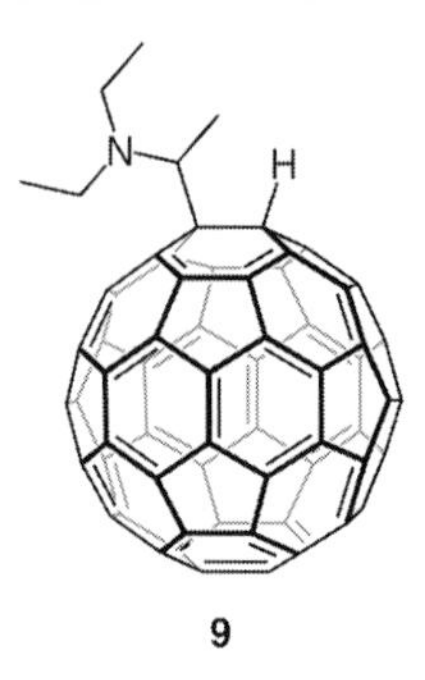

Figure 6.8 **9** can be obtained by photochemical reaction of NEt_3 with C_{60} under exclusion of O_2.

In the presence of oxygen or in air-saturated solutions, **11** is the major product (Scheme 6.9) [56, 59]. The formation of the cycloadduct follows a multistep process that probably includes several inter- or intramolecular electron-transfer steps. Suggestions include the electron transfer to singlet oxygen as an important part of the mechanism [58]. One possible mechanism is shown in Scheme 6.9. Oxygen plays the role of both an electron- and a proton-acceptor.

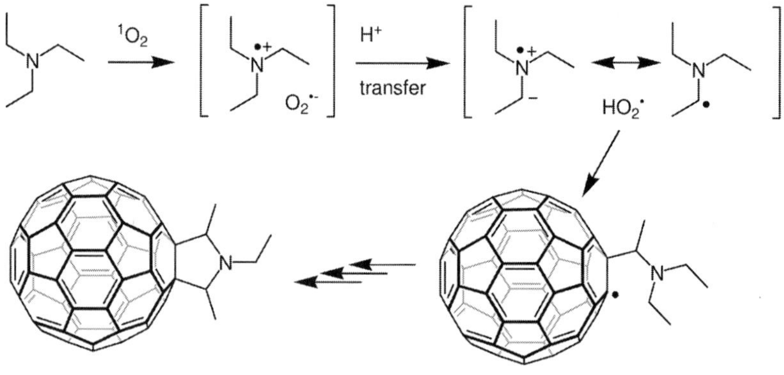

$^1O_2$ H⁺ transfer $O_2^{\bullet -}$ $HO_2^{\bullet}$

11 **10**

Scheme 6.9

For the formation of **15** (Figure 6.9) a different mechanism has been proposed [62]. Reaction of singlet oxygen with the corresponding amine (iminodiacetic methylester) leads – after two H-abstractions – to an azomethine ylide as a key intermediate. This ylide can add to the double bond via a [3+2]-cycloaddition related to the Prato reaction (Section 4.3.4).

Further products synthesized via photochemical addition of amines to C_{60} are shown in Figure 6.9 [54, 56–58]. The yields are usually in a range of ca. 15–25%. More complex addition products could be obtained in good yields by reaction with the alkaloids scandine, tazettine or gramine [60, 61].

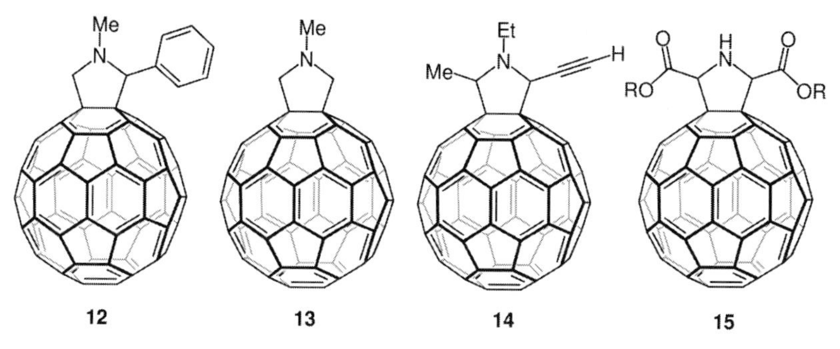

Figure 6.9 Different products of the photochemical addition of amines to C_{60}.

6.4 Photochemical Reaction with Silanes

Trialkylsilyl radicals prepared from the corresponding silanes add to C_{60} to form $R_3SiC_{60}^{\bullet}$ (Scheme 6.10) [63]. The Si–C_{60} bond is slightly longer than the C–C_{60} bond in an alkyl fullerenyl radical, and the rotation barrier is also lower. Hindered rotation can not be observed in the ESR spectrum. For $(^tBu)_3SiC_{60}^{\bullet}$, the hyperfine interaction of the unpaired electron with 27 equivalent protons can be seen [10, 63].

$$C_{60} + R_3Si^{\bullet} \longrightarrow R_3SiC_{60}^{\bullet} \quad (\mathbf{16}) \qquad R = Me, Et, {}^iPr, {}^tBu$$

Scheme 6.10

Further reaction of **16** with another trialkylsilyl radical $R_3Si^{\bullet}$ leads to the 1,16-bis-silylated adduct (**17**) as the major product (Scheme 6.11) [64–66]. Different alkyl groups (R) can be used and the yields are generally good (≈50%). If at least one of the silyl substituents is a phenyl group, another reaction pathway can be observed (Scheme 6.11) [64, 65]. The unusual 1,2-addition product **18** is probably generated by the mechanism shown in Scheme 6.11. For some silanes, especially with a higher number of phenyl groups, adduct **18** is the only product and can also be isolated in good yields.

Photochemical reaction with disiliranes leads to a [3+2]-cycloaddition (Section 4.3.9). Disilylcyclobutanes [67, 68] and cyclotetrasilanes [68, 69] react in a similar fashion. In both cases the four-membered ring is photolytically cleaved and a diradical is formed. This diradical adds in a formal cycloaddition to the [6,6] double bond of C_{60}. The cycloadduct **20** (R = 4-MeC_6H_4) can be obtained from **19** in 13% yield, but it is not the major product (Scheme 6.12). Rearrangement of an H atom leads to **21** in 46% yield. The product distribution depends strongly on R. Changing R from 4-MeC_6H_4 to phenyl leads to an exclusive formation of **21**.

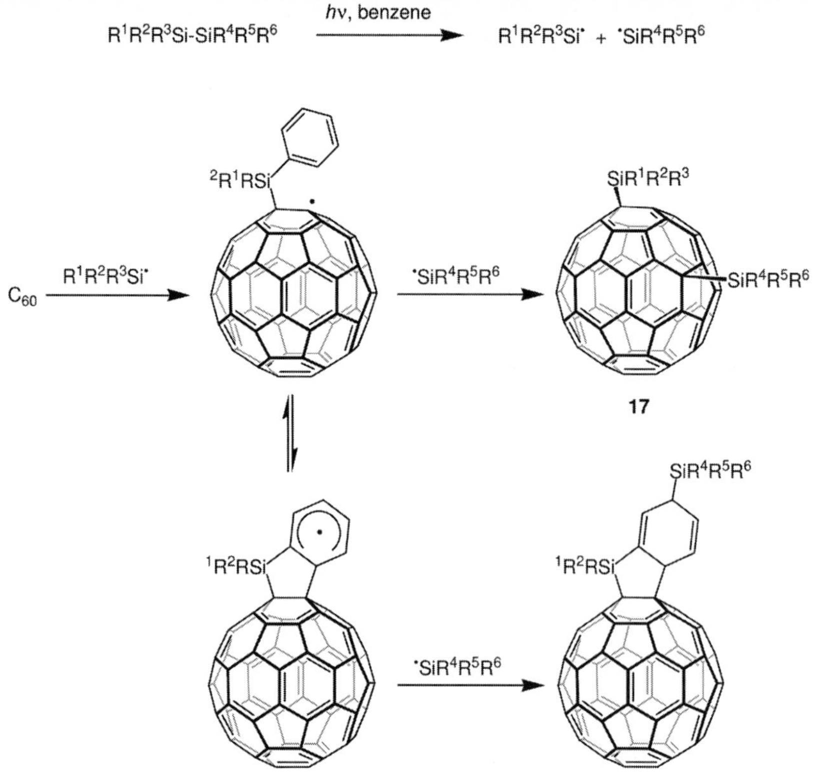

$$R^1R^2R^3Si\text{-}SiR^4R^5R^6 \xrightarrow{h\nu,\ \text{benzene}} R^1R^2R^3Si^\bullet + {}^\bullet SiR^4R^5R^6$$

C_{60} $\xrightarrow{R^1R^2R^3Si^\bullet}$ $^2R^1RSi$ (radical adduct) $\xrightarrow{{}^\bullet SiR^4R^5R^6}$ $SiR^1R^2R^3$ / $SiR^4R^5R^6$ **17**

$^1R^2RSi$ (cyclized radical) $\xrightarrow{{}^\bullet SiR^4R^5R^6}$ $SiR^4R^5R^6$ / $^1R^2RSi$ **18**

Scheme 6.11

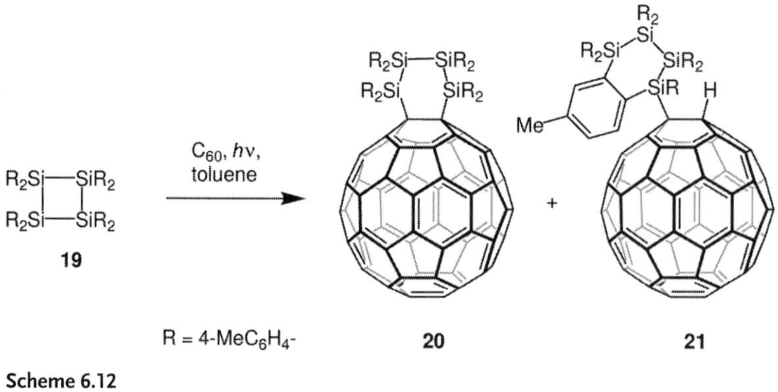

R_2Si—SiR_2 / R_2Si—SiR_2 **19** $\xrightarrow{C_{60},\ h\nu,\ \text{toluene}}$ R_2Si—SiR_2 / R_2Si SiR_2 **20** + R_2 Si R_2Si SiR_2 SiR H Me **21**

R = 4-MeC$_6$H$_4$-

Scheme 6.12

6.5 Metalation of C_{60} with Metal-centered Radicals

Photodissociation of the Re–Re bond of $Re_2(CO)_{10}$ has been used to generate the rhenium carbonyl radicals $^{\bullet}Re(CO)_5$, which add to C_{60} in benzene to form $C_{60}[Re(CO)_5]_2$ **22** (Scheme 6.13) [70]. This reaction has been followed quantitatively by IR spectroscopy, allowing the determination of the stoichiometry of this complex. The IR spectrum of **22** is characteristic of a species with the local structure X-$Re(CO)_5$ [71]. Alternatively, $C_{60}[Re(CO)_5]_2$ (**22**) can be synthesized by the reaction of (η^3-Ph_3C)$Re(CO)_4$ with C_{60} in the presence of carbon monoxide (Scheme 6.13) [70]. Thereby, $^{\bullet}Re(CO)_5$ radicals are generated in situ via homolysis of the weak Re–C bond in the $Ph_3CRe(CO)_5$ intermediate. Molecular modeling investigations on **22** suggest that a 1,4-addition mode takes place [70].

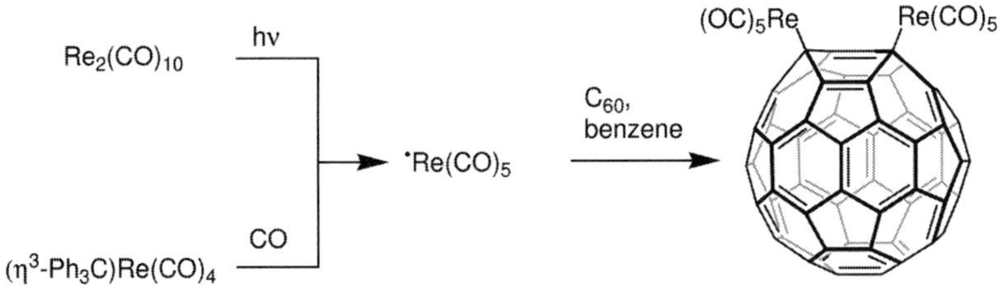

Scheme 6.13

The formation of the adduct **22** is reversible. Although the photochemical preparation of **22** proceeds in yields greater than 90%, solutions containing this complex decompose over a period of one day to regenerate C_{60} and $Re_2(CO)_{10}$ [70].

A moderately stable platinum containing fullerenyl radical **23** is accessible by reaction of the photolysis product of *cis*-$(CF_3)_2CF$-Hg-$Pt(PPh_3)_2$-$CH{=}CPh_2$ with C_{60} (Scheme 6.14) [72]. From the ESR spectrum a non-symmetrical arrangement of the Pt-ligands can be deduced. One PPh_3 ligand is located over a pentagon; the other PPh_3 is located over a hexagon. A large hyperfine splitting, caused by the interaction with ^{195}Pt, is evidence for the σ-bonding of platinum to the fullerene core. In addition to the radical adduct described, a broad line is recorded in the

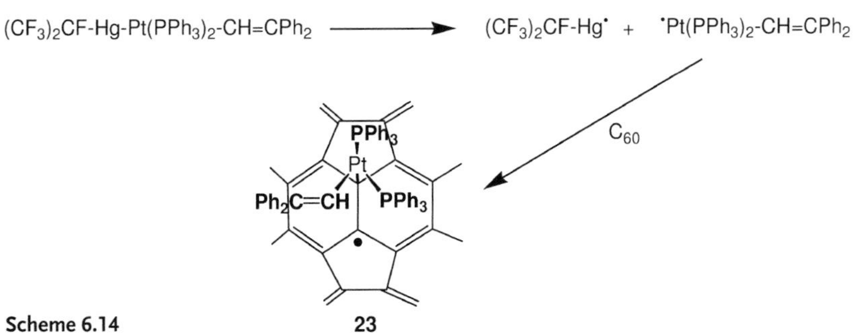

Scheme 6.14

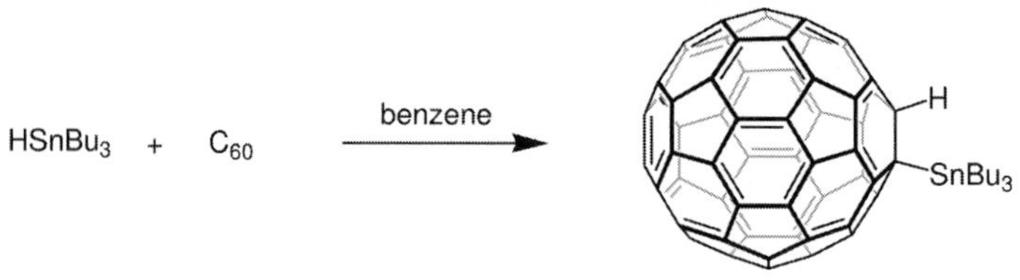

$$HSnBu_3 + C_{60} \xrightarrow{\text{benzene}} \mathbf{24}$$

Scheme 6.15

center of the spectrum, which is apparently related to the products of multiple addition of the carbon-centered radicals resulting from the homolytic cleavage of Hg-CF$(CF_3)_2$ [10, 72].

Refluxing benzene solutions of C_{60} in the presence of a 20-fold excess of Bu_3SnH leads to hydrostannylation (Scheme 6.15) [73]. Multiple additions can also take place. To maximize the yield of the monoadduct $C_{60}HSnBu_3$ (**24**), the time dependence of the reaction was followed quantitatively by HPLC. After about 4 h, the concentration of the monoadduct **24** reaches its maximum. Compound **24** can be isolated by preparative HPLC on a C_{18}-reversed-phase stationary phase with $CHCl_3$–CH_3CN (60 : 40, v/v) as eluent. The structure of $C_{60}HSnBu_3$ (**24**) was determined by ^1H NMR spectroscopy and other methods, showing that a 1,2-addition takes place regioselectively (Scheme 6.15) [73].

6.6
Addition of bis(Trifluoromethyl)nitroxide

The treatment of C_{60} with an excess of bis(trifluoromethyl)nitroxide [$(CF_3)_2NO^\bullet$ radicals] leads to multiple additions with an average of 18 $(CF_3)_2NO$ groups bound to the fullerene sphere (Scheme 6.16) [74]. The reaction has been carried out in a glass tube equipped with a Teflon valve. The resulting light brown solid decomposes above 158 °C.

$$(CF_3)_2NO^\bullet + C_{60} \longrightarrow C_{60}[(CF_3)_2NO]_n \qquad (n \approx 18)$$

Scheme 6.16

References

1 J. R. Morton, K. F. Preston, P. J. Krusic, S. A. Hill, E. Wasserman, *J. Phys. Chem.* **1992**, *96*, 3576.
2 J. R. Morton, K. F. Preston, P. J. Krusic, S. A. Hill, E. Wasserman, *J. Am. Chem. Soc.* **1992**, *114*, 5454.
3 J. R. Morton, K. F. Preston, P. J. Krusic, E. Wasserman, *J. Chem. Soc., Perkin Trans. 2:* **1992**, 1425.
4 P. J. Krusic, D. C. Roe, E. Johnston, J. R. Morton, K. F. Preston, *J. Phys. Chem.* **1993**, *97*, 1736.

5 J. R. Morton, K. F. Preston, P. J. Krusic, L. B. Knight, Jr., *Chem. Phys. Lett.* **1993**, *204*, 481.

6 P. N. Keizer, J. R. Morton, K. F. Preston, *J. Chem. Soc., Chem. Commun.* **1992**, 1259.

7 M. A. Cremonini, L. Lunazzi, G. Placucci, P. J. Krusic, *J. Org. Chem.* **1993**, *58*, 4735.

8 P. J. Fagan, B. Chase, J. C. Calabrese, D. A. Dixon, R. Harlow, P. J. Krusic, N. Matsuzawa, F. N. Tebbe, D. L. Thorn, E. Wasserman, *Carbon* **1992**, *30*, 1213.

9 B. L. Tumanskii, O. G. Kalina, V. V. Bashilov, *Mol. Cryst. Liq. Cryst.* **2000**, *13*, 23.

10 B. L. Tumanskii, O. G. Kalina, *Radical Reactions of Fullerenes and their Derivatives*, Vol. 2, 1st Ed., Kluwer Academic Publishers, New York, **2002**.

11 J. R. Morton, K. F. Preston, P. J. Krusic, *Hyperfine Interact.* **1994**, *86*, 763.

12 J. R. Morton, F. Negri, K. F. Preston, *Acc. Chem. Res.* **1998**, *31*, 63.

13 J. R. Morton, F. Negri, K. F. Preston, *Magn. Res. Chem.* **1995**, *33*, S20.

14 J. R. Morton, K. F. Preston, *Electron Spin Reson.* **1996**, *15*, 152.

15 R. Borghi, L. Lunazzi, G. Placucci, P. J. Krusic, D. A. Dixon, N. Matsuzawa, M. Ata, *J. Am. Chem. Soc.* **1996**, *118*, 7608.

16 I. V. Koptyug, A. G. Goloshevsky, I. S. Zavarine, N. J. Turro, P. J. Krusic, *J. Phys. Chem. A* **2000**, *104*, 5726.

17 R. G. Gasanov, B. L. Tumanskii, *Russ. Chem. Bull.* **2002**, *51*, 240.

18 B. L. Tumanskii, R. G. Gasanov, V. V. Bashilov, M. V. Tsikalova, N. N. Bubnov, V. I. Sokolov, V. P. Gubskaya, L. S. Berezhnaya, V. V. Zverev, I. A. Nuretdinov, *Russ. Chem. Bull.* **2003**, *52*, 1512.

19 R. G. Gasanov, V. V. Bashilov, B. L. Tumanskii, V. I. Sokolov, I. A. Nuretdinov, V. V. Zverev, V. P. Gubskaya, L. S. Berezhnaya, *Russ. Chem. Bull.* **2003**, *52*, 380.

20 K. Ohkubo, S. Fukuzumi, *Inorg. React. Mech.* **2000**, *2*, 147.

21 W. T. Ford, T. Nishioka, F. Qiu, F. D'Souza, J.-p. Choi, W. Kutner, K. Noworyta, *J. Org. Chem.* **1999**, *64*, 6257.

22 W. T. Ford, T. Nishioka, F. Qiu, F. D'Souza, J.-p. Choi, *J. Org. Chem.* **2000**, *65*, 5780.

23 H. Okamura, T. Terauchi, M. Minoda, T. Fukuda, K. Komatsu, *Macromolecules* **1997**, *30*, 5279.

24 B. L. Tumanskii, E. N. Shaposhnikova, V. V. Bashilov, S. P. Solodovnikov, N. N. Bubnov, S. R. Sterlin, *Russ. Chem. Bull.* **1997**, *46*, 1174.

25 M. Yoshida, D. Suzuki, M. Iyoda, *Chem. Lett.* **1996**, 1097.

26 M. Yoshida, F. Sultana, N. Uchiyama, T. Yamada, M. Iyoda, *Tetrahedron Lett.* **1999**, *40*, 735.

27 J. R. Morton, F. Negri, K. F. Preston, *Can. J. Chem.* **1994**, *72*, 776.

28 C. S. Johnson, Jr., *Adv. Magn. Reson.* **1965**, *1*, 33.

29 C. S. Johnson, Jr., *J. Chem. Phys.* **1964**, *41*, 3277.

30 P. J. Krusic, P. Meakin, J. P. Jesson, *J. Phys. Chem.* **1971**, *75*, 3438.

31 P. J. Fagan, P. J. Krusic, D. H. Evans, S. A. Lerke, E. Johnston, *J. Am. Chem. Soc.* **1992**, *114*, 9697.

32 J. R. Morton, F. Negri, K. F. Preston, G. Ruel, *J. Phys. Chem.* **1995**, *99*, 10114.

33 J. R. Morton, F. Negri, K. F. Preston, G. Ruel, *J. Chem. Soc., Perkin Trans. 2* **1995**, 2141.

34 J. R. Morton, F. Negri, K. F. Preston, *Chem. Phys. Lett.* **1995**, *232*, 16.

35 R. Borghi, B. Guidi, L. Lunazzi, G. Placucci, *J. Org. Chem.* **1996**, *61*, 5667.

36 R. Borghi, L. Lunazzi, G. Placucci, G. Cerioni, A. Plumitallo, *J. Org. Chem.* **1996**, *61*, 3327.

37 P. J. Krusic, E. Wasserman, B. A. Parkinson, B. Malone, E. R. Holler, Jr., P. N. Keizer, J. R. Morton, K. F. Preston, *J. Am. Chem. Soc.* **1991**, *113*, 6274.

38 B. L. Tumanskii, V. V. Bashilov, N. N. Bubnov, S. P. Solodovnikov, V. I. Sokolov, *Russ. Chem. Bull.* **1992**, 1938.

39 B. L. Tumanskii, V. V. Bashilov, N. N. Bubnov, S. P. Solodovnikov, A. A. Khodak, *Russ. Chem. Bull.* **1994**, 1671.

40 B. L. Tumanskii, V. V. Bashilov, N. N. Bubnov, S. P. Solodovnikov, V. I. Sokolov, *Mol. Cryst. Liq. Cryst.* **1996**, *8*, 61.

41 B. L. Tumanskii, V. V. Bashilov, E. N. Shaposhnikova, S. P. Solodovnikov, N. N. Bubnov, V. I. Sokolov, *Russ. Chem. Bull.* **1998**, *47*, 1823.

42 R. G. Gasanov, V. V. Bashilov, O. G. Kalina, B. L. Tumanskii, *Russ. Chem. Bull.* **2000**, *49*, 1642.

43 J. R. Morton, K. F. Preston, *Chem. Phys. Lett.* **1996**, *255*, 15.

44 J. M. McBridge, *Tetrahedron* **1974**, *30*, 2009.

45 R. Borghi, L. Lunazzi, G. Placucci, P. J. Krusic, D. A. Dixon, L. B. Knight, Jr., *J. Phys. Chem.* **1994**, *98*, 5395.

46 P. J. Krusic, E. Wasserman, P. N. Keizer, J. R. Morton, K. F. Preston, *Science* **1991**, *254*, 1183.

47 P. J. Krusic, J. K. Kochi, *J. Am. Chem. Soc.* **1968**, *90*, 7155.

48 N. Matsuzawa, D. A. Dixon, T. Fukunaga, *J. Phys. Chem.* **1992**, *96*, 7594.

49 B. L. Tumanskii, O. G. Kalina, V. V. Bashilov, A. V. Usatov, E. A. Shilova, Y. I. Lyakhovetskii, S. P. Solodovnikov, N. N. Bubnov, Y. N. Novikov, A. S. Lobach, V. I. Sokolov, *Russ. Chem. Bull.* **1999**, *48*, 1108.

50 P. J. Fagan, P. J. Krusic, C. N. McEwen, J. Lazar, D. H. Parker, N. Herron, E. Wasserman, *Science* **1993**, *262*, 404.

51 Y. I. Lyakhovetsky, E. A. Shilova, B. L. Tumanskii, A. V. Usatov, E. A. Avetisyan, S. R. Sterlin, A. P. Pleshkova, Y. N. Novikov, Y. S. Nekrasov, R. Taylor, *Fullerene Sci. Technol.* **1999**, *7*, 263.

52 B. Ma, G. E. Lawson, C. E. Bunker, A. Kitaygorodskiy, Y.-P. Sun, *Chem. Phys. Lett.* **1995**, *247*, 51.

53 G. E. Lawson, A. Kitaygorodskiy, B. Ma, C. E. Bunker, Y.-P. Sun, *J. Chem. Soc., Chem. Commun.* **1995**, 2225.

54 D. Zhou, H. Tan, C. Luo, L. Gan, C. Huang, J. Pan, M. Lu, Y. Wu, *Tetrahedron Lett.* **1995**, *36*, 9169.

55 L. Gan, D. Zhou, C. Luo, H. Tan, C. Huang, M. Lue, J. Pan, Y. Wu, *J. Org. Chem.* **1996**, *61*, 1954.

56 K.-F. Liou, C.-H. Cheng, *Chem. Commun.* **1996**, 1423.

57 L. Gan, J. Jiang, W. Zhang, Y. Su, Y. Shi, C. Huang, J. Pan, M. Lue, Y. Wu, *J. Org. Chem.* **1998**, *63*, 4240.

58 R. Bernstein, C. S. Foote, *J. Phys. Chem. A* **1999**, *103*, 7244.

59 G. E. Lawson, A. Kitaygorodskiy, Y.-P. Sun, *J. Org. Chem.* **1999**, *64*, 5913.

60 L.-W. Guo, X. Gao, D.-W. Zhang, S.-H. Wu, H.-M. Wu, Y.-J. Li, *Chem. Lett.* **1999**, 411.

61 L.-W. Guo, X. Gao, D.-W. Zhang, S.-H. Wu, H.-M. Wu, Y.-J. Li, S. R. Wilson, C. F. Richardson, D. I. Schuster, *J. Org. Chem.* **2000**, *65*, 3804.

62 H. Cheng, L. Gan, Y. Shi, X. Wei, *J. Org. Chem.* **2001**, *66*, 6369.

63 P. N. Keizer, J. R. Morton, K. F. Preston, P. J. Krusic, *J. Chem. Soc., Perkin Trans. 2* **1993**, 1041.

64 T. Kusukawa, W. Ando, *Organometallics* **1997**, *16*, 4027.

65 T. Kusukawa, W. Ando, *J. Organomet. Chem.* **1998**, *559*, 11.

66 T. Akasaka, T. Suzuki, Y. Maeda, M. Ara, T. Wakahara, K. Kobayashi, S. Nagase, M. Kako, Y. Nakadaira, M. Fujitsuka, O. Ito, *J. Org. Chem.* **1999**, *64*, 566.

67 T. Kusukawa, Y. Kabe, T. Erata, B. Nestler, W. Ando, *Organometallics* **1994**, *13*, 4186.

68 T. Kusukawa, A. Shike, W. Ando, *Tetrahedron* **1996**, *52*, 4995.

69 T. Kusukawa, Y. Kabe, W. Ando, *Organometallics* **1995**, *14*, 2142.

70 S. Zhang, T. L. Brown, Y. Du, J. R. Shapley, *J. Am. Chem. Soc.* **1993**, *115*, 6705.

71 K. Raab, W. Beck, *Chem. Ber.* **1985**, *118*, 3830.

72 B. L. Tumanskii, V. V. Bashilov, O. G. Kalina, V. I. Sokolov, *J. Organomet. Chem.* **2000**, *599*, 28.

73 A. Hirsch, T. Grösser, A. Skiebe, A. Soi, *Chem. Ber.* **1993**, *126*, 1061.

74 D. Brizzolara, J. T. Ahlemann, H. W. Roesky, K. Keller, *Bull. Soc. Chim. Fr.* **1993**, *130*, 745.

7
Transition Metal Complex Formation

7.1
Introduction

The electron deficiency of the fullerenes C_{60} and C_{70}, established by calculations, electrochemistry and the reaction behavior towards nucleophiles, has been independently demonstrated by investigations of fullerene transition metal complexes. Various single-crystal structures and spectroscopic studies show that the complexation of transition metals to the fullerene core proceeds similar to well-established reactions of electron-deficient olefins in a dihapto manner to one or more π-bonds or as hydrometalation reactions. Some of these complexation reactions are more or less reversible. The resulting thermodynamic control explains the remarkable regioselectivity observed at the formation of higher addition products, for example bisadducts or hexakisadducts. Packing effects in the solid-state structure can also play an important role for the exclusive formation of a specific regioisomer. One driving force for the formation of a certain solid-state structure of fullerene derivatives could be due to π–π interactions of ligand-bound electron-rich arene moieties and the electron-poor fullerene core itself. This is demonstrated by impressive examples for supramolecular arrangements in single crystals of fullerene transition metal complexes.

7.2
(η^2-C_{60}) Transition Metal Complexes

The question as to whether C_{60} behaves like an aromatic or an electron-deficient alkene was answered elegantly by Fagan [1–3]. The complex $[Cp^*Ru(CH_3CN)_3]^+$-$O_3SCF_3^-$ ($Cp^* = \eta^5$-$C_5(CH_3)_5$) reacts with electron-rich planar arenes upon displacement of the three coordinated acetonitrile ligands, resulting in a strong η^6-binding of ruthenium to the six-membered rings of the arenes [4]. Conversely, the reaction of this ruthenium complex with an electron-deficient alkene leads to the displacement of one acetonitrile ligand and the formation of a η^2-olefin complex. Upon the reaction with a ligand containing both an alkene functionality and an arene ring, the ruthenium exclusively binds to the arene ring [4]. This remarkable selectivity makes the ruthenium complex a good candidate for testing the transition metal

Fullerenes: Chemistry and Reactions. Andreas Hirsch and Michael Brettreich

ISBN: 3-527-30820-2

complex formation behavior of C_{60}. Allowing C_{60} to react with a 10-fold excess of $[Cp^*Ru(CH_3CN)_3]^+\ O_3SCF_3^-$ in CH_2Cl_2 at 25 °C for five days results in a brown precipitate of $[[Cp^*Ru(CH_3CN)_2]_3(C_{60})]^{3+}\ (O_3SCF_3^-)_3$, in which two acetonitrile ligands are retained on each ruthenium [2]. This suggests that each ruthenium is bound to one double bond of the C_{60} sphere. Therefore, the reactivity of C_{60} is also shown from this point of view to be that of an electron-deficient olefin.

Considering the geometry of the p-orbitals within a hexagon of the C_{60} framework, which are tilted away from the center of the ring, it is already obvious that a binding of C_{60} to a metal in a hexahapto fashion is not favorable because the orbital overlap will be weakened. This was confirmed in some theoretical calculations [5–8]. Using PM3(tm) or combined PM3(tm)-density functional theory studies the exchange of a η^6-bound benzene for C_{60} was calculated. With most transition metal complex fragments this exchange was found to be endothermal.

Coordination to two or three of the double bonds of a hexagon was found in various osmium, rhenium, iridium or ruthenium complexes, but these complexes do not exhibit a η^6-complexation. In fact metal clusters, not single metal atoms, bind to C_{60} such that the metals are η^2-bound to adjacent bonds of a C_6 face (Section 7.3).

The binding of C_{60} to transition metals in a η^2-fashion becomes clearly evident by the formation of various complexes [9] (Scheme 7.1) of platinum, palladium and nickel [1–3, 10–16], iridium, cobalt and rhodium [9, 17–30] iron, ruthenium and osmium [25, 31, 32], manganese [29, 33], titanium [34], rhenium and tantalum [25, 29] and molybdenum and tungsten [35–44]. Low-valent complexes of these metals easily undergo complexation with electron-deficient olefins [45, 46]. A typical structural aspect is the loss of planarity of the olefin coordinated to the transition metal, because the four groups bound to the olefin bend back away from the metal (Figure 7.1). In substituted ethylenes, C_2X_4, this deformation increases upon increasing electronegativity of X. In C_{60}, the arrangement around [6,6] double bonds is already preorganized. Therefore, the combination of both strain and electron deficiency of the [6,6] double bonds is an important driving force of η^2-binding of C_{60} with low-valent transition metals. The reaction of equimolar amounts of $(Ph_3P)_2Pt(\eta^2\text{-}C_2H_4)$ with C_{60} leads to a dark emerald green solution of $(Ph_3P)_2Pt(\eta^2\text{-}C_{60})$ (**4**, Scheme 7.1). Single-crystal X-ray structure analysis confirmed the η^2-binding of the platinum to a [6,6] fullerene double bond [1].

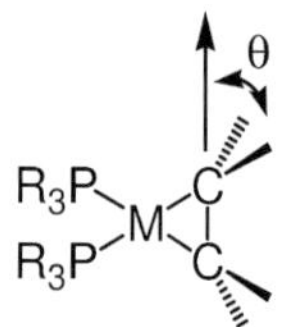

Figure 7.1 Coordination of an olefin to the transition metal fragment $M(PPh_3)_2$ (M = e.g. Ni, Pd, Pt) leads to a deviation from planarity of the olefin system. The angle θ is a measure of the deformation of the groups attached to the carbon–carbon double bonds from planarity.

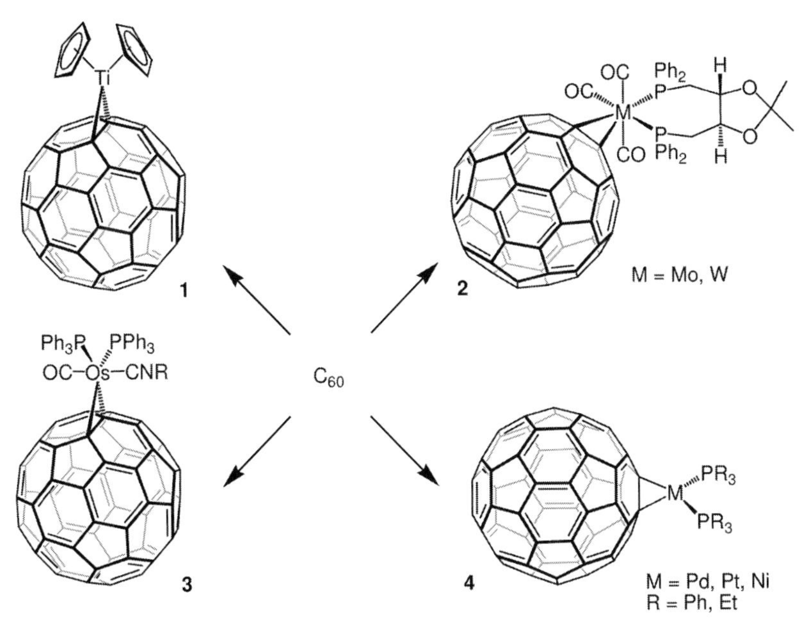

Scheme 7.1 Examples of η^2-C_{60} complexes.

Similar complex formation reactions are also possible with the metal reagents $M(PEt_3)_4$ (M = Ni, Pd) (Scheme 7.1) [3]. All these metal derivatives exhibit almost identical properties to $(Ph_3P)_2Pt(\eta^2\text{-}C_{60})$. The same compounds can also be synthesized by the reaction with the complexes $(Et_3P)_2Pd(\eta^2\text{-}CH_2{=}CHCO_2CH_3)$ [12]. Using stoichiometric amounts of reagents, these reactions are very selective in forming only monoadducts in a high yield rather than a mixture of polyadducts and unreacted C_{60}.

Heterobimetallic palladium- or platinum-[60]fullerene complexes (**5**) with a bisdentate bis-phosphinoferrocene ligand have been obtained via different routes [15, 16]. The palladium complex was prepared (Scheme 7.2) either electrochemically or in a one-pot reaction with another Pd complex, the ligand dppf [dppf = 1,1′-bis(bisdiphenylphosphino)ferrocene] and C_{60} [15]. Unfortunately it was not possible to obtain crystals for the complexes prepared via this route. With another procedure, which includes two reaction steps (Scheme 7.2), X-ray suitable crystals could be obtained for the corresponding Pt and Pd complexes [16]. In the first step the known complex $(Ph_3P)_2M(\eta^2\text{-}C_{60})$ was formed, whose phosphino ligands were exchanged in a second step with the ferrocene-containing bisdentate ligand.

Osmium forms various multinuclear complexes (Section 7.3) but a mononuclear η^2-complex of this member of the platinum family was missing. Recently it was possible to verify the existence of a mononuclear Os complex [32]. The *cis*-dihydride complex $[OsH_2(CO)(PPh_3)_3]$ was refluxed in toluene together with C_{60} and ${}^tBuNC$. Substitution of one PPh_3 ligand with ${}^tBuNC$ accompanied by elimination of dihydrogen yielded the new complex $[(\eta^2\text{-}C_{60})Os(CO)({}^tBuNC)(PPh_3)_2]$ (**3**) (Scheme 7.1). Molybdenum and tungsten form octahedral complexes with one ligand being C_{60}.

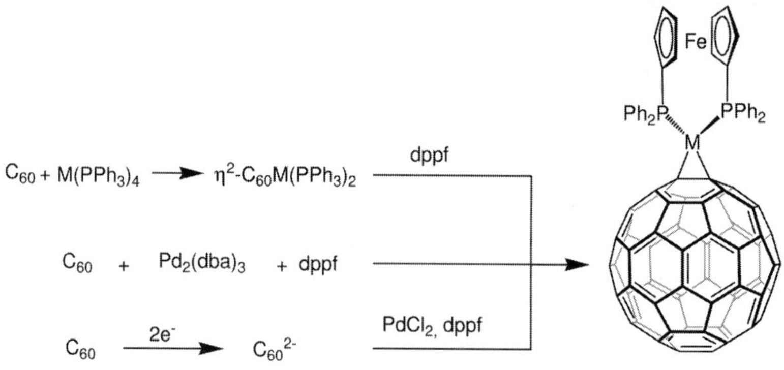

C_{60} + $M(PPh_3)_4$ → η^2-$C_{60}M(PPh_3)_2$ —dppf→

C_{60} + $Pd_2(dba)_3$ + dppf →

C_{60} —2e⁻→ C_{60}^{2-} —$PdCl_2$, dppf→

Fe; Ph_2P; PPh_2; M

5

Scheme 7.2

M = Pt, Pd

The other positions are usually occupied by three CO and two (mostly bridged) donor ligands, which can be phenanthroline [47, 48] or bridged bifunctional phosphines such as e.g. dppb [40, 42, 49], dppe [50] or dppf [41] (abbreviations see [51]). Some remarkably stable complexes could be prepared by photolysis of either $W(CO)_4(Ph_2PCH_2CH_2PPh_2)$ with C_{60} or $Mo(CO)_4(Ph_2PCH_2CH_2PPh_2)$ in dichloro- or chlorobenzene as solvent. The thereby generated complexes $(\eta^2\text{-}C_{60})W(CO)_3$-$(Ph_2PCH_2CH_2PPh_2)$ and $(\eta^2\text{-}C_{60})Mo(CO)_3(Ph_2PCH_2CH_2PPh_2)$, respectively, are relatively stable in solution against the loss of C_{60} even at elevated temperatures. Air-stable Mo and W complexes have been synthesized with phenanthroline as a ligand. Optically active Mo and W derivatives have been obtained by replacing the achiral diphosphine ligand with the chiral diphosphine DIOP [43], which is 2,3-O,O′-isopropylidene-2,3-dihydroxy-1,4-bis(diphenylphosphanyl)butane. It is formed for both metals as the *mer*-isomer **2** (Scheme 7.1).

The same ligands used successfully to synthesize stable C_{60} complexes of molybdenum, tungsten and chromium have been used to synthesize the corresponding C_{70} complexes. Some examples are $M(CO)_3(dppb)(\eta^2C_{70})$ [42] (with M = Mo, Cr, W), $Mo(CO)_3(dppe)(\eta^2C_{70})$ [50], $W(CO)_3(dppf)(\eta^2C_{70})$ [41] or Mo(CO)-(phen)dbm(η^2C_{70}) [47, 48] (abbreviations see [52]). As far as could be proven via X-ray spectroscopy the addition takes place at the poles of C_{70} at the 1,2-double bond. The same coordination site was also found for the brown-black Pd complex $(\eta^2C_{70})Pd(PPh_3)_2$ [53].

Unlike zirconium, the group IV metal titanium does not form the hydrometalation product but rather a $(\eta^2\text{-}C_{60})$-complex. The first titanium-fullerene complex **1** was prepared by reaction of the bis(trimethylsilyl)-acetylene complex of titanocene with equimolar amounts of C_{60} (Scheme 7.1).

As already described, complexations with metals of the platinum group (Ni, Pd, Pt) and of other metals mostly lead selectively to η^2-monoaddition. However, with platinum metals the reactions can be driven to the formation of air-sensitive hexaadducts $[(Et_3P)_2M]_6C_{60}$ by using a 10-fold excess of the metal reagent $M(PEt_3)_4$ (M = Ni, Pd, Pt) [2, 3, 10]. Each compound exists as a single structural isomer.

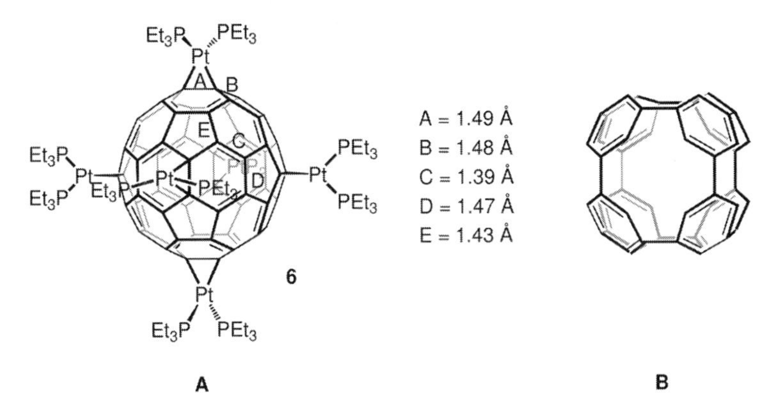

Figure 7.2 (A) Bond lengths for the five different sets of C–C bonds and favored electronic resonance structure of $[(Et_3P)_2Pt]_6C_{60}$ determined by X-ray crystal structure analysis. (B) 48-Electron π-system produced by the saturation of six double bonds in the octahedral sites of C_{60}, leaving eight linked benzenoid hexagons.

The X-ray crystal structure of $[(Et_3P)_2Pt]_6C_{60}$, for example, shows that the molecule has a C_{60} core octahedrally coordinated with six $(Et_3P)_2Pt$ groups (Figure 7.2). This results in an overall T_h point group symmetry, ignoring the ethyl groups. Each platinum is bound across a [6,6] double bond. It is very interesting that excluding the carbons bound to the platinum leads to a network of eight 1,3,5-linked benzene-like rings (Figure 7.2) (see also Chapter 10). Indeed, the crystal structure of $[(Et_3P)_2Pt]_6C_{60}$ shows that the bond alternation within the six octahedrally arranged "benzene rings" (bonds C and E) is reduced to 0.037 Å, half the value in C_{60}. Both bonds approach the typical value for arenes (ca. 1.395 Å). Consequently, the remaining π-system in the C_{60} core becomes more delocalized, which is corroborated by Hückel calculations [54].

This electronic argument, the formation of a more delocalized aromatic system (8 benzene-like rings) out of the more bond-localized non-aromatic C_{60} system, may be an additional driving force for the octahedral arrangement of six metal units on the fullerene sphere. However, steric factors must also be considered, because there is no room for a seventh $(Et_3P)_2Pt$ ligand and the octahedral array of metal fragments in $[(Et_3P)_2M]_6C_{60}$ is, sterically, the optimum situation.

An important requirement for the exclusive formation of octahedrally coordinated $[(Et_3P)_2M]_6C_{60}$ in these high yields is a certain degree of reversibility of the $(Et_3P)_2M$ addition, because not only the octahedral sites in C_{60} are available for complex formation. The complexation of the metal fragments in $[(Et_3P)_2Pt]_2C_{60}$ in non-octahedral sites was confirmed by ^{31}P NMR spectroscopy. The reversibility of the adduct formation was confirmed by substitution experiments as well as by electrochemistry studies [12]. The addition of one equiv. of diphenylacetylene to a solution of $[(Et_3P)_2Pt]_6C_{60}$ in benzene leads to an instantaneous reaction to produce the complexes $(Et_3P)_2Pt(\eta^2\text{-}C_2\text{-}Ph_2)$ and $[(Et_3P)_2Pt]_5C_{60}$ as major components

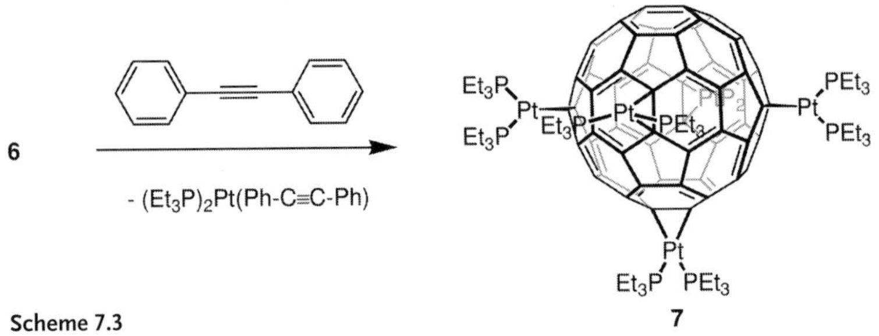

6 → $-(Et_3P)_2Pt(Ph\text{-}C{\equiv}C\text{-}Ph)$ → 7

Scheme 7.3

(Scheme 7.3) [3]. Because of the steric inaccessibility of the C_{60} bonds in the hexaadduct, it is unlikely that this is an associative reaction. Therefore, $[(Et_3P)_2Pt]_6C_{60}$ is concluded to be in an equilibrium with a small amount of the metal fragment $(Et_3P)_2Pt$ even in non-polar solvents.

The first reduction potentials of $(Et_3P)_2MC_{60}$ (M = Ni, Pd, Pt) are shifted by about 0.23–0.34 V to more negative potentials than for C_{60} [12]. Therefore, these organometallic complexes are about 0.1–0.2 V harder to reduce than organic monoadducts of the type $C_{60}RR'$ (R, R′ = organic group or H). The additional shift in the metal complexes relative to the organic derivatives was attributed to a higher inductive donation of electron density into the C_{60} moiety, which further lowers the electron affinity. This also explains the high selectivity for the monoadduct formation with these low-valent transition metal reagents. The tendency of a second metal fragment to add to a monoadduct is significantly reduced. Bisadducts $[(Et_3P)_2M]_2C_{60}$ react with C_{60} in solution and form the corresponding monoaddition products. With increasing amounts of metal complexation, the adducts become increasingly harder to reduce. This agrees with a decreasing tendency of further complexation of electron-rich metal fragments. The addends begin to equilibrate on and off the C_{60} framework most readily in the hexaadducts $[(Et_3P)_2M]_6C_{60}$, which allows the T_h symmetry to be achieved in these complexes.

Several effects can influence the electronic structure of C_{60} upon metal complex formation. One is the removal of one double bond from the remaining 29 fullerene double bonds. As in any polyene system, this decreased conjugation is expected to raise the energy of the LUMO and therefore decreases the electron affinity of the system. Conversely, the d-orbital backbonding transfers electron density from the metal into π^* orbitals of the remaining double bonds, which also decreases the electron affinity.

Coordination of C_{60} with the $(\eta^5\text{-}C_9H_7)Ir(CO)$ fragment lowers the first reduction potential only by 0.08 V relative to C_{60} [9, 22], indicating a weaker backbonding capability. The corresponding complex $(\eta^5\text{-}C_9H_7)Ir(CO)(\eta^2\text{-}C_{60})$ can be obtained by refluxing equimolar amounts of $(\eta^5\text{-}C_9H_7)Ir(CO)(\eta^2\text{-}C_8H_{14})$ and C_{60} in CH_2Cl_2 (Scheme 7.4). The coordinated C_{60} molecule in this complex can be substituted upon treatment with strongly coordinating ligands, for example with CO, $P(OMe)_3$ or PPh_3 (Scheme 7.4). Thereby, the solutions change from green to the characteristic purple of C_{60}. These reactions can be monitored quantitatively by UV/Vis and IR

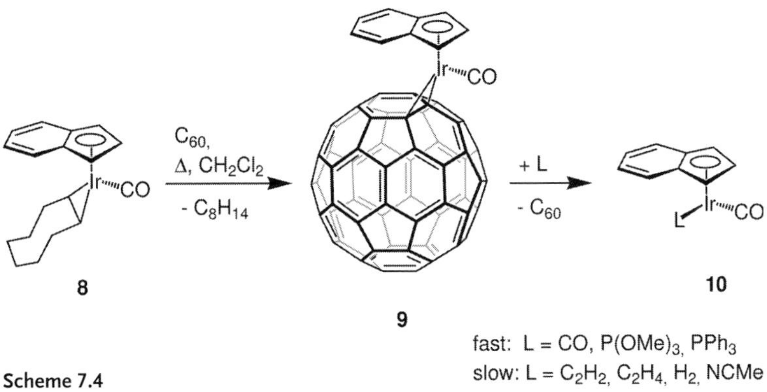

C_{60}, Δ, CH_2Cl_2 ; $- C_8H_{14}$; $+ L$; $- C_{60}$

8 9 10

fast: L = CO, $P(OMe)_3$, PPh_3
slow: L = C_2H_2, C_2H_4, H_2, NCMe

Scheme 7.4

spectroscopy. Significantly, the substitution of C_{60} in $(\eta^5\text{-}C_9H_7)Ir(CO)(\eta^2\text{-}C_{60})$ by weaker coordinating ligands, such as C_2H_4 or C_2H_2, is more than 100× slower than the reaction with carbon monoxide. This implies an associative pathway for the substitution reaction. The air-stable solutions of $(\eta^5\text{-}C_9H_7)Ir(CO)(\eta^2\text{-}C_{60})$ show an additional well-defined absorption band at 436 nm. This band is also observed in the monoadducts $C_{60}RR'$ (R, R′ = organic group or H), discussed in preceding chapters, and can, therefore, not be attributed to Ir-to-C_{60} charge-transfer transitions. It arises instead from a transition largely centered on the C_{60} ligand [55].

A more reversible adduct formation occurs upon addition of Vaska's complex $Ir(CO)Cl(PPh_3)_2$ (Scheme 7.5) [17]. This complex reacts with electron-deficient olefins, such as tetracyanoethylene, to form stable η^2 adducts [56]. The carbon monoxide stretching frequency in these complexes can be used as a measure of the electron-withdrawing influence of the ligands. In the tetracyanoethylene adduct of $Ir(CO)Cl(PPh_3)_2$, for example, the carbonyl stretching frequency is considerably stronger than in the C_{60} adduct $(\eta^2\text{-}C_{60})Ir(CO)Cl(PPh_3)_2$, demonstrating that C_{60} is less effective in electron withdrawing than tetracyanoethylene. This was also confirmed by Mössbauer spectroscopy [57]. The electron-withdrawing influence of C_{60} is similar to that of O_2 in $O_2Ir(CO)Cl(PPh_3)_2$. The reversibility of adduct formation was found to be correlated with the magnitude of the carbonyl stretching frequency. Therefore, both the C_{60} and the dioxygen complex fall into the category of "easily reversible". Single crystals of $(\eta^2\text{-}C_{60})Ir(CO)Cl(PPh_3)_2$ containing five

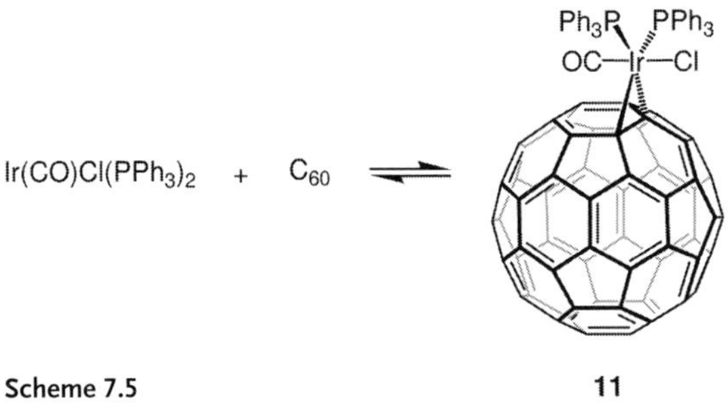

$Ir(CO)Cl(PPh_3)_2$ + C_{60} ⇌ 11

Scheme 7.5

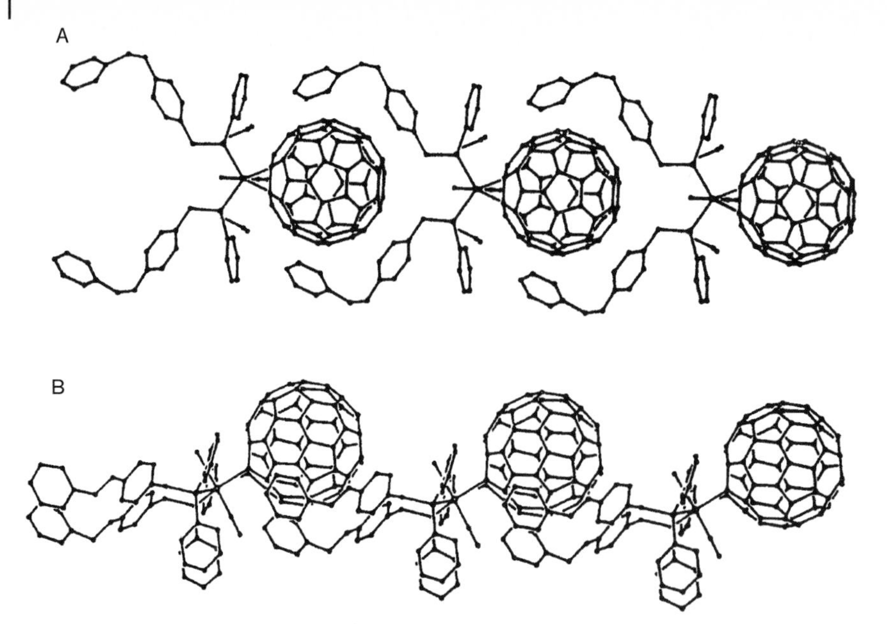

Figure 7.3 X-ray crystal structure of $(\eta^2\text{-}C_{60})Ir(CO)Cl(bobPPh_2)_2$ (bob = 4-$(PhCH_2O)C_6H_4CH_2$) [20]. The two views (A) and (B) show the supramolecular chelation architecture of this system.

molecules of benzene per formula unit can be obtained by mixing equimolar benzene solutions of the components. The reversible process takes place upon redissolving these crystals in CH_2Cl_2.

In addition to $Ir(CO)Cl(PPh_3)_2$, other Vaska-type iridium compounds have been added to C_{60} [18, 20]. By using the complex $Ir(CO)Cl(bobPPh_2)_2$ [bob = 4-$(PhCH_2O)$-$C_6H_4CH_2$], which contains two phenyl rings in each side chain, a supramolecular architecture is formed in the single crystals of $(\eta^2\text{-}C_{60})Ir(CO)Cl(bobPPh_2)_2$ (Figure 7.3). Each of the C_{60} spheres is chelated by the phenyl rings in the two side-arms of another molecule, which is a further example of a π–π interaction between an electron-rich moiety in the side chain of the addend and the electron-poor C_{60}. This attractive interaction is also reflected by the decrease in the P–Ir–P bond angle compared with that in $(\eta^2\text{-}C_{60})Ir(CO)Cl(PPh_3)_2$.

Replacing the PPh_3 ligand in the Vaska complex with alkylphoshine ligands leads to an increased reactivity with regard to oxidative addition. Thus the binding constant for oxidative addition to $Ir(CO)Cl(PMe_2Ph)$ is 200× larger than for $Ir(CO)Cl(PPh_3)$ [9, 58]. Multiple adducts of C_{60} have been obtained by using such modified Vaska complexes with the ligands PMe_2Ph [18], PMe_3 and PEt_3 [19]. The addition of $Ir(CO)Cl(PMe_2Ph)_2$ to C_{60} in benzene in different molar ratios leads to the formation of air-sensitive crystals of the bisadducts, which were identified by X-ray crystallography (Scheme 7.6) [18]. Two different conformational isomers were observed. In each case the Ir moieties are bound at the opposite ends of the C_{60} molecules in *trans*-1 positions. Both electronically and sterically, the formation of the *trans*-1

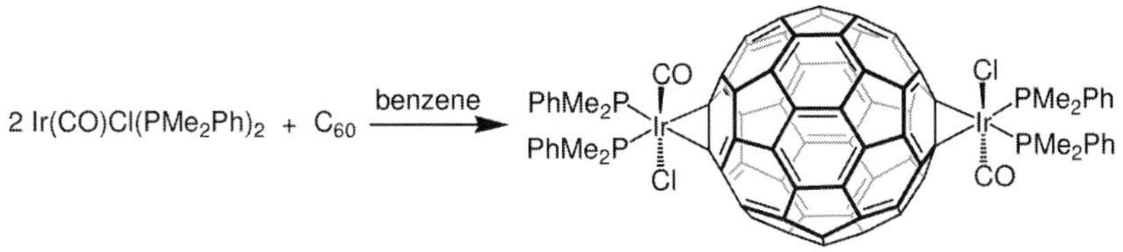

$$2\ Ir(CO)Cl(PMe_2Ph)_2 + C_{60} \xrightarrow{\text{benzene}} \mathbf{12}$$

Scheme 7.6

isomers is not expected to be favored over the *trans*-2, *trans*-3, *trans*-4 and *e*-isomers to the extent that the formation of the latter is completely suppressed. Since the additions of the iridium complexes to C_{60} are reversible, the low solubility of this bisadduct, which is characteristic for *trans*-1 isomers, as well as packing effects in the solid, can play a major role in the exclusive formation of this regioisomer.

The reversibility of oxidative addition of these Ir complexes to C_{60} was proven with temperature-dependent ^{31}P NMR measurements [19]. These measurements were carried out with $C_{60}[Ir(CO)Cl(PEt_3)_2]_2$ [19], or in another study with a dendritic complex derived from $Ir(CO)Cl(PPh_2R)_2$ with R being a Fréchet-type dendron of the first or second generation [21].

Refluxing C_{60} with the metallacyclic carboranyliridium dihydride complex **13** in a toluene–acetonitrile mixture yields the complex **14**, which contains two different polyhedral clusters as ligands (Scheme 7.7) [28]. The strong trans-influence of the σ-bonded carborane ligand leads to a distortion of the iridacyclopropane moiety. The Ir–C bond trans to the carborane is significantly longer than the other Ir–C bond (2.229 vs. 2.162 Å). This distortion can be utilized for the selective insertion

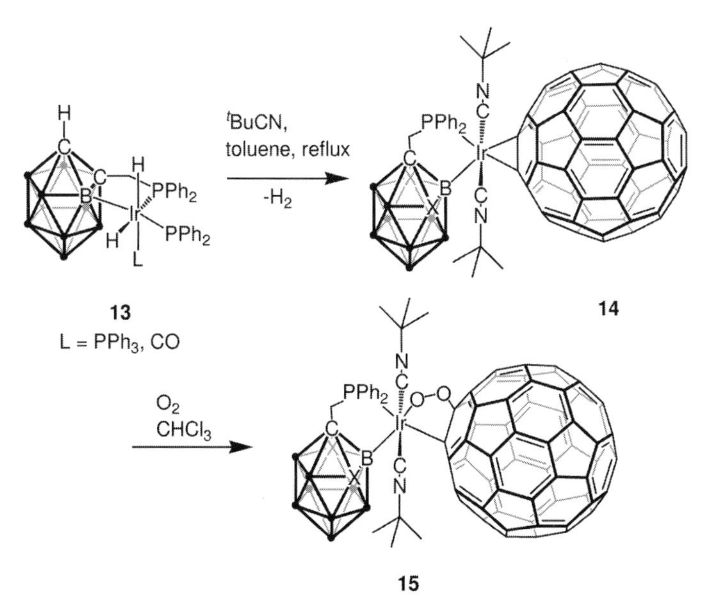

Scheme 7.7 Dots represent BH, X represents BH or CH.

of dioxygen into the longer Ir–C bond (Scheme 7.7) [59]. The result is the unusual σ-coordinated fullerene complex **15**, which is one of the rare examples of a non-η^2-coordination of C_{60}.

Upon treatment of C_{70} with Vaska's complex in benzene one regioisomer of $(\eta^2\text{-}C_{70})Ir(CO)Cl(PPh_3)_2$ is selectively formed, which was characterized by IR spectroscopy as well as by X-ray crystallography (Scheme 7.8) [60]. The addition of the Ir complex occurs in the 1,2-position at the poles. As with the C_{60} analogues, such as $(\eta^2\text{-}C_{60})Ir(CO)Cl(PPh_3)_2$, the two C atoms of the fullerene involved in the metal binding are pulled out from the surface. Therefore, the exo-double bonds of the pole pentagons are the most accessible for such coordination because the other [6,6] bonds in C_{70} have a more flattened local structure. The bond lengths of the [5]radialene subunits of C_{70} at the poles are almost the same as those in C_{60}.

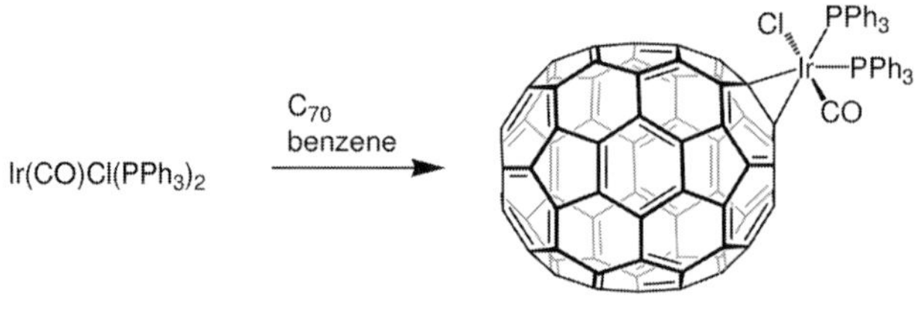

Ir(CO)Cl(PPh3)2 —(C_{70}, benzene)→ **16** (Cl, PPh3, PPh3, CO on Ir)

Scheme 7.8

The complex $Ir(CO)Cl(PPhMe_2)_2$ is more reactive in oxidative addition reactions than Vaska's complex [58]. This allows the synthesis of a bisadduct of C_{70} by treatment with a six- to twelve-fold excess of $Ir(CO)Cl(PPhMe_2)_2$ in benzene (Scheme 7.9) [61]. Whereas in solution several isomers exist in equilibrium, which can be shown by $^{31}P\{^1H\}$ NMR spectroscopy, the uniform and characteristic morphology of the single crystals of $\{(\eta^2\text{-}C_{70})[Ir(CO)Cl(PPhMe_2)_2]_2\}$ indicate that there is a single regioisomer present in the solid state. The molecule has a C_2-symmetry and the Ir atoms are bound on the *exo*-double bonds of the opposite pole pentagons, which electronically as well as sterically is the most favorable situation. Three isomers of "pole–opposite pole" bound bisadducts are in principle possible. The fact that this C_2-symmetrical isomer is formed exclusively was explained by packing effects of the solid state structure [61]. Arene–fullerene π–π interactions play an important role thereby. The phenyl rings of the phosphine ligand of one fullerene lie close over the surface of another C_{70} portion.

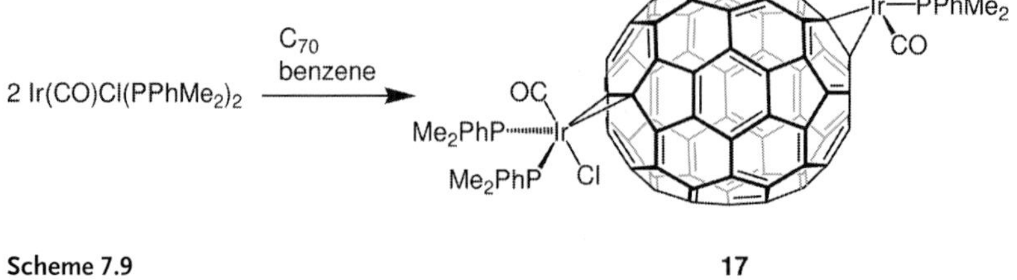

2 Ir(CO)Cl(PPhMe2)2 —(C_{70}, benzene)→ **17** (Cl, PPhMe2, PPhMe2, CO on one Ir; OC, Me2PhP, Me2PhP, Cl on the other Ir)

Scheme 7.9

The hydrogenation catalyst $RhH(CO)(PPh_3)_3$ undergoes hydrometalation reactions with electron-deficient olefins [62]. For the reaction with C_{60} or with C_{70}, however, a complexation in a dihapto manner takes place, leading to $(\eta^2\text{-}C_{60})[RhH(CO)(PPh_3)_2]$ and $(\eta^2\text{-}C_{70})[RhH(CO)(PPh_3)_2]$, respectively, rather than a hydrometalation (Scheme 7.10), as shown by X-ray structure analysis and NMR spectroscopy [23, 24, 26]. The green-black crystals of the C_{60} complex are formed in 75% yield. Unlike $(\eta^2\text{-}C_{60})Ir(CO)Cl(PPh_3)_2$, which dissociates into C_{60} and $Ir(CO)Cl(PPh_3)_2$, the green solution of $(\eta^2\text{-}C_{60})[RhH(CO)(PPh_3)_3]$ is stable.

The corresponding Ir complex was prepared similarly (Scheme 7.10) [24]. Multiple addition of the $MH(CO)(PPh_3)_2$ fragment with M = Ir or Rh leads to a mixture of at least five different bisadduct-isomers [24]. The analogous reaction with $RuH(Cl)(PPh_3)_3$ or $RuH_2(CO)(PPh_3)_3$ does not yield any product. A reaction similar to that with $MH(CO)(PPh_3)_3$ (M = Ir, Rh) only occurs with $RuH(NO)(PPh_3)_3$, yielding the complex $(\eta^2\text{-}C_{60})[RuH(NO)(PPh_3)_2]$, which is air stable in solution and as a solid [25, 63].

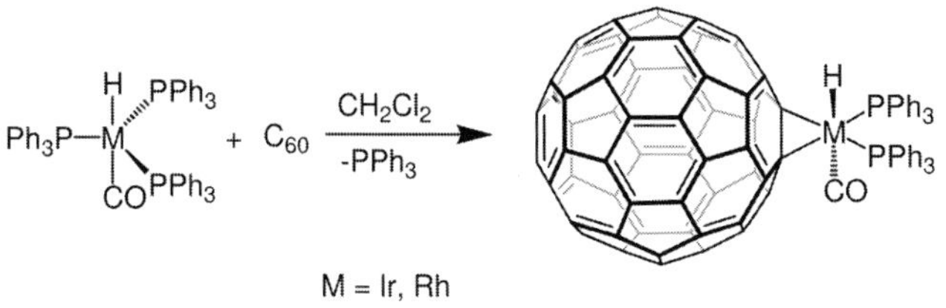

Scheme 7.10

7.3 Multinuclear Complexes of C_{60}

A binuclear addition product of C_{60} has been synthesized by the addition of two molecules $Ir_2(\mu\text{-}Cl)_2(1,5\text{-}COD)_2$ (Scheme 7.11) (1,5-COD = 1,5-cyclooctadiene) [64]. Single crystals of the air-stable complex $(\eta^2{:}\eta^2\text{-}C_{60})[Ir_2Cl_2(\eta^4\text{-}1,5\text{-}COD)_2]_2{\cdot}2C_6H_6$ were grown by slow diffusion of benzene solutions of the components. Complexation of $Ir_2Cl_2(1,5\text{-}COD)_2$ takes place without any leaving groups. In this complex, two molecules $Ir_2Cl_2(1,5\text{-}COD)_2$ bind to the opposite ends of the same C_{60} framework and the two Ir atoms of each $Ir_2Cl_2(1,5\text{-}COD)_2$ are bound cis-1 to the same hexagon of C_{60}, leading to a C_{2h} symmetry. Each 1,5-COD ligand is bound to the Ir with two η^2-bonds.

A similar complex is accessible by reaction of C_{60} with an equimolar mixture of the bridged binuclear ruthenium complexes $[(\eta^5\text{-}C_5Me_5)Ru(\mu\text{-}H)]_2$ and $[(\eta^5\text{-}C_5Me_5)Ru(\mu\text{-}Cl)]_2$ [65]. Heating a mixture of this complex with C_{60} in toluene yields green crystals of $C_{60}Ru_2(\mu\text{-}H)(\mu\text{-}Cl)(\eta^5\text{-}C_5Me_5)_2$. It shows the same addition pattern to C_{60} as with Ir complex, i.e. an addition to adjacent double bonds of the hexagon in a $\eta^2{:}\eta^2$-fashion. The distance between the two Ru-metals indicates the presence of a bond between the metals. This Ru–Ru can not be observed in the complex $(\eta^2{:}\eta^2\text{-}C_{60})Ru_2(\mu\text{-}Cl)_2(\eta^5\text{-}C_5Me_5)_2$ were H is replaced with the bigger

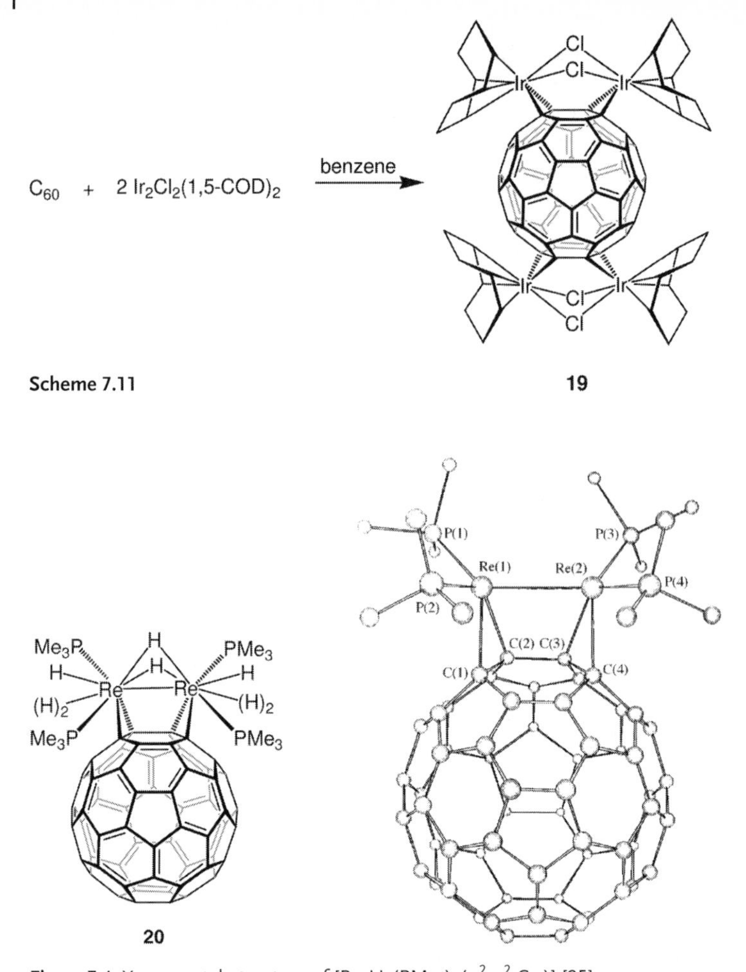

Scheme 7.11

Figure 7.4 X-ray crystal structure of $[Re_2H_8(PMe_3)_4(\eta^2{:}\eta^2\text{-}C_{60})]$ [25].

chlorine [65]. Another example with an observable M–M bond, which shows this rather unusual $\eta^2{:}\eta^2$-coordination mode, is the polyhydrido complex $(\eta^2{:}\eta^2\text{-}C_{60})Re_2H_8(PMe_3)_4$ **20** [25]. Reaction of C_{60} with $Re_2H_8(PMe_3)_4$ gives brown, air-stable crystals in high yield. The X-ray structure of this complex is shown in Figure 7.4 [25], whereas the positions of the hydrogens can only be concluded from NMR spectroscopy.

Theoretical studies found that a η^6-coordination to one hexagon of pristine C_{60} may be possible [6, 8]. The fragments $Co(\eta^3\text{-}C_3H_3)$ or $Rh(\eta^3\text{-}C_3H_3)$ were suggested as good candidates for a coordination to the three double bonds of the 6-ring. Nonetheless, to date, no η^6-complex could be proved. This is one more indication for the low aromaticity of the C_{60} hexagons. As expected they react more like a cyclohexatriene unit, as shown by C_{60}'s ability to bind to various cluster frameworks via a face-capping bonding mode to give $\mu_3\text{-}\eta^2{:}\eta^2{:}\eta^2\text{-}C_{60}$-complexes (Figure 7.5).

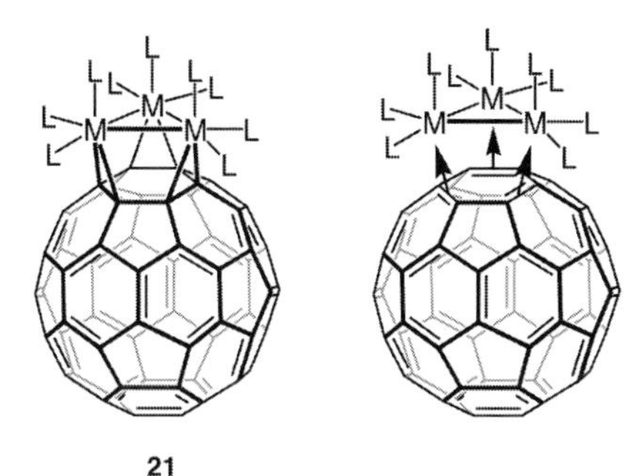

Figure 7.5 μ_3-η^2:η^2:η^2-binding mode of metalcluster-C_{60}-complexes.

Owing to the curvature of C_{60} the exohedral π-orbitals are oriented about 10° away from the perpendicular to the face of the C_6-ring [66]. Thus sufficient overlap with one η^6-bound metal is not given but the overlap of π-orbitals with the metal-triangle of the metal-cluster shown in Figure 7.5 can be very effective. Molecular orbital calculations on the complex of $Ru_3(CO)_9$ [7] with either C_{60} or benzene showed that, compared with the corresponding binding of a cluster fragment to benzene, C_{60} can bind via many more fragment orbitals, allowing the fullerene to form many more bonding interactions. The metal–carbon bonds in $Ru_3(CO)_9(\mu_3$-η^2:η^2:η^2-$C_{60})$ are therefore stronger than in $Ru_3(CO)_9(\mu_3$-η^2:η^2:η^2-$C_6H_6)$ [7].

$Ru_3(CO)_9(\mu_3$-η^2:η^2:η^2-$C_{60})$ can be synthesized and isolated as red crystals by reacting two equiv. of $Ru_3(CO)_{12}$ with C_{60} in refluxing hexanes, or in higher yield by refluxing in chlorobenzene [67, 68]. Chlorobenzene turned out to be the best solvent for the preparation of this fullerene-cluster complex [66]. The bonds in the coordinated hexagon are elongated compared with C_{60} but the bond-length alternation is still observable. The Ru_3-triangle is positioned centrally over the coordinated hexagon of the C_{60}-framework whereas the Ru_3-triangle and the C_6-ring are essentially planar (Figure 7.6) [67].

Figure 7.7 shows some more examples of these trinuclear (μ_3-η^2:η^2:η^2-C_{60})-complexes. The rhenium-hydrido complex $Re_3(\mu$-$H)_3(CO)_{11}(\mu_3$-η^2:η^2:η^2-$C_{60})$ (**24**) can be prepared by reaction of $Re_3(\mu$-$H)_3(CO)_{11}(NCMe)$ with C_{60} in refluxing chlorobenzene in 50% yield [69]. The osmium cluster **23** is accessible by decarbonylation of $Os_3(CO)_9(\mu_3$-η^2:η^2:η^2-$C_{60})$ with Me_3NO–MeCN in the presence of PMe_3 [70, 71]. Instead of PMe_3, other ligands may be used. $Os_3(CO)_9(\mu_3$-η^2:η^2:η^2-$C_{60})$ can be synthesized either by thermolysis of $Os_3(CO)_{10}(NCMe)(\eta^2$-$C_{60})$ in refluxing chlorobenzene or via direct reaction of $Os_3(CO)_{11}(NCMe)$ with C_{60} also in refluxing chlorobenzene. The η^2-precursor complex is made by reaction of the same $Os_3(CO)_{11}(NCMe)$ complex with C_{60} but in refluxing toluene. The Re_3 and Os_3 fullerene-cluster complexes have structures similar to the already described structure of the Ru_3-compound. Various complexes have been synthesized by varying the ligands. The most common ligands are CO, phosphines and MeCN, and the possibility of ligand fluctuation was studied [68, 72]. NMR measurements showed that fluctuation takes place at elevated temperatures but is obviously rather slow at room temperature.

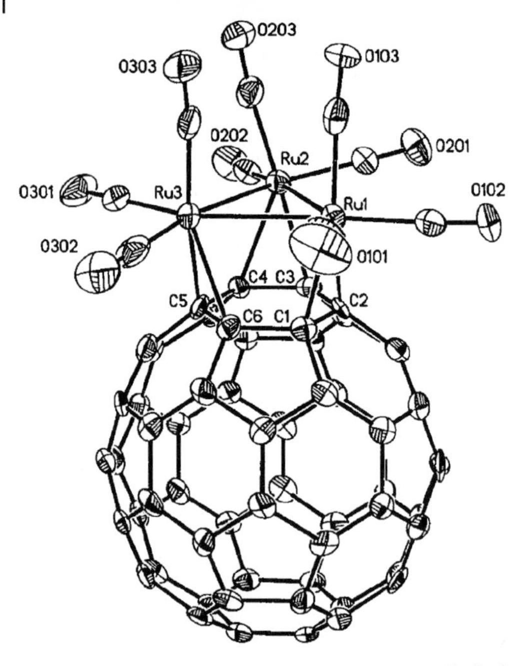

Figure 7.6 X-ray crystal structure of $Ru_3(CO)_9(\mu_3\text{-}\eta^2{:}\eta^2{:}\eta^2\text{-}C_{60})$ [67].

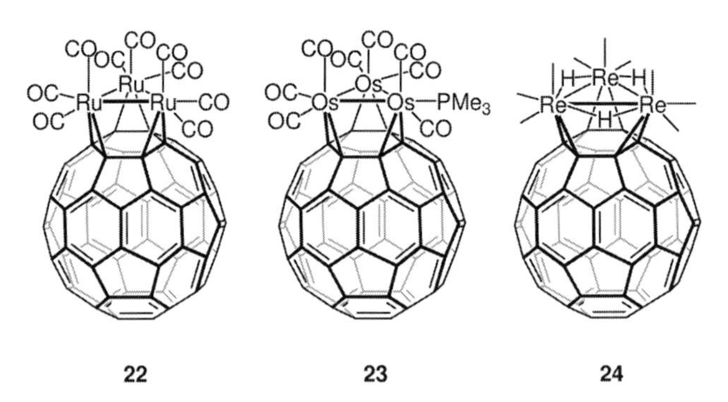

Figure 7.7 Examples of metal-C_{60} clusters. CO ligands omitted from **24** for clarity.

The trinuclear Os complex **25** undergoes some interesting conversions that are induced by ligands such as benzyl isocyanide or triphenylphosphine [66, 73, 74]. Treatment of complex **25** with an excess of $PhCH_2NC$ leads to an insertion of this molecule into an Os–Os bond and at the same time induces a change of the $(\mu_3\text{-}\eta^2{:}\eta^2{:}\eta^2)$ bonding mode to a $(\mu_3\text{-}\eta^1{:}\eta^2{:}\eta^1)$ bonding mode (Scheme 7.12). This new complex is one of the rare examples for a M–C_{60} σ-bond. The two M–C_{60} σ-bonds are found in the 1,4-position of the C_6-ring, leaving a cyclohexadiene fragment, which in the X-ray structure clearly shows a boat shape [73, 74].

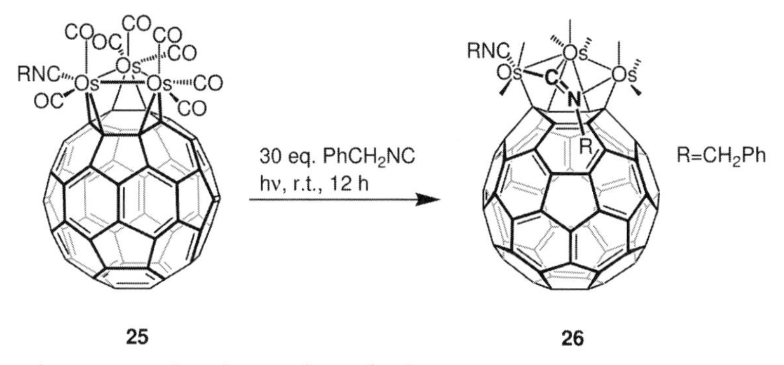

Scheme 7.12 CO ligands omitted in **26** for clarity.

Heating **26** with an excess of PPh_3 under pressure in chlorobenzene leads to the loss of one coordination site. One more Os–Os bond is broken upon formation of the complex $Os_3(CO)_6(CNR)(\mu_3\text{-}CNCH_2C_6H_4)\text{-}(PPh_3)(\mu\text{-}PPh_2)(\mu\text{-}\eta^2{:}\eta^2\text{-}C_{60})$ [74]. The $\eta^2{:}\eta^2$ binding mode observed in this complex is very unusual for C_{60}. One further example, where C_{60} can act as a four-electron ligand is the fullerene carbido pentaosmium cluster complex **28** shown in Scheme 7.13 [75, 76]. The $\eta^2{:}\eta^2$-complex **28** is formed in a mixture with the "normal" $\eta^2{:}\eta^2{:}\eta^2$ carbido pentaosmium complex **27**. The two complexes can be converted into one another at elevated temperatures.

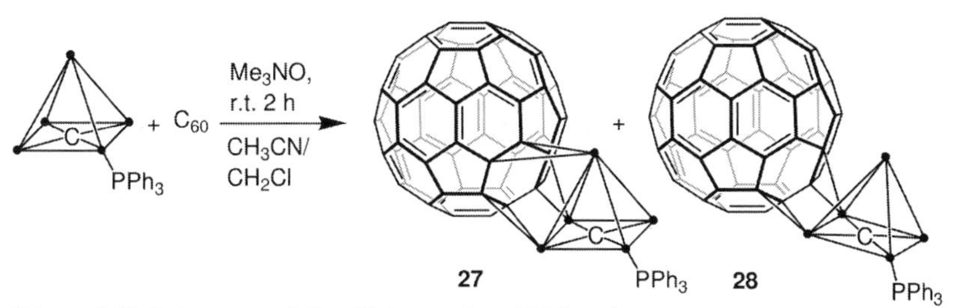

Scheme 7.13 Dots represent Os with two or three CO ligands.

Penta- and hexanuclear clusters of the metals osmium and ruthenium coordinate with the same $\eta^2{:}\eta^2{:}\eta^2$- binding mode as the trinuclear clusters to C_{60}. Such complexes are known for the clusters Os_5C [75, 76], Ru_5C [77–79], Ru_6C [78], $PtRu_5C$ [77] and Rh_6 [80]. In this collection of metal clusters rhenium plays a special role, because it forms a new fullerene-metal sandwich complex, where two C_{60} are bound to one cluster.

7.4
Hydrometalation Reactions

Whereas the reaction of C_{60} with $RhH(CO)(PPh_3)_3$ results in complex formation in a dihapto fashion, a hydrometalation is possible with the stronger nucleophilic reagent $Cp_2Zr(H)Cl$ ($Cp = \eta^5\text{-}C_5H_5$) [81] (Scheme 7.14) [82]. Upon treatment of

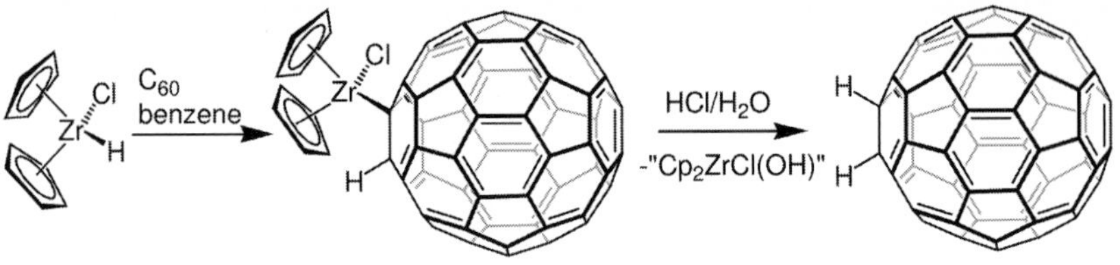

Scheme 7.14

C_{60} with this Zr complex, a red solution is formed, unlike the green solution of η^2 transition metal complexes of C_{60}. The structure of the air-sensitive $Cp_2ZrClC_{60}H$ was confirmed by ^1H NMR spectroscopy. The hydrogen transferred from the Zr to C_{60} resonates at $\delta = 6.09$, a typical value for fullerenyl protons [83]. Hydrolysis of $Cp_2ZrClC_{60}H$ with aqueous HCl provides access to the simplest C_{60} hydrocarbon $C_{60}H_2$ (**30**, Scheme 7.14). Spectroscopic characterization of $C_{60}H_2$ showed that the compound is the isomerically pure 1,2-addition product.

7.5 Organometallic Polymers of C_{60}

Organometallic polymers [9] of C_{60} can be obtained by the reaction of C_{60} solutions with the palladium or platinum complex $M_2(dba)_3{\cdot}CHCl_3$ [84] or the platinum complex $Pt(dba)_2$ [85] (dba = dibenzylideneacetone). The resulting replacement of the dba ligands leads to a dark brown precipitate of $C_{60}Pd_n$. Depending on the C_{60}:Pd ratio these polymers form linear "pearl necklace" or two- and three-dimensional arrangements. The Pd:C_{60} ratio can be changed from 1 : 1 to 3 : 1 upon heating a suspension of the initially formed polymer in toluene under reflux with regeneration of free C_{60}. These materials exhibit catalytic activity in the hydrogenation of olefins and acetylenes at room temperature [86]. The mode of the binding of the metal to the fullerene core has not been conclusively proven but it is believed that the fullerene cages are linked via a η^2-binding mode with the metal M (Figure 7.8).

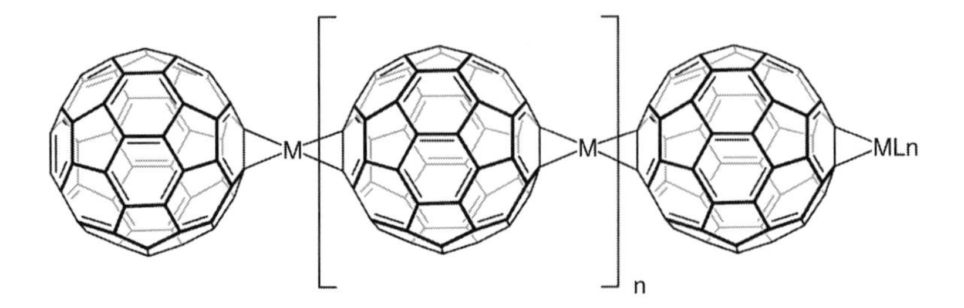

Figure 7.8 Organometallic polymers of C_{60} are probably linked via a η^2-binding-mode.

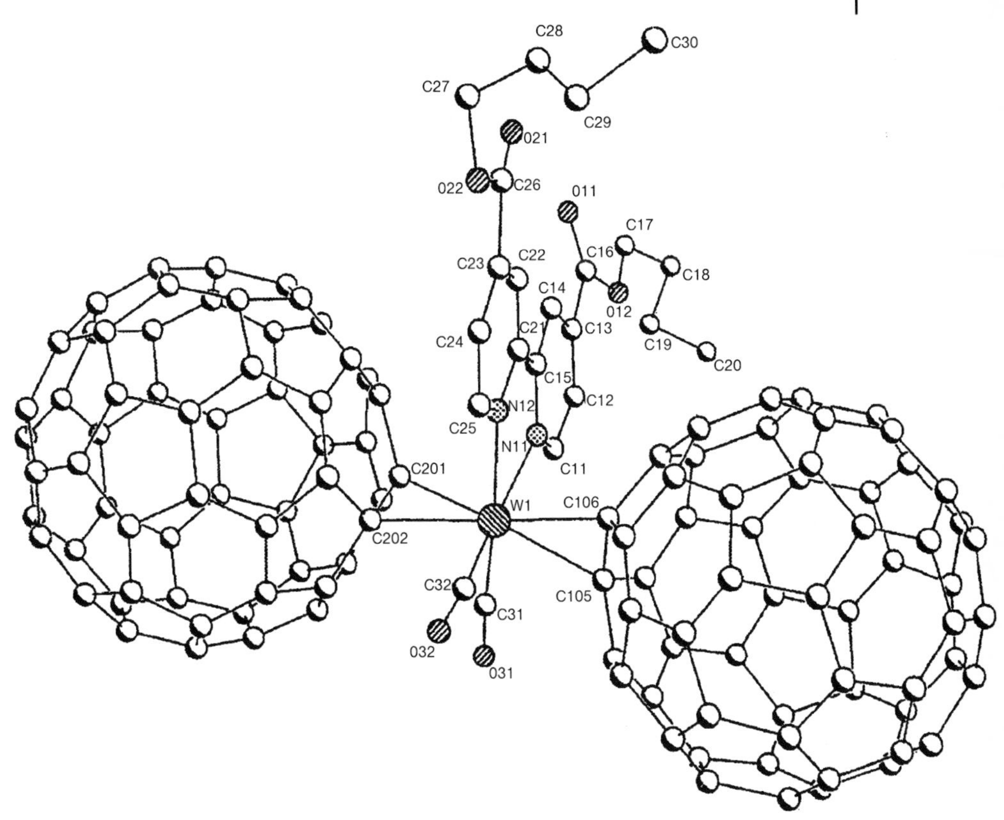

Figure 7.9 X-ray crystal structure of $[W(\eta^2\text{-}C_{60})_2(CO)_2(\text{dbcbipy})]$ (dbcbipy = 4,4′-di(butyl carboxyl)-2,2′-bipyridine) [87].

The possibility of this binding mode was shown recently with the synthesis and structural analysis of the dumbbell-shaped bis-fullerene complexes of molybdenum and tungsten. The X-ray crystal structure proves the $(\eta^2\text{-}C_{60})_2M$ binding mode (Figure 7.9) [87].

The polymer $C_{60}Pt_n$ has also been made by using Pt(1,5-cyclooctadiene)$_2$ [88]. Various $C_{60}M$ polymer films have been prepared by electrochemical reduction of transition metal complexes of the central metal Pd(II), Pt(II), Ir(I), Rh(I), Rh(II), Rh(III) and Au(I) [89, 90]. The metal complex in solution is first electrochemically reduced to the zerovalent metal, which then forms polymer films containing M_nC_{60} on the electrode.

References

1. P. J. Fagan, J. C. Calabrese, B. Malone, *Science* **1991**, *252*, 1160.
2. P. J. Fagan, J. C. Calabrese, B. Malone, *ACS Symp. Ser.* **1992**, *481*, 177.
3. P. J. Fagan, J. C. Calabrese, B. Malone, *Acc. Chem. Res.* **1992**, *25*, 134.
4. P. J. Fagan, M. D. Ward, J. C. Calabrese, *J. Am. Chem. Soc.* **1989**, *111*, 1698.
5. J. R. Rogers, D. S. Marynick, *Chem. Phys. Lett.* **1993**, *205*, 197.
6. E. D. Jemmis, M. Manoharan, P. K. Sharma, *Organometallics* **2000**, *19*, 1879.
7. M. A. Lynn, D. L. Lichtenberger, *J. Cluster Sci.* **2000**, *11*, 169.
8. S.-K. Goh, D. S. Marynick, *J. Comput. Chem.* **2001**, *22*, 1881.
9. A. L. Balch, M. M. Olmstead, *Chem. Rev.* **1998**, *98*, 2123.
10. P. J. Fagan, J. C. Calabrese, B. Malone, *J. Am. Chem. Soc.* **1991**, *113*, 9408.
11. B. Chase, P. J. Fagan, *J. Am. Chem. Soc.* **1992**, *114*, 2252.
12. S. A. Lerke, B. A. Parkinson, D. H. Evans, P. J. Fagan, *J. Am. Chem. Soc.* **1992**, *114*, 7807.
13. F. J. Brady, D. J. Cardin, M. Domin, *J. Organomet. Chem.* **1995**, *491*, 169.
14. L. Pandolfo, M. Maggini, *J. Organomet. Chem.* **1997**, *540*, 61.
15. V. V. Bashilov, T. V. Magdesieva, D. N. Kravchuk, P. V. Petrovskii, A. G. Ginzburg, K. P. Butin, V. I. Sokolov, *J. Organomet. Chem.* **2000**, *599*, 37.
16. L.-C. Song, G.-F. Wang, P.-C. Liu, Q.-M. Hu, *Organometallics* **2003**, *22*, 4593.
17. A. L. Balch, V. J. Catalano, J. W. Lee, *Inorg. Chem.* **1991**, *30*, 3980.
18. A. L. Balch, J. W. Lee, B. C. Noll, M. M. Olmstead, *J. Am. Chem. Soc.* **1992**, *114*, 10984.
19. A. L. Balch, J. W. Lee, B. C. Noll, M. M. Olmstead, *Inorg. Chem.* **1994**, *33*, 5238.
20. A. L. Balch, V. J. Catalano, J. W. Lee, M. M. Olmstead, *J. Am. Chem. Soc.* **1992**, *114*, 5455.
21. V. J. Catalano, N. Parodi, *Inorg. Chem.* **1997**, *36*, 537.
22. R. S. Koefod, M. F. Hudgens, J. R. Shapley, *J. Am. Chem. Soc.* **1991**, *113*, 8957.
23. A. L. Balch, J. W. Lee, B. C. Noll, M. M. Olmstead, *Inorg. Chem.* **1993**, *32*, 3577.
24. A. V. Usatov, S. M. Peregudova, L. I. Denisovich, E. V. Vorontsov, L. E. Vinogradova, Y. N. Novikov, *J. Organomet. Chem.* **2000**, *599*, 87.
25. A. N. Chernega, M. L. H. Green, J. Haggitt, A. H. H. Stephens, *J. Chem. Soc., Dalton Trans.* **1998**, 755.
26. A. V. Usatov, K. N. Kudin, E. V. Vorontsov, L. E. Vinogradova, Y. N. Novikov, *J. Organomet. Chem.* **1996**, *522*, 147.
27. J. W. Lee, M. M. Olmstead, J. S. Vickery, A. L. Balch, *J. Cluster Sci.* **2000**, *11*, 67.
28. A. V. Usatov, E. V. Martynova, F. M. Dolgushin, A. S. Peregudov, M. Y. Antipin, Y. N. Novikov, *Eur. J. Inorg. Chem.* **2002**, 2565.
29. D. M. Thompson, M. Bengough, M. C. Baird, *Organometallics* **2002**, *21*, 4762.
30. L.-C. Song, P.-C. Liu, Q.-M. Hu, G.-L. Lu, G.-F. Wang, *J. Organomet. Chem.* **2003**, *681*, 264.
31. M. Rasinkangas, T. T. Pakkanen, T. A. Pakkanen, *J. Organomet. Chem.* **1994**, *476*, C6.
32. A. V. Usatov, E. V. Martynova, I. S. Neretin, Y. L. Slovokhotov, A. S. Peregudov, Y. N. Novikov, *Eur. J. Inorg. Chem.* **2003**, 2041.
33. M. N. Bengough, D. M. Thompson, M. C. Baird, G. D. Enright, *Organometallics* **1999**, *18*, 2950.
34. V. V. Burlakov, A. V. Usatov, K. A. Lyssenko, M. Y. Antipin, Y. N. Novikov, V. B. Shur, *Eur. J. Inorg. Chem.* **1999**, 1855.
35. H. F. Hsu, J. R. Shapley, *Proc. Electrochem. Soc.* **1995**, *95-10*, 1087.
36. L.-C. Song, Y.-H. Zhu, Q.-M. Hu, *Polyhedron* **1998**, *17*, 469.
37. J. R. Shapley, Y. Du, H. F. Hsu, J. J. Way, *Proc. – Electrochem. Soc.* **1994**, *94-24*, 1255.
38. L.-C. Song, Y.-H. Zhu, Q.-M. Hu, *Polyhedron* **1997**, *16*, 2141.
39. K. Tang, S. Zheng, X. Jin, H. Zeng, Z. Gu, X. Zhou, Y. Tang, *J. Chem. Soc., Dalton Trans.* **1997**, 3585.

40 L.-C. Song, J.-T. Liu, Q.-M. Hu, L.-H. Weng, *Organometallics* **2000**, *19*, 1643.
41 L.-C. Song, J.-T. Liu, Q.-M. Hu, G.-F. Wang, P. Zanello, M. Fontani, *Organometallics* **2000**, *19*, 5342.
42 L.-C. Song, J.-T. Liu, Q.-M. Hu, *J. Organomet. Chem.* **2002**, *662*, 51.
43 L.-C. Song, P.-C. Liu, J.-T. Liu, F.-H. Su, G.-F. Wang, Q.-M. Hu, P. Zanello, F. Laschi, M. Fontani, *Eur. J. Inorg. Chem.* **2003**, 3201.
44 D. M. Thompson, M. Jones, M. C. Baird, *Eur. J. Inorg. Chem.* **2003**, 175.
45 S. D. Ittel, J. A. Ibers, *Adv. Organomet. Chem.* **1976**, *14*, 33.
46 T. Yoshida, T. Matsuda, S. Otsuka, *Inorg. Synth.* **1990**, *28*, 122.
47 C. Liu, G. Zhao, Q. Gong, K. Tang, X. Jin, P. Cui, L. Li, *Opt. Commun.* **2000**, *184*, 309.
48 P. Zanello, F. Laschi, A. Cinquantini, M. Fontani, K. Tang, X. Jin, L. Li, *Eur. J. Inorg. Chem.* **2000**, 1345.
49 P. Zanello, F. Laschi, M. Fontani, L.-C. Song, Y.-H. Zhu, *J. Organomet. Chem.* **2000**, *593–594*, 7.
50 H.-F. Hsu, Y. Du, T. E. Albrecht-Schmitt, S. R. Wilson, J. R. Shapley, *Organometallics* **1998**, *17*, 1756.
51 dppb = 1,2-bis(diphenylphosphino)-benzene; dppe = 1,2-bis(diphenyl-phosphino)ethane; dppf = 1,1′-bis-(diphenylphosphino)-ferrocene; phen = 1,10-phenanthroline; dbm = dibutylmaleate.
52 dppb, dppe, dppf, phen and dbm are as in Ref. [51].
53 M. M. Olmstead, L. Hao, A. L. Balch, *J. Organomet. Chem.* **1999**, *578*, 85.
54 P. W. Fowler, D. J. Collins, S. J. Austin, *J. Chem. Soc., Perkin Trans. 2:* **1993**, 275.
55 Y. Zhu, R. S. Koefod, C. Devadoss, J. R. Shapley, G. B. Schuster, *Inorg. Chem.* **1992**, *31*, 3505.
56 L. Vaska, *Acc. Chem. Res.* **1968**, *1*, 335.
57 A. Vertes, M. Gal, F. E. Wagner, F. Tuczek, P. Gütlich, *Inorg. Chem.* **1993**, *32*, 4478.
58 A. J. Deeming, B. L. Shaw, *J. Chem. Soc. A* **1969**, 1802.
59 A. V. Usatov, E. V. Martynova, F. M. Dolgushin, A. S. Peregudov, M. Y. Antipin, Y. N. Novikov, *Eur. J. Inorg. Chem.* **2003**, 29.
60 A. L. Balch, V. J. Catalano, J. W. Lee, M. M. Olmstead, S. R. Parkin, *J. Am. Chem. Soc.* **1991**, *113*, 8953.
61 A. L. Balch, J. W. Lee, M. M. Olmstead, *Angew. Chem.* **1992**, *104*, 1400; *Angew. Chem. Int. Ed. Engl.* **1992**, *31*, 1356.
62 G. Yagupsky, C. K. Brown, G. Wilkinson, *J. Chem. Soc. A* **1970**, 1392.
63 M. L. H. Green, A. H. H. Stephens, *Chem. Commun.* **1997**, 793.
64 M. Rasinkangas, T. T. Pakkanen, T. A. Pakkanen, M. Ahlgren, J. Rouvinen, *J. Am. Chem. Soc.* **1993**, *115*, 4901.
65 I. J. Mavunkal, Y. Chi, S.-M. Peng, G.-H. Lee, *Organometallics* **1995**, *14*, 4454.
66 K. Lee, H. Song, J. T. Park, *Acc. Chem. Res.* **2003**, *36*, 78.
67 H.-F. Hsu, J. R. Shapley, *J. Am. Chem. Soc.* **1996**, *118*, 9192.
68 H. F. Hsu, J. R. Shapley, *J. Organomet. Chem.* **2000**, *599*, 97.
69 H. Song, Y. Lee, Z.-H. Choi, K. Lee, J. T. Park, J. Kwak, M.-G. Choi, *Organometallics* **2001**, *20*, 3139.
70 J. T. Park, H. Song, J.-J. Cho, M.-K. Chung, J.-H. Lee, I.-H. Suh, *Organometallics* **1998**, *17*, 227.
71 H. Song, K. Lee, J. T. Park, M.-G. Choi, *Organometallics* **1998**, *17*, 4477.
72 H. Song, K. Lee, J. T. Park, H. Y. Chang, M. G. Choi, *J. Organomet. Chem.* **2000**, *599*, 49.
73 H. Song, C. H. Lee, K. Lee, J. T. Park, *Organometallics* **2002**, *21*, 2514.
74 H. Song, K. Lee, M.-G. Choi, J. T. Park, *Organometallics* **2002**, *21*, 1756.
75 K. Lee, C. H. Lee, H. Song, J. T. Park, H. Y. Chang, M.-G. Choi, *Angew. Chem.* **2000**, *112*, 1871; *Angew. Chem. Int. Ed. Engl.* **2000**, *39*, 1801.
76 K. Lee, Z.-H. Choi, Y.-J. Cho, H. Song, J. T. Park, *Organometallics* **2001**, *20*, 5564.
77 K. Lee, J. R. Shapley, *Organometallics* **1998**, *17*, 3020.
78 K. Lee, H.-F. Hsu, J. R. Shapley, *Organometallics* **1997**, *16*, 3876.
79 A. J. Babcock, J. Li, K. Lee, J. R. Shapley, *Organometallics* **2002**, *21*, 3940.
80 K. Lee, H. Song, B. Kim, J. T. Park, S. Park, M.-G. Choi, *J. Am. Chem. Soc.* **2002**, *124*, 2872.
81 J. Schwartz, *Pure Appl. Chem.* **1980**, *52*, 733.

82 S. Ballenweg, R. Gleiter, W. Krätschmer, *Tetrahedron Lett.* **1993**, *34*, 3737.
83 A. Hirsch, T. Grösser, A. Skiebe, A. Soi, *Chem. Ber.* **1993**, *126*, 1061.
84 H. Nagashima, A. Nakaoka, Y. Saito, M. Kato, T. Kawanishi, K. Itoh, *J. Chem. Soc., Chem. Commun.* **1992**, 377.
85 H. Nagashima, Y. Kato, H. Yamaguchi, E. Kimura, T. Kawanishi, M. Kato, Y. Saito, M. Haga, K. Itoh, *Chem. Lett.* **1994**, 1207.
86 H. Nagashima, A. Nakaoka, S. Tajima, Y. Saito, K. Itoh, *Chem. Lett.* **1992**, 1361.
87 X. Jin, X. Xie, K. Tang, *Chem. Commun.*, **2002**, 750–751.
88 M. Van Wijnkoop, M. F. Medine, A. G. Avent, A. D. Darwish, H. W. Kroto, R. Taylor, D. R. M. Walton, *J. Chem. Soc., Dalton Trans.: Inorg. Chem.* **1997**, 675.
89 K. Winkler, A. De Bettencourt-Dias, A. L. Balch, W. Kutner, K. Noworyta, *Proc. Electrochem. Soc.* **2000**, *2000-10*, 31.
90 K. Winkler, A. De Bettencourt-Dias, A. L. Balch, *Chem. Mater.* **2000**, *12*, 1386.

8
Oxidation and Reactions with Electrophiles

8.1
Introduction

Although the reduction of the fullerenes is far more facile than their oxidation, a large variety of oxidative functionalizations as well as electrophilic additions have been carried out. Controlled oxygenations and osmylations lead to defined adducts. Iridium complexes of oxygenated C_{60} and both mono- and polyadducts of osmylated fullerenes have been completely characterized by X-ray crystallography and NMR spectroscopy. Investigation of the all-carbon cations $C_{60}{}^{n+}$ has made good progress. Recently, a rather stable $C_{60}{}^{+}$ salt was isolated and the redox potentials of the fullerene cations $C_{60}{}^{2+}$ and $C_{60}{}^{3+}$ could be measured. With the synthesis of the $C_{60}{}^{+}$salt, measurements of the cation properties were cleared of the uncertainty as to whether the cation or a degradation product is measured. Nitration and oxidative sulfonation leads to polyadducts that serve as starting material for interesting products. Hydrolysis of these species with water provides access to the polar, water-soluble fullerenols, aminolysis leads to polyamino-adducts that have been used as linkers in polymers or as cores in dendrimers. The interaction of fullerenes with Lewis acids has been used to separate C_{60} from C_{70} as well as to carry out fullerenations of aromatics or the addition of halogenated hydrocarbons to C_{60}.

8.2
Electrochemical Oxidation of C_{60} and C_{70}

Whereas C_{60} and C_{70} are easy to reduce, their oxidation occurs at comparatively high anodic potentials [1, 2]. Theoretical investigations predict the first oxidation potential of C_{60} to be comparable to that of naphthalene [3]. Anodic electrochemistry with fullerenes has been carried out with C_{60} films [4] as well as in solution [5–7]. Cyclic voltammetry of C_{60} in a 0.1 M solution of Bu_4NPF_6 in trichloroethylene (TCE) at room temperature exhibits a chemically reversible, one-electron oxidation wave at +1.26 V vs Fc/Fc^+ (Figure 8.1) [7]. Under identical conditions, a one-electron, chemically reversible oxidation is also observed for C_{70}. The oxidation of C_{70} occurs 60 mV more negative than that of C_{60} at +1.20 V vs Fc/Fc^+. The energy difference between the first oxidation and the first reduction potential is a measure of the

Fullerenes: Chemistry and Reactions. Andreas Hirsch and Michael Brettreich

ISBN: 3-527-30820-2

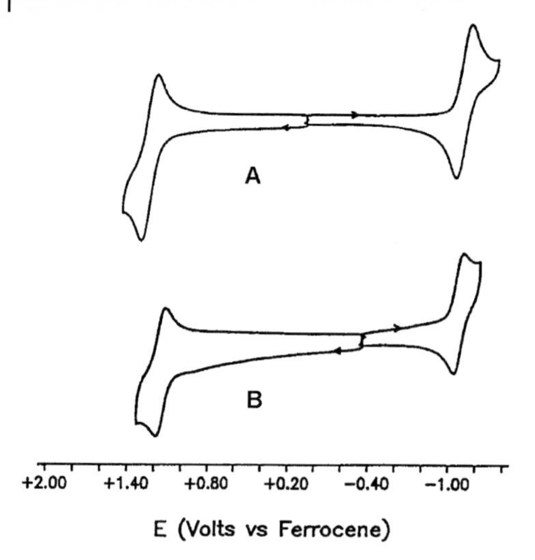

Figure 8.1 CV for C_{60} and C_{70} in 0.1 M (TBA)PF_6/TCE at room temperature; CVs were run at a scan rate of 100 mV s^{-1}. CV with C_{60} (A) and C_{70} (B), no ferrocene added. Cathodic and anodic cyclic scans were performed separately [7].

energy difference between the HOMO and the LUMO in solution and was determined to be 2.32 V for C_{60} and 2.22 V for C_{70}. The cationic radical species generated electrochemically are relatively long-lived at room temperature. The life-time of the radicals was estimated to be > 0.5 min. For C_{70}, a second chemically irreversible, one-electron oxidation ($C_{70}^{\bullet+} \rightarrow C_{70}^{2+}$) was observed at 1.75 V.

The stability of the C_{60}^{+}-cation can be increased by adding trifluoromethane-sulfonic acid CF_3SO_3H to the TCE/(TBA)PF_6 solutions [8]. Under these conditions the life-time of the cation is estimated to be at least some hours. C_{60}^{+} is probably stabilized by CF_3SO_3H because of acid's ability to scavenge water and nucleophiles that would react readily with the cation. The elevated lifetime of the cation made it possible to prepare C_{60}^{+} in a bulk-electrolysis and to analyze the oxidized C_{60} with UV/VIS/NIR spectroscopy and ESR spectroscopy. The UV/VIS/NIR spectrum shows a sharp peak at 983 nm and a broad peak at 846 nm. These two absorbances are attributed to allowed NIR-transitions and these values are consistent with spectra of the cation obtained with other methods [2]. EPR spectroscopy of C_{60}-cations, produced by different methods, leads to a broad distribution of measured g-values. These differences are caused by the short lifetime of the cation, the usually low signal-noise ratio and the uncertainty of the purity. The most reliable value until now is probably the one obtained by Reed and co-workers for the salt $C_{60}^{+}(CB_{11}H_6Cl_6)^{-}$ ($g = 2.0022$) [2, 9] (see also Section 8.5). Ex situ ESR spectroscopy of above-mentioned bulk electrolysis solutions led to a g-value of 2.0027 [8], which is very close to that of the salt, whereas the ESR spectra of this electrolytically formed cation shows features not observed earlier. The observed splitting of the ESR signal at lower modulation amplitudes was assigned to a rhombic symmetry of the cation radical at lower temperatures (5–200 K).

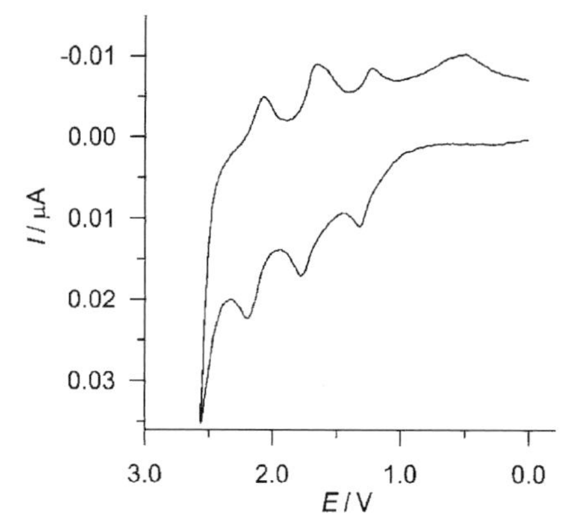

Figure 8.2 Cyclic voltammetric curve of a solution of 0.15 mM (saturated) C_{60}, 0.05 M $TBAAsF_6$ in dichloromethane [Pt electrode, scan rate: 100 mV s^{-1}, $T = -55$ °C, E (V) vs ferrocene] [10].

The C_{60}-radical cation reacts immediately with any nucleophile. C_{60}-cations with higher oxidation states are exceedingly reactive species. Their generation and spectroscopic proof requires carefully selected conditions. The electrochemical oxidation of C_{60}^{+} to C_{60}^{2+} and C_{60}^{3+} was possible by using ultra-dry dichloromethane as a solvent and $TBAAsF_6$ (tetrabutylammoniumarsenhexafluoride) as electrolyte [10]. $TBAAsF_6$ is commonly used as an electrolyte with high oxidation resistance and very low nucleophilicity. The anodic peaks have halfwave potentials of $E_{½} = 1.27$, 1.71 and 2.14 V for the three subsequent oxidations of C_{60} (Figure 8.2). As expected, the second and third oxidation potentials are not chemically reversible.

8.3 Oxygenation

Oxygenated fullerenes $C_{60}O_n$ and $C_{70}O_n$ can be found in the fullerene mixture generated by graphite vaporization [11, 12]. The formation of these oxides is due to a small amount of molecular oxygen present in the fullerene reactor. $C_{70}O$ was isolated from the fullerene extract by preparative HPLC [11]. Mixtures of $C_{60}O_n$ (n up to 4) can also be generated by electrochemical oxidation of C_{60} [13] or by photolysis of the crude fullerene extract [12]. More drastic conditions, such as UV irradiation in hexane [14] or heating in the presence of oxygen [15, 16] lead to an extensive oxidation or fragmentation of C_{60}.

$C_{60}O$ (**1**) can be obtained in 7% yield, in addition to negligible amounts of higher oxides, upon photooxygenation of C_{60} by irradiating an oxygenated benzene solution for 18 h at room temperature and subsequent purification by flash chromatography followed by semipreparative HPLC (Scheme 8.1) [15, 17]. The short lifetime of $C_{60}O_2$ under UV-radiation is responsible for the absence of higher oxides [18]. Photochemical formation of $C_{60}O$ from C_{60} probably proceeds via reaction of singlet oxygen with the lowest triplet state of C_{60} [19].

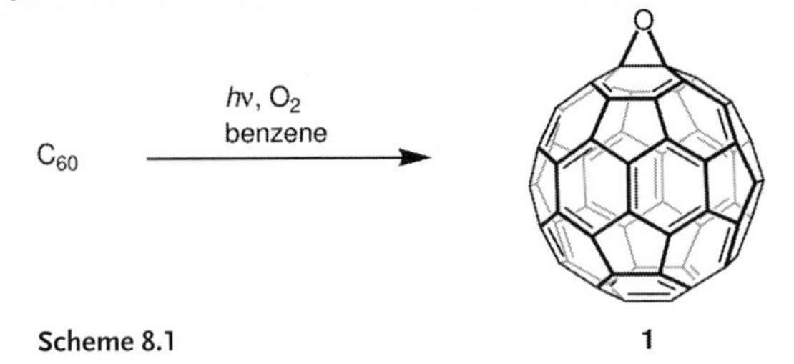

Scheme 8.1

Better yields of the mono-oxygenated product and the concurrent generation of higher oxides can be achieved by oxidation with *m*-chloroperoxybenzoic acid (MCPBA) (Scheme 8.2) [20–24]; C_{60} and usually about 10 to 30 equiv. of the peroxy acid are stirred in toluene at 80 °C [21]. The product distribution can be influenced by reaction time and the amount of acid. The existence of the $C_{60}O$-isomer **1** – accessible in 30% yield – and the *cis*-1-$C_{60}O_2$-isomer **2** – accessible in 8% yield – could be proven (Scheme 8.2) [21], but higher oxides $C_{60}O_n$ up to $n = 12$ can also be formed [24]. The main products are oxides with $n = 1$–3.

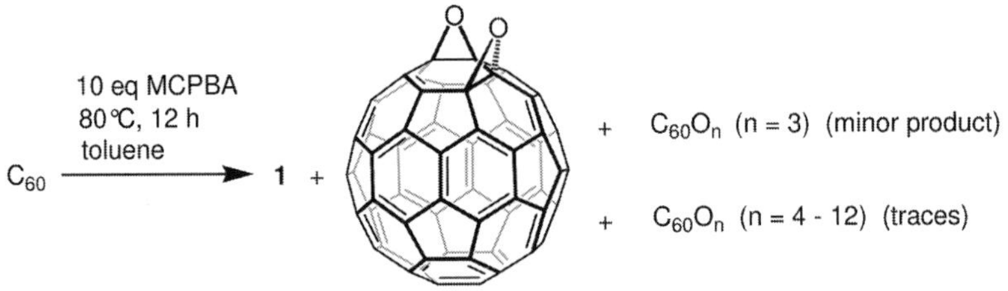

10 eq MCPBA
80 °C, 12 h
toluene

C_{60} → 1 + 2 + $C_{60}O_n$ (n = 3) (minor product) + $C_{60}O_n$ (n = 4 - 12) (traces)

Scheme 8.2

For one of the $C_{60}O_3$ isomers the C_{3v}-symmetrical structure **5** was proposed (Figure 8.3) [24]. Experimentally as well as theoretically [25, 26] the *cis*-1-addition pattern for oxygenated C_{60} turns out to be the most favorable. The three structures shown in Figure 8.3 are calculated to be the most stable isomers out of a possible 47 [25]. C_{60} reacts with hydrogen peroxide and methyltrioxorhenium in very similar fashion to the MCPBA oxidation but gives slightly higher yields [27]. This smooth reaction can be carried out at room temperature.

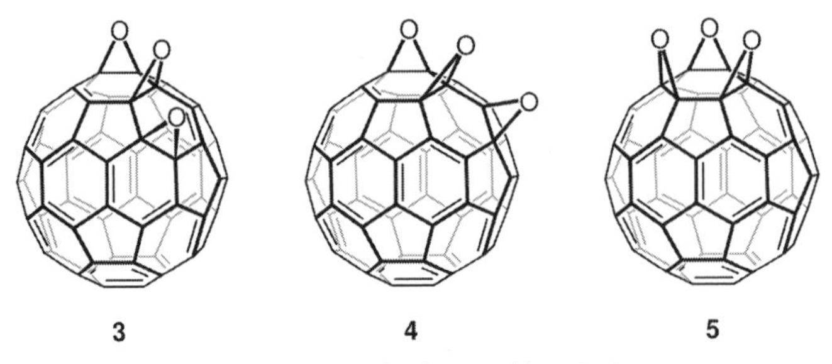

Figure 8.3 Most stable structures for the trisadduct $C_{60}O_3$.

$C_{60}O$ (**1**) can also be prepared by allowing toluene solutions of C_{60} to react with dimethyldioxirane (Scheme 8.3) [28]. The so-obtained product is identical to that prepared by photochemical epoxidation [15]. Upon treatment of C_{60} with dimethyldioxirane, a second product is formed simultaneously (Scheme 8.3), which was identified to be the 1,3-dioxolane **6**. Upon heating **6** in toluene for 24 h at 110 °C, no decomposition could be observed by HPLC, implying that **1** and **6** are formed by different pathways. Replacement of dimethyldioxirane with the more reactive methyl(trifluoromethyl)dioxirane allows much milder reaction conditions [29]. At 0 °C and a reaction time of only some minutes this reaction renders a C_{60} conversion rate of more than 90% and higher yields for $C_{60}O$ as well as for the higher oxides. The smooth reaction conditions may explain why no 1,3-dioxolane **6** can be found. $C_{60}O$ can be isolated in 20% yield, the combined $C_{60}O_2$ isomers in 35% yield and the $C_{60}O_3$ isomers in 17% yield. Also, under these reaction conditions *cis*-1-$C_{60}O_2$ is the most abundant bisadduct. Of the $C_{60}O_3$ isomers, two could be isolated as a mixture. The observed ^{13}C NMR spectrum strongly suggests that the given mixture consists of the isomers **3** and **4** (Figure 8.3).

Scheme 8.3 R = CH_3 or CF_3.

Spectroscopic analysis of **1** demonstrates that it exhibits the 1,2-bridged epoxide structure as shown in Scheme 8.1. This becomes especially evident by the appearance of a peak at $\delta = 90.18$ in the ^{13}C NMR spectrum [15], which is due to the sp^3 bridgehead carbons of the C_{60} cage. Further proof for this epoxide structure is given by the X-ray structure determination of a (η^2-$C_{60}O$)-iridium complex and the corresponding (η^2-$C_{60}O_2$) complex [20–22]. In these complexes the epoxide functionality is clearly present. The "O-addend" has a *cis*-1-directing influence on the addition of the transition metal. Another possible isomer of $C_{60}O$ could be the ring-opened 1,6-bridged structure **8** (Scheme 8.4). Interestingly, ab initio calculations of **1** and **8** at the LSD/DZP level [30, 31] and calculations at the AM1 level [26] predict **8** to be more stable, although its formation in photooxygenation, oxidation with peroxides or with dioxirane has never been observed. The missing [5,6]-open $C_{60}O$ is also not formed under the usual conditions of ozonolysis. But it is accessible by first forming the ozonide **7** at low temperatures in the dark [32]. The purified ozonide was then exposed to light at room temperature for several minutes [33]. These conditions were sufficient to convert **7** into **8** with a yield of nearly 100%. Thermal ozonolysis yields **1** in 20% yield and higher oxides [34–36].

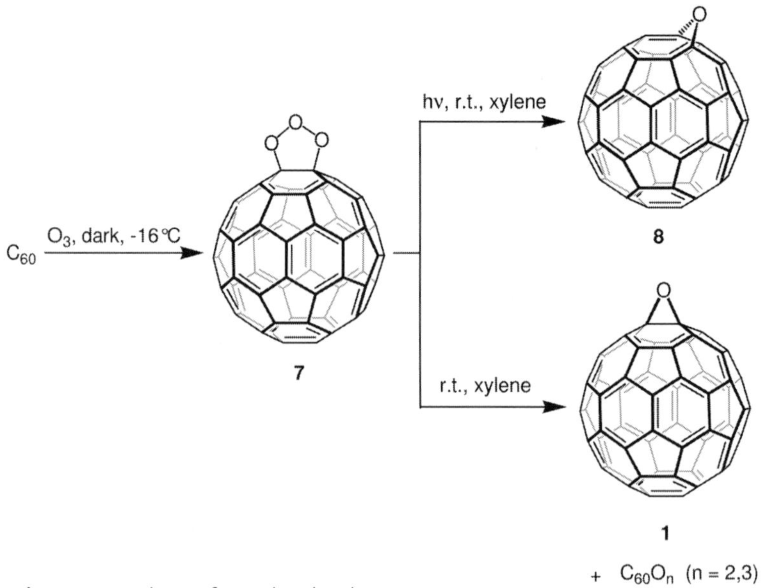

C_{60} $\xrightarrow{O_3,\ dark,\ -16\,°C}$ 7 → hν, r.t., xylene → 8 ; r.t., xylene → 1 + $C_{60}O_n$ (n = 2,3)

Scheme 8.4 Synthesis of [6,6]-closed and [5,6]-open $C_{60}O$.

Chromatography of $C_{60}O$ (**1**) on neutral alumina leads to an efficient conversion into C_{60} in 91% yield [15]. The same conversion is observed during electrochemical reduction of $C_{60}O$. Cyclovoltammetry of $C_{60}O$ shows three reduction waves [37, 38], whereas $C_{60}O$ is – as expected – a stronger electron acceptor than C_{60}. The transfer of the first electron is assigned to a cage reduction. $C_{60}O^-$ is not stable and decomposes to C_{60}. The transfer of the second electron is believed to initiate a polymerization process of $C_{60}O^{2-}$.

Studies on crystalline $C_{60}O$ [39] using calorimetry and high-resolution X-ray powder diffraction show a face centered cubic lattice (a = 14.185 Å) with an orientational disorder at room temperature. An orientational ordering transition occurs at 278 K, upon which a simple cubic phase develops. At 19 K this phase, which is similar to the orientational ordered phase of C_{60} itself, shows additional randomness due to a distribution of orientation of the oxygens in $C_{60}O$.

A light-induced oxygen incision of C_{60} was reported to lead to an opening of the fullerene cage [40, 41]. For more examples of ring-opening reactions see Chapter 11.

Oxygenation of C_{60} not only takes place in targeted syntheses, but also unsolicited by exposure to light and air [42–44]. This leads to an almost ubiquitous contamination of C_{60}. The main impurity was determined as the $C_{120}O$ dimer **9** (Figure 8.4) [42]. It probably originates from reaction of C_{60} and the reactive C_{60}-oxide $C_{60}O$ in a [2+2]-cycloaddition. The same reaction path was chosen to synthesize the dimer in preparative amounts [34, 45, 46]. C_{60} and $C_{60}O$ (e.g. prepared by ozonolysis) is diluted in CS_2. After removal of the solvent the homogenous mixture is heated in vacuum at 200 °C for some hours. $C_{120}O$ can be isolated in pure form and in good yields via HPLC purification. Side products are, probably, higher dimer-oxides such as $C_{120}O_n$ (n > 1) [34].

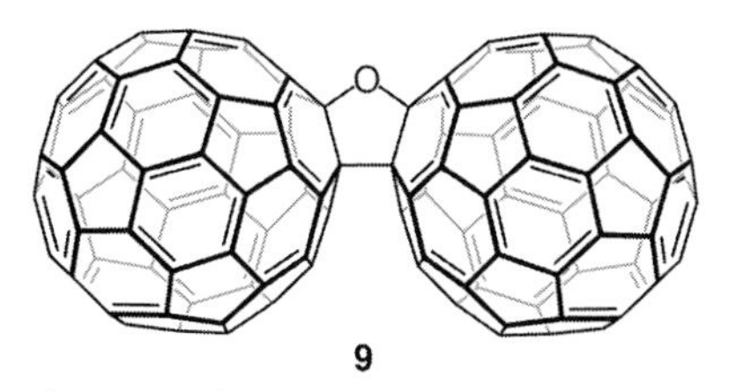

Figure 8.4 The dimer $C_{120}O$ can often be observed as an impurity in C_{60}-samples.

Highly oxygenated $C_{60}O_n^-$ anions with $n \leq 30$ are formed by corona discharge ionization in the gas phase in the presence of traces of oxygen [47]. The higher oxides were analyzed by mass spectrometry. Until now they could not be separated, but this method could be a new route for the synthesis of highly oxygenated C_{60}.

Different isomers of $C_{70}O$ have been prepared by photooxygenation [48], by MCPBA-oxidation [48], by ozonolysis [49] or they were extracted from fullerene soot [11, 50]. Isolation from fullerene soot and analysis of the product of photooxygenation and thermal ozonolysis yields only [6,6]-closed epoxide structures. As already observed for $C_{60}O$, ozonolysis and subsequent photolysis of the ozonide $C_{70}O_3$ gives different [5,6]-open oxidoannulene structures [49].

8.4 Osmylation

The addition of the strong, selective oxidizing reagent osmium tetroxide, carried out by Hawkins, provided the first fully characterized C_{60} derivatives (Scheme 8.5) [51–55]. The pyridine-accelerated addition of two equiv. of OsO_4 in toluene leads predominantly to the precipitation of the bisadducts **10** as mixtures of regioisomers. The monoadduct **11** can be obtained in high yields by osmylation in the absence of pyridine followed by the addition of pyridine or by using a stoichiometric amount of OsO_4 in the presence of pyridine (Scheme 8.5) [52]. The osmylation product fully reverted back to C_{60} when heated under vacuum [51].

The pyridine molecules in **10** or **11** can be exchanged by other coordinating ligands, such as 4-*tert*-butylpyridine (**12**) (Scheme 8.6) [52]. This ligand-exchange reaction has been used to increase the solubility of the osmylated fullerenes, which allows their spectroscopic characterization as well as the growth of high quality single crystals [52].

The soccer ball shaped framework of C_{60} was confirmed for the first time by the X-ray crystal structure of **13** [52]. The bridging of the OsO_4 unit occurs in the characteristic 1,2-mode to a [6,6] double bond of the fullerene core. The 17 sets of carbons in the C_{2v}-symmetric **13** were assigned using the 2D NMR INADEQUATE technique on the basis of their connectivities [53]. In these experiments ^{13}C-enriched material was used. The coupling constants $^1J_{CC}$ fall into three categories, 48, 54–57 and 65–71 Hz. These values can be attributed to the three types of C–C bonds present in **13**, namely $C(sp^2)$–$C(sp^3)$, the longer [5,6], and the shorter [6,6] bonds, respectively [53].

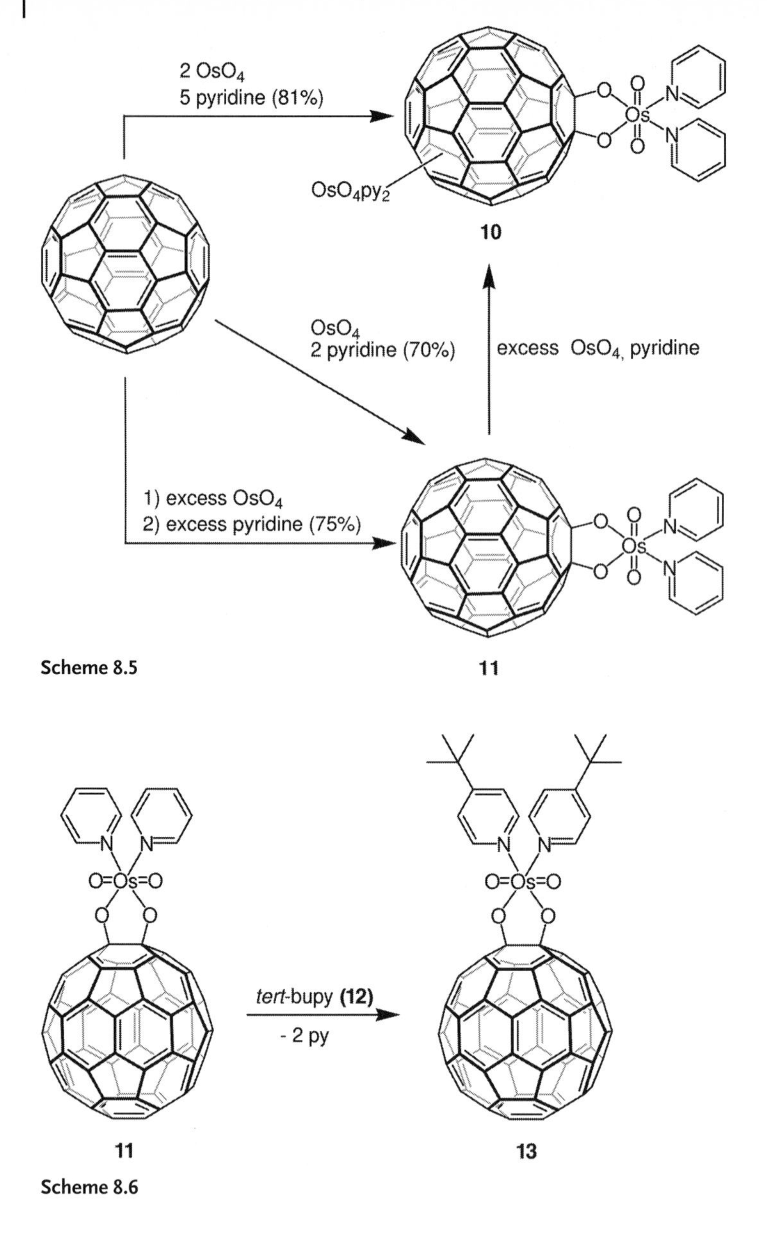

Scheme 8.5

Scheme 8.6

In a detailed investigation of the regiochemistry of the bisosmylation of C_{60}, five regioisomers of $C_{60}[OsO_4(py)_2]_2$ (**14**, not shown) have been isolated by preparative HPLC [55]. After their isolation, these regioisomers were converted into their 4-*tert*-butylpyridine analogues and analyzed by NMR spectroscopy. As with the regioisomers of $C_{60}(COOEt)_4$ (Chapter 10) [56], the least polar *trans*-1 regioisomer is much less soluble than the other isomers. For **14** [55], the solubility of the least retentive regioisomer is so low that it could not be analyzed. The symmetry and the elution order of the other 4 regioisomers of **14** allow a tentative assignment of the *trans*-2, *trans*-3, *trans*-4 isomers and an unambiguous assignment of the *e*-isomer [55] Three sets of peaks for the bridge head fullerene C-atoms as well as for the 4-*tert*-butylpyridine protons are found in the *e*-isomer, which is consistent with the positioning of one of the osmyl groups across the mirror plane. The same order of elution and symmetry was observed in the regioisomers of the bisadducts $C_{62}(COOEt)_4$ [56], which further corroborated the assignment.

The C_2-symmetric *trans*-2 and *trans*-3 isomers are chiral. Bisosmylation in the presence of chiral N-donor ligands L* provides an asymmetric induction for the formation of enantiomerically enriched *trans*-2 and *trans*-3 isomers of **10** with optical rotations up to 3700° [57]. These asymmetric bisosmylations are carried out by treating C_{60} with 2 equiv. of OsO_4 and 5 equiv. of a chiral ligand L* followed by exchange of the chiral ligands for pyridine. Five regioisomers of **10** in varying ratios can be isolated after the exchange reactions.

The osmylation of C_{70} with 0.75 equiv of OsO_4 in pyridine–toluene at 0 °C leads to the two regioisomers **15** (major product) and **16** (minor) (Scheme 8.7) [58]. The same regioselectivity was observed in the hydrogenation (Chapter 5) [59], nucleophilic additions (Chapter 3) [60] as well as in Ir-adduct formations (Chapter 7) [61] of C_{70} and is in line with the predicted stabilities [59, 62] of C_{70} adducts. The reaction occurs at the most reactive [6,6] double bonds, with the closest geometric similarity to the [6,6] double bonds of C_{60}. These are the double bonds between C-1/C-2 (bond 1, Scheme 8.7) and C-5/C-6 (bond 2, Scheme 8.7) respectively.

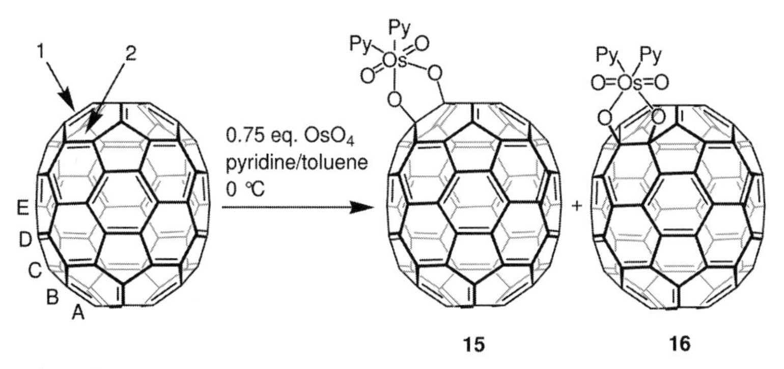

Scheme 8.7

The chiral fullerene C_{76} was also asymmetrically osmylated using the chiral ligands L^1 and L^2 (Scheme 8.8) [63]. In this way an optically active allotrope of a pure element was prepared. C_{76} contains 15 different types of [6,6] bonds. The pronounced regioselectivity of C_{70} towards osmylation [58] suggests that specific bonds in C_{76} may be favored for an attack by OsO_4. An analysis of the ab initio calculated curvature of C_{76} shows that two of the five pyracylene-type bonds are particularly distorted, which could enhance their reactivity [64]. Indeed HPLC analysis of $C_{76}[OsO_4L^*]$ shows that two regioisomers are predominantly formed upon osmylation of C_{76} [63].

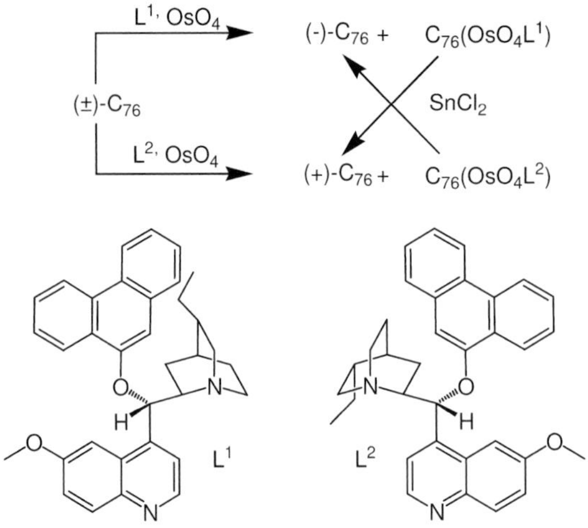

Scheme 8.8

During the kinetically controlled asymmetric osmylation, one enantiomer reacts faster with the asymmetric reagent than the other. The starting material becomes enriched with the discriminated enantiomer of C_{76}, and the reaction products $C_{76}[OsO_4L^*]$ are predominantly formed with the more reactive enantiomer. The C_{76} fraction, recovered by HPLC from the treatment with OsO_4 and the chiral alkaloid ligands L^1 or L^2, shows a maximum specific rotation $[\alpha]_D$ of about –4000°. The circular dichroism spectrum corresponds with the UV/Vis spectrum [63]. C_{76} can be regenerated from $C_{76}[OsO_4L^*]$ by treatment with $SnCl_2$ (Scheme 8.8). In this procedure the opposite enantiomer can also be generated. The two enantiomers of C_{76} show mirror image circular dichroism spectra [63]. The pronounced chiral discrimination observed in these asymmetric osmylation reactions were interpreted in terms of diastereotropic attractive π–π interactions between the phenanthryl units of L^1 and L^2 and the reactive part with the greatest local curvature of the fullerene surface [63]. Also, the enantiomeric pairs of D_3-C_{78} and D_2-C_{84} were kinetically resolved with this method [65].

8.5
Reactions with Strong Oxidizing Reagents and Acids

The treatment of C_{60} with powerful oxidizing reagents such as SbF_5/SO_2ClF [66, 67], $SbCl_5$ [68], magic acid (FSO_3H, SbF_5) [69] or a mixture of fuming sulfuric acid and SO_2ClF [70] gives, in most cases, a dark green solution. On the basis of NMR, ESR or NIR spectroscopy the formation of radical cations of C_{60} was assumed. Oxidation with chlorine dioxide leads to the instant formation of a brown precipitate [71], which also shows an ESR signal indicative for the radical cation.

A problem in all these oxidations is the extreme instability of C_{60} cations under the applied conditions. C_{60}^{+} and the higher cations are susceptible to oxidative degradation and nucleophilic attack by the anions of superacids (e.g. HSO_4^-, $SbCl_6^-$, SbF_6^-) or other present nucleophiles. This may be one reason for the wide range of observed ESR signals or NIR-λ_{max}s [2].

Oxidation of C_{60} to C_{60}^{+} can be facilitated by photoinduced electron transfer (PET). Photoinduced oxidation of C_{60} has been achieved by electron transfer from excited $^3C_{60}^*$ to a strong electron acceptor such as *p*-chloranil [72, 73], *p*-benzoquinone [73], tetracyano-*p*-quinodimethane (TCNQ) or tetracyanoethylene (TCNE). This electron transfer proceeds efficiently only by addition of promoters such as $Sc(OTf)_3$ or triflic acid, both of which strongly enhance the electron-transfer process [72, 73]. Another possibility to produce the cation is the electron transfer from C_{60} to the singlet excited state of a strong electron acceptor such as *N*-methylacridinium hexafluorophosphate (NMA^+) [74, 75], triphenylpyriliumtetrafluoroborate (TPP^+) [74, 75] or 9,10-dicyanoanthracene [76]. The latter electron transfer can also be mediated by the biphenyl radical cation produced by PET from biphenyl (BP) to photosensitizers (NMA^+) (Scheme 8.9) [74, 75]. C_{60} cations that are formed via PET are intermediates in various addition reactions, such as the addition of alcohols, ethers, aldehydes or hydrocarbons (RH in Scheme 8.9) [75–77].

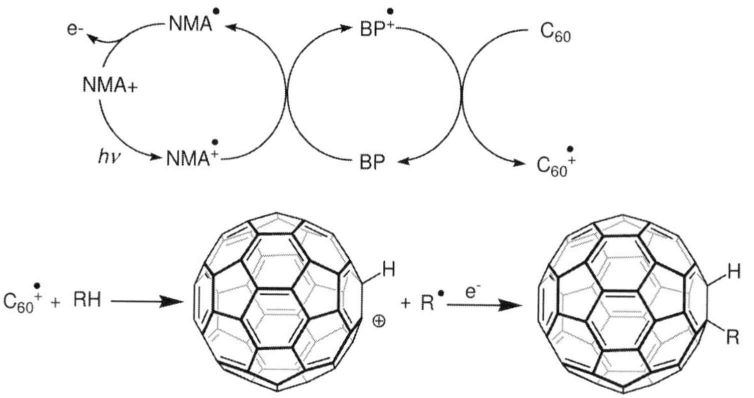

Scheme 8.9

The first synthesis of an isolated and sufficiently characterized fullerene radical cation was established by Reed and co-workers [2, 9]. The main problem of cation synthesis is the nucleophilicity of the oxidant and/or the counter-ions that causes subsequent reactions and degradation. This was overcome by using a very strong oxidant and an inert counter-anion, both with extremely low nucleophilicity. The potential of the stable hexabrominated carbazole radical **18**$^+$ (Figure 8.5) is high enough to oxidize C_{60}. The boron-cluster anion **17**$^-$ (Figure 8.5) is exceptionally stable due to the three-dimensional σ-aromaticity of the CB_{11} framework. The basicity and nucleophilicity of this boron cluster is less than that of common halocarbon solvents [2, 9, 78]. Addition of the salt **17**$^-$**18**$^+$ to a solution of C_{60} in *o*-dichlorobenzene produces a stable, dark red solution. The $[C_{60}^{+}][CB_{11}H_6Cl_6^-]$ salt co-crystallizes with C_{60}, which is probably formed in a disproportionation reaction.

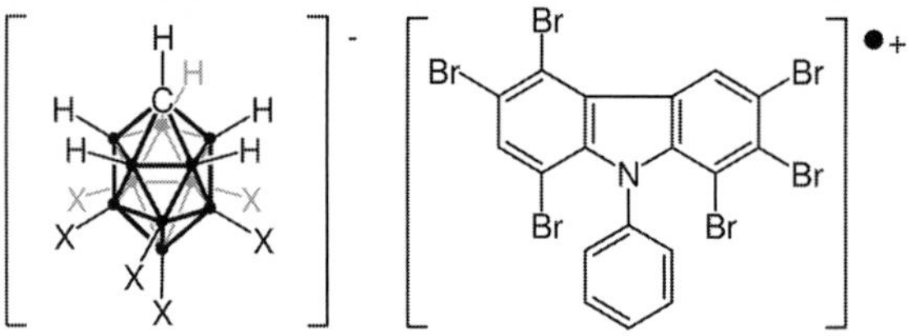

17^- 18^+

Figure 8.5 X = Cl; dots represent B.

The EPR spectrum shows a signal at $g = 2.002$ with a line width ΔH_{pp} of 3–8 G. In the NIR spectrum the typical C_{60}^+ absorption at 980 nm is accompanied by a weaker broad band at 850 nm [2, 9]. Under similar conditions stable salts of C_{70} were obtained [79].

Treatment of C_{60} with the usual superacids, such as magic acid, leads to degradation, while reaction with acids, such as nitric acid or sulfuric acid, yields polyaddition products (see below). With the superacid $[CB_{11}H_6Cl_6H]$ derived by protonation of the above-mentioned carborane $[CB_{11}H_6Cl_6^-]$, proton transfer onto C_{60} is possible and gives access to characterizable $C_{60}H^+$ [9]. Analytically pure samples of $[HC_{60}^+]$ $[CB_{11}H_6Cl_6^-]$ can be produced by adding $[CB_{11}H_6Cl_6H]$ to C_{60} in *o*-dichlorobenzene and removing the solvent under vacuum. The salt was characterized by IR and NMR spectroscopy [9, 80].

C_{60} reacts with sulfuric/nitric acid, and subsequent hydrolysis of the intermediates with aqueous base affords fullerenols (Scheme 8.10) [81–84]. Either nitric acid itself or nitric acid produced in situ from potassium nitrate and fuming sulfuric acid can be used.

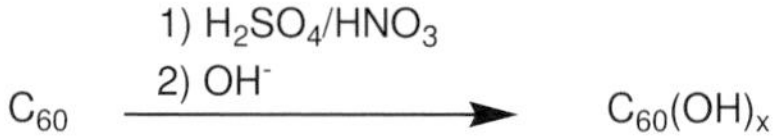

$$C_{60} \xrightarrow[\text{2) } OH^-]{\text{1) } H_2SO_4/HNO_3} C_{60}(OH)_x$$

Scheme 8.10

The reaction with fuming sulfuric acid is significantly enhanced by addition of P_2O_5 [83] or metal oxides [85]. The chemical composition of the water-soluble fullerenols depends on the conditions but, usually, about 6–15 hydroxy groups are attached to the C_{60}. In a similar approach, fullerene derivatives with polar functional groups, namely polyhydroxy polyacyloxy C_{60}, were synthesized by the reaction with nitronium tetrafluoroborate and carboxylic acids [86]. After an initial electrophilic attack of the nitronium ion, a subsequent nucleophilic 1,3- or 1,5-substitution of acyloxy occurs. Treatment with water and hydrolysis of these derivatives in hydroxide solutions affords the corresponding water-soluble fullerenols, consisting on average of 18–20 hydroxy groups per C_{60} molecule [86].

8.6
Reactions with Lewis Acids and Fullerylation of Aromatics and Chloroalkanes

The reaction of fullerenes with the Lewis acid BH_3 is discussed in Chapter 5. These hydroborations followed by hydrolysis provide access to hydrogenated fullerenes [87]. Also, other Lewis acids have been allowed to react with C_{60} and C_{70} [66, 88–91]. If these reactions are carried out in CS_2, fullerene–Lewis acid complexes precipitate [91]. The complexation behavior of C_{60} and C_{70} is different, with C_{70} complexing much more strongly than C_{60}. This phenomenon was taken advantage of in separating the fullerene mixtures. Other Lewis acids that form CS_2-insoluble complexes are $AlBr_3$, $TiCl_4$, $SnCl_4$ and $FeCl_3$. The parent fullerenes can be recovered from the Lewis acid complexes by reaction with ice water [92].

If solutions of C_{60} in aromatic hydrocarbons are treated with Lewis acids, such as $AlCl_3$, $AlBr_3$, $FeBr_3$, $FeCl_3$, $GaCl_3$ or $SbCl_5$, then a fullerylation of aromatics takes place (Scheme 8.11) [66, 88, 90, 91]. In this case, the Lewis acid serves as a catalyst and increases the electrophilicity of the fullerene. Mixtures of polyarylated fullerenes are obtained. Depending on the reaction conditions and the aromatic used for the fullerylation, up to 16 aryl groups are covalently bound to the fullerene core [66, 88, 90]. With the phenyl addition to C_{60}, the highest intensity molecular ion peak in the mass spectra was that for $C_{60}Ph_{12}$, indicating that this compound exhibits an enhanced stability [66, 91].

$$C_{60} \xrightarrow{AlCl_3,\ benzene} C_{60}(HC_6H_5)_n$$

Scheme 8.11

The reaction takes place only with relatively strong Lewis acids as catalysts. No reaction was observed with weak Lewis acids, such as $SnCl_4$ and $TiCl_4$, and $GeCl_4$, BF_3 and BCl_3 were also inactive [66, 91]. The order of Lewis acid activity was found to be $AlBr_3 > AlCl_3 > FeCl_3 > GaCl_3 > SbCl_5$ [66]. The polyarylated fullerenes appear to be unstable as, on standing, partial decomposition was observed [90].

Electrophilic addition of polychloroalkanes such as, e.g., chloroform or 1,1,2,2-tetrachloroethane to C_{60} with $AlCl_3$ in a 100-fold excess gives the monoadduct with a 1,4-addition pattern (Scheme 8.12) [93, 94]. The reaction proceeds via a $C_{60}R^+$ cation (**19**, Scheme 8.12) that is stabilized by the coordination of a chlorine atom to the cationic center. The cation is trapped by Cl^- to give the product **20**. The chloroalkyl fullerenes can be readily hydrolyzed to form the corresponding fullerenol **21**. This fullerenol can be utilized as a proper precursor for the cation, which is easily obtained by adding triflic acid. The stability of $C_{60}R^+$ is similar to tertiary alkyl cations such as the *tert*-butyl-cation [95].

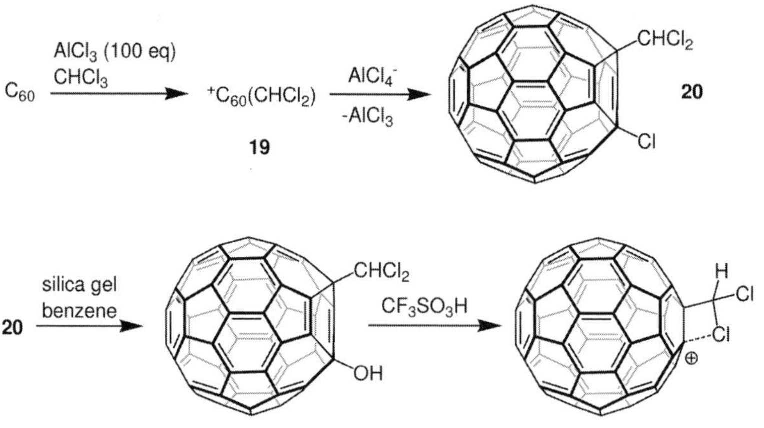

Scheme 8.12

References

1 L. Echegoyen, L. E. Echegoyen, *Acc. Chem. Res.* **1998**, *31*, 593.

2 C. A. Reed, R. D. Bolskar, *Chem. Rev.* **2000**, *100*, 1075.

3 R. C. Haddon, L. E. Brus, K. Raghavachari, *Chem. Phys. Lett.* **1986**, *125*, 459.

4 C. Jehoulet, A. J. Bard, F. Wudl, *J. Am. Chem. Soc.* **1991**, *113*, 5456.

5 D. Dubois, K. M. Kadish, S. Flanagan, L. J. Wilson, *J. Am. Chem. Soc.* **1991**, *113*, 7773.

6 K. Meerholz, P. Tschuncky, J. Heinze, *J. Electroanal. Chem.* **1993**, *347*, 425.

7 Q. Xie, F. Arias, L. Echegoyen, *J. Am. Chem. Soc.* **1993**, *115*, 9818.

8 R. D. Webster, G. A. Heath, *Phys. Chem. Chem. Phys.* **2001**, *3*, 2588.

9 C. A. Reed, K.-C. Kim, R. D. Bolskar, L. J. Müller, *Science* **2000**, *289*, 101.

10 C. Bruno, I. Doubitski, M. Marcaccio, F. Paolucci, D. Paolucci, A. Zaopo, *J. Am. Chem. Soc.* **2003**, *125*, 15738.

11 F. Diederich, R. Ettl, Y. Rubin, R. L. Whetten, R. Beck, M. Alvarez, S. Anz, D. Sensharma, F. Wudl, K. C. Khemani, A. Koch, *Science* **1991**, *252*, 548.

12 J. M. Wood, B. Kahr, S. H. Hoke II, L. Dejarme, R. G. Cooks, D. Ben-Amotz, *J. Am. Chem. Soc.* **1991**, *113*, 5907.

13 W. A. Kalsbeck, H. H. Thorp, *J. Electroanal. Chem. Interfacial Electrochem.* **1991**, *314*, 363.

14 R. Taylor, J. P. Parsons, A. G. Avent, S. P. Rannard, T. J. Dennis, J. P. Hare, H. W. Kroto, D. R. M. Walton, *Nature* **1991**, *351*, 277.

15 K. M. Creegan, J. L. Robbins, W. K. Robbins, J. M. Millar, R. D. Sherwood, P. J. Tindall, D. M. Cox, J. P. McCauley, Jr., D. R. Jones, A. B. Smith III, R. T. Gallagher, *J. Am. Chem. Soc.* **1992**, *114*, 1103.

16 A. M. Vassallo, L. S. K. Pang, P. A. Cole-Clarke, M. A. Wilson, *J. Am. Chem. Soc.* **1991**, *113*, 7820.

17 J. O. Escobedo, A. E. Frey, R. M. Strongin, *Tetrahedron Lett.* **2002**, *43*, 6117.

18 D. Heymann, L. P. F. Chibante, *Chem. Phys. Lett.* **1993**, *207*, 339.

19 D. I. Schuster, P. S. Baran, R. K. Hatch, A. U. Khan, S. R. Wilson, *Chem. Commun.* **1998**, 2493.

20 A. L. Balch, D. A. Costa, J. W. Lee, B. C. Noll, M. M. Olmstead, *Inorg. Chem.* **1994**, *33*, 2071.

21 A. L. Balch, D. A. Costa, B. C. Noll, M. M. Olmstead, *J. Am. Chem. Soc.* **1995**, *117*, 8926.

22 A. L. Balch, D. A. Costa, B. C. Noll, M. M. Olmstead, *Inorg. Chem.* **1996**, *35*, 458.

23 S. G. Penn, D. A. Costa, A. L. Balch, C. B. Lebrilla, *Int. J. Mass Spectrom. Ion Processes* **1997**, *169/170*, 371.

24 Y. Tajima, K. Takeuchi, *J. Org. Chem.* **2002**, *67*, 1696.

25 N. P. Curry, B. Doust, D. A. Jelski, *J. Cluster Sci.* **2000**, *12*, 385.

26 D. L. Kepert, B. W. Clare, *Inorg. Chim. Acta* **2002**, *327*, 41.

27 R. W. Murray, K. Iyanar, *Tetrahedron Lett.* **1997**, *38*, 335.

28 Y. Elemes, S. K. Silverman, C. Sheu, M. Kao, C. S. Foote, M. M. Alvarez, R. L. Whetten, *Angew. Chem.* **1992**, *104*, 364; *Angew. Chem. Int. Ed. Engl.* **1992**, *31*, 351.

29 C. Fusco, R. Seraglia, R. Curci, V. Lucchini, *J. Org. Chem.* **1999**, *64*, 8363.

30 K. Raghavachari, C. Sosa, *Chem. Phys. Lett.* **1993**, *209*, 223.

31 K. Raghavachari, *Chem. Phys. Lett.* **1992**, *195*, 221.

32 D. Heymann, S. M. Bachilo, R. B. Weisman, F. Cataldo, R. H. Fokkens, N. M. M. Nibbering, R. D. Vis, L. P. F. Chibante, *J. Am. Chem. Soc.* **2000**, *122*, 11473.

33 R. B. Weisman, D. Heymann, S. M. Bachilo, *J. Am. Chem. Soc.* **2001**, *123*, 9720.

34 S. Lebedkin, S. Ballenweg, J. Gross, R. Taylor, W. Krätschmer, *Tetrahedron Lett.* **1995**, *36*, 4971.

35 J.-P. Deng, C.-Y. Mou, C.-C. Han, *J. Phys. Chem.* **1995**, *99*, 14907.

36 R. D. Beck, C. Störmer, C. Schulz, R. Michel, P. Weis, G. Bräuchle, M. M. Kappes, *J. Chem. Phys.* **1994**, *101*, 3243.

37 K. Winkler, D. A. Costa, A. L. Balch, W. R. Fawcett, *J. Phys. Chem.* **1995**, *99*, 17431.

38 T. Suzuki, Y. Maruyama, T. Akasaka, W. Ando, K. Kobayashi, S. Nagase, *J. Am. Chem. Soc.* **1994**, *116*, 1359.

39 G. B. M. Vaughan, P. A. Heiney, D. E. Cox, A. R. McGhie, D. R. Jones, R. M. Strongin, M. A. Cichy, A. B. Smith III, *Chem. Phys.* **1992**, *168*, 185.

40 C. Taliani, G. Ruani, R. Zamboni, R. Danieli, S. Rossini, V. N. Denisov, V. M. Burlakov, F. Negri, G. Orlandi, F. Zerbetto, *J. Chem. Soc., Chem. Commun.* **1993**, 220.

41 J. W. Arbogast, A. P. Darmanyan, C. S. Foote, F. N. Diederich, R. L. Whetten, Y. Rubin, M. M. Alvarez, S. J. Anz, *J. Phys. Chem.* **1991**, *95*, 11.

42 R. Taylor, M. P. Barrow, T. Drewello, *Chem. Commun.* **1998**, 2497.

43 P. Paul, K.-C. Kim, D. Sun, P. D. W. Boyd, C. A. Reed, *J. Am. Chem. Soc.* **2002**, *124*, 4394.

44 P. Paul, R. D. Bolskar, A. M. Clark, C. A. Reed, *Chem. Commun.* **2000**, 1229.

45 A. L. Balch, D. A. Costa, W. R. Fawcett, K. Winkler, *J. Phys. Chem.* **1996**, *100*, 4823.

46 M. P. Barrow, N. J. Tower, R. Taylor, T. Drewello, *Chem. Phys. Lett.* **1998**, *293*, 302.

47 H. Tanaka, K. Takeuchi, Y. Negishi, T. Tsukuda, *Chem. Phys. Lett.* **2004**, *384*, 283.

48 A. B. Smith III, R. M. Strongin, L. Brard, G. T. Furst, J. H. Atkins, W. J. Romanow, M. Saunders, H. A. Jimenez-Vazquez, K. G. Owens, R. J. Goldschmidt, *J. Org. Chem.* **1996**, *61*, 1904.

49 D. Heymann, S. M. Bachilo, R. B. Weisman, *J. Am. Chem. Soc.* **2002**, *124*, 6317.

50 V. N. Bezmelnitsin, A. V. Eletskii, N. G. Schepetov, A. G. Avent, R. Taylor, *J. Chem. Soc., Perkin Trans. 2* **1997**, 683.

51 J. M. Hawkins, T. A. Lewis, S. D. Loren, A. Meyer, J. R. Heath, Y. Shibato, R. J. Saykally, *J. Org. Chem.* **1990**, *55*, 6250.

52 J. M. Hawkins, A. Meyer, T. A. Lewis, S. Loren, F. J. Hollander, *Science* **1991**, *252*, 312.

53 J. M. Hawkins, S. Loren, A. Meyer, R. Nunlist, *J. Am. Chem. Soc.* **1991**, *113*, 7770.

54 J. M. Hawkins, *Acc. Chem. Res.* **1992**, *25*, 150.

55 J. M. Hawkins, A. Meyer, T. A. Lewis, U. Bunz, R. Nunlist, G. E. Ball, T. W. Ebbesen, K. Tanigaki, *J. Am. Chem. Soc.* **1992**, *114*, 7954.

56 A. Hirsch, I. Lamparth, H. R. Karfunkel, *Angew. Chem.* **1994**, *106*, 453; *Angew. Chem. Int. Ed. Engl.* **1994**, *33*, 437.

57 J. M. Hawkins, A. Meyer, M. Nambu, *J. Am. Chem. Soc.* **1993**, *115*, 9844.

58 J. M. Hawkins, A. Meyer, M. A. Solow, *J. Am. Chem. Soc.* **1993**, *115*, 7499.
59 C. C. Henderson, C. M. Rohlfing, P. A. Cahill, *Chem. Phys. Lett.* **1993**, *213*, 383.
60 A. Hirsch, T. Grösser, A. Skiebe, A. Soi, *Chem. Ber.* **1993**, *126*, 1061.
61 A. L. Balch, V. J. Catalano, J. W. Lee, M. M. Olmstead, S. R. Parkin, *J. Am. Chem. Soc.* **1991**, *113*, 8953.
62 H. R. Karfunkel, A. Hirsch, *Angew. Chem.* **1992**, *104*, 1529; *Angew. Chem. Int. Ed. Engl.* **1992**, *31*, 1468.
63 J. M. Hawkins, A. Meyer, *Science* **1993**, *260*, 1918.
64 J. R. Colt, G. E. Scuseria, *J. Phys. Chem.* **1992**, *96*, 10265.
65 J. M. Hawkins, M. Nambu, A. Meyer, *J. Am. Chem. Soc.* **1994**, *116*, 7642.
66 G. A. Olah, I. Bucsi, R. Aniszfeld, G. K. S. Prakash, *Carbon* **1992**, *30*, 1203.
67 J. W. Bausch, G. K. S. Prakash, G. A. Olah, D. S. Tse, D. C. Lorents, Y. K. Bae, R. Malhotra, *J. Am. Chem. Soc.* **1991**, *113*, 3205.
68 L. Gherghel, M. Baumgarten, *Synth. Methods* **1995**, *70*, 1389.
69 G. P. Miller, C. S. Hsu, H. Thomann, L. Y. Chiang, M. Bernardo, *Mater. Res. Soc. Symp. Proc.* **1992**, *247*, 293.
70 H. Thomann, M. Bernardo, G. P. Miller, *J. Am. Chem. Soc.* **1992**, *114*, 6593.
71 V. I. Sokolov, V. V. Bashilov, Q. K. Timerghazin, E. V. Avzyanova, A. F. Khalizov, N. M. Shishlov, V. V. Shereshovets, *Mendeleev Commun.* **1999**, 54.
72 L. Biczok, H. Linschitz, *J. Phys. Chem. A* **2001**, *105*, 11051.
73 S. Fukuzumi, H. Mori, H. Imahori, T. Suenobu, Y. Araki, O. Ito, K. M. Kadish, *J. Am. Chem. Soc.* **2001**, *123*, 12458.
74 L. Dunsch, F. Ziegs, C. Siedschlag, J. Mattay, *Chem. Eur. J.* **2000**, *6*, 3547.
75 C. Siedschlag, H. Luftmann, C. Wolff, J. Mattay, *Tetrahedron* **1999**, *55*, 7805.
76 G. Lem, D. I. Schuster, S. H. Courtney, Q. Lu, S. R. Wilson, *J. Am. Chem. Soc.* **1995**, *117*, 554.
77 I. G. Safonov, S. H. Courtney, D. I. Schuster, *Res. Chem. Intermed.* **1997**, *23*, 541.
78 C. A. Reed, *Acc. Chem. Res.* **1998**, *31*, 133.
79 K.-C. Kim, C. A. Reed, *Abstr. Paper. – Am. Chem. Soc.* **2001**, *221*, INOR-087.
80 L. J. Müller, D. W. Elliott, K.-C. Kim, C. A. Reed, P. D. W. Boyd, *J. Am. Chem. Soc.* **2002**, *124*, 9360.
81 L. Y. Chiang, J. W. Swirczewski, C. S. Hsu, S. K. Chowdhury, S. Cameron, K. Creegan, *J. Chem. Soc., Chem. Commun.* **1992**, 1791.
82 L. Y. Chiang, R. B. Upasani, J. W. Swirczewski, S. Soled, *J. Am. Chem. Soc.* **1993**, *115*, 5453.
83 B.-H. Chen, J.-P. Huang, L. Y. Wang, J. Shiea, T.-L. Chen, L. Y. Chiang, *J. Chem. Soc., Perkin Trans. 1* **1998**, 1171.
84 L. Y. Chiang, L.-Y. Wang, J. W. Swirczewski, S. Soled, S. Cameron, *J. Org. Chem.* **1994**, *59*, 3960.
85 B.-H. Chen, T. Canteenwala, S. Patil, L. Y. Chiang, *Synth. Commun.* **2001**, *31*, 1659.
86 L. Y. Chiang, R. B. Upasani, J. W. Swirczewski, *J. Am. Chem. Soc.* **1992**, *114*, 10154.
87 C. C. Henderson, P. A. Cahill, *Science* **1993**, *259*, 1885.
88 S. H. Hoke, II, J. Molstad, G. L. Payne, B. Kahr, D. Ben-Amotz, R. G. Cooks, *Rapid Commun. Mass Spectrom.* **1991**, *5*, 472.
89 G. A. Olah, I. Bucsi, C. Lambert, R. Aniszfeld, N. J. Trivedi, D. K. Sensharma, G. K. S. Prakash, *J. Am. Chem. Soc.* **1991**, *113*, 9387.
90 R. Taylor, G. J. Langley, M. F. Meidine, J. P. Parsons, A. A. K. Abdul-Sada, T. J. Dennis, J. P. Hare, H. W. Kroto, D. R. M. Walton, *J. Chem. Soc., Chem. Commun.* **1992**, 667.
91 G. A. Olah, I. Bucsi, D. S. Ha, R. Aniszfeld, C. S. Lee, G. K. S. Prakash, *Full. Sci. Technol.* **1997**, *5*, 389.
92 G. A. Olah, personal communication.
93 T. Kitagawa, H. Sakamoto, K. i. Takeuchi, *J. Am. Chem. Soc.* **1999**, *121*, 4298.
94 T. Kitagawa, K. i. Takeuchi, *Bull. Chem. Soc. Jap.* **2001**, *74*, 785.
95 T. Kitagawa, Y. Lee, M. Hanamura, H. Sakamoto, H. Konno, K. i. Takeuchi, K. Komatsu, *Chem. Commun.* **2002**, 3062.

9
Halogenation

9.1
Introduction

Inspired by the fascinating properties of Teflon, one of the first products that scientists had in mind when C_{60} was discovered was a Teflon-like ball. Thus, halogenations and, in particular, fluorinations were among the first reactions carried out with C_{60}. Fluorinated and, even more so, the chlorinated and brominated fullerenes are not stable against hydrolysis and temperature, making the synthesis of a Teflon ball unreachable. The instability of halogenated fullerenes derives from their lower C–X bond energy compared with alkylhalogenides. One reason for the decreased bond energy is the increased eclipsing interactions, which increase with the level of halogenation and with the size of the halogen. Another reason is the need to introduce [5,6] double bonds, which is not favorable. Even if the stability of their halogenated hydrocarbons can not be reached, halofullerenes can still be isolated and various derivatives with interesting properties could be synthesized and fully characterized by NMR spectroscopy and X-ray crystallography. Valuable information about the regiochemistry and aromaticity of C_{60} and C_{60}-derivatives can be obtained from chlorinations, brominations and fluorinations. To date, iodination of C_{60} could not be achieved due to the very weak C_{60}–iodine bond and the bulkiness of iodine.

9.2
Fluorination

Calculations on fluorinated fullerenes [1–5] predict the fluorination to be very exothermic. The large stabilizations are due to the formation of strong C–F bonds and the breaking of weak F–F bonds. Similar to the hydrogenations of C_{60}, it is predicted that fluorinations in a 1,2-addition mode to form $C_{60}F_n$ are favored over the 1,4-addition mode [4, 5]. Of the eight possible regioisomers of $C_{60}F_4$, the isomer in which the addition to the second [6,6] double bond occurs at the *e* position relative to the first addition has the lowest MNDO or AM1 heat of formation [4, 6]. Nevertheless, the $C_{60}F_4$-isomer **2** with the same *cis*-1-addition-pattern as already observed for the main $C_{60}H_4$-isomer is the only isolated isomer so far [7]. All known

Fullerenes: Chemistry and Reactions. Andreas Hirsch and Michael Brettreich

ISBN: 3-527-30820-2

isomers of $C_{60}F_n$ that have been isolated and characterized show a contiguous addition pattern [7, 8]. They were formed by subsequent 1,2-additions exclusively to [6,6] double bonds. The only exception is the $C_{60}F_{24}$ isomer **7** with only non-adjacent fluorine atoms (see Section 9.2.2) [9]. Contrary to all other products, which derive from high-temperature fluorinations of pristine C_{60}, this isomer was derived from the room temperature conversion of $C_{60}Br_{24}$ into $C_{60}F_{24}$ while conserving the structure of the bromine derivative. So far, a series of isomers of $C_{60}F_n$ with an established structure have been isolated [7, 8]. They are shown in Figure 9.1.

Comparing these structures with those of the hydrogenation products (see Chapter 5) reveals the obvious similarity between these two types of C_{60} adducts. In polyhydrofullerenes as well as in polyfluorofullerenes the formation of isolated benzene rings leads to an increased aromaticity relative to the fullerene precursor and, therefore, to increased stability [8]. Compared with C_{60}, fluorofullerenes have – as expected – higher electron affinities and higher redox potentials, making them good oxidants or electron acceptors. For example, the reduction potential of $C_{60}F_{48}$ is 1.38 V more positive than that of C_{60} [10]. They are reasonably stable under exposure of air and light but sensitive to hydrolysis. Most common solvents for fluorofullerenes are aromatic or halogenated aromatic hydrocarbons such as benzene, toluene, chlorobenzene and hydrocarbons such as pentane or hexane.

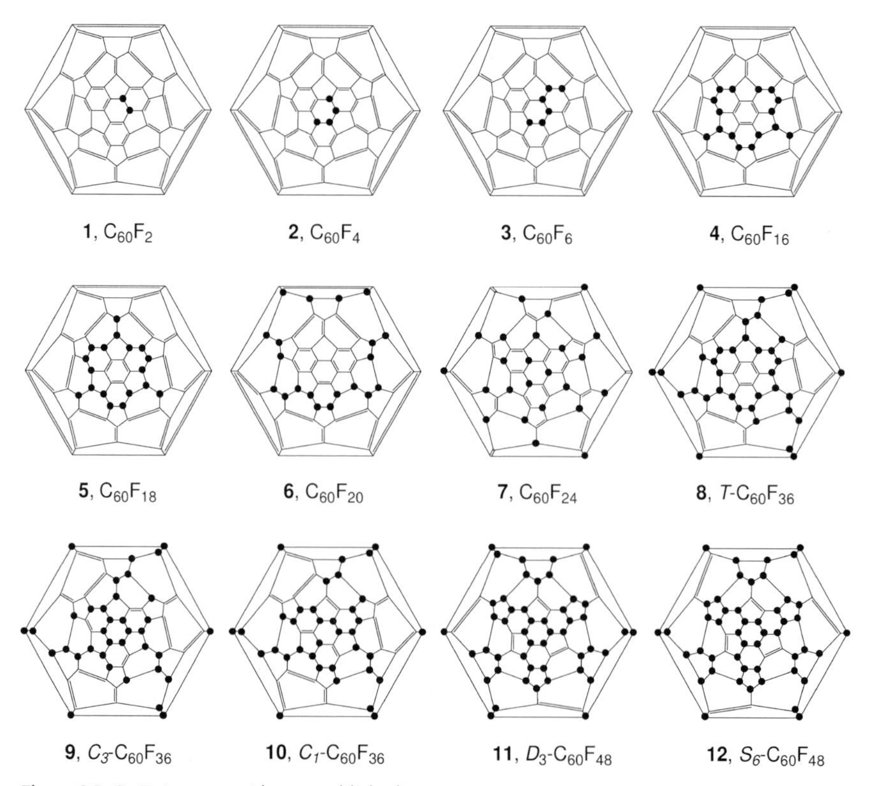

Figure 9.1 $C_{60}F_n$ isomers with an established structure.

In aliphatic solvents fluorofullerenes exhibit a much higher solubility than C_{60}. Solvents such as THF, acetone or methanol completely degrade the fluorinated C_{60}-derivatives [7, 8] even though in carefully dried THF, e.g., $C_{60}F_{18}$ is stable for weeks [7]. In the absence of any organic solvent, $C_{60}F_n$ does not react with boiling water because of its insolubility in this medium [11]. However, a very exothermic reaction together with the formation of HF takes place when THF is added to the suspension [12, 13].

Fluorofullerenes have been prepared by reaction with halogen fluorides, by direct fluorination with F_2 or by fluorination with noble gas fluorides [7, 8, 14]. The reaction with high valence metal fluorides is the most versatile route for the synthesis of $C_{60}F_n$ [7].

9.2.1
Direct Fluorination with F_2

Allowing fluorine gas at low pressure to react with the solid fullerene (Scheme 9.1) [11, 15, 16] was among the first halogenation reactions performed with C_{60}. After the first promising results, direct fluorination [7, 8, 17] was carried out under various conditions [18–23]. Different pressures and temperatures were applied, the reaction was carried out under static conditions or with a stream of fluorine and different reaction vessels such as quartz, steel or Monel were used. Monel is a nickel-copper alloy commonly used for fluorination reactions due to its stability against this extremely reactive gas.

$$C_{60} \xrightarrow{F_2} C_{60}F_n$$

Scheme 9.1

The fluorine uptake was monitored by measuring the weight gain with different methods, by mass spectrometry or by XPS spectrometry. As expected, the results of these fluorinations strongly depend on the conditions, but nevertheless some generally valid statements can be deduced from all these attempts to directly fluorinate C_{60} with F_2 gas. In all reactions a mixture of $C_{60}F_n$ with a wide range of compositions was found. Fluorofullerenes with $n = 2$ up to $n = 102$ [19] were detected but due to the high reactivity of fluorine gas the derivatives with a lower fluorine content are usually not formed. In most cases the distribution of the fluorine content of $C_{60}F_n$ lies approximately within the limit $n = 30$ to 50. Derivatives with a notably higher occurrence are $C_{60}F_{36}$ and $C_{60}F_{48}$. Disruption of the fullerene framework probably exists in compounds with $n > 48$ but must exist when n exceeds 60 [17, 19]. $C_{60}F_{48}$ is the only product synthesized by direct fluorination that can be isolated [7, 24–28]. Of the isolated and characterized fluorofullerenes it is the derivative with the highest fluorine content. Fluorofullerenes with higher fluorine content were detected by mass spectrometry but never isolated. Thus $C_{60}F_{48}$ may be considered as the final product of direct fluorination of C_{60}. Products with lower fluorine contents represent incomplete fluorination [7].

By using severe fluorinating conditions $C_{60}F_{48}$ is accessible by either a one- or two-stage process in good yields and in high compositional purity (Scheme 9.2) [7]. The two-stage process involves fluorination at 250 °C for 20 h and a subsequent fluorination of the obtained and purified intermediate product at 275 °C for 30 h [24]. The intermediate product is mainly $C_{60}F_{46}$. The one-stage process affords the desired fluorofullerene by heating C_{60} and fluorine gas in a flow reactor to 315–355 °C for 2–3 h [17, 29].

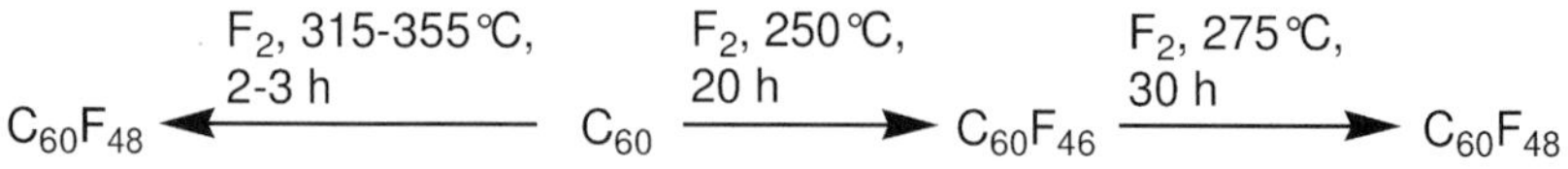

$$C_{60}F_{48} \xleftarrow[\text{2-3 h}]{F_2,\ 315\text{-}355\,°C,} C_{60} \xrightarrow[\text{20 h}]{F_2,\ 250\,°C,} C_{60}F_{46} \xrightarrow[\text{30 h}]{F_2,\ 275\,°C,} C_{60}F_{48}$$

Scheme 9.2

Under all applied conditions only $C_{60}F_{48}$ is formed, probably as three different optical isomers [24–26] (Figure 9.1). Gakh and co-workers determined the correct structure on the basis of ^{19}F NMR spectroscopy [24]. The structure was confirmed by resolution of the X-ray crystal structure [25, 26]. This shows that, probably, all three optical isomers – namely the two enantiomeric forms with D_3-symmetry and the mesoform with S_6-symmetry – can be found in the crystal (Figure 9.1, structures **11** and **12**).

In this structure two groups, each consisting of three double bonds, cause a significant indendation, leading to an effective shielding of the double bonds (Figure 9.2) [8, 25]. This may inhibit a further attack of fluorine and thus make $C_{60}F_{48}$ the "final" product in direct fluorination of C_{60}. $C_{60}F_{48}$ is a stable compound, potentially valuable as a highly concentrated source of fluorine [8, 30, 31], which can be released on heating, and it also forms colored charge-transfer complexes with aromatic solvents [8, 32].

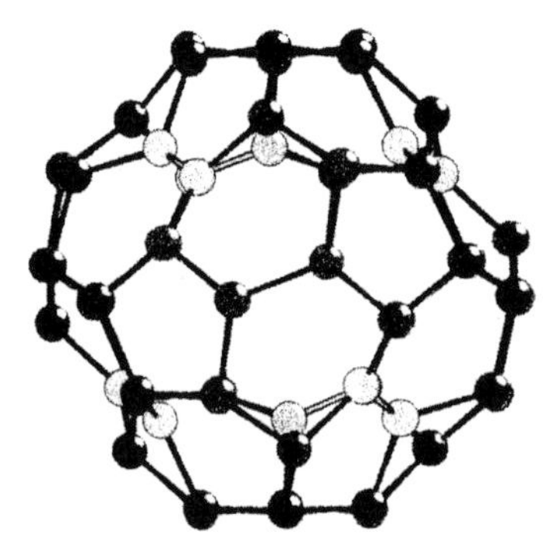

Figure 9.2 View of the D_3 isomer of $C_{60}F_{48}$ along the noncrystallographic twofold axis. Black carbons are sp^3, grey carbons are sp^2 [25].

9.2.2
Fluorination with Noble Gas Fluorides and Halogen Fluorides

The noble gas fluorides xenon difluoride (XeF_2) or krypton difluoride (KrF_2) are more powerful fluorinating agents than fluorine gas. C_{60} can be fluorinated either by treating dichloromethane solutions with XeF_2 [15] or by treatment with KrF_2 in anhydrous HF (Scheme 9.3). KrF_2 is more reactive than XeF_2 and yields products with a higher degree of fluorination [33]. Until now, fluorination with noble gas fluorides could not be established as a standard procedure for C_{60} because only complex, inseparable mixtures were obtained.

$$C_{60} \xrightarrow{XeF_2,\ CH_2Cl_2,\ r.t.} C_{60}F_{6\text{-}44}$$

$$C_{60} \xrightarrow{KrF_2,\ HF,\ r.t.} C_{60}F_{36\text{-}78}$$

Scheme 9.3 [14].

What was said about the noble gas fluorides is also valid for halogen fluorides such as ClF_3 or BrF_5. They are more reactive than F_2 but fluorination of C_{60} does not result in identifiable products [7, 34]. While fluorination of C_{60} with noble gas fluorides did not lead to defined products, one example for a selective product formation is known. The conversion of T_h-$C_{60}Br_{24}$ into $C_{60}F_{24}$ has been accomplished with XeF_2 in anhydrous HF [9]. The structure of T_h-$C_{60}Br_{24}$ is preserved in $C_{60}F_{24}$, which was proven by ^{19}F NMR spectroscopy and comparison of the experimental and calculated IR and Raman spectra. T_h-$C_{60}F_{24}$ (Figure 9.1) is the first example for a non-contiguous addition pattern of a fluorofullerene.

9.2.3
Reactions with Metal Fluorides

The main disadvantage of gaseous fluorine and noble gas fluorides is their high reactivity, which results in a low selectivity. Thus only one defined product can be obtained via this route (see Section 9.2.1). This constraint was overcome by the introduction of inorganic metal fluorides as milder fluorination agents [7, 14, 35, 36]. Such reactions are carried out in a Knudsen effusion cell reactor [36] under vacuum and at high temperatures of 300–600 °C. Various high-valence transition metal and rare-earth metal fluorides were mixed with C_{60}. The solid-state reaction, in a setup with a Knudsen cell associated with a mass spectrometer and a product collector, allows the simultaneous synthesis and analysis of the fluorofullerenes (Scheme 9.4). This technique facilitates significantly the search for optimized conditions for the selective preparation of fluorofullerenes with a certain degree of fluorination [36].

Fluorination of C_{60} leads to products with a higher volatility than C_{60} itself [37]. Reaching a certain degree of fluorination, which is dependent on the applied

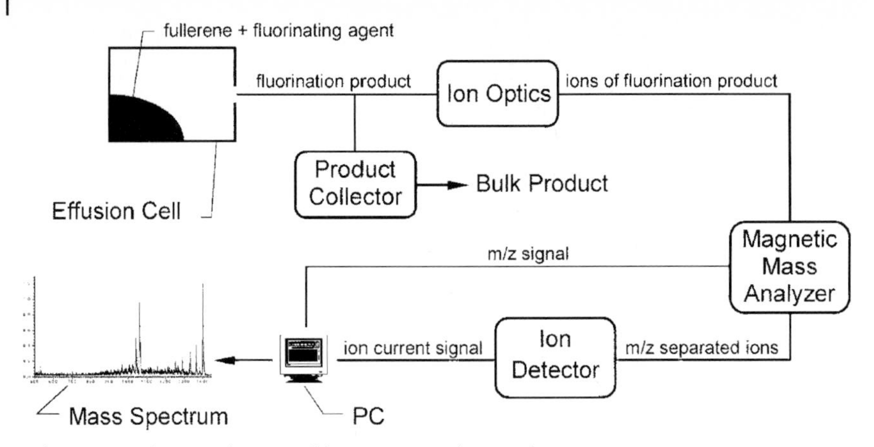

Scheme 9.4 Schematic drawing of the experimental set-up for simultaneous in situ Knudsen-cell synthesis (fluorination) and mass spectral characterization of fullerenes [36].

temperature, leads to sublimation of the product and resublimation on a specially designed collection platform. Some of the metal fluorides used with this method are shown in Scheme 9.5 [7, 14, 38].

All these fluorides are binary fluorides and are listed in their order of reactivity [38]. TbF_4, CeF_4, CoF_3, AgF_2 and MnF_3 belong to the group of very reactive metal fluorides and can produce highly fluorinated fullerenes in a temperature range of 300–400 °C. MnF_3 leads selectively to the formation of $C_{60}F_{36}$ [39]. Another selective reaction can be carried out with the less reactive AgF at 420–480 °C [38]. Exclusively, $C_{60}F_{18}$ can be isolated. The very mild fluorinating agents CuF_2 and FeF_3 allow generation of the smallest fluorofullerene $C_{60}F_2$ in the gas phase [40]. As well as the binary metal fluorides the use of ternary fluorides was established. $KPtF_6$ has a medium reactivity and requires temperatures of 450–520 °C to reach a good selectivity in forming $C_{60}F_{18}$ [41, 42]. Other used complex fluorides are of composition $M_{2-3}PbF_{6-7}$ (M = alkaline metal) or $MPbF_6$ (M = earth alkaline metal) [43]. Depending on the metal M, a high selectivity for either $C_{60}F_{18}$ or $C_{60}F_{36}$ can be reached. Of the above-mentioned reactions, the K_2PtF_6 reaction has a high selectivity affording > 90% purity of the single isomer of $C_{60}F_{18}$ [42]. However, in most cases several isomers of major products and side products are formed, which can be isolated by HPLC [8].

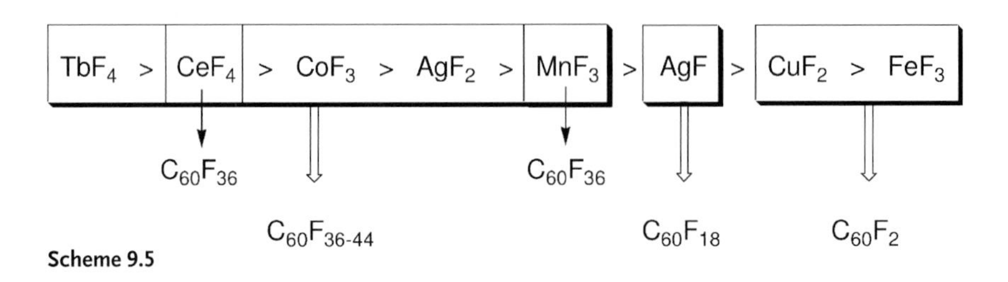

Scheme 9.5

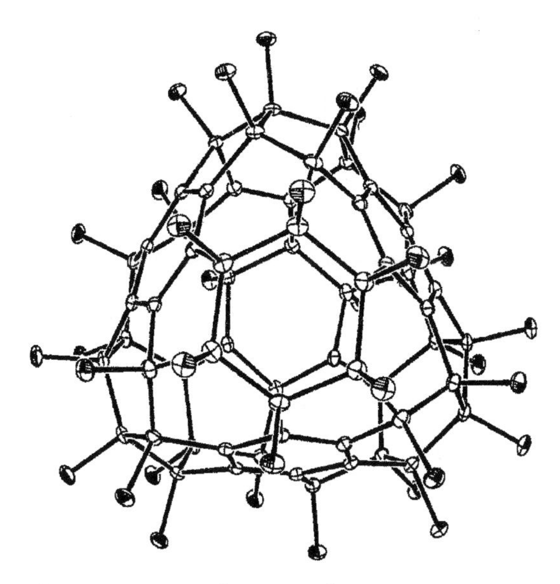

Figure 9.3 X-ray crystal structure of C_1-$C_{60}F_{36}$ [47].

$C_{60}F_{36}$ was synthesized with MnF_3 (350 °C) [39, 44, 45], CeF_4 (330 °C), [46] CoF_3 (350 °C) [46] or with ternary lead fluorides [43] in different yields and selectivities. The best method turned out to be the fluorination with MnF_3 and CeF_4 [7]. Separation from other fluorofullerenes such as $C_{60}F_{20}$ or $C_{60}F_{18}$ formed during the reaction was performed with HPLC. HPLC purification requires several steps that involve elution with toluene, toluene–hexane and finally pure hexane [44]. Three isomers of $C_{60}F_{36}$ have been characterized. Reaction with MnF_3 yields a C_3-isomer as the major product (> 90%) and two minor isomers with T-symmetry (< 10%) and C_1-symmetry [44, 47]. Schlegel diagrams of the structures of the three-isomers, which were proven with ^{19}F and ^{13}C NMR spectroscopy [44, 48], are shown in Figure 9.1. The correct assignment of the structures of T- and C_1-$C_{60}F_{36}$ was supported by X-ray crystallography [47, 49]. Even though the C_3-isomer is the most abundant product its crystal structure is not yet reported. The crystal structure depicted in Figure 9.3 shows clearly the severe distortion from the spherical C_{60}-shape. To obtain sufficient quantities of the C_1-isomer for X-ray structure analysis the synthesis was modified to improve the yield of this isomer by using a mixture of MnF_3 and K_2NiF_6 at 480 °C for the fluorination [47].

K_2PtF_6 or ternary lead fluorides have been utilized to reach reasonable yields of $C_{60}F_{18}$ [41–43, 50]. The binary fluoride AgF, heated from 420 to 480 °C, also gives mainly this compound [38]. Characterization of this fluorofullerene was somewhat easier than for the $C_{60}F_{36}$ isomer, since only one isomer is selectively formed. The single-crystal X-ray structure of $C_{60}F_{18}$ was the first to be reported for a fluorofullerene [50]. In this C_{3v}-symmetrical structure all fluorine atoms are bound on one hemisphere of the C_{60}, leaving one aromatic benzene ring surrounded by fluorine on one side and an almost intact C_{60} moiety on the other side (Figures 9.1

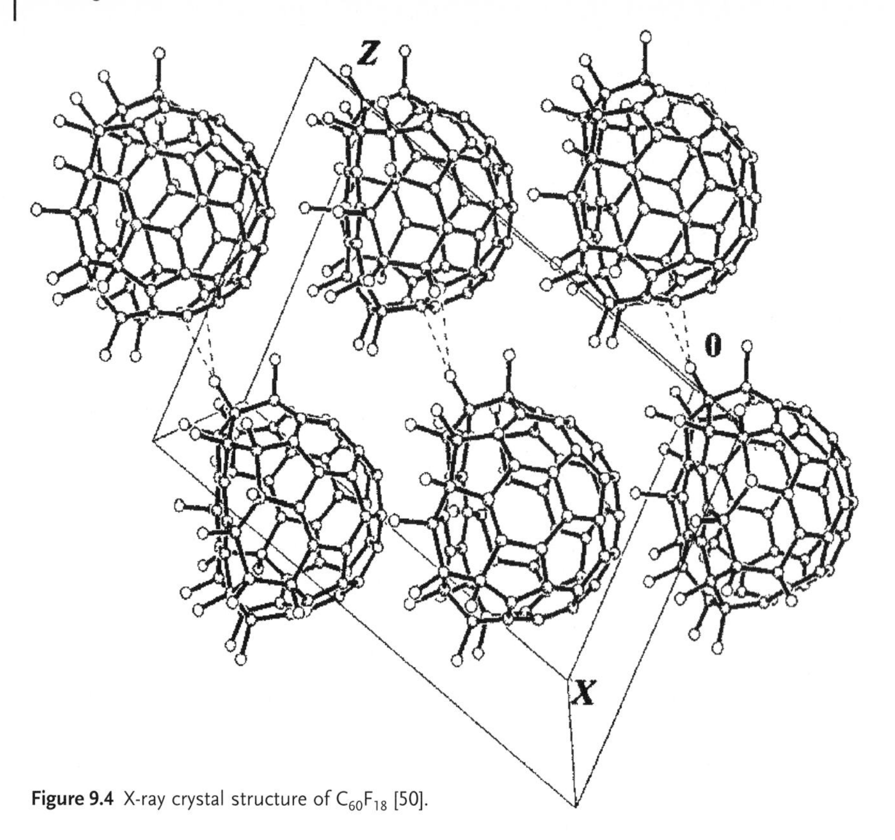

Figure 9.4 X-ray crystal structure of $C_{60}F_{18}$ [50].

and 9.4). This flattened fullerene structure might well be considered as a hexa-substituted benzene, making this the first fullerene derivative with a truly benzenoid ring on the C_{60} surface [50].

One of the numerous minor products formed during the synthesis of $C_{60}F_{18}$ is the fluorofullerene $C_{60}F_{20}$ [51]; its very small yield could be improved by deactivating the fluorinating agent MnF_3, which usually leads to $C_{60}F_{36}$, with KF. Reaction of C_{60} with an equimolar amount of MnF_3 and KF yielded sufficient amounts for a characterization of $C_{60}F_{20}$ [51]. Surprisingly, its structure is completely different to that of $C_{60}F_{18}$. It can not be derived by simply adding two fluorine atoms, thus $C_{60}F_{18}$ is not an intermediate in the formation of $C_{60}F_{20}$. ^{19}F NMR unveiled the highly symmetrical structure. Only one single signal in the spectrum was observed, leading to the conclusion that all twenty fluorine atoms are equivalent. The proposed structure can be described as two dehydrocorannulenes held together by a $(CF)_{20}$ chain (Figures 9.1 and 9.5). This appearance gave the compound the name "saturnene" [51].

Contrary to $C_{60}F_{20}$ the structure of $C_{60}F_{16}$ can easily be derived from the derivative with 18 fluorine atoms by removing two of them (Figure 9.1). Scaling up the preparation of $C_{60}F_{18}$ led to the discovery of the minor product $C_{60}F_{16}$ (about 2% of the $C_{60}F_{18}$ amount) [52]. Next to $C_{60}F_{16}$ the characterization of additional adducts

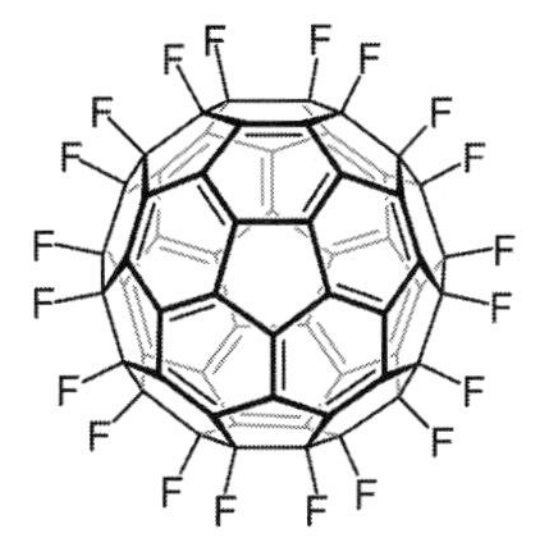

Figure 9.5 Structure of $C_{60}F_{20}$ [51].

that are intermediates on the way from C_{60} to $C_{60}F_{18}$ by contiguous addition of F_2-units could be characterized. This also provided further evidence for the proposed mechanism that subsequent addition takes place via activation of adjacent double bonds. The lower fluorofullerenes are difficult to obtain and characterize, since the stability of fluorinated C_{60} decreases with decreasing number of attached fluorines. Structures of $C_{60}F_2$ [53], $C_{60}F_4$ [54] and $C_{60}F_6$ [54] have been proposed on the basis of ^{19}F NMR spectroscopy (Figure 9.1). The originally proposed structure of $C_{60}F_8$ [54] was recently proven not to be the most favorable; the structure with non-contiguous F-pattern, which is consistent with NMR data and theoretical data on relative stabilities, was suggested instead [55].

In most of the fluorination reactions oxygenated fluorofullerenes [54, 56–63] and products such as $C_{60}F_nCF_2CF_3$ or $C_{60}F_nCF_3$ [54, 64, 65] are often observed as side products. Mass spectrometric analysis of the product mixture of fluorinations with metal fluorides [54, 59, 60] or elemental fluorine [57] gas shows oxides with one or more oxygens for almost any fluorofullerene. Some of them could be isolated and characterized with ^{19}F NMR spectroscopy ($C_{60}F_nO$ with $n = 2, 4, 6, 8, 16$ and 18) [7, 54, 59, 60]. These oxygenated fluorofullerenes are probably formed during the synthesis due to the presence of traces of oxygen or water [7]. In all characterized structures, not the earlier assumed epoxide structure [57] – as observed in oxofullerenes – but an intramolecular ether structure was found. For the oxahomo fullerene $C_{60}F_{18}O$ the insertion of “O” into a C-sp$_3$–C-sp$_3$ bond of the fullerene cage could be proven via X-ray crystallography [56]. The structure of this first oxahomo fullerene derivative is shown in Figure 9.6.

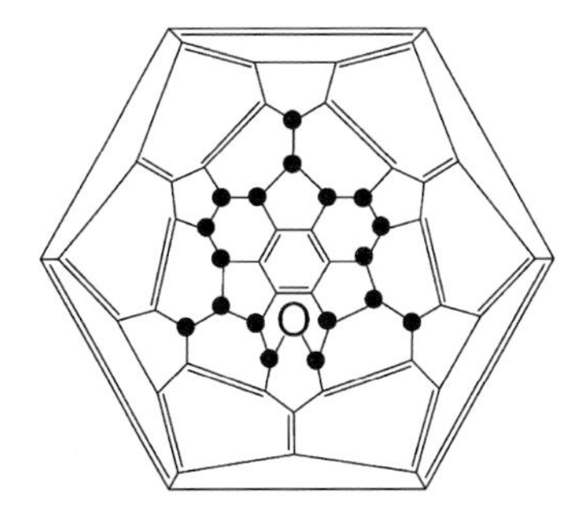

Figure 9.6 The first oxahomofullerene derivative $C_{60}F_{18}O$. The dots represent CF-units.

Another isomer of the fluorofullerene ether $C_{60}F_{18}O$ has been found recently [61]. This fluoroxyfluoro fullerene $C_{60}F_{17}OF$ is generated by insertion of "O" into a C–F bond.

The CF_3-substituted fluorofullerenes are formed via fragmentation of $C_{60}F_n$ during the reaction giving CF_3 radicals or difluoromethylene species [64]. The formation of trifluoromethyl derivatives of fluorofullerenes is significantly enhanced by performing the K_2PtF_6 reaction in the presence of CF_3COOAg, the latter serving as a source of CF_3 radicals formed upon heating [36]. Alternatively, CF_2 is supposed to insert into the most accessible C–F bond to give $C_{60}F_{n-1}CF_3$ [54, 64, 65]. Another insertion into this newly-created C–C bond leads to $C_{60}F_{n-1}CF_2CF_3$. In the X-ray crystal structure analysis of $C_{60}F_{17}CF_3$ three different isomers, one C_s-symmetrical isomer and two C_1-symmetrical enantiomeric isomers, can be seen [64].

Fluorination of C_{70} yields complex mixtures of highly fluorinated products ($C_{70}F_n$), with major peaks observed with mass-spectrometry for $n \approx 36$–52 [16, 34, 66]. Some of the $C_{70}F_n$ species, with $n = 34, 36, 38, 40, 42$ and 44, and a few oxygenated species could be isolated and characterized via mass spectrometry and ^{19}F NMR spectroscopy [66]. They were obtained by reaction of C_{70} with MnF_3 at 450 °C and separated via HPLC. So far no X-ray structure of any of these isomers has been obtained.

9.2.4 Reactions of Fluorofullerenes

Fluorofullerenes are good oxidants, and they should be powerful electrophiles and very susceptible to nucleophilic attack [7, 8]. The number of fluorofullerenes is currently limited, and few are accessible in weighable amounts. Thus, chemistry carried out with fluorofullerenes as reactants or substrates is still in its infancy. Nonetheless, some very promising reactions with $C_{60}F_{18}$ [7, 8, 67–70], one of the most easily accessible compounds, have been performed, leading to interesting products. $C_{60}F_{18}$ was chosen because of its almost intact spherical C_{60}-like moiety, which is, sterically, not hindered for an attack of a nucleophile or a diene. It was subjected to some reactions that are well established for pristine C_{60}, i.e. the Bingel reaction [67, 71, 72], the Prato reaction [7, 8] and the cycloaddition of anthracene [69] and tetrathiafulvalene [70] or electrophilic substitution of aromatic compounds [68].

One modification of the Bingel reaction involves DBU as a base. Initial experiments showed that exposure of $C_{60}F_{18}$ to DBU leads to a complete loss of fluorine [71]. Fortunately this defluorination does not take place in the presence of diethyl bromomalonate. Reaction of $C_{60}F_{18}$ with DBU and diethylbromomalonate in toluene at room temperature yields three products and unreacted $C_{60}F_{18}$ [67, 71]. The three products are the result of mono-, bis- and tris-substitution; their relative yield depends on how fast the base is added. The reaction does not – as in the usual Bingel reaction – proceed via a nucleophilic addition reaction to one double bond but as a nucleophilic substitution with a S_N2'-mechanism. The bromomalonate anion attacks a double bond in δ-position to one of the least sterically hindered fluorine atoms, which leads to the loss of this fluorine. The bromine atom is not replaced (Scheme 9.6).

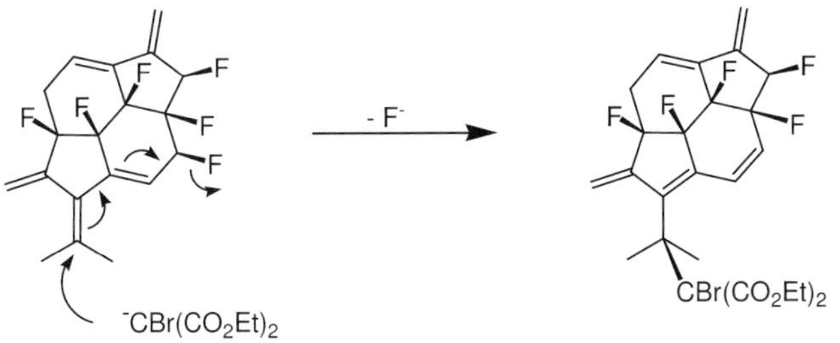

Scheme 9.6 Mechanism of the "Bingel" reaction with $C_{60}F_{18}$ (only a fragment of $C_{60}F_{18}$ is shown).

Such substitution can take place three times, whereas the threefold addition leaves a C_{60} derivative with a very interesting structure. It contains an equatorial 18π-annulene belt. The aromaticity of this first annulene-fullerene with its high level of conjugation is responsible for the emerald green color of this compound. Bond-length alternation, which was derived from the single X-ray crystal structure [67], is very low, indicating the high delocalization. The all-trans arrangement of the constituent bonds of the belt defines $C_{60}F_{15}[CBr(CO_2Et)_2]_3$ (**13**) as a trannulene (Scheme 9.7).

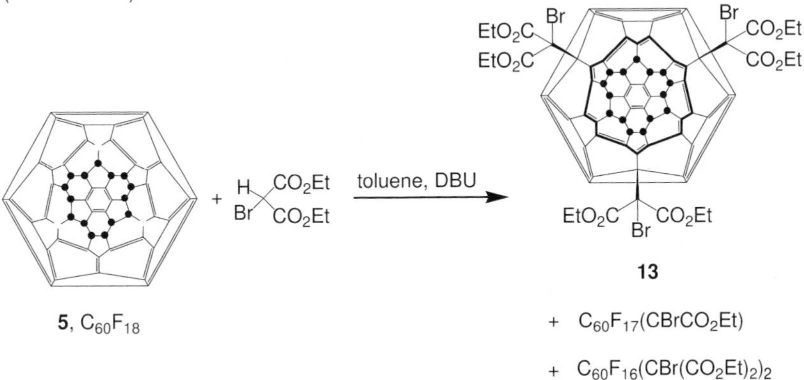

Scheme 9.7 The annulene belt is drawn with bold bonds.

The all-trans annulenes can also be obtained by performing the Bingel reaction with malonates such as $XCH(CO_2Et)_2$ ($X = NO_2$ or SO_2CH_2Ph) [72].

Under the influence of a Lewis acid, C_{60} can be arylated via a Friedel–Crafts reaction (see Section 8.6). A similar reaction occurs with $C_{60}F_{18}$ where *ipso* substitution of fluorine atoms takes place [68]. The fluorofullerene is treated with $FeCl_3$ in benzene for two weeks (Scheme 9.8). Electrophilic substitution leads to arylated $C_{60}F_{18-n}Ph_n$ with $n = 1$–3. Probably due to steric reasons the δ-substitution is not observed in this arylation.

Four different sorts of double bonds are accessible for a [4+2]-cycloaddition. Anthracene reacts with $C_{60}F_{18}$ in refluxing toluene mainly by addition to two of the four possible positions, giving the adducts **15** and **16** (Scheme 9.9) [69]. Seemingly, the least sterically hindered position is attacked. Furthermore, the less stable $C_{60}F_{18}$-anthracene adduct **16** can rearrange to the sterically more stable isomer **15**.

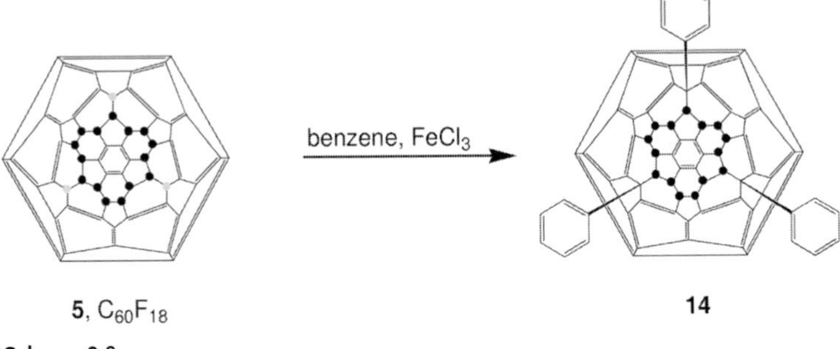

Scheme 9.8

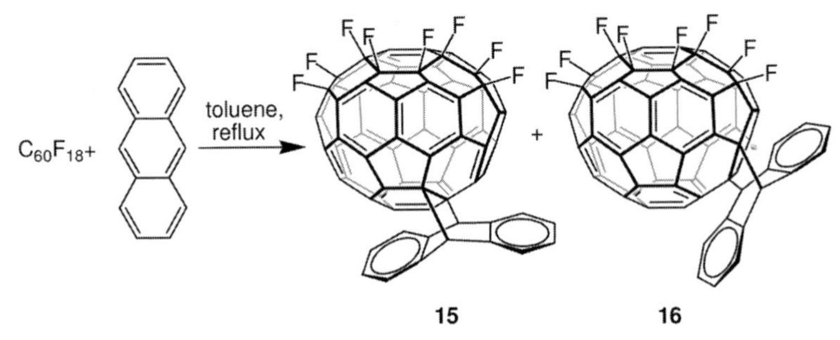

Scheme 9.9 Diels–Alder adduct of anthracene with $C_{60}F_{18}$. Only the fluorine atoms directed towards the viewer are shown.

An unusual six-electron cycloaddition could be observed in the reaction of $C_{60}F_{18}$ with tetrathiafulvalene (TTF) [70]. TTF is not added to one of the free double bonds but replaces two of the fluorine atoms in a way that leaves the $C_{60}F_{16}$ moiety, whose structure is described above (Figure 9.1). Under loss of F_2 the product $C_{60}F_{16}$(TTF) is formed (Figure 9.7).

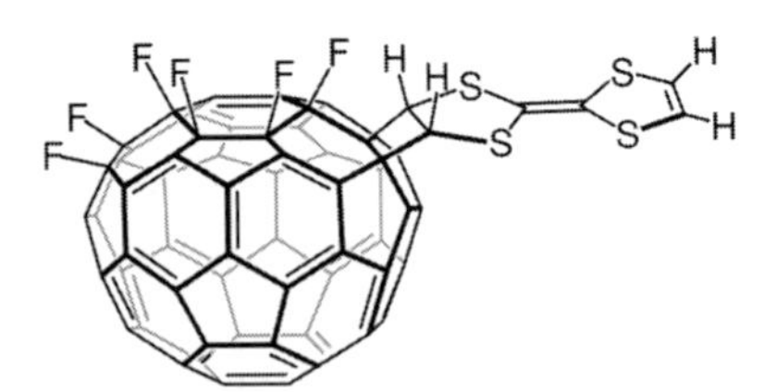

Figure 9.7 TTF reacts with $C_{60}F_{18}$ under loss of F_2 to form $C_{60}F_{16}$(TTF). Only the fluorine atoms directed towards the viewer are shown.

9.3 Chlorination

9.3.1 Synthesis and Properties of Chlorofullerenes

Whereas the treatment of C_{60} with chlorine in organic solvents does not lead to any detectable reaction, a polychlorination [73] to $C_{60}Cl_n$ can be achieved by allowing a slow stream of chlorine gas to react with C_{60} in a hot glass tube at between 250 and 400 °C (Scheme 9.10) [74]. By this method an average of 24 chlorine atoms can be attached to C_{60}. The polychlorofullerenes are obtained as a mixture of products, which are light orange and soluble in many organic solvents [74]. Chlorofullerenes $C_{60}Cl_n$ with a lower degree of chlorination ($n \approx 6$) can be synthesized by the treatment of solid C_{60} with liquid chlorine at −35 °C (Scheme 9.10) [75]. These reactions are completed in about one day after passing liquid chlorine continuously over C_{60} on a glass filter. By reaction of ICl, ICl_3, $KICl_4$ under different conditions chlorofullerenes $C_{60}Cl_n$ with average $n = 6$, 8, 10, 12, 14, 26 could be synthesized and characterized by IR and NMR spectroscopy and MALDI-TOF mass spectrometry, but only as a mixture. Except for one isomer ($C_{60}Cl_6$, see below) they could not be separated [76].

$$C_{60} \xrightarrow{Cl_2} C_{60}Cl_n \xrightarrow{400\,°C \text{ or } PPh_3 \text{ or } e^-} C_{60}$$

Scheme 9.10

The chlorofullerenes are less stable than their fluorine analogues. Ions deduced to be from $C_{60}Cl_n$ could be observed in the MALDI [77] of FAB [78] mass spectra, but with extensive fragmentation down to C_{60}. The instability of the chlorofullerenes can also be seen from their behavior towards thermal, chemical and electrochemical manipulations. Heating $C_{60}Cl_n$ at 400 °C under argon results in dechlorination and the parent fullerenes are recovered (Scheme 9.10) [74]. The dechlorination starts at 200 °C [75]. Cyclic voltammetry of $C_{60}Cl_n$ ($n \approx 6$) indicates that, after reduction, chloride anions dissociate from the chlorinated fullerene [75]. The treatment of toluene solutions of $C_{60}Cl_n$ with triphenylphosphine leads to a dechlorination yielding 80% C_{60} (Scheme 9.10) [75]. Similar to their fluorinated analogues, the polychlorofullerenes also undergo substitution reactions with nucleophiles (see Section 9.2.2) [74].

Isomerically pure chlorofullerene $C_{60}Cl_6$ has been reported to be the predominant product of the reaction of C_{60} with an excess of iodine monochloride in benzene or toluene at room temperature (Scheme 9.11) [79]. The product is very soluble in benzene, carbon disulfide and tetrachloromethane. Deep orange crystals can be obtained by recrystallization from pentane. The synthesis of $C_{60}Cl_6$ using toluene as a solvent proceeds more slowly than with benzene, indicating that radicals are involved and are scavenged by the toluene [79].

The structure of $C_{60}Cl_6$ as 1,2,4,11,15,30-hexachloro[60]fullerene (Scheme 9.11) was deduced from its ^{13}C NMR spectrum, which shows 32 lines for 54 sp^2-hybridized

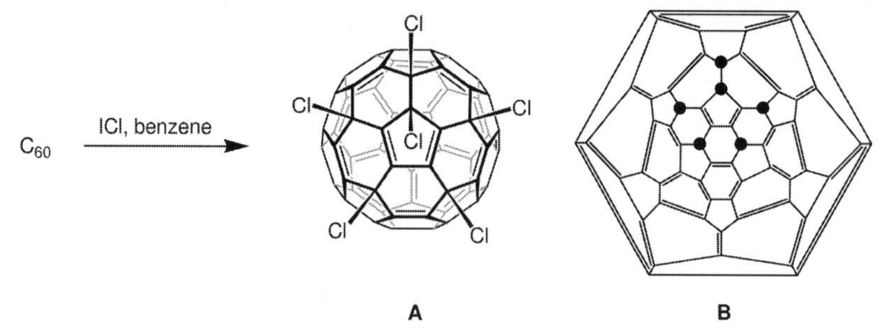

Scheme 9.11 Schematic representation of the structure of $C_{60}Cl_6$ obtained by the reaction of C_{60} with ICl. (A) front view and (B) Schlegel diagram.

and 6 sp^3-hybridized C atoms, demonstrating the C_s-symmetry. In this structure of $C_{60}Cl_6$, the same building principles seem to be important that also direct the formation of the radical $C_{60}R_5^{\bullet}$, as shown by ESR spectroscopy [80]. In $C_{60}R_5^{\bullet}$ the additions of the radicals $R^{\bullet}$ also occurred at positions 1, 4, 11, 15 and 30. The regiochemistry of the chlorine addition leading to $C_{60}Cl_6$ therefore indicates that a radical mechanism is also valid in this case. For the chlorine addition to C_{60}, in contrast to that of hydrogen and fluorine, a 1,4-mode is predicted to be more favorable than a 1,2-mode [4]. This difference is mainly due to the enhanced steric requirement of the chlorine.

The only structurally characterized chloro[70]fullerene ($C_{70}Cl_{10}$) was also synthesized with iodine monochloride in benzene solution [81]. Its proposed structure is shown in Figure 9.8.

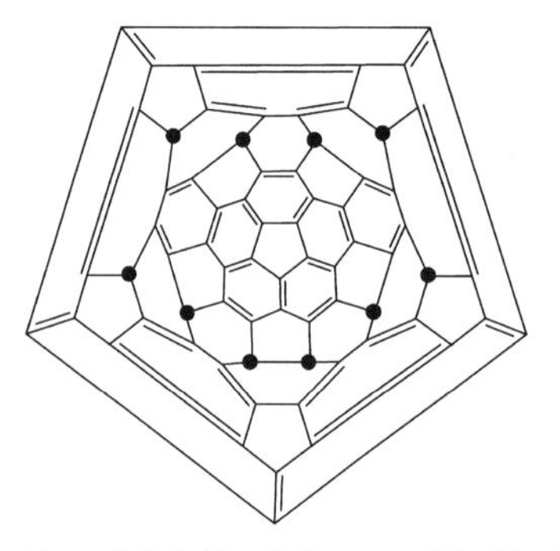

Figure 9.8 Schlegel diagram of $C_{70}Cl_{10}$ [81].

9.3.2 Reactions of Chlorofullerenes

Chlorofullerenes are sensitive to hydrolysis and readily react with nucleophiles [73]. A mixture of polychlorofullerenes reacts with potassium methanolate or benzene to form polymethoxylated or polyphenylated fullerenes in which all of the chlorine atoms are replaced [74]. So far, only $C_{60}Cl_6$ has been prepared in high purity [82], and thus defined reactions of chlorofullerenes are limited to reactions with this compound. Nucleophilic substitutions with $C_{60}Cl_6$ have been accomplished with alkoxide [83], MeLi [84], benzene and some benzene derivatives [85, 86] or allyltrimethylsilane [87]. The main products of these reactions are isostructural to $C_{60}Cl_6$, which means they are derived by successive *ipso* substitution of chlorine [84].

A Friedel–Crafts-type reaction can be carried out with hexachlorofullerene and, e.g., benzene, fluorobenzene, anisol or toluene, but due to steric hindrance not with mesitylene [85, 86]. Upon stirring benzene solutions of $C_{60}Cl_6$ with ferric chloride for 20 min, $C_{60}Ph_5Cl$ can be obtained (Scheme 9.12). Replacement of the sixth chlorine is prevented by steric hindrance. This hindrance can be overcome by addition of the more reactive toluene to $C_{60}Ph_5Cl$ to give $C_{60}Ph_5(C_6H_4CH_3)$ [86].

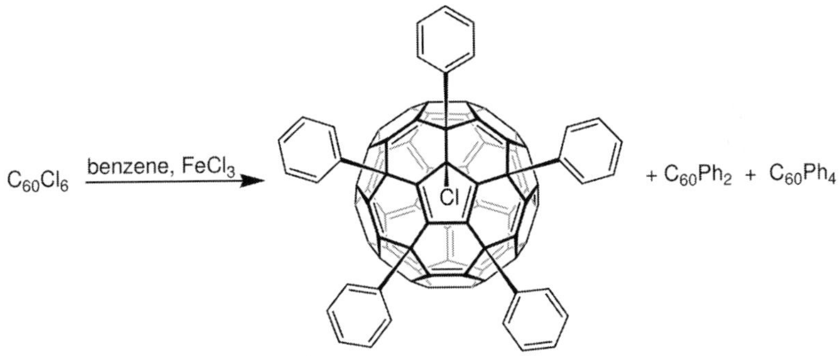

$C_{60}Cl_6$ $\xrightarrow{\text{benzene, FeCl}_3}$ [structure: Cl] + $C_{60}Ph_2$ + $C_{60}Ph_4$

Scheme 9.12 Arylation of $C_{60}Cl_6$.

Minor products of the phenylation of hexachlorofullerene are the two lower arylated compounds $C_{60}Ph_4$ and $C_{60}Ph_2$ [85, 86]. By treatment with PPh_3 the chlorine in $C_{60}Ph_5Cl$ is readily replaced by hydrogen, derived from traces of water in the solvent [86]. In the presence of water, treatment with $AlCl_3$ also leads to the hydrochlorofullerene $C_{60}Ph_5H$ [73]. This replacement proceeds via the formation of the $C_{60}Ph_5^+$-cation [88]. Unlike C_{60}^+ the fullerene-cations $C_{60}Ph_5^+$ and also $C_{60}(C_6H_4F)^+$ are relatively stable and can be observed with ^{13}C and ^3He NMR-spectroscopy if no water is present [88]. $C_{60}Ph_5H$ and $C_{60}Ph_5Cl$ spontaneously oxidize under exposure to air. Under loss of one chlorine or hydrogen and the loss of ortho-hydrogen of the phenyl ring, a benzofuranyl fullerene derivative is formed [89].

An excess of allyltrimethylsilane reacts with $C_{60}Cl_6$ in the presence of $TiCl_4$ to give the hexasubstitution product $C_{60}(CH_2CH{=}CH_2)_6$ in good yields [87]. Steric hindrance encountered during the arylation is not critical in this reaction and all chlorines can be substituted. The fivefold substituted compound $C_{60}(CH_2CH{=}CH_2)_5Cl$ can be isolated as a minor product.

Treatment of $C_{60}Cl_6$ with sodium methanolate or sodium ethanolate under reflux respectively at room temperature for some days leads to $C_{60}(OR)_5Cl$ in moderate yields [83]. A by-product of the reaction with EtO^- is 1,4-$(EtO)_2C_{60}$, showing that chlorine elimination accompanies substitution. Reaction under loss of chlorine is the main reaction when using the alcohols and not the alkoxylates as nucleophiles. Heating with either methanol or isopropanol under reflux for 140 h yields 1,4-$(MeO)_2C_{60}$ and 1,4-$(^iPrO)_2C_{60}$.

Substitution of $C_{60}Cl_6$ with methyllithium in THF at room temperature yields a mixture of compounds all showing the same addition pattern [84]. Among the main products are the fully alkylated derivative $C_{60}Me_6$ and the chloroalkylfullerene $C_{60}Me_5Cl$. Additionally, various oxygenated byproducts are formed.

Friedel–Crafts reaction of $C_{70}Cl_{10}$ with benzene and $FeCl_3$ proceeds in similar fashion to the reaction with $C_{60}Cl_6$ [90]. $C_{70}Ph_{10}$ and $C_{70}Ph_8$ are produced as major products. $C_{70}Ph_n$ with $n = 2, 4, 6$ and $C_{70}Ph_9OH$ [91] are formed as by-products.

9.4
Bromination [73]

Bromination of C_{60} is predicted to be less exothermic than the chlorination or fluorination, and in general 1,4-additions should be favored over 1,2-additions [4]. C_{60} reacts with liquid bromine to form the bromofullerene $C_{60}Br_{24}$, which is a yellow-orange crystalline compound and is obtained as a bromine solvate $C_{60}Br_{24}(Br_2)_x$ (Scheme 9.13) [92, 93]. This compound has a simple IR spectrum, indicating a highly symmetrical structure. Upon heating to 150 °C, all the bromine is eventually lost [93].

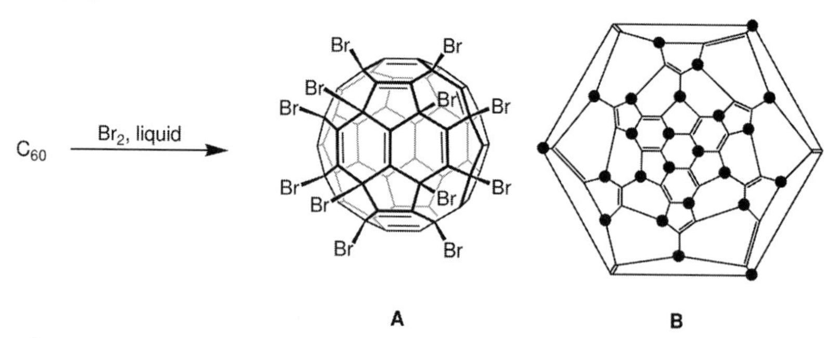

Scheme 9.13 Schematic representation of $C_{60}Br_{24}$. (A) front view and (B) Schlegel diagram. Only the bromine atoms directed towards the viewer are shown in (A).

An X-ray crystal structure analysis shows that $C_{60}Br_{24}$ has T_h-symmetry [92]. This is the highest symmetry available for a brominated fullerene, $C_{60}Br_n$. The addition pattern of the bromine atoms on the C_{60} surfaces arises from 1,4-additions to fused pairs of six-membered rings with the closest Br,Br placement being 1,3 (Scheme 9.13). Bromines attached to C atoms in the 1 and 4 positions cause 12 of the hexagons to adopt boat conformations and the remaining eight hexagons to adopt chair conformations [92]. The addition of another two bromine atoms would require locations in 1,2-positions, which is very unfavorable for these bulky atoms [4]. The bromination of C_{60} in CS_2 leads to the formation of dark brown crystals in 80%

yield after 24 h (Scheme 9.14) [93]. The single-crystal X-ray diffraction analysis showed the crystals to be $C_{60}Br_8$ (Scheme 9.14). The same compound can also be obtained in 58% yield by bromination in chloroform (Scheme 9.14). The octabromide is not very soluble in common organic solvents but it is soluble in bromine.

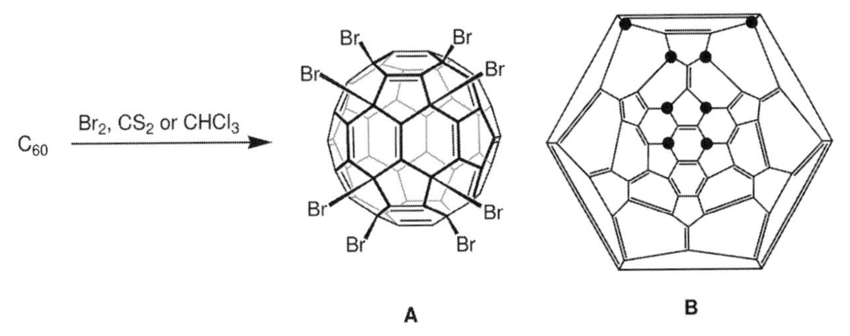

Scheme 9.14 Schematic representation of $C_{60}Br_8$. (A) front view and (B) Schlegel diagram.

If the bromination of C_{60} is carried out in either benzene or tetrachloromethane, another bromide, $C_{60}Br_6$, crystallizes in magenta plates in 54 and 92% yields respectively (Scheme 9.15) [93]. The structure of $C_{60}Br_6$ was also determined by single-crystal X-ray analysis. This bromide is isostructural to $C_{60}Cl_6$ [79].

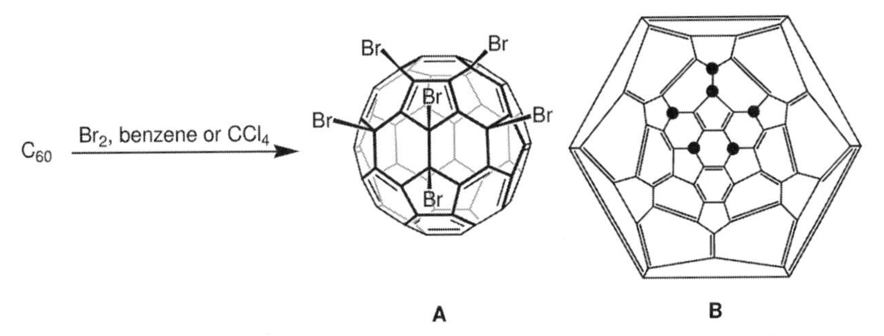

Scheme 9.15 Schematic representation of $C_{60}Br_6$. (A) front view and (B) Schlegel diagram.

$C_{60}Br_6$ is moderately soluble in organic solvents. Each of the bromo derivatives is unstable when heated, and loses bromine [93, 94]. The most stable bromide is $C_{60}Br_{24}$. Upon heating in tetrachloromethane or benzene, $C_{60}Br_6$ disproportionates to give C_{60} and $C_{60}Br_8$ [93]. The higher stability of $C_{60}Br_8$ than that of $C_{60}Br_6$ may be explained by the lack of the eclipsing interactions in $C_{60}Br_8$ that are present in $C_{60}Br_6$. During the formation of $C_{60}Br_8$ from $C_{60}Br_6$, a sequence of 1,3-allylic bromine shifts may be involved [93]. The general instability of the bromides, especially $C_{60}Br_6$ and $C_{60}Br_8$, is related to the introduction of [5,6] double bonds. Different bromides are formed in different solvents due to the crystallization of insoluble bromofullerenes, such as $C_{60}Br_6$ in benzene, which inhibits further bromination.

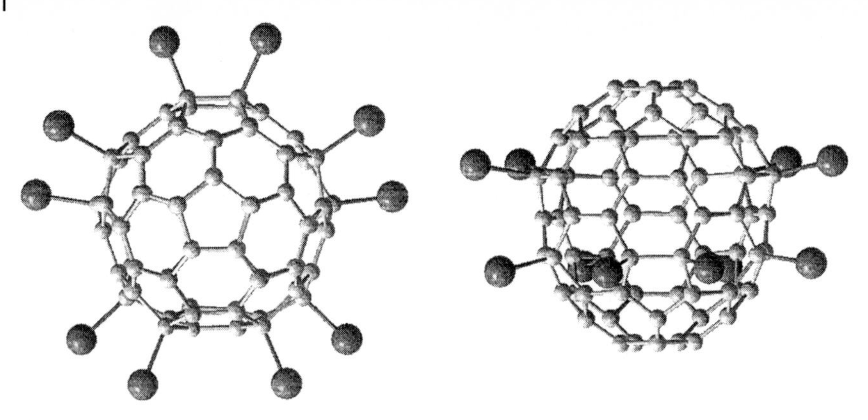

Figure 9.9 Top and side views of $C_{70}Br_{10}$ [95].

A few attempts to brominate C_{70} were undertaken by reaction with bromine either neat or in solution. The only derivative whose composition could be confirmed is $C_{70}Br_{10}$. A single X-ray structure of this compound revealed its structure (Figure 9.9), which is identical to that of $C_{70}Cl_{10}$ (see Figure 9.8) [95].

9.5
Reaction with Iodine

Iodination of C_{60} has not yet been observed. This may be attributed to the weak carbon–iodine bond and to the bulkiness of iodine. The size of iodine requires that the addends are well separated, which would result in substantial bond reorganization with double bonds being placed in pentagonal rings, which is unfavorable [73]. A reaction of C_{60} with iodine was observed upon irradiating a C_{60}/I_2 mixture with Hg light at 130 °C for about 20 h [96]. X-ray analysis of the resulting simple hexagonal structure revealed a $C_{60}I_2$ stoichiometry and an intercalation of iodine atoms [96, 97]. In contrast to alkali metal doped C_{60}, such as K_3C_{60}, no transition into a superconducting phase has been observed [96–98]. Some more examples of intercalated $C_{60}I_n$ have been observed and their properties studied [99–101].

References

1 G. E. Scuseria, *Chem. Phys. Lett.* **1991**, *176*, 423.
2 J. Cioslowski, *Chem. Phys. Lett.* **1991**, *181*, 68.
3 B. I. Dunlap, D. W. Brenner, J. W. Mintmire, R. C. Mowrey, C. T. White, *J. Phys. Chem.* **1991**, 95, 5763.
4 N. Matsuzawa, T. Fukunaga, D. A. Dixon, *J. Phys. Chem.* **1992**, *96*, 10747.
5 D. A. Dixon, N. Matsuzawa, T. Fukunaga, F. N. Tebbe, *J. Phys. Chem.* **1992**, *96*, 6107.
6 B. W. Clare, D. L. Kepert, *Theochem* **2003**, *621*, 211.
7 O. V. Boltalina, S. Strauss, H., in *Encyclopedia of Nanoscience and Nanotechnology* (Eds.: J. A. Schwartz, C. I. Contescu, K. Putyera), Dekker, New York, **2004**, p. 3200.
8 R. Taylor, *Chem. Eur. J.* **2001**, *7*, 4074.
9 N. I. Denisenko, S. I. Troyanov, A. A. Popov, V. Kuvychko Igor, B. Zemva, E. Kemnitz, H. S. Strauss Steven, O. Boltalina, V., *J. Am. Chem. Soc.* **2004**, *126*, 1619.
10 F. Zhou, G. J. Van Berkel, B. T. Donovan, *J. Am. Chem. Soc.* **1994**, *116*, 5485.
11 K. Kniaz, J. E. Fischer, H. Selig, G. B. M. Vaughan, W. J. Romanow, D. M. Cox, S. K. Chowdhury, J. P. McCauley, R. M. Strongin, A. B. Smith III, *J. Am. Chem. Soc.* **1993**, *115*, 6060.
12 R. Taylor, J. H. Holloway, E. G. Hope, A. G. Avent, G. J. Langley, T. J. Dennis, J. P. Hare, H. W. Kroto, D. R. M. Walton, *J. Chem. Soc., Chem. Commun.* **1992**, 665.
13 R. Taylor, G. J. Langley, A. K. Brisdon, J. H. Holloway, E. G. Hope, H. W. Kroto, D. R. M. Walton, *J. Chem. Soc., Chem. Commun.* **1993**, 875.
14 O. V. Boltalina, *J. Fluorine Chem.* **2000**, *101*, 273.
15 H. Holloway John, E. G. Hope, R. Taylor, G. J. Langley, A. G. Avent, T. J. Dennis, J. P. Hare, H. W. Kroto, D. R. M. Walton, *J. Chem. Soc., Chem. Commun.* **1991**, 966.
16 H. Selig, C. Lifshitz, T. Peres, J. E. Fischer, A. R. McGhie, W. J. Romanow, J. P. McCauley, Jr., A. B. Smith III, *J. Am. Chem. Soc.* **1991**, *113*, 5475.
17 O. V. Boltalina, N. A. Galeva, *Russ. Chem. Rev.* **2000**, *69*, 609.
18 A. A. Tuinman, P. Mukherjee, J. L. Adcock, R. L. Hettich, R. N. Compton, *J. Phys. Chem.* **1992**, *96*, 7584.
19 A. A. Tuinman, A. A. Gakh, J. L. Adcock, R. N. Compton, *J. Am. Chem. Soc.* **1993**, *115*, 5885.
20 F. Okino, H. Touhara, K. Seki, R. Mitsumoto, K. Shigematsu, Y. Achiba, *Fullerene Sci. Technol.* **1993**, *1*, 425.
21 F. Okino, S. Yajima, S. Suganuma, R. Mitsumoto, K. Seki, H. Touhara, *Synth. Methods* **1995**, *70*, 1447.
22 A. Hamwi, C. Fabre, P. Chaurand, S. Della-Negra, C. Ciot, D. Djurado, J. Dupuis, A. Rassat, *Fullerene Sci. Technol.* **1993**, *1*, 499.
23 A. Hamwi, C. Latouche, V. Marchand, J. Dupuis, R. Benoit, *J. Phys. Chem. Solids* **1996**, *57*, 991.
24 A. A. Gakh, A. A. Tuinman, J. L. Adcock, R. A. Sachleben, R. N. Compton, *J. Am. Chem. Soc.* **1994**, *116*, 819.
25 S. I. Troyanov, P. A. Troshin, O. V. Boltalina, I. N. Ioffe, L. N. Sidorov, E. Kemnitz, *Angew. Chem.* **2001**, *113*, 2345; *Angew. Chem. Int. Ed.* **2001**, *40*, 2285.
26 I. S. Neretin, K. A. Lyssenko, M. Y. Antipin, Y. L. Slovokhotov, *Russ. Chem. Bull.* **2002**, *51*, 754.
27 A. A. Gakh, A. A. Tuinman, *J. Phys. Chem. A* **2000**, *104*, 5888.
28 A. A. Gakh, A. A. Tuinman, *Tetrahedron Lett.* **2001**, *42*, 7137.
29 V. F. Bagryantsev, A. S. Zapol'skii, O. V. Boltalina, N. A. Galeva, L. N. Sidorov, *Z. Neorgan. Khim.* **2000**, *45*, 1121.
30 O. V. Boltalina, L. N. Sidorov, E. V. Sukhanova, I. D. Sorokin, *Chem. Phys. Lett.* **1994**, *230*, 567.
31 A. A. Gakh, A. A. Tuinman, J. L. Adcock, R. N. Compton, *Tetrahedron Lett.* **1993**, *34*, 7167.
32 O. V. Boltalina, L. N. Sidorov, V. F. Bagryantsev, V. A. Seredenko, A. S. Zapol'skii, J. M. Street, R. Taylor, *J. Chem. Soc., Perkin Trans. 2* **1996**, 2275.

33 O. V. Boltalina, A. a. K. Abdul-Sada, R. Taylor, *J. Chem. Soc., Perkin Trans. 2* **1995**, 981.
34 H. Selig, K. Kniaz, G. B. M. Vaughan, sJ. E. Fischer, A. B. Smith III, *Macromol. Symp.* **1994**, *82*, 89.
35 O. V. Boltalina, V. Y. Markov, A. Y. Lukonin, T. V. Avakjan, D. B. Ponomarev, I. D. Sorokin, L. N. Sidorov, *Proc. – Electrochem. Soc.* **1995**, *95-10*, 1395.
36 O. V. Boltalina, A. A. Goryunkov, V. Y. Markov, I. N. Ioffe, L. N. Sidorov, *Int. J. Mass Spectrom.* **2003**, *228*, 807.
37 O. V. Boltalina, V. Y. Markov, A. Y. Borschevskii, N. A. Galeva, L. N. Sidorov, G. Gigli, G. Balducci, *J. Phys. Chem. B* **1999**, *103*, 3828.
38 A. A. Goryunkov, V. Y. Markov, O. V. Boltalina, B. Zemva, A. K. Abdul-Sada, R. Taylor, *J. Fluorine Chem.* **2001**, *112*, 191.
39 O. V. Boltalina, A. Y. Borschevskii, L. N. Sidorov, J. M. Street, R. Taylor, *Chem. Commun.* **1996**, 529.
40 O. V. Boltalina, D. B. Ponomarev, A. Y. Borschevskii, L. N. Sidorov, *J. Phys. Chem. A* **1997**, *101*, 2574.
41 O. V. Boltalina, V. Y. Markov, R. Taylor, M. P. Waugh, *Chem. Commun.* **1996**, 2549.
42 I. y. V. Goldt, O. V. Boltalina, L. N. Sidorov, E. Kemnitz, S. I. Troyanov, *Solid State Sci.* **2002**, *4*, 1395.
43 P. A. Troshin, O. V. Boltalina, N. V. Polyakova, Z. E. Klinkina, *J. Fluorine Chem.* **2001**, *110*, 157.
44 O. V. Boltalina, J. M. Street, R. Taylor, *J. Chem. Soc., Perkin Trans. 2* **1998**, 649.
45 N. S. Chilingarov, A. V. Nikitin, J. V. Rau, I. V. Golyshevsky, A. V. Kepman, F. M. Spiridonov, L. N. Sidorov, *J. Fluorine Chem.* **2002**, *113*, 219.
46 O. V. Boltalina, A. Y. Lukonin, A. A. Gorjunkov, V. K. Pavlovich, A. N. Rykov, V. M. Seniavin, L. N. Sidorov, R. Taylor, *Proc. – Electrochem. Soc.* **1997**, *97-14*, 257.
47 A. G. Avent, B. W. Clare, P. B. Hitchcock, D. L. Kepert, R. Taylor, *Chem. Commun.* **2002**, 2370.
48 A. A. Gakh, A. A. Tuinman, *Tetrahedron Lett.* **2001**, *42*, 7133.
49 P. B. Hitchcock, R. Taylor, *Chem. Commun.* **2002**, 2078.
50 I. S. Neretin, K. A. Lyssenko, M. Y. Antipin, Y. L. Slovokhotov, O. V. Boltalina, P. A. Troshin, A. Y. Lukonin, L. N. Sidorov, R. Taylor, *Angew. Chem.* **2000**, *112*, 3411; *Angew. Chem. Int. Ed.* **2000**, *39*, 3273.
51 O. V. Boltalina, V. Y. Markov, P. A. Troshin, A. D. Darwish, J. M. Street, R. Taylor, *Angew. Chem.* **2001**, *113*, 809; *Angew. Chem. Int. Ed.* **2001**, *40*, 787.
52 A. G. Avent, O. V. Boltalina, A. Y. Lukonin, J. M. Street, R. Taylor, *J. Chem. Soc., Perkin Trans. 2* **2000**, 1359.
53 O. V. Boltalina, A. Y. Lukonin, J. M. Street, R. Taylor, *Chem. Commun.* **2000**, 1601.
54 O. V. Boltalina, A. D. Darwish, J. M. Street, R. Taylor, X.-W. Wei, *J. Chem. Soc., Perkin Trans. 2* **2002**, 251.
55 J. P. B. Sandall, P. W. Fowler, *Org. Biomol. Chem.* **2003**, *1*, 1061.
56 O. V. Boltalina, P. A. Troshin, B. de La Vaissiere, P. W. Fowler, J. P. B. Sandall, P. B. Hitchcock, R. Taylor, *Chem. Commun.* **2000**, 1325.
57 R. Taylor, G. J. Langley, J. H. Holloway, E. G. Hope, A. K. Brisdon, H. W. Kroto, D. R. M. Walton, *J. Chem. Soc., Perkin Trans. 2* **1995**, 181.
58 O. Boltalina, J. H. Holloway, E. G. Hope, J. M. Street, R. Taylor, *J. Chem. Soc., Perkin Trans. 2* **1998**, 1845.
59 A. G. Avent, O. V. Boltalina, P. W. Fowler, A. Y. Lukonin, V. K. Pavlovich, J. P. B. Sandall, J. M. Street, R. Taylor, *J. Chem. Soc., Perkin Trans. 2* **1998**, 1319.
60 O. V. Boltalina, A. Y. Lukonin, A. G. Avent, J. M. Street, R. Taylor, *J. Chem. Soc., Perkin Trans. 2* **2000**, 683.
61 A. D. Darwish, A. a. K. Abdul-Sada, A. G. Avent, J. M. Street, R. Taylor, *J. Fluorine Chem.* **2003**, *121*, 185.
62 O. V. Boltalina, B. de La Vaissiere, P. W. Fowler, A. Y. Lukonin, A. a. K. Abdul-Sada, J. M. Street, R. Taylor, *J. Chem. Soc., Perkin Trans. 2* **2000**, 2212.
63 O. V. Boltalina, B. de La Vaissiere, A. Y. Lukonin, P. W. Fowler, A. a. K. Abdul-Sada, J. M. Street, R. Taylor, *J. Chem. Soc., Perkin Trans. 2* **2001**, 550.

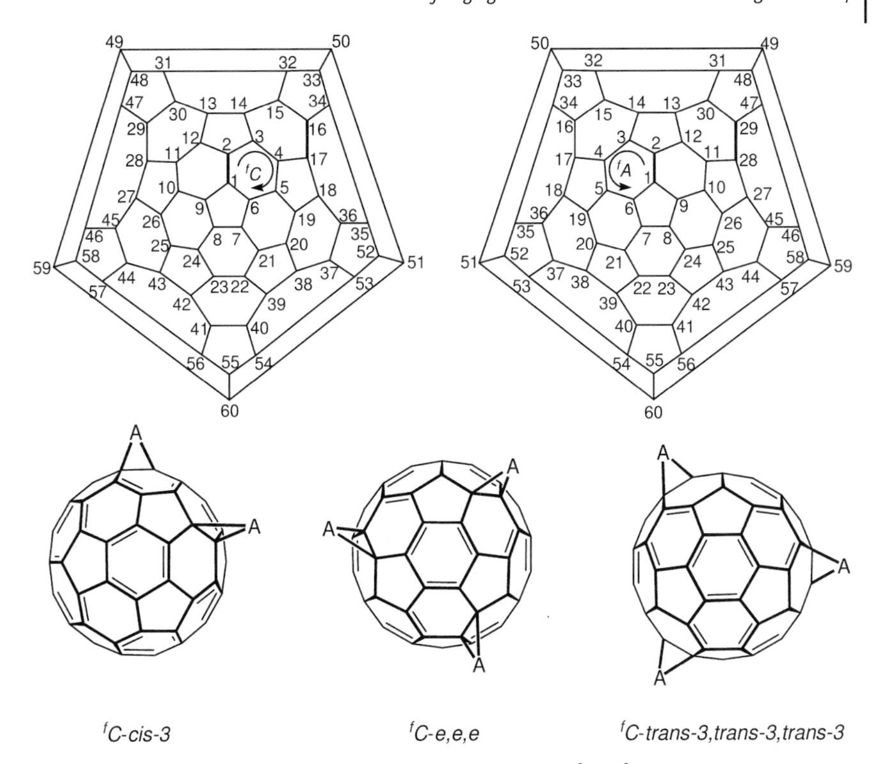

Figure 10.3 Scheme for determining the absolute configuration (^{f}C vs ^{f}A) of chiral C_{60} adducts and assignment of the absolute configuration of adducts.

10.2.1
Subsequent Cycloadditions to [6,6]-double Bonds

Cycloadditions to [6,6]-double bonds of C_{60} are among the most important reactions in fullerene chemistry. For a second attack to a [6,6]-bond of a C_{60} monoadduct nine different sites are available (Figure 10.1). For bisadducts with different but symmetrical addends nine regioisomeric bisadducts are, in principle, possible. If only one type of symmetrical addends is allowed, eight different regioisomers can be considered, since attack to both *e*′- and *e*″-positions leads to the same product. Two successive cycloadditions mostly represent the fundamental case and form the basis for the regioselectivity of multiple additions. In a comprehensive study of bisadduct formations with two identical as well as with two different addends, nucleophilic cyclopropanations, Bamford–Stevens reactions with dimethoxybenzophenone–tosylhydrazone and nitrene additions have been analyzed in detail (Scheme 10.1) [3, 9, 10].

The results can be summarized as: (1) Product distributions (Figure 10.4) are not statistical (in principle: one possibility for a *trans*-1 attack, two possibilities for *e*′- or *e*″-attacks and each of four possibilities for attack to the other *trans*- or *cis*-positions); (2) in most cases *e*-isomers followed by the *trans*-3-isomers are the

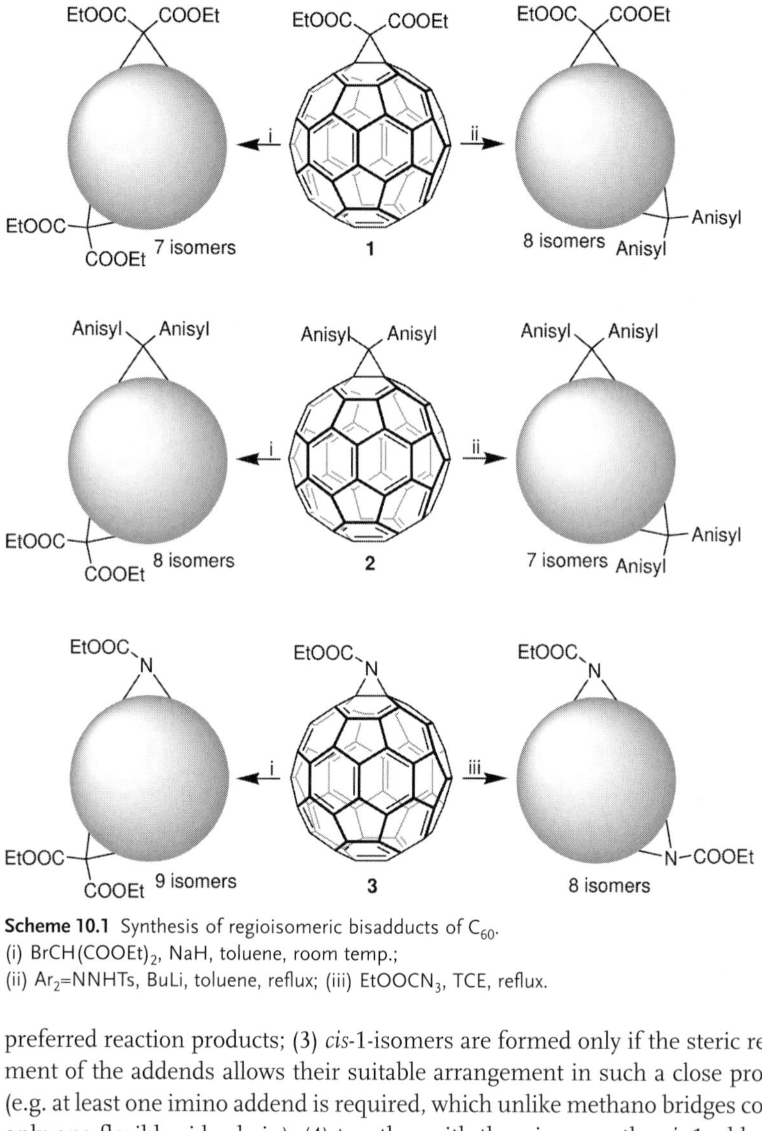

Scheme 10.1 Synthesis of regioisomeric bisadducts of C_{60}.
(i) $BrCH(COOEt)_2$, NaH, toluene, room temp.;
(ii) Ar_2=NNHTs, BuLi, toluene, reflux; (iii) $EtOOCN_3$, TCE, reflux.

preferred reaction products; (3) *cis*-1-isomers are formed only if the steric requirement of the addends allows their suitable arrangement in such a close proximity (e.g. at least one imino addend is required, which unlike methano bridges contains only one flexible side chain); (4) together with the *e*-isomers the *cis*-1-adducts are the major products if their formation is possible at all; (5) an attack to an *e*″-position is slightly preferred over an attack to an *e*′-position; and (6) the regioselectivity of bisadduct formation is less pronounced if more drastic reaction conditions are used [e.g. less regioselectivity for nitrene additions in refluxing 1,1,2,2-tetrachloroethane (TCE) compared with reactions with diethylbromomalonate at room temperature]. Similar product distributions were observed, for example, for twofold additions of diamines [11], for bisosmylations [12], for twofold addition of azomethine ylides [13], for the formation of the tetrahydro[60]fullerenes [14] and for the twofold addition of benzyne [15].

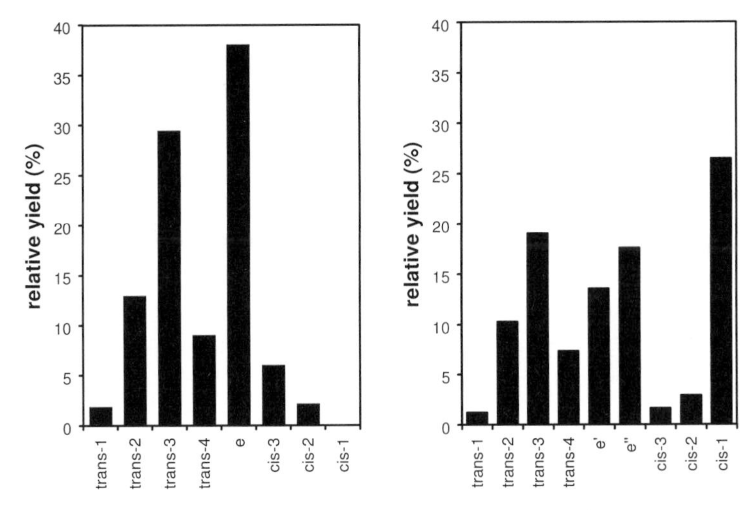

Figure 10.4 Relative yields of isolated regioisomeric bisadducts of $C_{62}(COOEt)_4$ (left) and $C_{61}(COOEt)_2(NCOOEt)$ (right).

Interpretation of these experimental results requires distinction between the properties of the reaction products and their precursor monoadducts [1]. Comparison of AM1-calculated energies of typical series of bisadducts (Table 10.1) shows that (1) in bisadducts with two sterically requiring addends such as dialkoxycarbonylmethylene or diarylmethylene groups *cis*-1-adducts are energetically forbidden; (2) the opposite case is observed in bisadducts with at least one imino addend where the *cis*-1-isomers are the most stable; and (3) all other isomers exhibit very similar stabilities, with the *e*-isomers being slightly stabilized and the *cis*-isomer slightly destabilized. In all cases the *trans*-isomers have about the same calculated heat of formation. As can be seen from space filling models, the instability of a *cis*-1-isomer such as *cis*-1-C_{62}(anisyl)$_4$ (Scheme 10.1) is due to the pronounced steric repulsion of the addends leading to considerable deformations of typical bond angles [10]. Conversely, a strain-free situation is provided if at least one imino addend is present, since in low energy invertomers unfavorable interactions between the addends are avoided. Analogous behavior occurs for the various regioisomers of $C_{60}H_4$ [14] where, except for eclipsing H-interactions, no additional strain due to the addends is present. The corresponding *cis*-1-adduct is the major product formed by a twofold hydroboration followed by hydrolysis. AM1-calculations predict the *cis*-isomers to be somewhat less stable than the *trans*- and *e*-isomers. According to ab initio calculations (HF/3-21G) the *cis*-1- followed by the e-isomer is the most stable. These considerations show that the thermodynamic properties of the bisadducts alone cannot explain the observed regiochemistry [1].

What is the influence of the structural and electronic properties of the precursor adducts on the product distribution? Analysis of various experimental and calculated [6,6] monoadduct structures shows that the bond length alternation between [5,6]- and [6,6]-bonds is preserved [1]. Significantly, independently of the nature of the

Table 10.1 Relative stabilities ($AM1_{HOF}$) in kcal mol^{-1} of the possible regioisomers of bis-adducts with two identical and two different addends.

Positional relationship	$C_{62}(COOMe)_4$	$C_{62}(phenyl)_2$	$C_{60}(NCOOMe)_2$	$C_{61}(COOMe)_2(NCOOMe)$
trans-1	0.2	0.2	4.4	1.3
trans-2	0.2	0.3	4.5	1.2
trans-3	0.1	0.2	4.3	1.1
trans-4	0.0	0.1	4.4	1.1
e′	0.0 [a]	0.0 [a]	4.2 [a]	0.8 [b]
e″	0.0 [a]	0.0 [a]	4.2 [a]	0.8 [c]
cis-3	1.3	2.3	5.9	3.7
cis-2	1.8	3.3	6.7	3.8
cis-1	17.7	24.9	0.0	0.0

a) *e*′- and *e*″-isomers are identical.
b) *e*′-isomer referred to $C_{61}(COOEt)_2$ as precursor molecule.
c) *e*″-isomer referred to $C_{61}(COOEt)_2$ as precursor molecule.

addend, the *cis*-1 bonds are considerably shorter than those of parent C_{60}. A less pronounced contraction is observed for the *e*″-bonds. Another trend to emerge is that the *cis*-2- and *cis*-3-bonds are somewhat elongated and that the opposite hemisphere is less disturbed [1]. Similarly to the geometrical distortions the polarizations (AM1-Mulliken-charges) of the C-framework in these monoadducts is somewhat enhanced in the neighborhood of the first addend but essentially zero in the opposite hemisphere. Computational analysis of frontier orbitals in monoadducts such as $C_{61}(COOEt)_2$ (**1**) revealed the following characteristics [1]: (1) In a first approximation the distribution of the MO coefficients to specific sites within monoadducts is totally independent of the nature of the addend. (2) The distribution of MO coefficients to specific sites within monoadducts is related to that within free C_{60}, as can be seen from the diagrams correlating the HOMOs and LUMOs of C_{60} with those of a monoadduct (Figure 10.5).

The lowest lying HOMO-4 of the monoadduct correlates with the h_u orbital of C_{60} having the highest coefficients in two opposing [6,6]-bonds (HOMO). Relative to this HOMO three of the others have high coeffiecients in the *equatorial* sites and not in the opposing I and I* (*trans*-1) sites (*trans*-1 effect). In the HOMO of the monoadduct, highest coefficients are located in the *cis*-1 and *e*″ site and in the HOMO-1 in the *e*′ followed by the *trans*-3 and *cis*-2 sites. The LUMO of C_{60} with high coefficients in opposing [6,6]-bonds correlates with the LUMO+2 of the monoadduct. Due to the perpendicular orientation of the LUMOs of C_{60} and their correlation with the LUMOs of the monoadduct there are no coefficients in *trans*-1 but preferably in the *e*-positions of the LUMO and LUMO+1 (*trans*-1 effect). In the LUMO, pronounced coefficients are also found in *trans*-3 and *cis*-2. Only in LUMO+2 do the *trans*-1-, *trans*-4- and *cis*-3-bonds also exhibit enhanced coefficients. The differences in [6,6]-bond lengths are influenced by the HOMO coefficients since

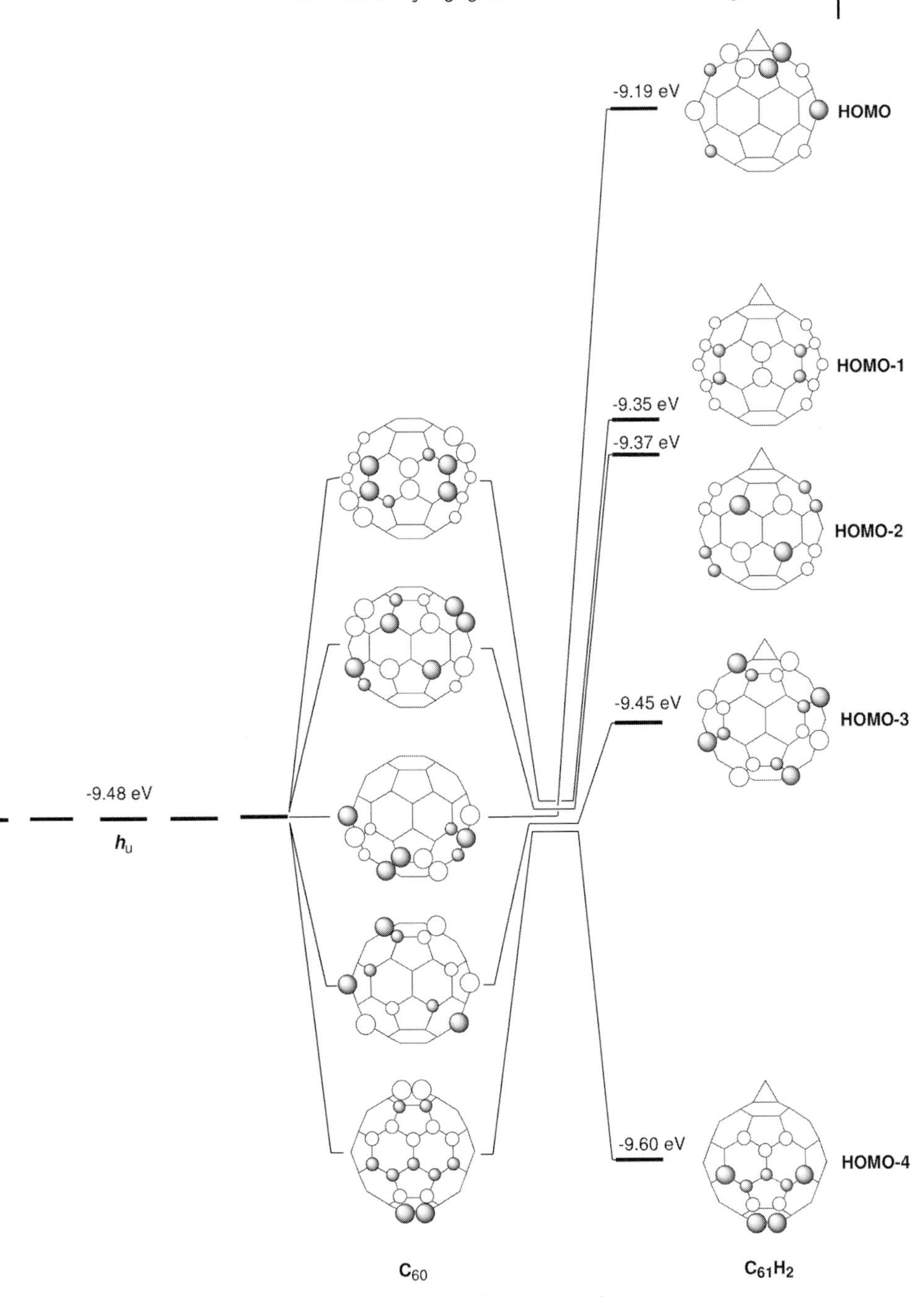

Figure 10.5 (a) Correlation of the HOMOs of C_{60} with the HOMOs of $C_{61}H_2$.

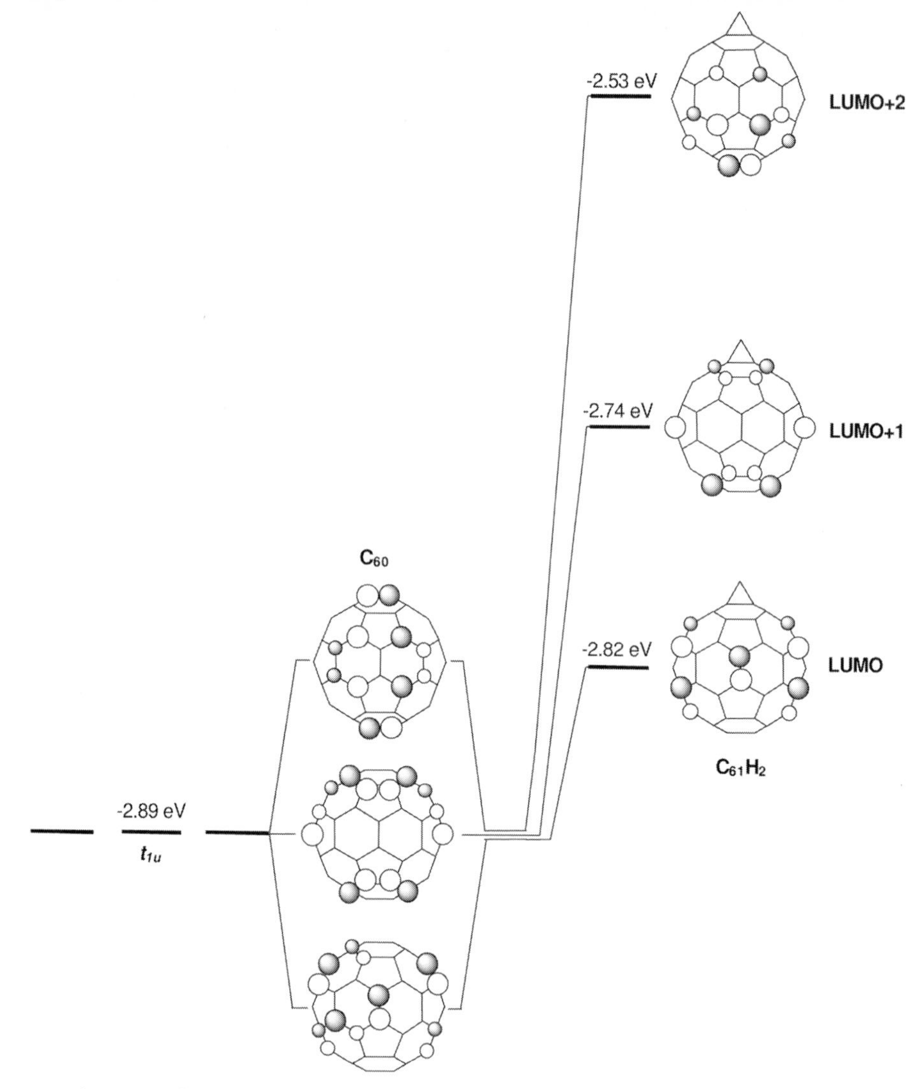

Figure 10.5 (b) Correlation of the LUMOs of C_{60} with the LUMOs of $C_{61}H_2$.

the shortest [6,6]-bonds are those with the highest coefficients in the HOMO. Pronounced bonding interactions at a binding site in general cause a contraction of the bond length. The fact the *cis*-1 bond is shorter than the *e''* bond, although the coefficients in the HOMO are comparable, could be due to the removal of cyclic conjugation within the six-membered ring involving the cis-1 bonds. For electrophilic attacks the highest lying HOMOs (e.g. HOMO and HOMO-1) and for nucleophilic attacks the lowest lying LUMOs (e.g. LUMO and LUMO-1) of the precursor adducts are the most important, and the distribution of their orbital coefficients affects the regioselectivty of a bisaddition. In conclusion, the typical product distributions of twofold additions to [6,6]-bonds, especially the preferred attacks of *e*-sites for sterically demanding and of *e*- and *cis*-1 sites for sterically less demanding addends, correlate with enhanced frontier orbital coefficients in the monoadduct precursors. At the same time the lengths of the corresponding monoadduct double bonds are the shortest, implying pronounced reactivity. The preference of *cis*-1 attacks leading, for example, to *cis*-1- $C_{61}(COOEt)_2(NCOOEt)$ (Scheme 10.1) indicates that not only the nitrene additions but also the cyclopropanations should be mainly HOMO-C_{60} controlled. Here, the cyclopropanation via malonates would be due to addition or carbenes. If it is due to an initial nucleophilic attack, then the preferred formation of *e*-adducts would also be reflected by the distribution of the coefficients of the LUMO and LUMO+1 in the precursor adducts. To explain the pronounced formation of *cis*-1 adducts, however, additional factors have then to be considered, which could be thermodynamic arguments, since for sterically non-demanding addends the *cis*-1-adducts followed by the e-adducts are the most stable. Another driving force could simply be that *cis*-1 bonds are the shortest, having the most double bond character. The extent of pyramidalization (curvature) at a given site, which is a predominant factor governing the preferred bond of an attack, for example, to C_{70} ([6,6]-bonds at the poles, Chapter 13), does not play a role for successive additions to C_{60}. This can be clearly seen from the fact that the reactive *cis*-1 sites are the least pyramidalized in a monoadduct.

For higher adducts of C_{60} the number of possible regioisomers increases dramatically with an increasing number of addends. At the same time it becomes even more difficult to experimentally identify a specific addition pattern. There are few systematic investigations to date of three successive additions of C_{2v}- or C_2 symmetrical addends to [6,6]-double bonds of C_{60} [3, 16–20]. For trisadducts, in principle, 46 different regioisomers are possible. The number of regioisomers that can theoretically be formed starting from a given bisadduct depends on its existing addition pattern. For example, 14 different trisadducts can be produced from a precursor with the addends bound in *e*-positions. However, for many addends, such as malonates, the number of preferably formed regioisomers is considerably smaller, since, for example, additions in *cis*-positions can be neglected. For bis-malonates or related adducts with either *e*- or *trans*-positional relationships, only attacks into *e*- or *trans*-positions to addends already bound have to be considered [20]. The number of allowed addition patterns is then reduced to ten (Table 10.2 see page 300).

Seven of these were found after cyclopropanation of *e*- and *trans-n*-$C_{62}(COOEt)_4$ ($n = 2$–4) (4–7) with diethylmalonate (Figure 10.6) [20]. In a few cases, such as the C_3-symmetric adduct **8** (e,e,e-addition pattern), the structure can be assigned based on NMR spectroscopy alone. NMR spectroscopy allows for the determination of the point group of the adduct. The e,e,e-addition pattern is the only one that has C_3-symmetry and as a consequence the assignment is unambiguous. The same is true for the D_3-symmetrical adduct **9**. Conversely, C_2-, C_s- or C_1-symmetry of trisadducts can arise from different addition patterns. The structural assignment of such adducts requires the additional analysis of their possible formation pathways.

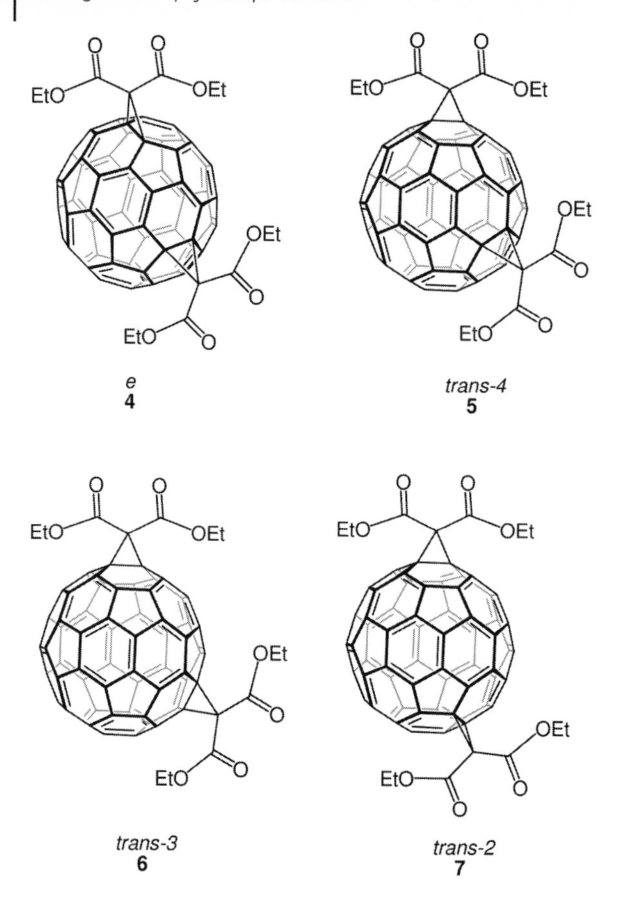

EtO OEt

^{f}C-I,eI,eII
8

^{f}C-I,III*,III*
9

Figure 10.6 Trisadducts obtained from cyclopropanation of bisadducts **4–7**.

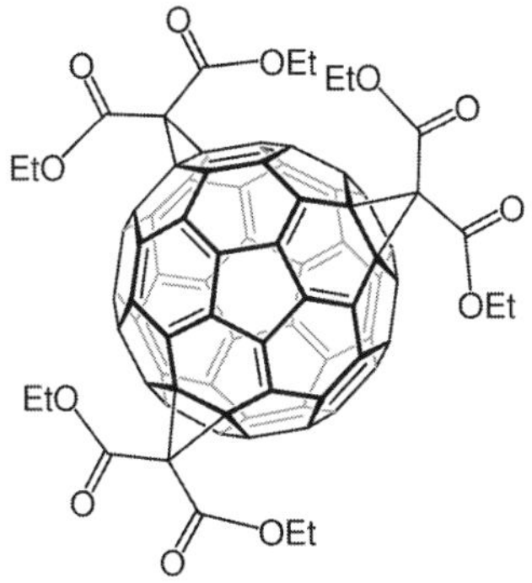

^{f}C-I,eI,III*
10

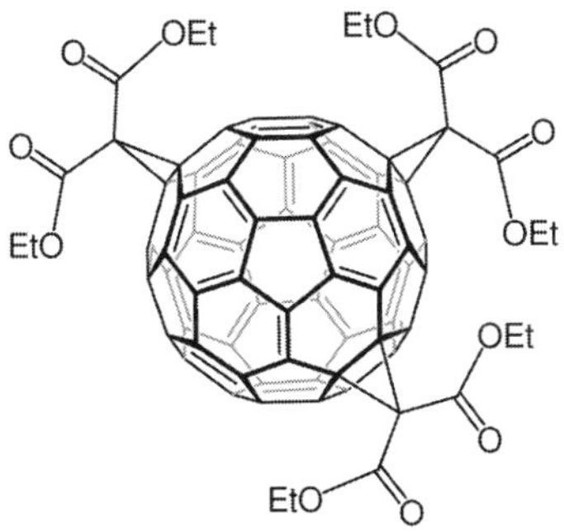

^{f}C-I,eI,IV*
11

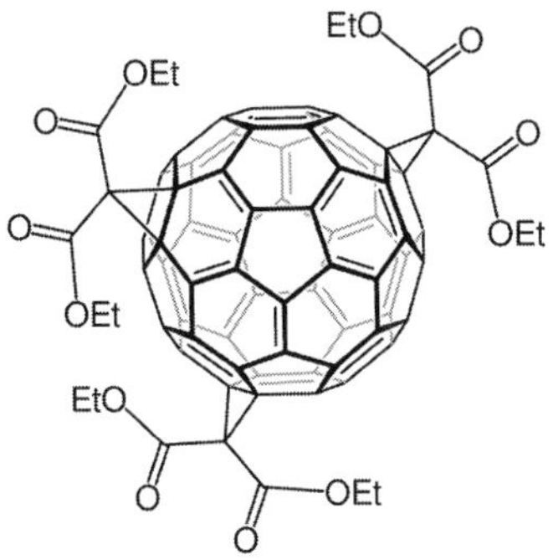

^{f}C-I,eII,III*
12

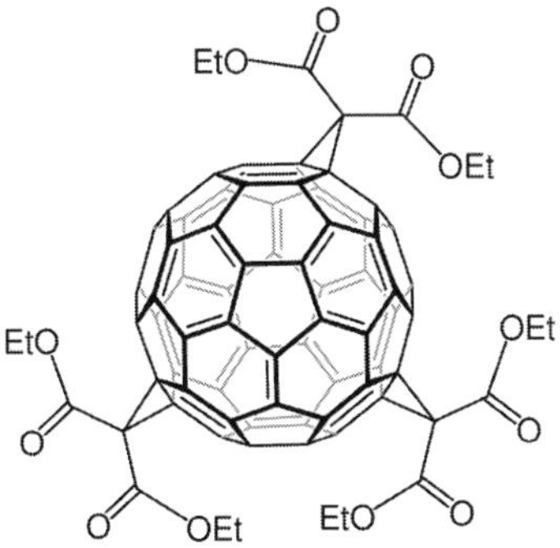

^{f}C-I,IV^{*1},IV^{*3}
13

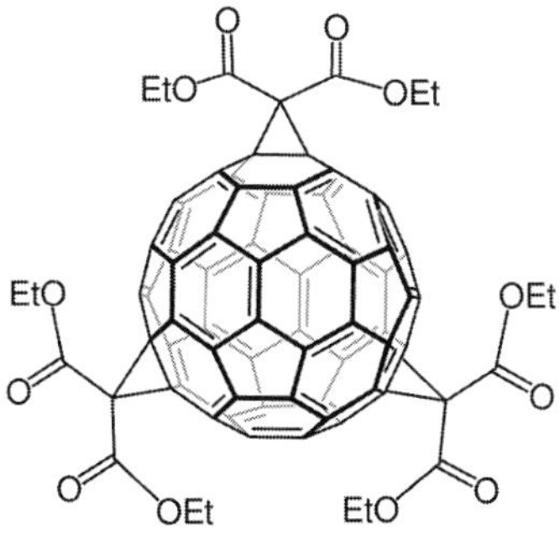

I,IV*,III*
14

Figure 10.6 (continued)

Table 10.2 Relative and absolute positional relationships, symmetry and number of possible formation pathways of trisadducts that can be formed out of *e* and *trans-n* ($n = 1–4$) bisadducts (first row) neglecting cis additions.

trans-1 ***I, I****	***trans-2*** ***I, II****	***trans-3*** ***I, III****	***trans-4*** ***I, IV****	***e*** ***I, eI***
e, e, **t-1** (I) [a)] I, eI, I* C_s 2	e, t-4, **t-2** I, eI, IV* C_1 2	e, t-4, **t-3** I, eI, III* C_1 2	e, **t-4**, t-2 I, eI, IV* C_1 2	**e, e, e** I, eI, eII C_3 2
e, e, **t-1** (II) I, eII, I* C_s 2	e, t-3, **t-2** I, eII, III* C_1 2	e, **t-3**, t-2 I, eII, III* C_1 2	e, **t-4**, t-3 I, eI, III* C_1 2	**e**, t-4, t-2 I, eI, IV* C_1 2
	t-4, t-4, **t-2** I, IV*^{1}, IV*^{3} C_2 1	t-4, **t-3, t-3** I, IV*, III* C_s 1	**t-4, t-4**, t-2 I, IV*^{1}, IV*^{3} C_2 2	**e, e**, t-1 (I) I, eI, I* C_s 1
		t-3, t-3, t-3 I, III*, III* D_3 1	**t-4, t-4,t-4** I, IV*^{1}, IV*^{4} C_{3v} 1	**e**, t-4, t-3 I, eI, III* C_1 2
			t-4, t-3, t-3 I, IV*, III* C_s 1	**e**, t-3, t-2 I, eII, III* C_1 2
				e, e, t-1 (II) I, eII, I* C_s 1

a) t = trans.

For example, adduct **12** was formed from *e*-**4**, *trans-3*-**6** and *trans-2*-**7**. Consequently, it must involve these three positional relationships. Therefore, its structural assignment is unambiguous. Similarly, various trisadducts carrying C_2-symmetrical bis(oxazoline) addends could be isolated and structurally assigned [19, 20]. The regioselectivities of these cyclopropanations strongly depend on the precursor bisadduct. Whereas, for example, all possible trisadducts **9**, **10**, **12**, **14** that can be obtained from *trans-3*-**6** were formed in about equal amounts the cyclopropanation of *trans-4*-**5** is more selective. Among the four isolated isomers **10** was the most abundant. The fifth isomer, with a C_{3v}–symmetrical *trans*-4, *trans*-4, *trans*-4- addition pattern, was not found.

Only three of the six possible trisadducts were obtained upon cyclopropanation of *e*-**4**. Especially, the adducts with the addition patterns *e,e,*-trans-1 (I) and *e,e,*-*trans*-1 (II) did not form [20]. Access to corresponding adducts requires *trans*-1 precursors and/or tether strategies (Section 10.3.3). Although the relative yield of *e,e,e*-**8** constitutes about 35–40% of the trisadducts formed and the yield of *e,trans-3,trans-2*-**12** is higher, the preferred mode of addition is *e* relative to the bound addends. Under the reaction conditions, substantial amounts of tetrakis- and pentakisadducts were formed by further additions into e-positions. The overall yield of such adducts formed by successive attacks into e-positions amounts to about 50%. The preferred cyclopropanation of *e*-**4** leading to *e,e,e*-**8** and *e, trans-3, trans-2*-**12** correlated well with enhanced coefficients of both the HOMO and the LUMO in *e*- and *trans*-3 positions.

A $C_{60}H_6$ isomer with a D_3 symmetrical all *trans*-3 addition pattern was found as the preferred adduct upon the hexahydrogenation of C_{60} via Zn-Cu acid reduction [21].

Further cyclopropanation of *e,e,e*-**8** leads to the C_s-symmetrical tetrakisadduct **15** and the C_{2v}-symmetrical pentakisadduct **16** and the T_h-symmetrical hexakisadduct **17** (Figure 10.7) [16].

Significantly, next to **15** only one other tetrakisadduct with C_1-symmetry is formed. The relative yield of **15** is 64%. Cyclopropanations of **15** and **16** are even more regioselective. The regioisomers **16** and **17** are the only pentakis- and hexakisadducts formed. The more addends bound in *e*-positions (octahedral sites) the more regioselective is an attack. Finally, the all-over regioselectivity of subsequent cyclopropanations is surprisingly high considering that in principle 316 regio-isomeric hexakisadducts without *cis*-1 positions can be formed. Again all precursor compounds **8**, **15**, and **16** exhibit pronounced frontier orbital coefficients at those *e*-positions that were preferably cyclopropanated [1]. Moreover, the very high regioselectivity leading to **16** and **17** is also due to thermodynamic effects.

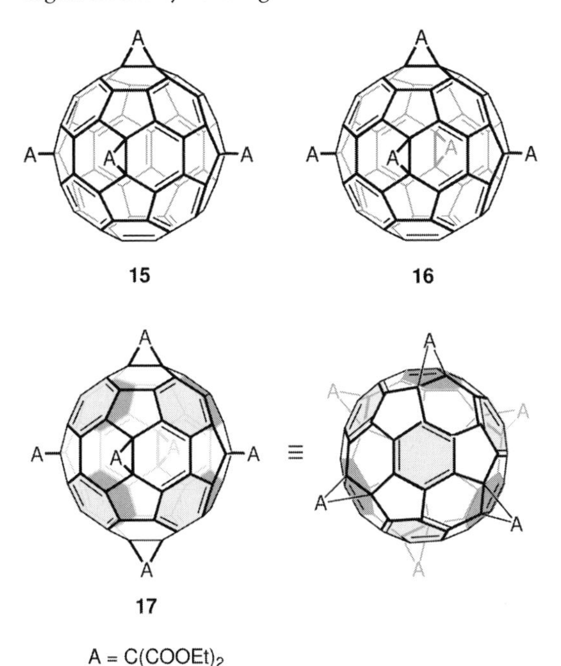

Figure 10.7 Tetrakis-, pentakis- and hexakisadducts of C_{60} formed by successive cyclopropanation of octahedral [6,6]-double bonds of the *e,e,e*-trisadduct **8**.

The regioisomer **17** is stablilized by at least 5 kcal mol^{-1} (AM1) compared with all other 43 hexakis-adducts, which in principle can be formed out of **8** without introducing unfavorable *cis*-1 positions [16].

Adduct **17** was the first purely organic prototype of such hexakisadducts involving a T_h-symmetry, which is unprecedented in organic chemistry. T_h-Symmetrical hexaadducts have also been synthesized via sixfold Diels–Alder reactions in comparatively high yields by allowing an access of dienes to react with C_{60} [22]. However, the latter results imply that reversible reactions are involved, allowing for a thermodynamic control. The sterically demanding tetramethyl azomethine ylide, generated from dry acetone and 2,2-dimethylglycine in chlorobenzene, showed high selectivity in the stepwise series of [3+2] cyclizations leading to the T_h symmetrical hexakisadduct **18** in 12% yield [23]. This compound gives rise to an intense, bright orange phosphorescence, with a lifetime of 4.4 s, when this solution was cooled to 77 K [23, 24].

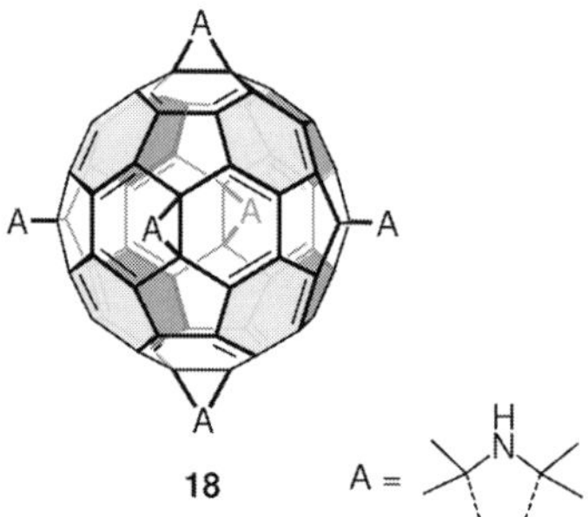

A A A A A A

18 A = H N

The single crystal structure of **17** [25] confirms computational results. Very remarkable is the bonding in the remaining π-electron system – a new type of an oligocyclophane (Figure 10.8). The alternation of bond lengths between [6,6]- and [5,6]-bonds within the benzenoid rings is reduced by half to approximately 0.03 Å compared with the parent C_{60}. Relative to solutions of C_{60} (purple) or its adducts $C_{61}(COOEt)_2$ to $C_{65}(COOEt)_{10}$ (red to orange) the light yellow solutions of **17** show only weak absorptions in the visible region. Hexakisadducts with other addends and mixed adducts have similar optical properties [26].

10.2.2
Adducts with an Inherently Chiral Addition Pattern

The I_h-symmetry of C_{60} is reduced if one or several addends are attached, for example, to the [6,6]-double bonds [1]. The addition pattern of the resulting adduct always corresponds to a subgroup of the I_h point group. Among the possible subgroups of I_h several are chiral, for example the D_3, C_3, C_2 and C_1 point groups (Figure 10.9). An addition pattern of a C_{60} adduct is inherently chiral if it belongs to one of these point groups [3, 8, 19, 27–29].

Various bis-and tris-adducts discussed so far, such as **6–13**, belong to one of these subgroups. However, when achiral addends such as malonates are used the

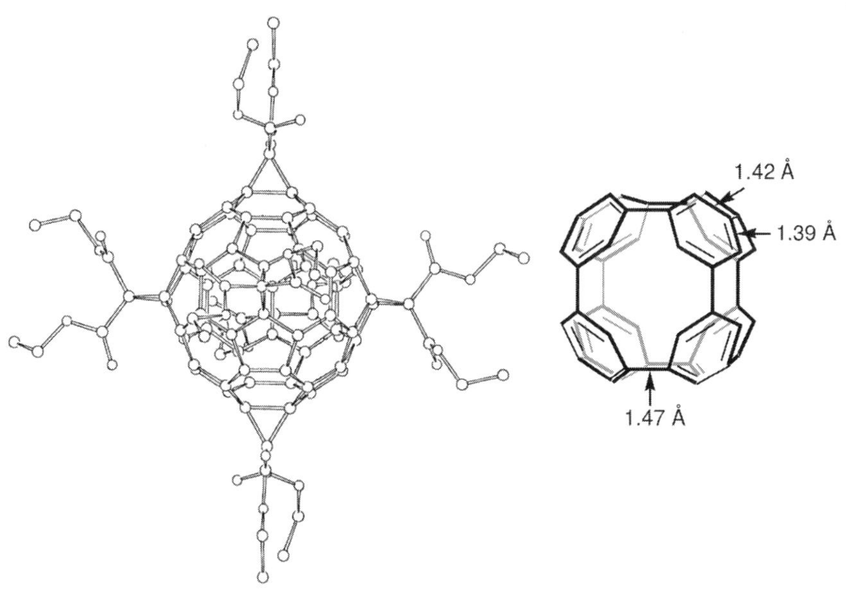

Figure 10.8 Single-crystal X-ray analysis of T_h-symmetrical $C_{66}(COOEt)_{12}$ (**17**) and cyclophane substructure of the remaining π-system of eight benzenoid rings [25].

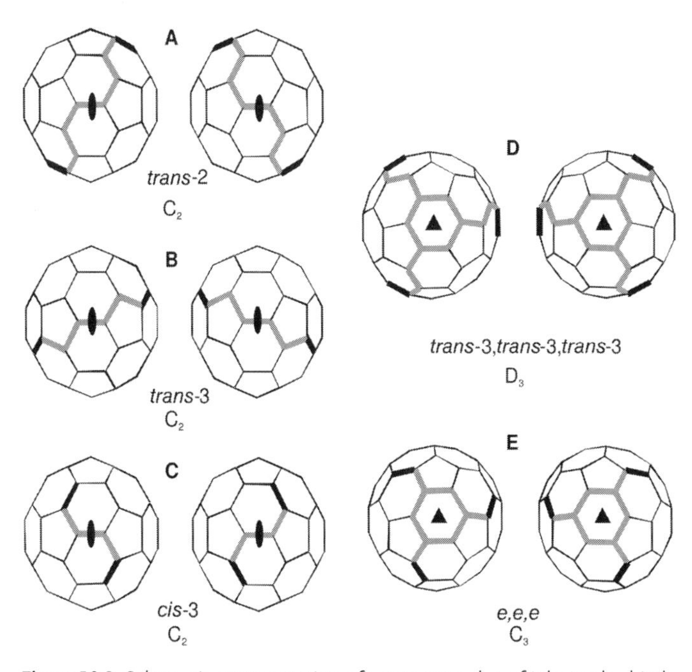

Figure 10.9 Schematic representation of some examples of inherently chiral bis- and tris-addition patterns of C_{60} adducts.

corresponding adducts are isolated as racemic mixtures. Enantiomer separation of various adducts was successful by HPLC using the chiral Welk-O1 phase on a semipreparative scale [30]. In a more efficient approach cyclopropanation of C_{60} or C_{60}-adducts was carried out with C_2-symmetrical bis(oxazolines) [19, 20]. Here, inherently chiral bis- or trisadducts such as **19** and **20** were formed as pairs of diastereomers.

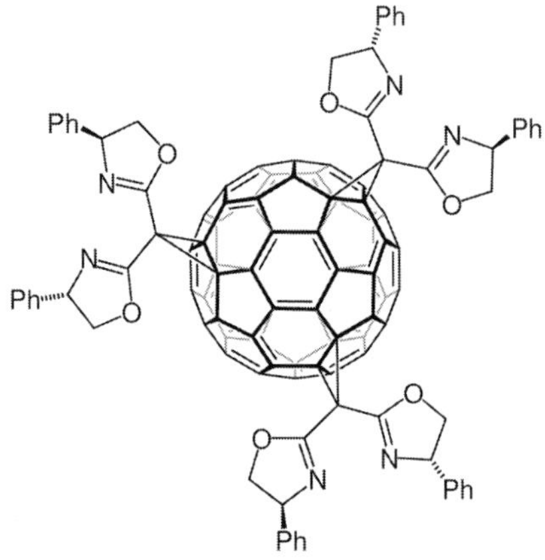

^{f}C - I, e I, e II
19

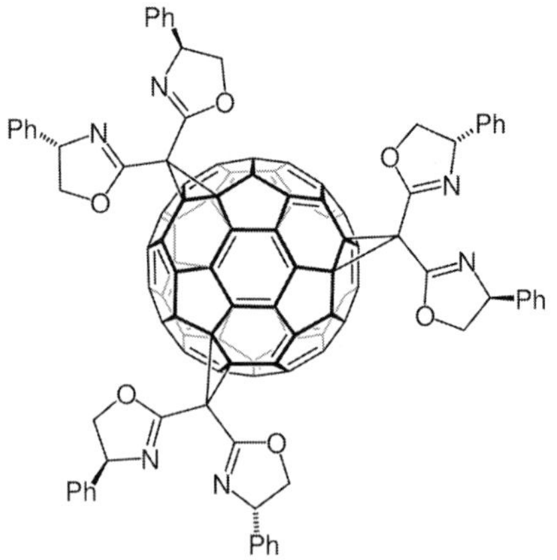

^{f}A - I, e I, e II
19

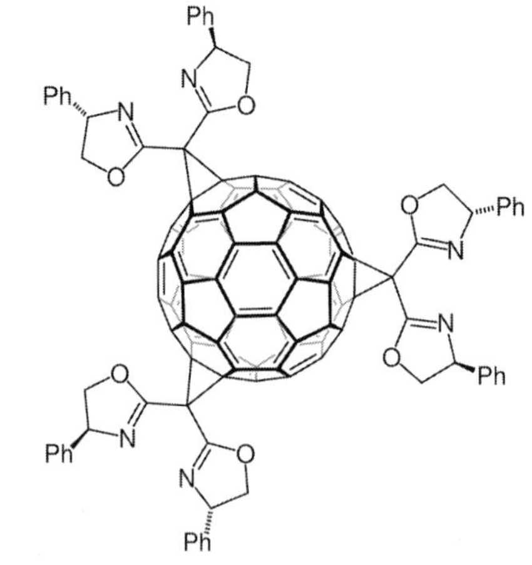

^{f}C-I,III*,III*
20

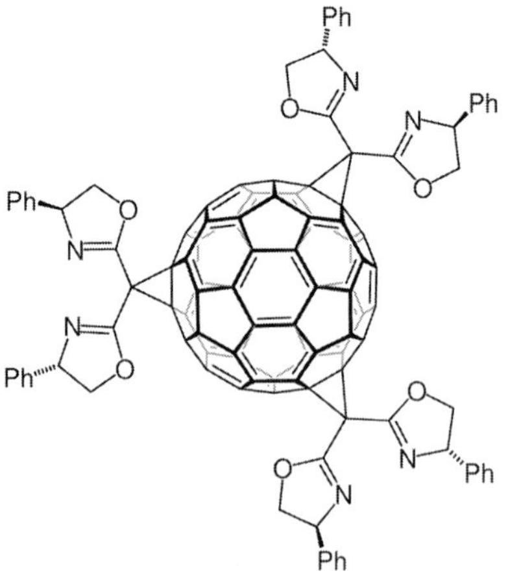

^{f}A-I,III*,III*
20

Their isolation by flash chromatography on silica gel was comparatively easy. The CD spectra of related pairs of diastereomers whose addition pattern represent pairs of enantiomers, reveal pronounced Cotton effects and mirror image behavior. It is the chiral arrangement of the conjugated π-electron system within the fullerene core that predominantly determines the chiroptical properties. Adducts with a C_2-

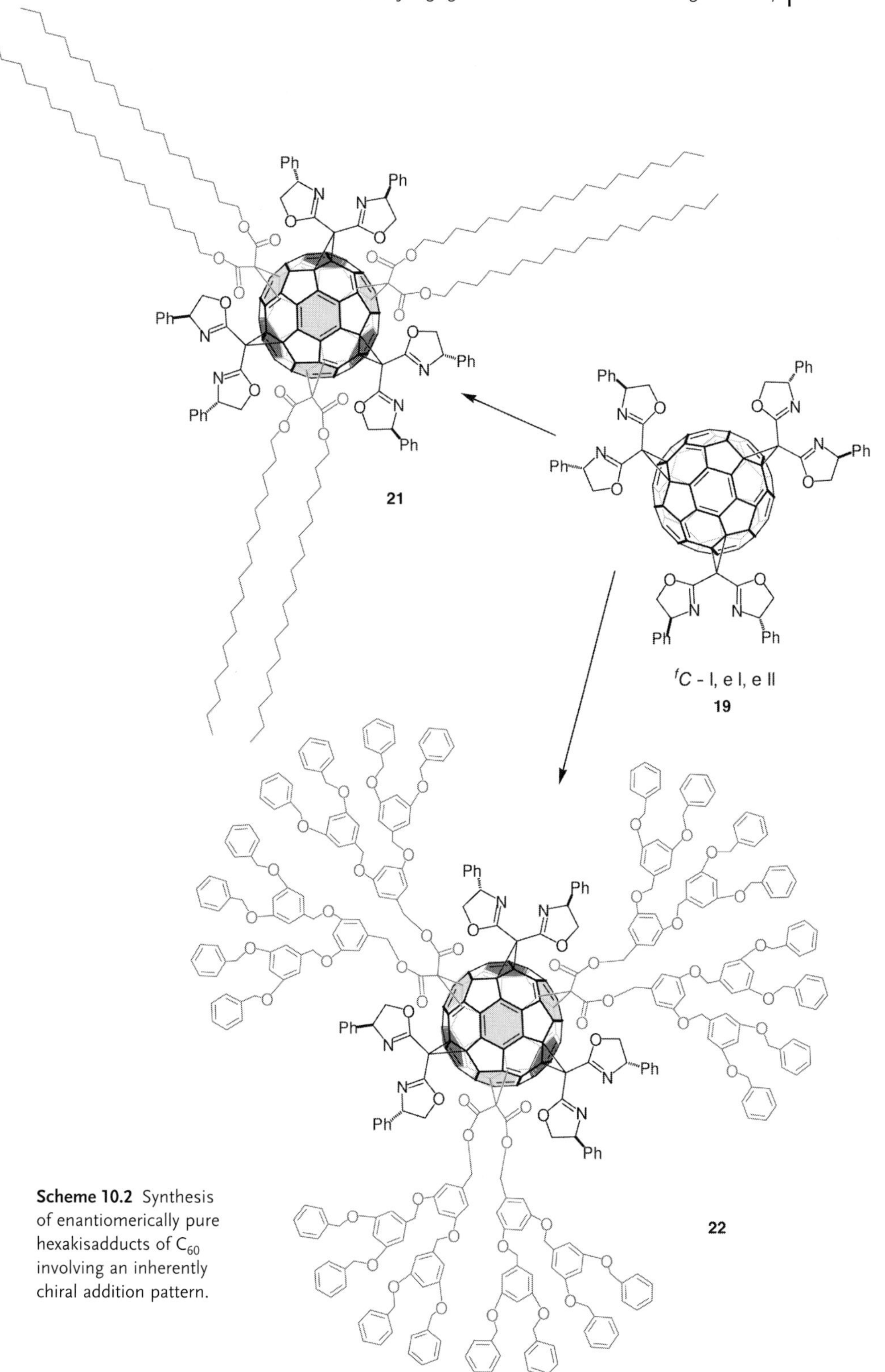

Scheme 10.2 Synthesis of enantiomerically pure hexakisadducts of C_{60} involving an inherently chiral addition pattern.

symmetrical addition pattern show significantly larger Cotton effects than those with more symmetrical addition patterns (C_3, D_3). The absolute configuration of a large variety of enantiomerically pure adducts with an inherently chiral addition pattern [27] have been determined either by comparison of the calculated and experimental CD spectra of the bis(oxazoline) adducts or with knowledge of the absolute configuration of the chiral bis-adduct precursors containing bis(oxazoline) addends [20]. Enantiomerically pure *e,e,e*-adducts with known absolute configuration such as **19** have been used as precursors for the synthesis of fullerene dendrimers and lipofullerenes with an inherently chiral addition pattern (Scheme 10.2) [31].

The relative configuration within the precursor trisadducts is retained in the inherently chiral C_3-symmetrical hexakisadducts, while the absolute configuration has changed. The Cotton effects of such hexakisadducts are much less pronounced than those of their precursor tris-adducts, since the local symmetry of the fullerene core is very close to T_h.

10.2.3
Vinylamine Mode

The introduction of two [5,6]-aza bridges shows a remarkable regioselectivity even if segregated alkylazides are used [17]. The diazabishomofullerenes **23** (Scheme 10.3) are by far the major products and only traces of one other bisadduct with unidentified structure are found if, for example, a twofold excess of azide is allowed to react with C_{60} at elevated temperatures [17]. To obtain clues on the mechanism of this most regioselective bisadduct formation process in fullerene chemistry a concentrated solution of an azahomofullerene precursor **24** was treated with an alkyl azide at room temperature.

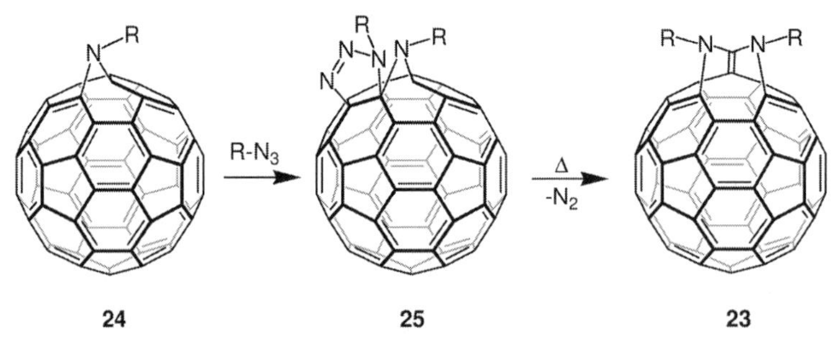

Scheme 10.3 Regioselective formation of diazabishomofullerenes **23**.

Under these conditions only one mixed [6,6]-triazoline-azahomofullerene isomer **25** was formed, which is explained by the fact that **24** behaves as a strained electron-poor vinylamine. The significantly highest Mulliken charge of 0.06 (AM1) is located at C-1 and C-6, and the lowest of –0.07 at C-2 and C-5 (Figure 10.10). The most negatively polarized N atom of the azide (AM1) is that bearing R. A kinetically controlled attack of the azide, therefore, leads predominantely to **25**.

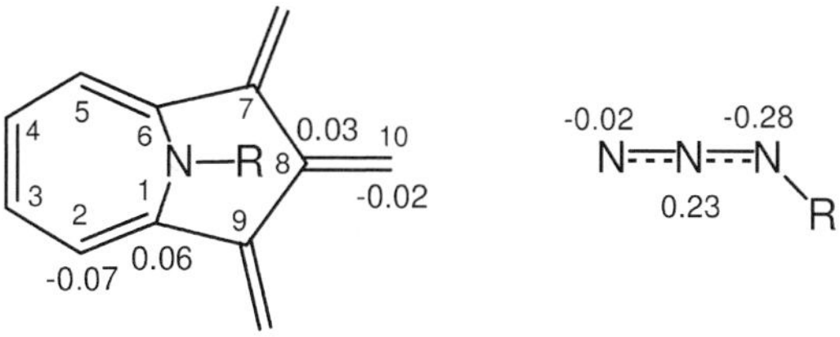

Figure 10.10 AM1 Mulliken charges near the imino bridge of **24** and in N_3R ($R = CH_2COOMe$).

The further ring-expanded doubly bridged diazabishomofullerenes **23** exhibit three seven-membered and one eleven-membered rings within the fullerene cage. One C atom is already halfway decoupled from the spherical carbon core. The pronounced relative reactivity of the vinylamine type double bonds within [5,6]-bridged azahomofullerenes was also demonstrated by the facial addition of mild nucleophiles such as water and amines. For example, the first stable fullerenol was synthesized by treating a toluene solution of **24** in the presence of water and neutral alumina in an almost quantitative reaction [32, 33]. The finding that the reactivity of the vinylamine type [6,6]-double bonds in **24** is dramatically enhanced over the remaining [6,6]-bonds turned out to be a key for further cluster-opening reactions and the formation of nitrogen heterofullerenes (Chapter 12). Subsequent attacks to [6,6]-bonds within [6,6]-adducts and also to [6,6]-bonds within the [5,6]-bridged methanofullerenes are by far less regioselective [1, 34]. Twofold addition of a bisazide bridged with a flexible oligoethylene glycol or alkyl tether afforded bridged systems involving the same addition pattern [35, 36].

Diazabishomofullerenes with another addition pattern can only be obtained in good yields if a tether directed synthesis is applied. If the tether between two azide groups is sufficiently rigid a second addition is forced to occur at specific regions of the fullerene cage [36, 37].

10.2.4
Cyclopentadiene Mode

C_{60} reacts with free radicals [38, 39] amines [40, 41], iodine chloride [42] or bromine [43], specific organocopper [44] and organolithium reagents [45] to give predominantly, or almost exclusively, substituted 1,4,11,14,15,30-hexahydro[60]fullerenes (Figure 10.11). In these hexa- or pentaadducts of C_{60} the addends are bound in five successive 1,4- and one 1,2-positions. The corresponding corannulene substructure of the fullerene core contains an integral cyclopentadiene moiety whose two [5,6]-double bonds are decoupled from the remaining conjugated π-electron system of the C_{60} core.

This characteristic structure type is completely different from those polyadducts of C_{60} described in Section 10.2.1. In many cases the detailed mechanisms of these polyadduct formations remain unclear; however, presumably, sequential additions of radicals or nucleophiles such as the cuprate Ph_2Cu^- are involved. It is very reasonable to assume that the subsequent addition of nucleophiles is accompanied

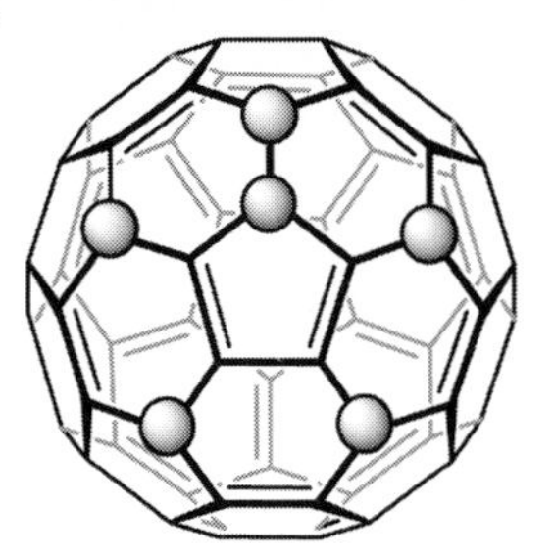

Figure 10.11 Schematic representation of the VB-structure of a 1,4,11,14,15,30-hexahydro[60]fullerene derivative involving a cyclopentadiene substructure (cyclopentadiene addition pattern).

by oxidation of the negatively charged intermediates by air oxygen. Valuable insight of multiple additions of free radicals came from the ESR spectroscopic investigations of benzyl radicals, ^{13}C-labeled at the benzylic positions (see Chapter 6) [38, 39]. The mechanism of the stepwise addition is shown in Scheme 10.4.

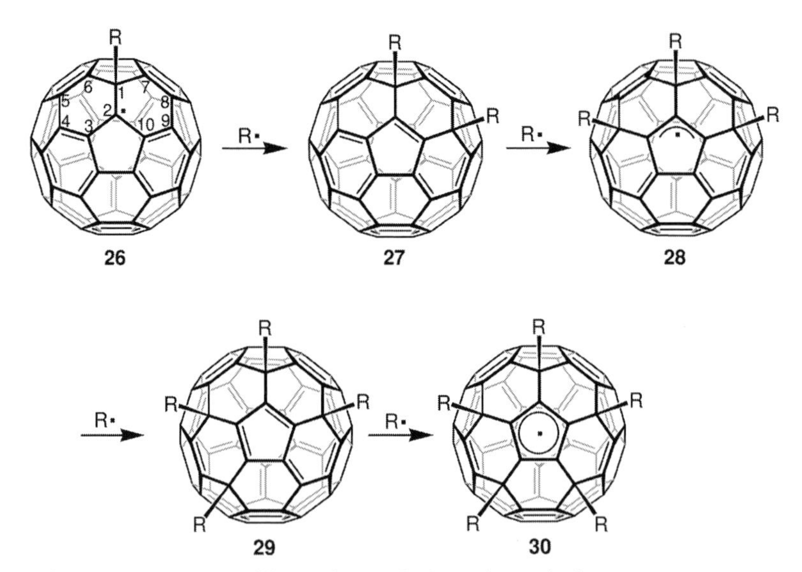

Scheme 10.4 Successive addition of sterically demanding radicals to C_{60} – the cyclopentadiene mode.

A stable representative **31** of an intermediate **29** was obtained in 40% yield by Komatsu upon reacting C_{60} with potassium fluorenide in the presence of neutral fluorene without rigorous exclusion of air (Scheme 10.5) [45].

The single-crystal structure of **31** clearly reflects the presence of an integral fulvene-type π-system on the spherical surface. The average bond length for the [5,6]-bond between C1-C2 and C3-C4 is 1.375 Å, which is considerably shorter than a typical [5,6]-bond in C_{60}. In contrast, the bond between C2 and C3 (1.488 Å) is notably

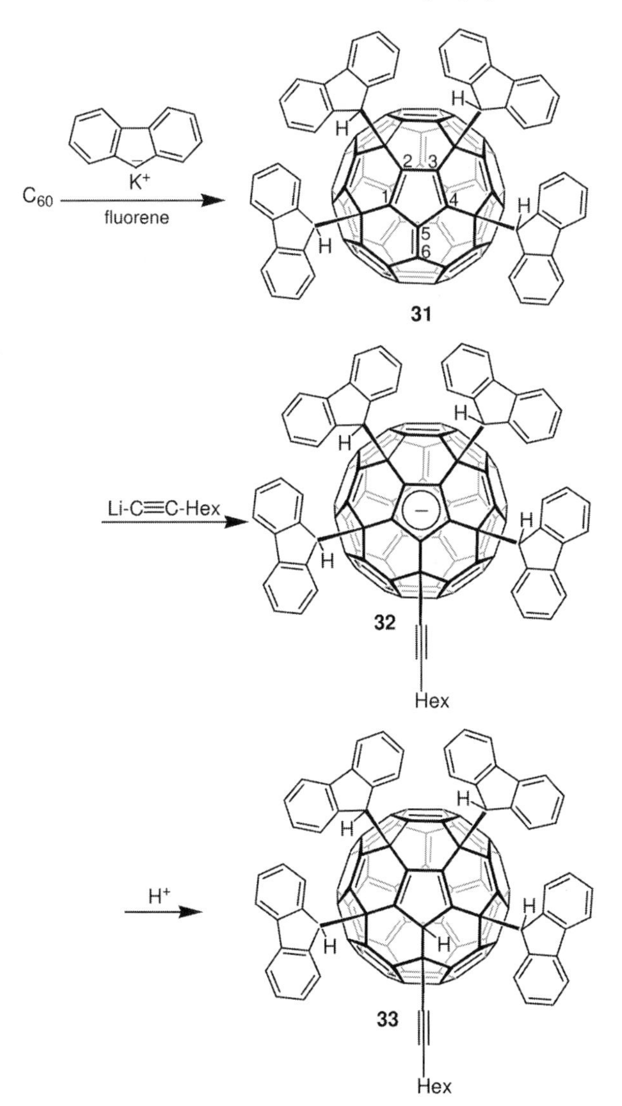

Scheme 10.5 Selective formation of C_{60} adducts with integral fulvene, cyclopentadienide and cyclopentadiene moieties.

longer. The fulvene C5–C6 bond length (1.350 Å) is the shortest reported for C_{60} derivatives. This tendency for the bonds to alternate in length is comparable to that observed in fulvene derivatives. Typical fulvene reactivity towards nucleophiles was demonstrated with the reaction of **31** with 1-octynyllithium. After quenching with acid, hexaadduct **33** with the *cyclopentadiene* addition pattern was obtained. The reaction intermediate is the cyclopentadienide **32** (Scheme 10.5). The results of this beautiful preparative fullerene chemistry support the earlier ESR investigations (Chapter 6).

A pentahaptofullerene metal complex (**34**) has been obtained upon the reaction of the appropriate hexakisadduct with TlOEt, demonstrating the acidity of the *cyclopentadiene* (Scheme 10.6) [44]. A single-crystal investigation confirmed the cyclopentadienide character of the complexed five-membered ring. The [6,6]- and [5,6]-bond lengths in the remaining C_{50} moiety are similar to those of monofunctionalized C_{60} derivatives. Next to **34** the transition metal-cyclopentadienide complex $Rh(\eta^5$-$C_{60}Me_5)(CO)_2$ has also been synthesized and characterized by X-ray crystallography [46]. Hybrids of ferrocene and fullerene have been synthesized by treating $C_{60}HMe_5$ [47] or $C_{70}HMe_3$ with $[FeCp(CO)_2]_2$ [47]. Recently, a whole series of adducts involving the same addition pattern have been synthesized, some of which exhibit remarkable supramolecular and materials properties [48–54].

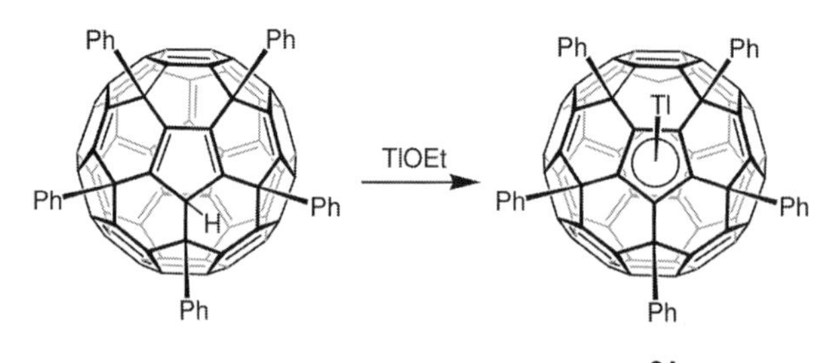

Scheme 10.6 Formation of a cyclopentadienide complex of a C_{60} derivative with Tl.

Upon reaction of C_{60} with MeLi a bis-epoxide bridged derivative of this cyclopentadiene adduct type has been isolated and characterized by single-crystal X-ray analysis [55]. Both epoxide groups bridge [5,6]-bonds in the same pentagon.

Penta- or hexaalkyl derivatives involving the cyclopentadiene-addition pattern have also been synthesized via electrophilic substitutions by $C_{60}Cl_6$ [56]. For electrophilic substitutions of aromatics, pentaaryl addended derivatives, $C_{60}Ar_5Cl$ [56, 57], were usually obtained as major reaction products. However, $C_{60}Cl_6$ reacts with allyltrimethylsilane in the presence of $TiCl_4$ to afford the hexaallyl addended $C_{60}(CH_2CH{=}CH_2)_6$ as the major reaction product. The analogous *heterocyclopentadiene* addition pattern was obtained by reacting the heterofullerene monoadducts with ICl (see Chapter 12).

10.3 Concepts for Regio- and Stereoselective Multiple Functionalization of C_{60}

10.3.1 Template Mediation Approaches

10.3.1.1 T_h-Symmetrical Hexakisadducts

To improve the yield of six-times cyclopropanated adducts such as T_h-$C_{66}(COOEt)_{12}$ (**17**) a very efficient one-pot method has been developed (Scheme 10.7) [1, 25, 26, 58, 59]. The lynchpin of this strategy was the discovery that 9,10-dimethylanthracene

(DMA) binds reversibly to C_{60}. A similar effect was observed for 2,6-dimethoxy-anthracene [60]. Use of, for example, a ten-fold excess of DMA results in an equilibrium between the various $C_{60}DMA_n$ adducts, with *e,e,e*-$C_{60}DMA_3$ as the main component. Hence, synergetic combination of kinetic and thermodynamic control results in the generation of templates such as *e,e,e*-$C_{60}DMA_3$, with incomplete octahedral addition patterns. Since (a) attack of irreversibly binding addends onto such templates occurs with highly pronounced regioselectivity at free octahedral sites, (b) facile rearrangement of DMA addends is possible in *wrong* intermediates, resulting in the formation of an octahedral isomer, and (c) the reversibly bound DMA molecules can easily be replaced by the desired addends, the yields of hexakisadducts such as **17** can be as high as 50%. Another important improvement was the in situ formation of the bromomalonates, by DBU-initiated reaction between the corresponding parent malonate and CBr_4 [59]. Consequently, a broad variety of easily available malonates can be used directly to synthesize large quantities of T_h-$C_{66}(COOR)_{12}$ [26]. A few examples of such adducts with remarkable properties deserve to be mentioned.

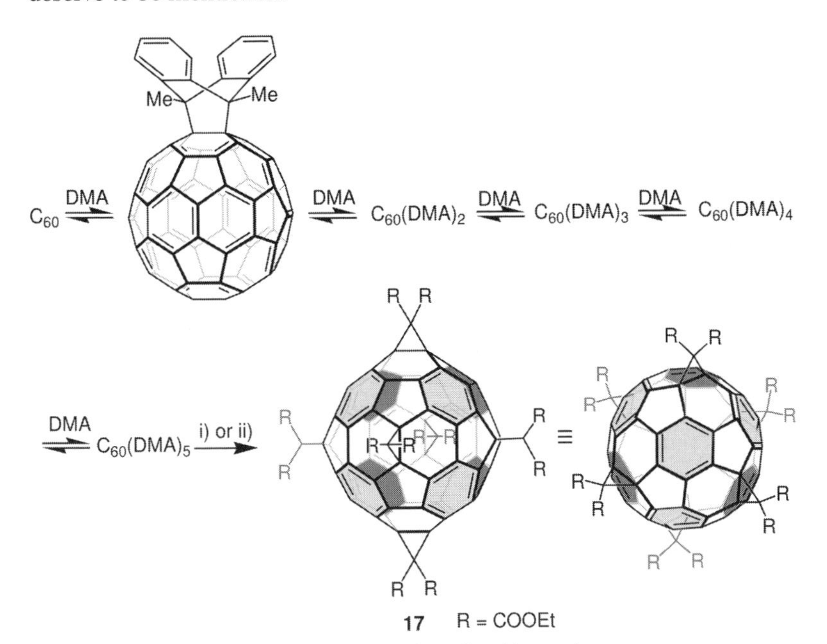

Scheme 10.7 Template mediated synthesis of hexakisadducts of C_{60} involving an octahedral addition pattern using DMA as equilibrating addend. (i) Diethyl bromomalonate in the presence of DBU; (ii) in situ formation of bromomalonate using diethyl malonate and CBr_4–DBU.

Lipofullerenes such as **35–37** self-assemble within lipid bilayers into rod-like structures of nanoscopic dimensions [61, 62]. These anisotropic superstructures may be important for future membrane technology. Significantly, lipofullerenes **35** and **37** have very low melting points, 22 and 67 °C (DSC, heating scan), respectively, with **35** being the first liquid fullerene derivative at room temperature.

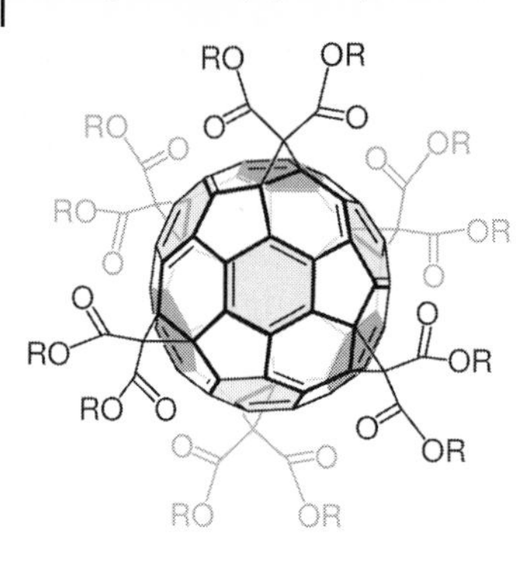

35 R =
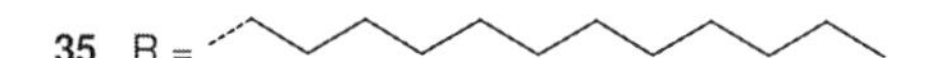

36 R =

37 R =

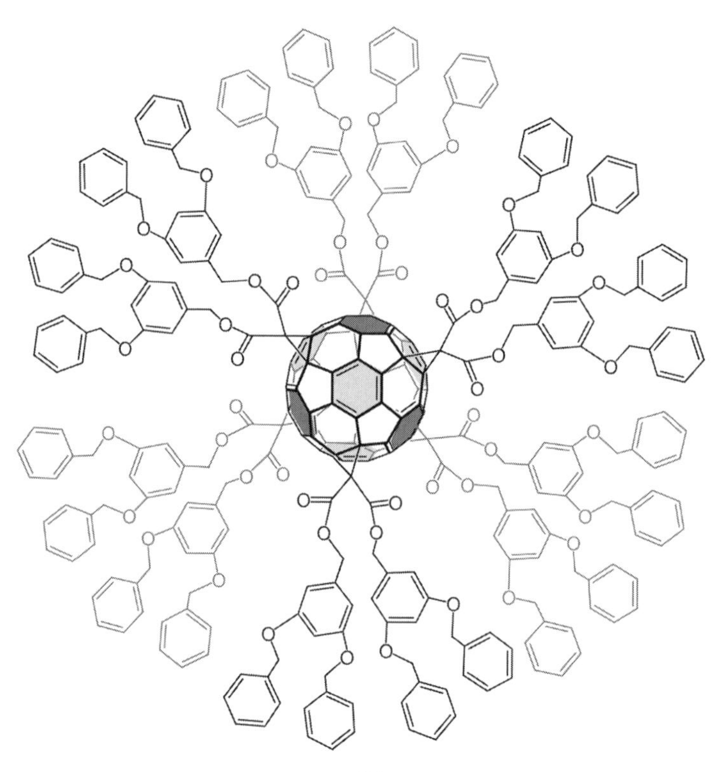

38

The T_h-symmetrical hexaaddition pattern also represents an attractive core tecton for dendrimer chemistry [26, 31, 63–67]. Examples for such dendrimers, involving a core branching multiplicity of 12, are **38** and **39** [63, 64]. Addition of six mesotropic cyanobiphenyl malonate addends produced the spherical thermotropic liquid crystal **40** [65]. DSC and POM investigations revealed a smectic A phase between 80 and 133 °C. Interestingly, this spherical and highly symmetrical compound gives rise to liquid crystallinity despite the absence of molecular anisotropy.

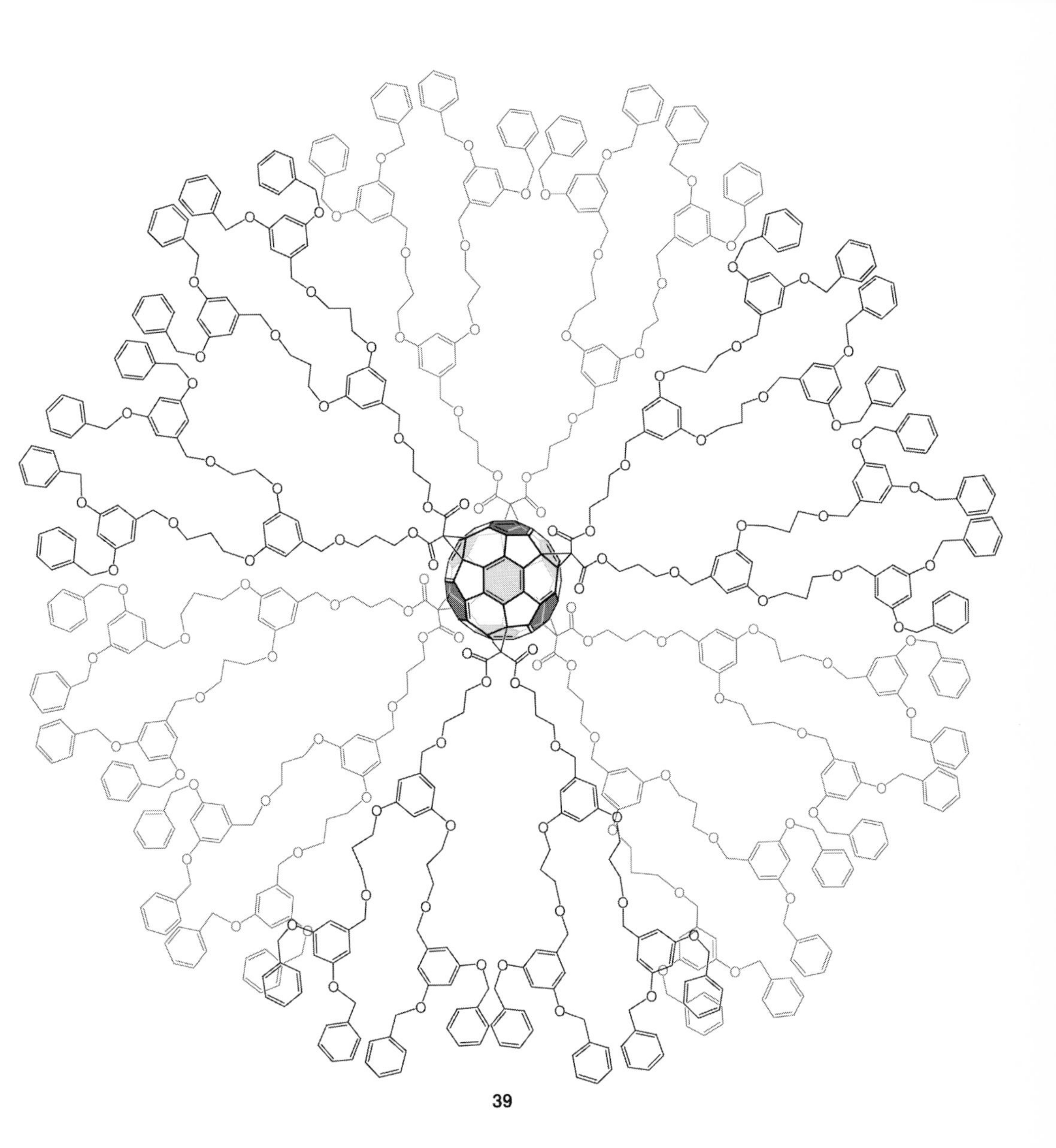

39

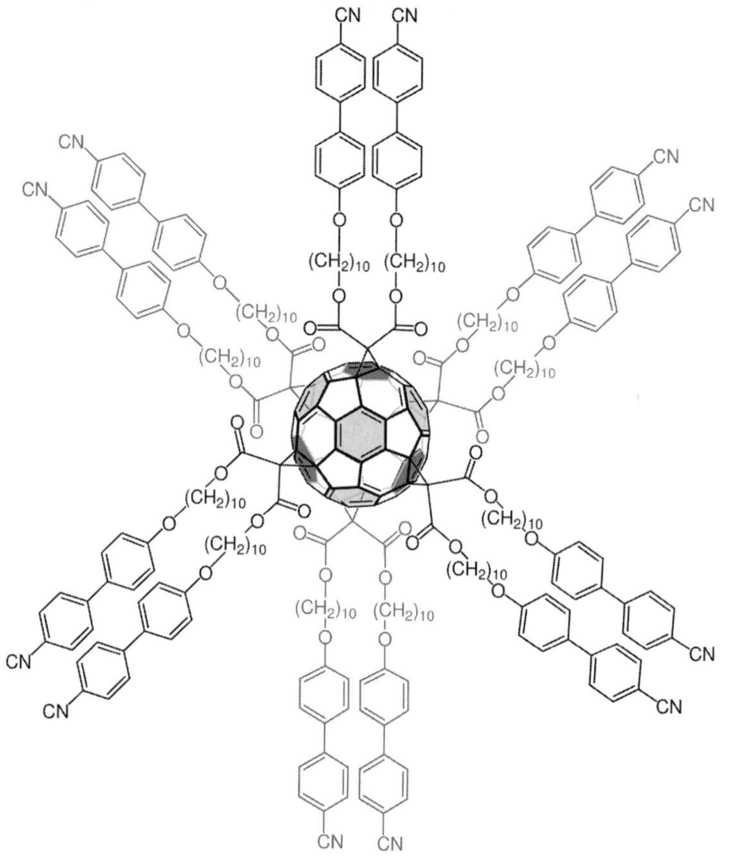

40

Fullerene malonates $C_{66}(COOR)_{12}$ can serve as valuable starting materials for further side-chain modifications, as demonstrated, for example, with the synthesis of the highly water-soluble hexamalonic acid derivative $C_{66}(COOH)_{12}$ (**41**) by hydrolysis of **17** [25]. A more efficient synthetic route to water-soluble fullereno-derivatives is cyclopropanation of C_{60} with bis(3-*tert*-butoxycarbonyl)propyl malonate **42** to afford the *tert*-butyl ester **43** (Scheme 10.8) [26]. Subsequent cleavage of the *tert*-butyl protecting group leads to the spherical dodecacarboxylic acid **44**, which can then be transformed into dodecaglycine adduct **45**.

10.3.1.2 Mixed Hexakisadducts

Using the template mediation technique, mixed hexakisadducts with different types of addends are also easily available [26]. Here the precursor is not C_{60} but a C_{60} adduct with an incomplete octahedral addition pattern. The possible structures of mixed hexakisadducts with two different addends in octahedral positions are depicted in Figure 10.12.

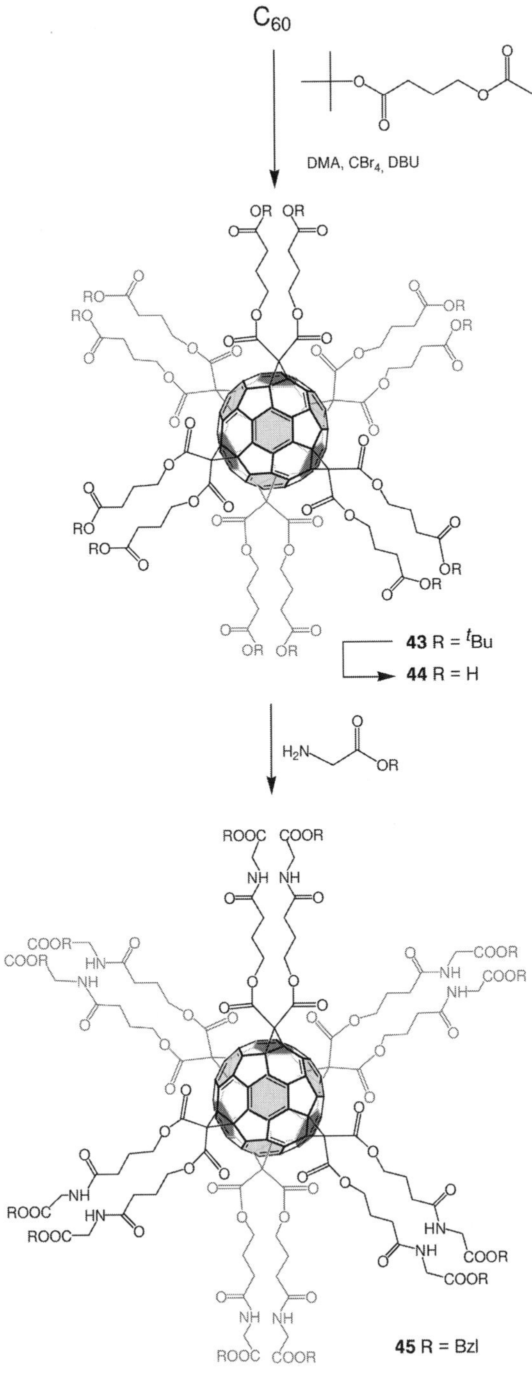

Scheme 10.8 Synthesis of highly water-soluble dodecacarboxylic acid **44** and protected glycine derivative **45**.

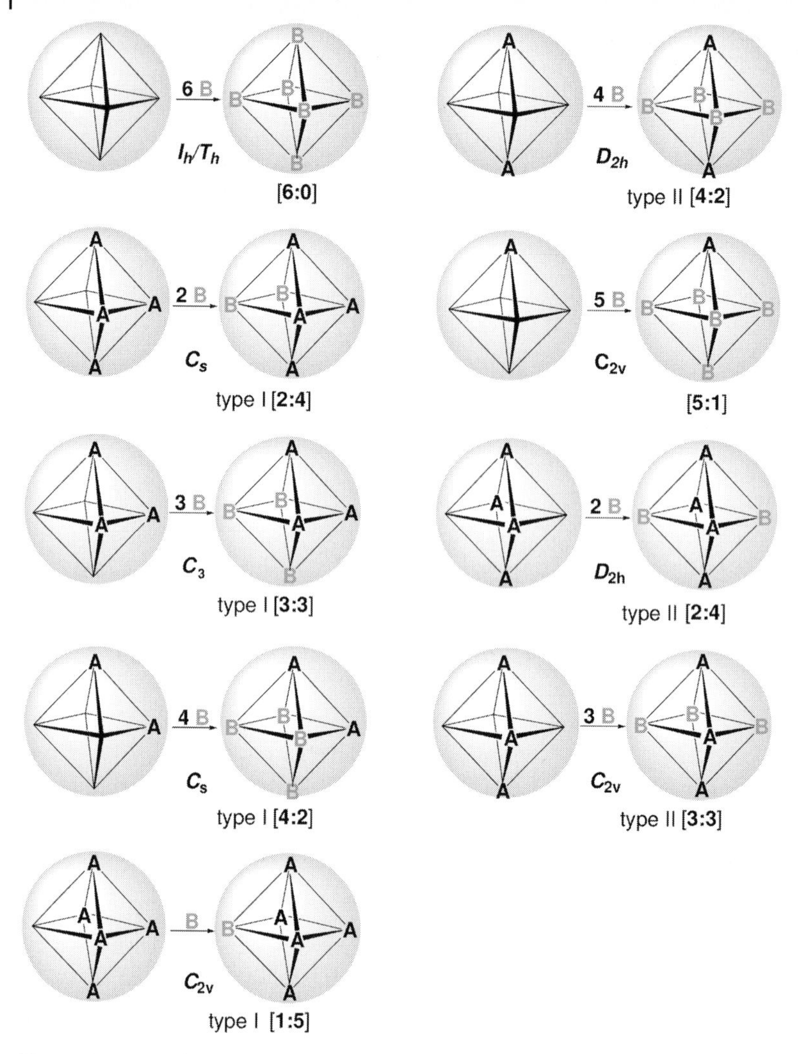

Figure 10.12 Complete series of octahedral addition patterns in hexakis-adducts of C_{60} with one or two different types of addends and their precursor adducts; type I adducts are derived from precursors obtained from successive e-additions, type II adducts from precursors synthesized by other means.

The most important aspect is the synthesis of precursor adducts possessing an incomplete octahedral addition pattern with one type of addend. Whereas mono-adducts or *e*-bisadducts are easily available, the production of *trans*-1-bisadducts or higher adducts with incomplete addition requires more effort. However, production of the mixed hexakisadducts from all the precursors is generally straightforward, since advantage can be taken of either the effective template mediation technique or the highly pronounced *e*-regioselectivity characteristic of higher adducts to complete the octahedral addition pattern [1, 26].

Mixed [5:1]-hexakisadducts

This addition pattern is accessible by starting either from a pentakisadduct with one unchanged octahedral site or from a monoadduct with five unoccupied sites. However, the stepwise synthesis of pentakisadducts such as **16** (Figure 10.7), with a C_{2v}-symmetrical addition pattern, is very time-consuming and the overall yield is unsatisfactory. For a convenient synthesis of this adduct type, an effective protection–deprotection strategy was developed (Scheme 10.9) [58]. The reaction sequence starts with the synthesis of the triazoline **46** by [3+2] cycloaddition of methyl azidoacetate to a [6,6]-double bond of C_{60}. After exhaustive template mediated cyclopropanation to **47** and thermally induced [3+2] cycloreversion, the pentakisadduct **16** was obtained in good overall yield. This pentakisadduct is a very valuable starting material, because attack at the remaining octahedral [6,6]-double bonds proceeds with quantitative regioselectivity. An example of a [5:1]-hexakisadduct originating from **16** is the dendrimer **48**.

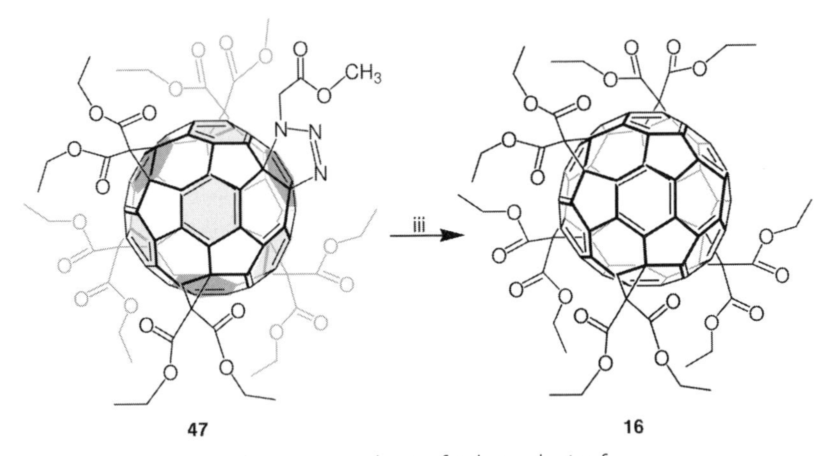

Scheme 10.9 Protection–deprotection technique for the synthesis of *e*-pentakisadduct **16**. (i) Methyl azidoacetate, 1-chloronaphthalene, 60 °C; (ii) 10 equiv. DMA, bromomalonate, DBU, toluene, room temp.; (iii) toluene, reflux.

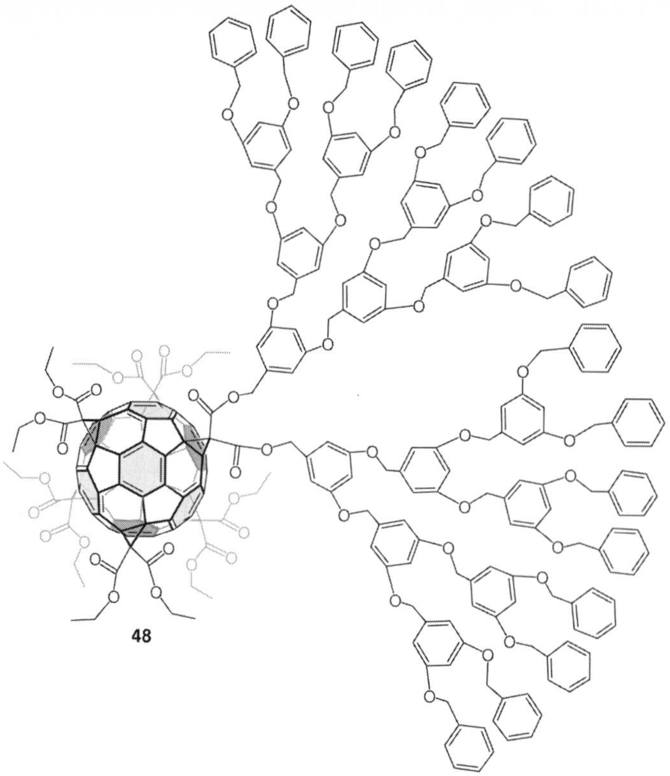

48

The inverse reaction sequence starting from easily available [6,6]-monoadducts, which are subsequently transformed into [1:5]-hexakisadducts using the template mediation technique, is even more efficient. This has been demonstrated in the synthesis of triazoline **47** and several related compounds [26, 58, 63, 64, 66–70]. Examples are functional dendrimers such as **49** [64] and **50** [66]. The globular amphiphile **50** dissolves in water, forming unilamellar vesicles with diameters typically between 100 and 400 nm. Stable monolayers of **50** on the air–water interface have been produced by the Langmuir–Blodgett technique [71]. An example of a biofunctional fullerene derivative that can intercalate into a DPPC bilayer is the biotinatyled lipofullerene **51** [68]. This molecule can be used as a transmembrane anchor for proteins located outside the membrane.

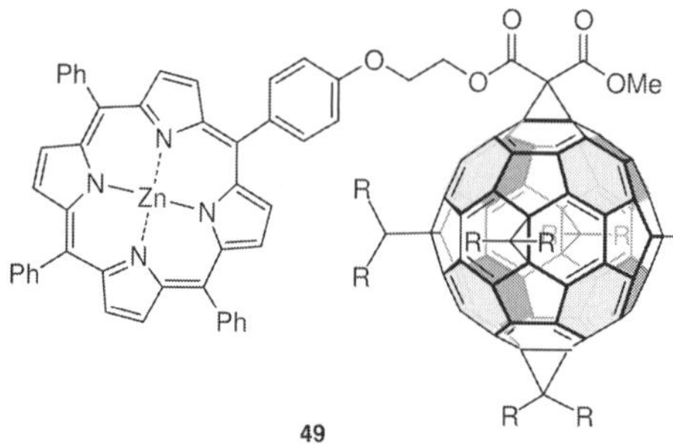

49

R =

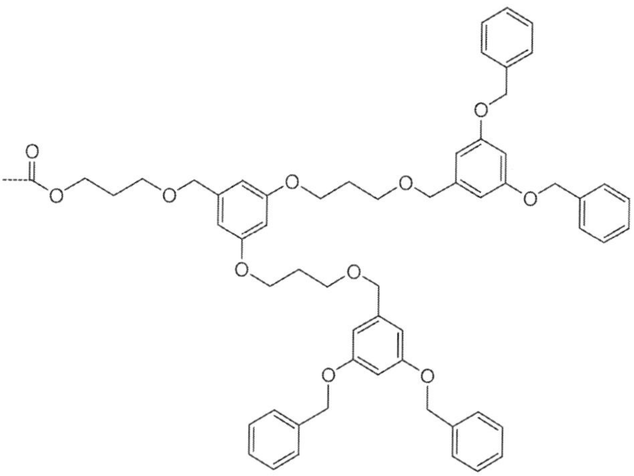

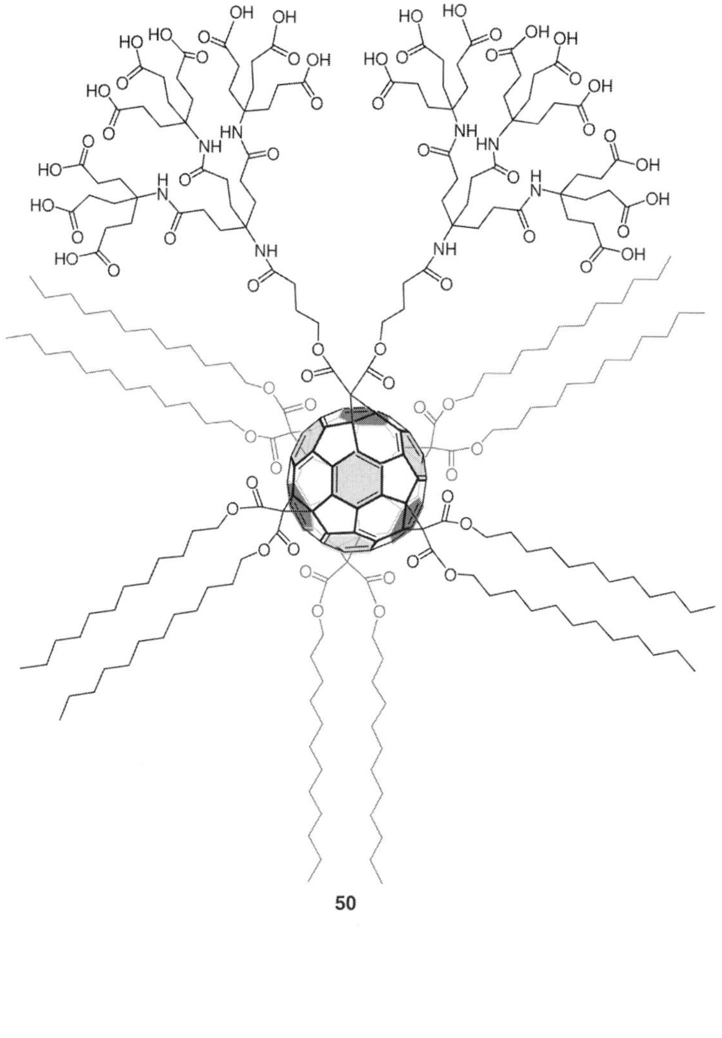

50

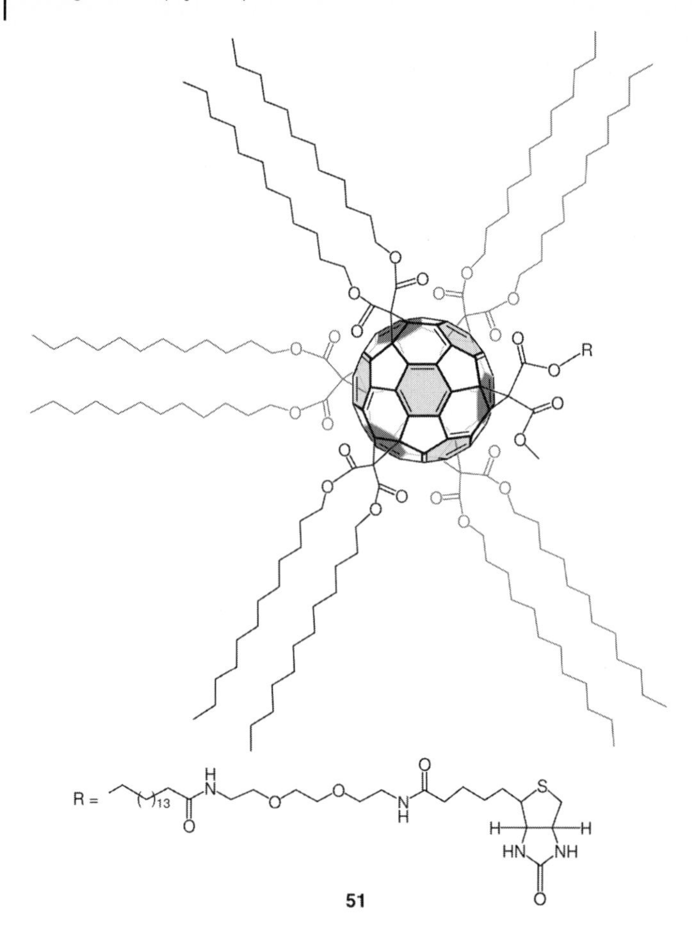

Diederich and co-workers transformed **52** into the supramolecular cyclophane **53** in quantitative yield by mixing equimolar amounts of **52** with *cis*-[Pt-$(PEt_3)_2$-$(OTf)_2$] [70].

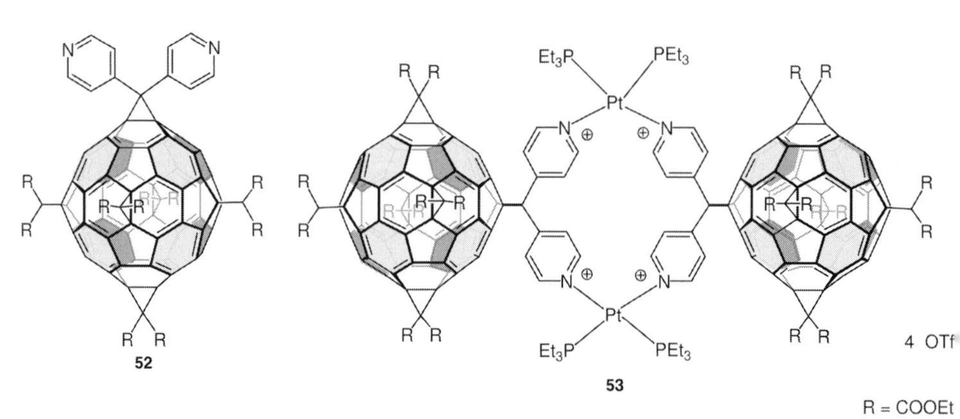

Mixed [4:2]-hexakisadducts

As a suitable starting material for the synthesis of [4:2] mixed hexakisadducts the tetrakisadduct **15** (Figure 10.7) was used [16, 26, 67]. Double cyclopropanation of this precursor core with the second generation (G2) Fréchet-dendron bromomalonate in the presence of DBU afforded C_S-symmetrical $C_{66}(COOEt)_8(COOG_2)_4$ **54** in 75% yield as a yellow powder (Scheme 10.10) [67]. The inverse [2:4] addition pattern can be obtained by successive fourfold cyclopropanation of the e-bisadduct **4** with second generation (G2) dendron bromomalonates to give $C_{66}(COOEt)_4$-$(COOG_2)_8$ **55** in 73% yield, also as a bright yellow powder (Scheme 10.11) [67].

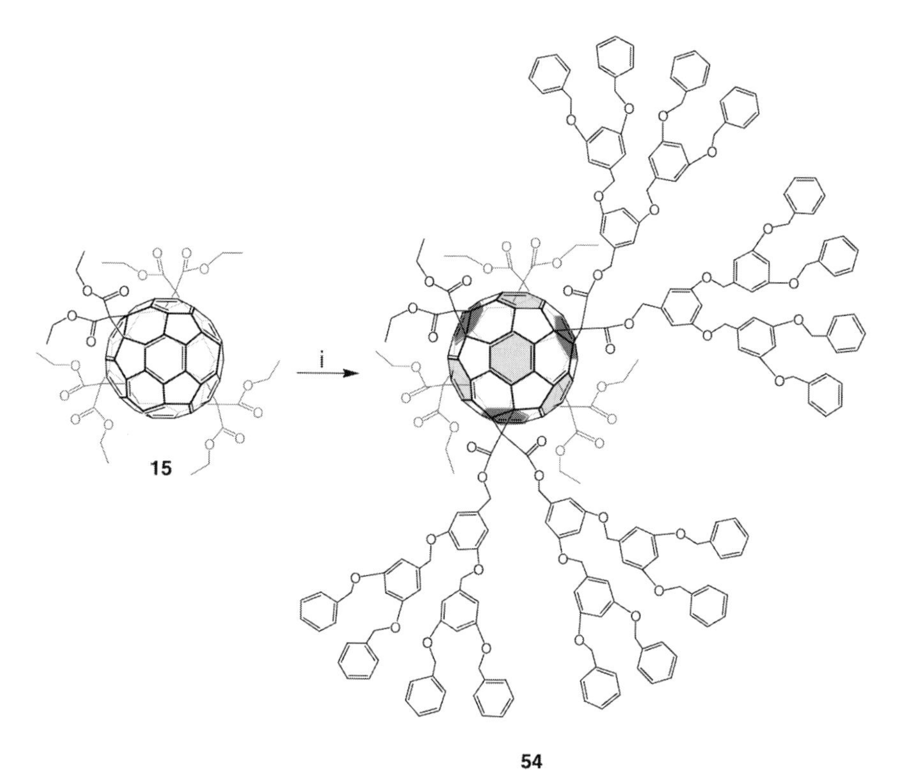

Scheme 10.10 Double nucleophilic cyclopropanation of all-e tetrakisadduct **15** with second generation (G2) bromomalonates yields dendrimeric [4:2] hexakisadduct **54**. (i) $CHBr(COOG2)_2$, DBU, toluene–CH_2Cl_2, 3 d, room temp.

In a similar fashion, **58** was synthesized starting from the e-bisadduct **56** with four protected terminal carboxylic functions (Scheme 10.12) [26]. Regioselective approaches to [4:2] hexakisadducts of type II are presented below (Sections 10.3.2 and 10.3.3).

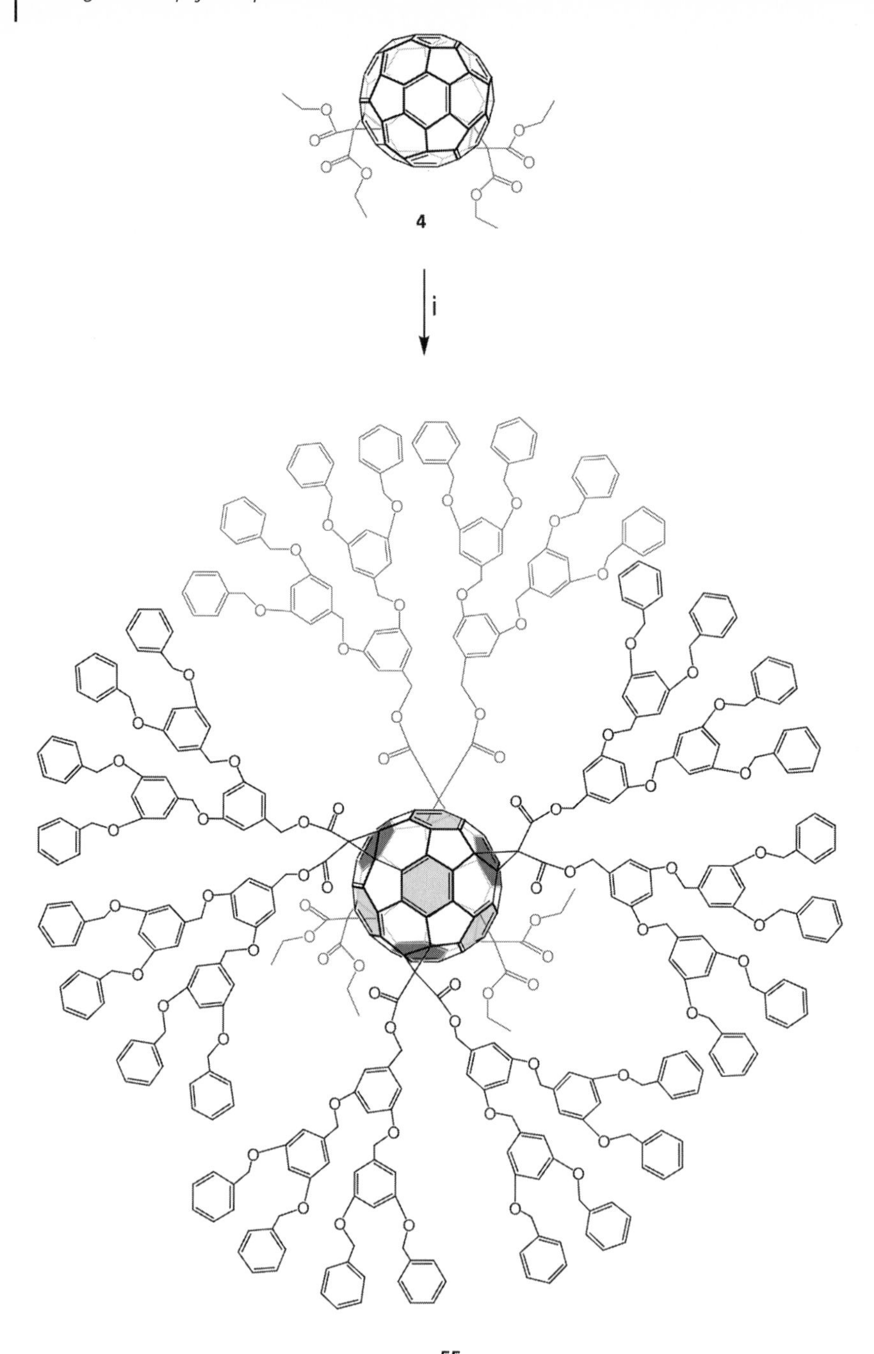

55

Scheme 10.11 Template-mediated fourfold cyclopropanation of *e*-bis-adduct **4**, leading to hexakisadduct **55**. (i) DMA, second generation (G2) bromomalonate $CHBr(COOG2)_2$, DBU, toluene–CH_2Cl_2, 3d, room temp.

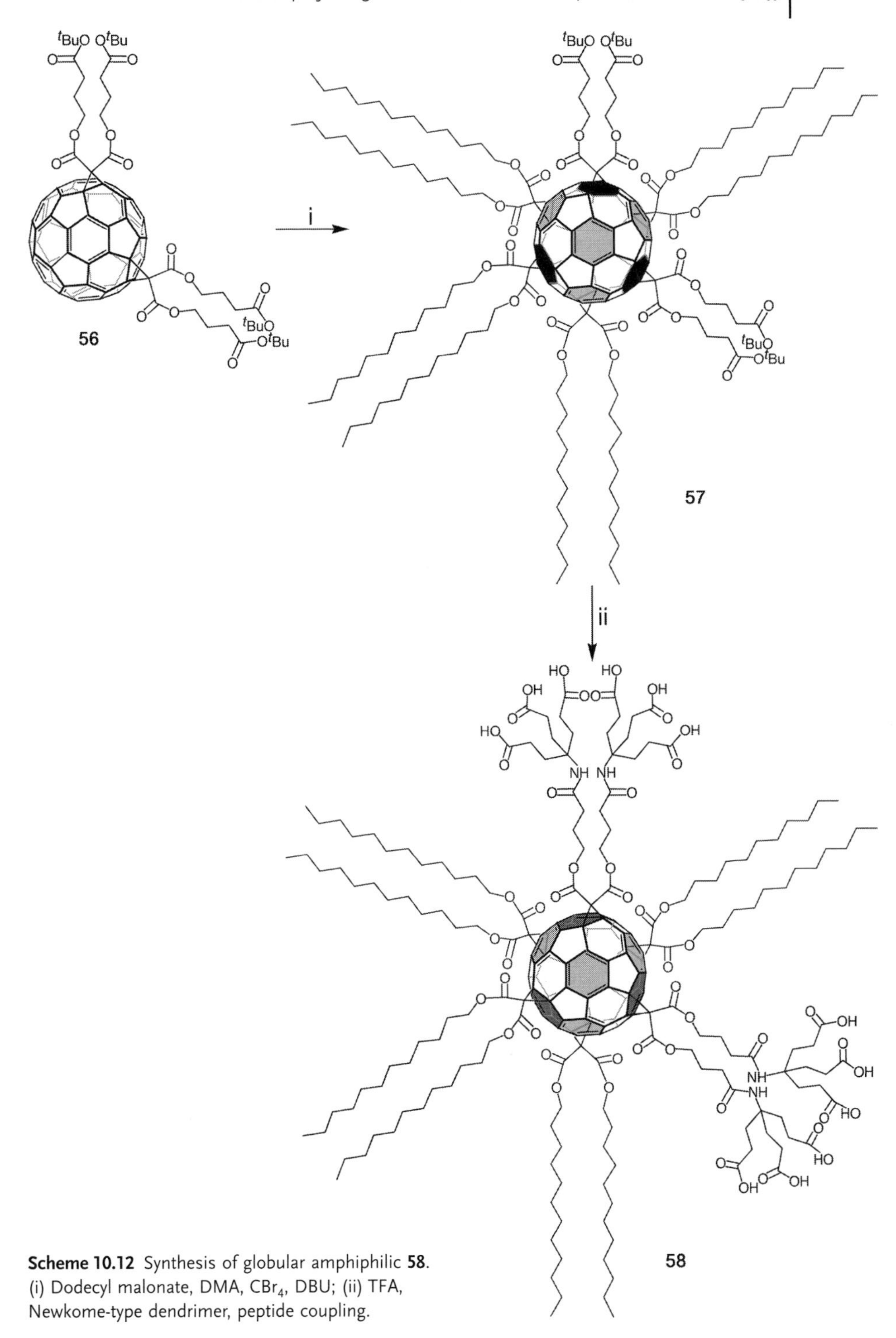

Scheme 10.12 Synthesis of globular amphiphilic **58**. (i) Dodecyl malonate, DMA, CBr_4, DBU; (ii) TFA, Newkome-type dendrimer, peptide coupling.

Mixed [3:3]-hexakisadducts

The synthesis of mixed [3:3] hexakisadducts of type I (Figure 10.12) requires trisadduct precursors with an e,e,e-addition pattern. A few examples for the synthesis of enantiomerically pure [3:3]-hexakisadducts have been discussed in Section 10.2.2. An additional example is the dendritic adduct **59** [67]. This was obtained as a racemic mixture by threefold cyclopropanation of the trisadduct ±**8**. Regioselective approaches to [3:3]-hexakisadducts of type II are presented below (Section 10.3.3).

59

10.3.2
Topochemically Controlled Solid-state Reactions

Kräutler and co-workers have developed a topochemically controlled, solid-state group-transfer synthesis to obtain the *trans*-1-bisanthracene adduct **61** (Scheme 10.13) [72–74]. The crystal packing provided the molding effect characteristic of a template. Heating a sample of crystalline monoadduct **60** at 180 °C for 10 min afforded 48% each of C_{60} and bisadduct **61** resulting from a regioselective intermolecular anthracene-transfer reaction. Bisadduct **61** can serve as starting material for the synthesis of type II [4:2]-addition pattern compounds, in which two addends are bound at the poles and four are attached at the equatorial belt. The two anthracene addends of **61** served to direct four bromomalonate addends regiospecifically into *e*-positions, giving hexakisadduct **62** in 95% yield. Subsequent thermal removal of the two polar anthracene molecules led to a tetrakis-adduct with an equatorial belt on the carbon sphere [73], representing a valuable tecton for further specific functionalization.

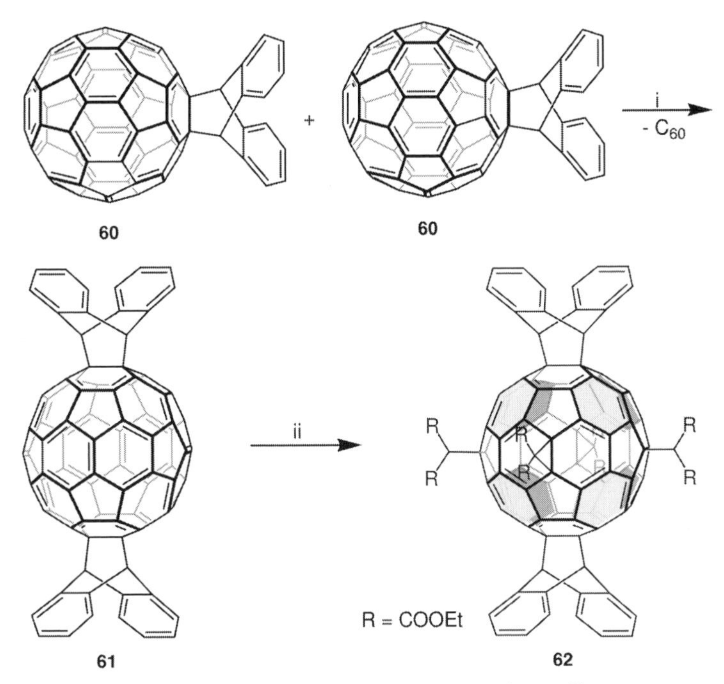

Scheme 10.13 Topochemically controlled solid-state synthesis of bis-adduct **61** and subsequent cyclopropanation to give [2:4]-hexakisadduct **62**. (i) 180 °C, 10 min; (ii) 40 equiv. diethyl bromomalonate, 40 equiv. DBU.

10.3.3 Tether-directed Remote Functionalization of C_{60}

In 1994, Diederich and co-workers reported a very important approach for the regioselective formation of multiple adducts of C_{60} by tether-directed remote functionalization [75]. This technique allows for the synthesis of fullerene derivatives with addition patterns that are difficult to obtain by thermodynamically or kinetically controlled reactions with free untethered addends. This important subject has been extensively reviewed [26, 76, 77].

10.3.3.1 Higher Adducts with the Addends Bound in Octahedral Sites

The first example of a tether remote functionalization was the synthesis of the tris-**64** (Scheme 10.14) [75, 78]. For this purpose the computer-aided chemically designed addend **63** was allowed to undergo a successive nucleophilic cyclopropanation/[4+2] cycloaddition sequence yielding the tris-adduct **64** in 60% yield with complete regioselectivity. Subsequent cyclopropanation of the remaining octahedral sites with a large excess of diethyl bromomalonate and DBU afforded the hexakisadduct **65**

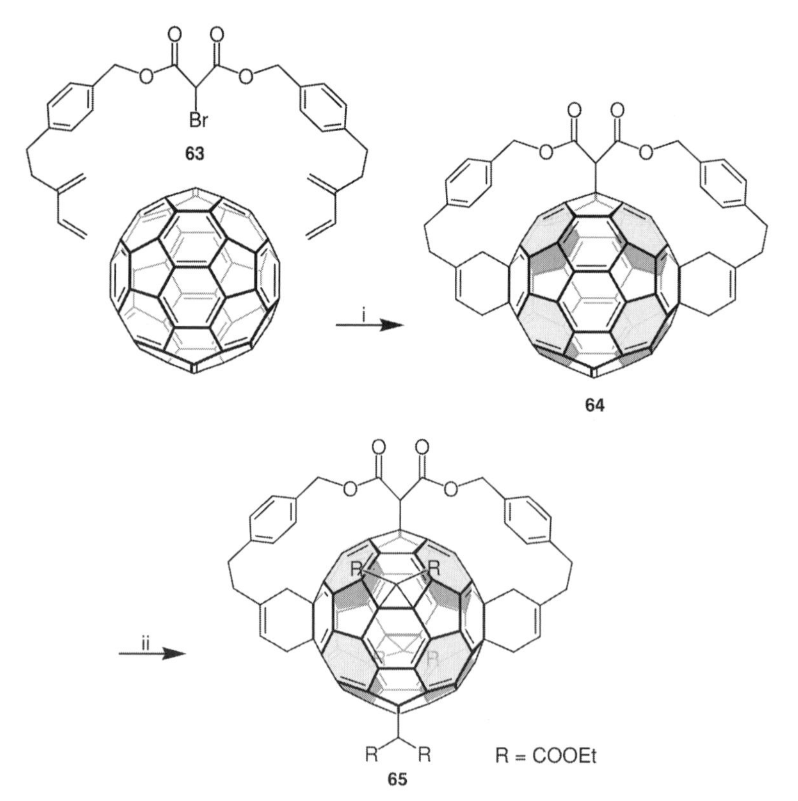

Scheme 10.14 Tether-mediated synthesis of trisadduct **64** and subsequent transformation into the [1:2:3]-hexakisadduct **65**. (i) DBU, toluene, room temp. 12 h, reflux 16 h; (ii) 10 equiv. diethyl bromomalonate, 10 equiv. DBU.

in 73% yield. Stepwise malonate additions to **64** have also been carried out, leading, for example, to pentakisadducts such as **66** (Scheme 10.15). In such pentakisadducts, with an incomplete octahedral addition pattern, the final addition to the remaining octahedral [6,6]-double bond proceeds in very good yields [26]. This was also demonstrated with the synthesis of **68** [79, 80]. Subsequent Eglinton–Glaser macrocyclization of **68** lead to soluble carbon-rich scaffolds in **69** and **70**, which are solubilized derivatives of C_{195} and C_{260}, two members of a new class of fullerene-acetylene hybrid carbon allotropes (Scheme 10.15) [79]. Repeated treatment of pentakisadducts **66** with diazomethane allowed for the synthesis of a mixture of

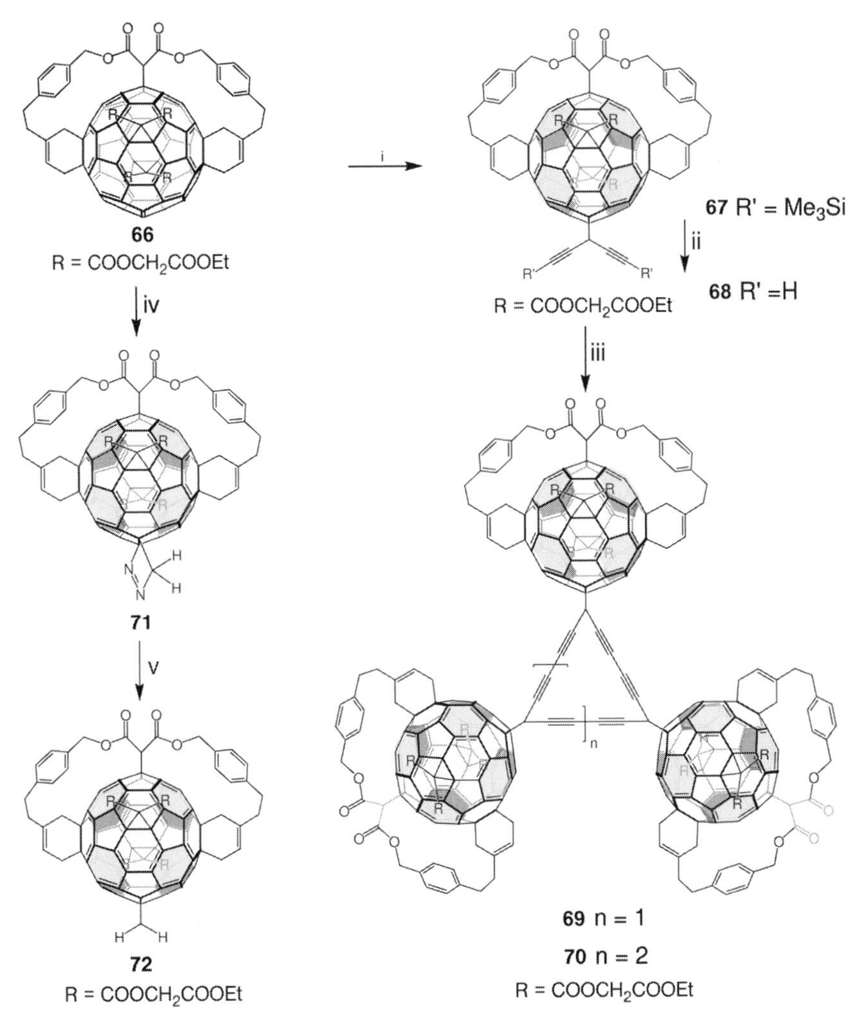

Scheme 10.15 Synthesis of fullerene appended macrocycles **69** and **70** and the [1:2:3] hexakisadduct **72**. (i) 10 equiv. 3-bromo-1,5-bis(trimethylsilyl)-1,4-pentadiyne (**75**), 10 equiv. DBU; (ii) $Bu_2NF(SiO_2)$; (iii) 200 equiv. $Cu(OAc)_2$, molecular sieves; (iv) 60 equiv. CH_2N_2; (v) $h\nu$.

mixed hexakis-, heptakis- and octakisadducts (Scheme 10.15) [81, 82]. Along with adducts with methano groups bridging the remaining octahedral [6,6]-bond such as **72**, adducts with bridged open [5,6]-bonds are also formed during diazomethane addition followed by extrusion of N_2.

Removal of the e-directing cyclohexene rings together with the *p*-xylylene tethers in hexakisadduct **65** was possible by a procedure originally introduced by Rubin [83]. When a solution of **65** containing C_{60} as $^1O_2$ sensitizer was irradiated while a stream of O_2 was bubbled through, an isomeric mixture of allylic hydroperoxides was obtained due to an $^1O_2$ ene reaction [84, 85]. Subsequent reduction to the corresponding allylic alcohols, dehydration (using TosOH) to the bis(cyclohexa-1,3-diene) derivative and a Diels–Alder retro-Diels–Alder sequence afforded the mixed tetrakisadduct **73** (Scheme 10.16). Transesterification yielded the octakis(ethyl ester) **74**. Tetrakisadduct **74** contains two reactive [6:6]-bonds at the pole activated by the four malonate addends in *e*-positions at the equator. Addition of 2 equiv. of dialkynyl bromide **75** and subsequent deprotection afforded hexakisadduct **76** as a useful building block for further molecular scaffolding.

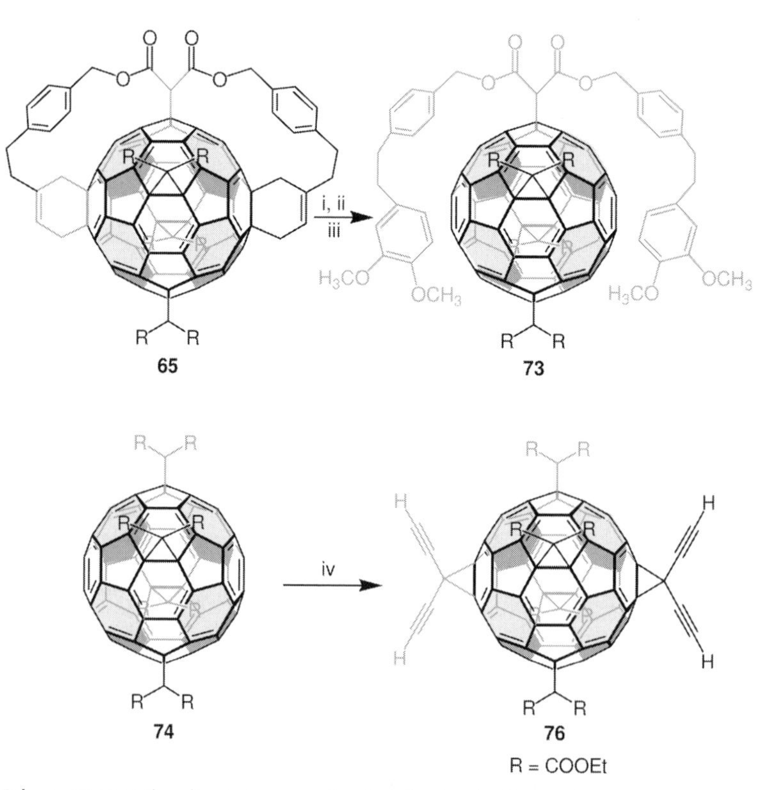

Scheme 10.16 Tether-directed remote functionalization leading to the tetraethynylated [2:4] hexakisadduct **76**. (i) O_2, $h\nu$, PhCl; (ii) PPh_3, PhCl; (iii) TosOH, ethyl acetylenedicarboxylate; (iv) 3-bromo-1,5-bis(trimethylsilyl)-1,4-pentadiyne (**75**).

As pointed out by Diederich [77] it is important to investigate fullerene properties as a function of addition pattern and the nature of the addend. Various pronounced correlations and trends can be deduced. In general, with increasing reduction in the conjugated fullerene π-chromophore, (1) the HOMO–LUMO gap increases; (2) the number of reversible one-electron reductions decreases, and the first reduction potential becomes increasingly negative; (3) the reactivity of the fullerene toward nucleophiles and carbenes or as dienophiles in cycloadditions decreases; and (4) the capacity for photosensitization of $^1O_2$ formation decreases [86]. Concerning the nature of the addend, C_{60} properties are retained more in cyclopropanated adducts than, for example, in Diels–Alder adducts. This is because cyclopropane fusion represents a smaller perturbation of the fullerene core. The smallest perturbation of the fullerene π-chromophore by multiple additions occurs when all-methano addends are introduced along an equatorial belt, as in **74** [85, 87].

10.3.3.2 Multiple Cyclopropanation of C_{60} by Tethered Malonates

A very simple, versatile method for the preparation of bisadducts of C_{60} with high regio- and diastereoselectivity is the macrocyclization between C_{60} and bismalonate derivatives in a double cyclopropanation reaction (Scheme 10.17) [76, 77]. As a general synthetic protocol, diols are transformed, for example, into bis(ethyl malonyl) derivatives and the one-pot reaction of C_{60}, bismalonate, I_2 or CBr_4 (to prepare the corresponding halomalonate in situ), and DBU in toluene generates the macrocyclic bisadducts in yields usually between 20 and 40% with high regio- and diastereoselectivity. With the exception of *cis*-1, examples for all possible addition patterns have been synthesized by this method; selected examples (**77–84**) are presented in Scheme 10.17 [28, 29, 76, 77, 88–91].

Each macrocyclization could, in principle, lead to a mixture of diastereoisomers, depending on how the EtOOC residues at the two methano bridge C atoms are oriented with respect to each other (in-in, in-out, and out-out stereoisomerism) [92]. Usually, only out-out stereoisomerism has been observed so far. One exception is the in-out isomer **81**.

Higher adducts of C_{60} can also be synthesized regiospecifically using this tether approach. For example, a clipping reaction sequence leads to the tetrakisadduct **87** with an all-*cis*-2 addition pattern along an equatorial belt (Scheme 10.18) [29].

Using a route that involves two sequential tether-directed remote functionalization steps hexakisadduct (±)-**91**, which features six addends in a distinct helical array along an equatorial belt, was prepared according to Scheme 10.19 [93–95]. In this new class of hexakisadducts, π-electron conjugation between the two unsubstituted poles of the carbon sphere is maintained. As a consequence, (±)-**91** exhibits different chemical and physical properties than hexakisadducts with a pseudo-octahedral addition pattern. For example, its reduction under cyclic voltammetric conditions is significantly facilitated (by 570 mV). Moreover, it readily undergoes additional, electronically favored cyclopropanation reactions at the two sterically well-accessible central polar [6:6]-bonds. The addition reactions lead to the heptakis- and octakisadducts, (±)-**92** and (±)-**93**, respectively (Scheme 10.20). Whereas hexakisadducts with an octahedral addition pattern are bright yellow (±)-**91** is shiny red.

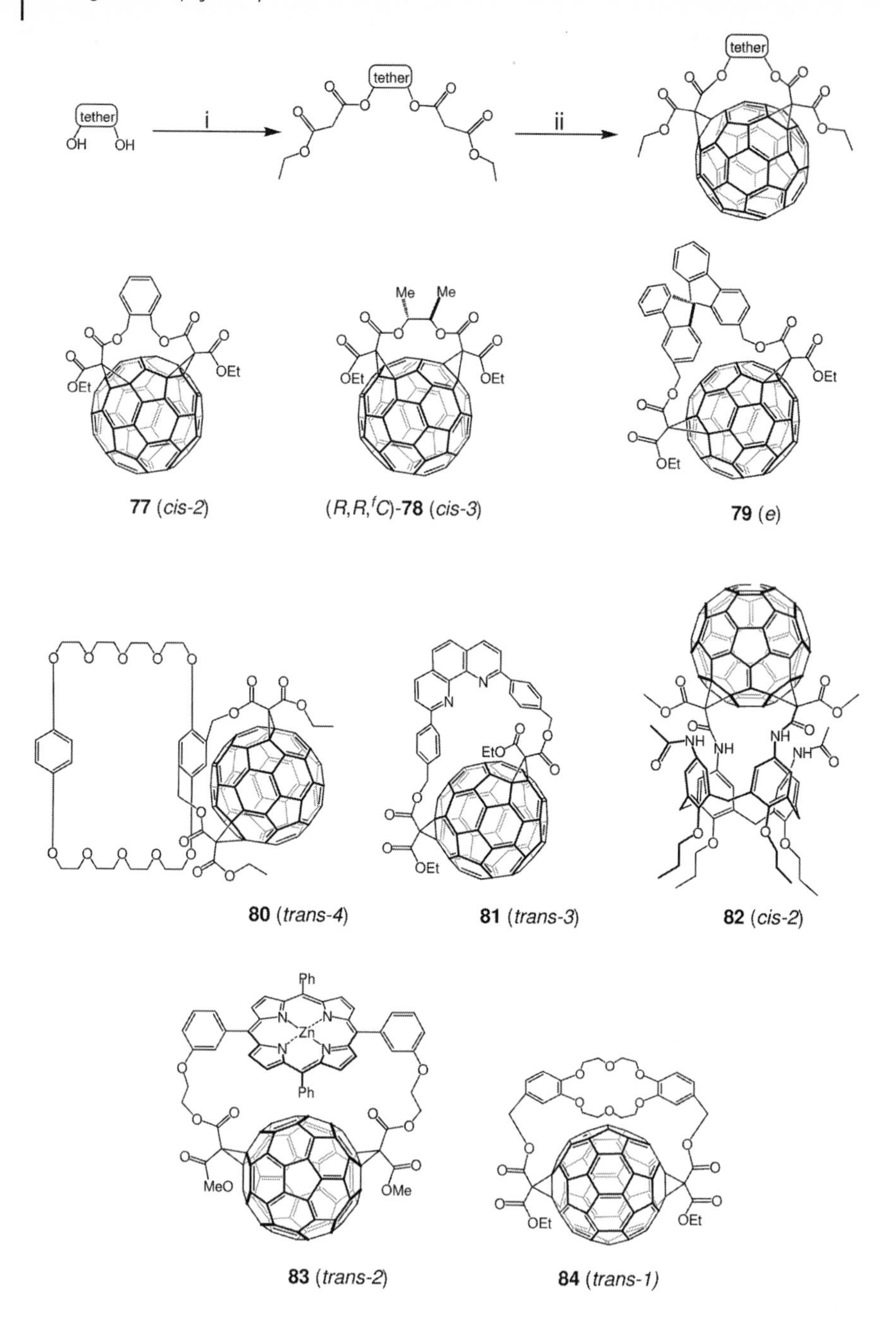

i) EtO_2CCH_2COCl, C_5H_5N, CH_2Cl_2, 0-20 °C, 60-70%
ii) C_{60}, DBU, I_2, PhMe, 20 °C

Scheme 10.17 Tether-controlled regio- and stereoselective preparation of bis(cyclopropanated) C_{60} derivatives and examples of bisadducts produced.

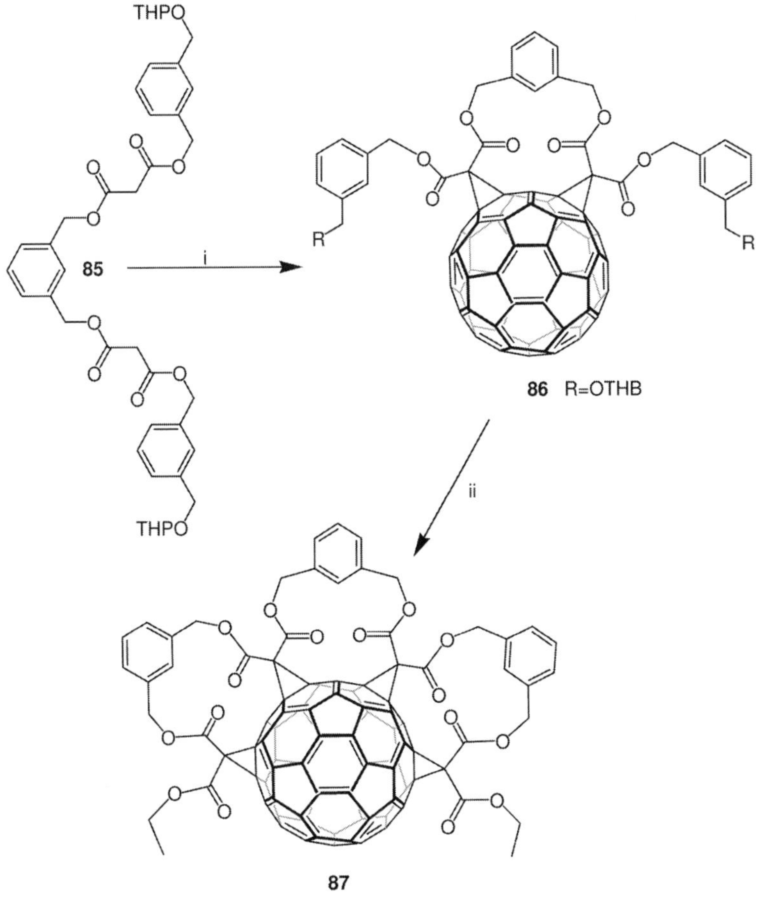

THPO

85

i

86 R=OTHB

ii

THPO

87

i) C_{60}, DBU, I_2, PhMe, r.t.
ii) 1. $TsOH.H_2O$, EtOH, PhMe, 80 °C
2. EtO_2CCH_2COCl, C_5H_5N, H_2Cl_2, 0 °C -> r.t.
3. DBU, I_2, PhMe, r.t.

Scheme 10.18 Regioselective synthesis of tetrakisadduct **87**.

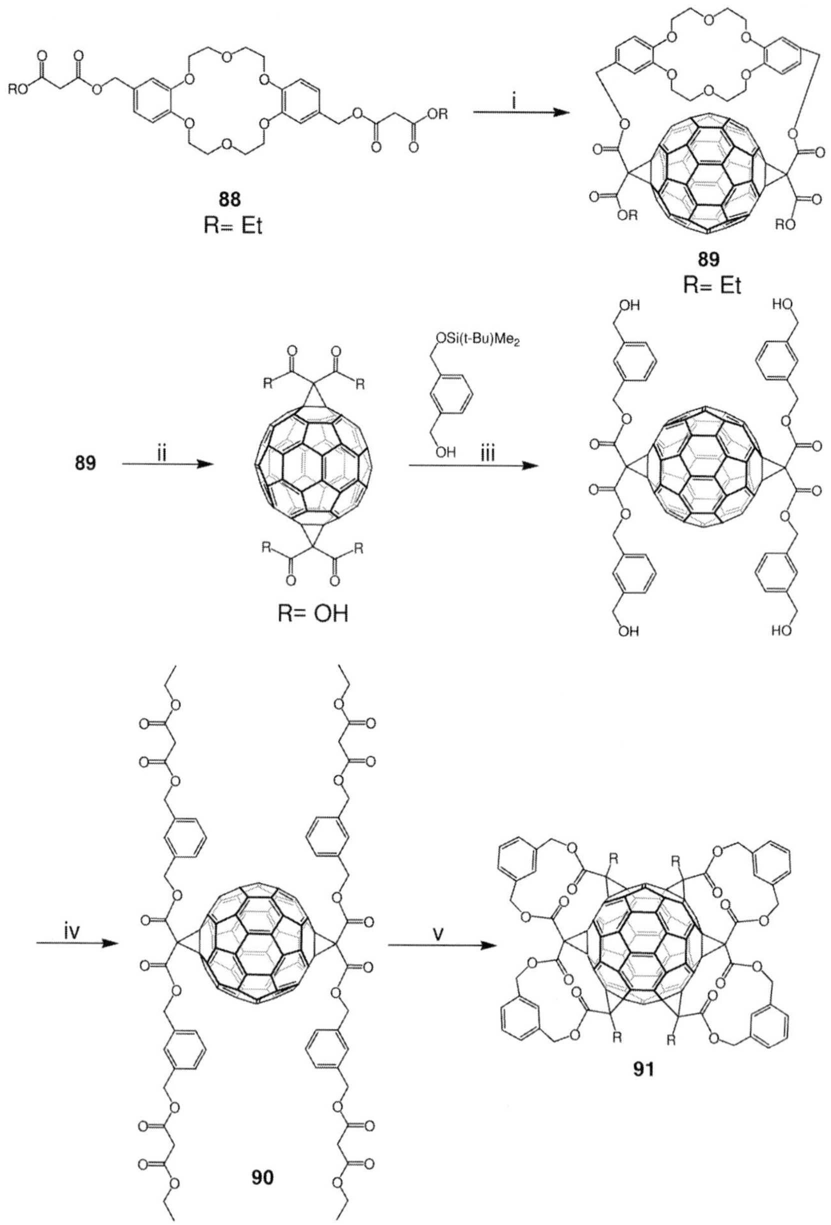

88 R= Et

89 R= Et

R= OH

OSi(t-Bu)Me$_2$

90

91

i) C_{60}, KPF_6, PhMe/MeCN, 20°C; 54% ((±)-13), 50% ((±)-16)
ii) TsOH·H_2O, PhMe, Δ, 12h
iii) $(COCC)_2$, Pyridine, CH_2Cl_2, 20°C, 12h; 21%(from (±)13)
iv) HF·pyridine, CH_2Cl_2, 0°C, 1h; 92%
v) I_2, DBU, PhMe/Me_2SO, 20°C,12h; 10%

Scheme 10.19 Synthesis of hexakisadduct **91**.

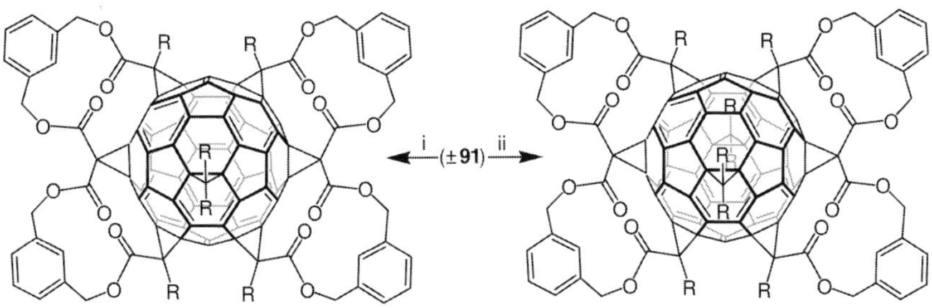

92 93

i) $(EtOOC)_2CHBr$ (2 equiv.), DBU, PhMe/Me_2SO 1:1, 20 °C, 12h; ca. 40%
ii) $(EtOOC)_2CHBr$ (20 equiv.), DBU, PhMe/Me_2SO 1:1, 20 °C, 12h; ca. 80%

Scheme 10.20 Synthesis of heptakisadduct **92** and octakisadduct **93**.

Transesterification of the fullerene crown ether conjugate (±)-**89** allowed for the removal of the tether, leading to the D_{2h}-symmetrical bisadduct **94** (34%) [93], which is a very useful building block for further addition chemistry.

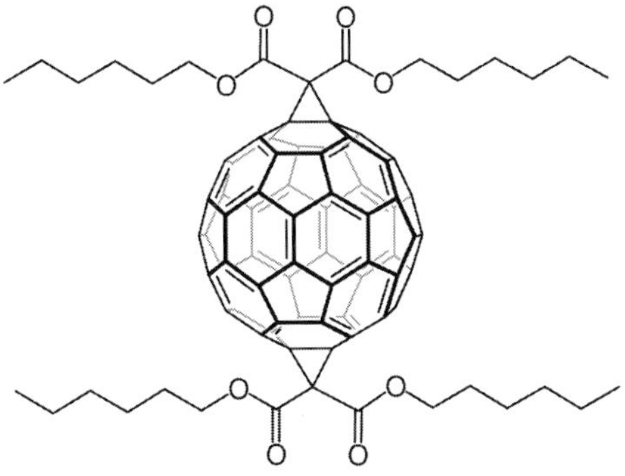

94

An enantioselective synthesis of optically active bisadducts in which the chirality results from the C_2-symmetrical addition pattern only has been carried out using a sequence of a highly diastereoselective tethered cyclopropanation followed by transesterification [8, 27–29]. For example, starting from (*R,R*)-**95** and (*S,S*)-**95**, which were prepared from the corresponding optically pure diols, the two enantiomeric *cis-3* bisadducts (*R,R,*^{f}*A*)-**96** and (*S,S,*^{f}*C*)-**96** were obtained with high diastereoselectivity (Scheme 10.21). Transesterification of (*R,R,*^{f}A)-**96** and (*S,S,*^{f}C)-**96** yielded the cis-3 tetraethyl esters (^{f}*A*)-**97** and (^{f}*C*)-**97** with an enantiomeric excess (ee) higher than 99% [(^{f}*A*)-2] and 97% [(^{f}*C*)-2] (HPLC), reflecting the ee values of the corresponding commercial starting diols.

Determination of the absolute configurations of these optically active fullerene derivatives was possible by comparison of their experimental and calculated circular dichroism (CD) spectra [96]. Tether controlled bis-cyclopropanation reactions have been extensively used to synthesize extended functional architectures such as dyads,

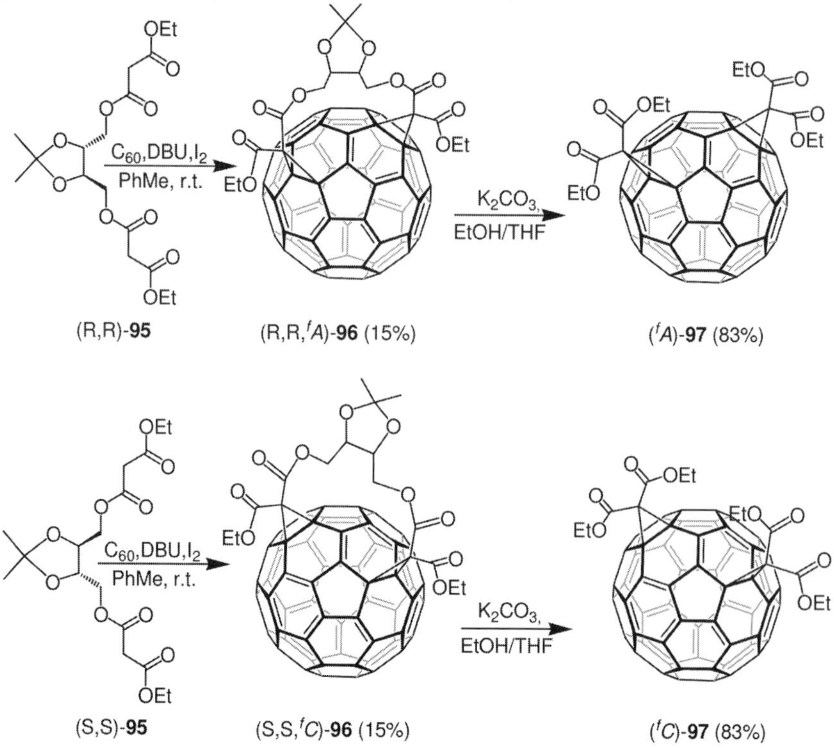

Scheme 10.21 Enantioselective synthesis of ^{f}A-**97** and ^{f}C-**97**.

dendrimers, amphiphiles and supramolecular systems [29, 76, 77, 97, 98]. Direct threefold cyclopropanation of C_{60} was possible using a C_3-symmetrical cycloveratrylene tether [99]. A mixture of the corresponding C_3-symmetrical trisadducts having *trans*-3,*trans*-3,*trans*-3 and *e,e,e* structures were obtained in one step in 11 and 9% isolated yields, respectively.

10.3.3.3 Highly Regioselective Cyclopropanation of C_{60} with Cyclo-[*n*]-alkylmalonates

The tether directed cyclopropanations of C_{60} discussed so far take advantage of specifically designed open-chain malonates with quite rigid spacers. This is a particularly suitable method for the synthesis of bisadducts. However, for an entry to tris- and higher adducts the design of a suitable steric arrangement becomes more and more difficult.

This difficulty has been circumvented by the introduction of a modified concept [100]. Here, flexible alkyl chains are used as tethers that are incorporated in macrocyclic cyclo-[*n*]-malonates. In this way, the regioselectivity is not based on steric preorganization but on the avoidance of unequal strain in the alkyl chains. This has very important consequences for the symmetry of the product addition pattern and dictates selection rules: (1) Cyclo-[*n*]-malonates containing alkyl chains

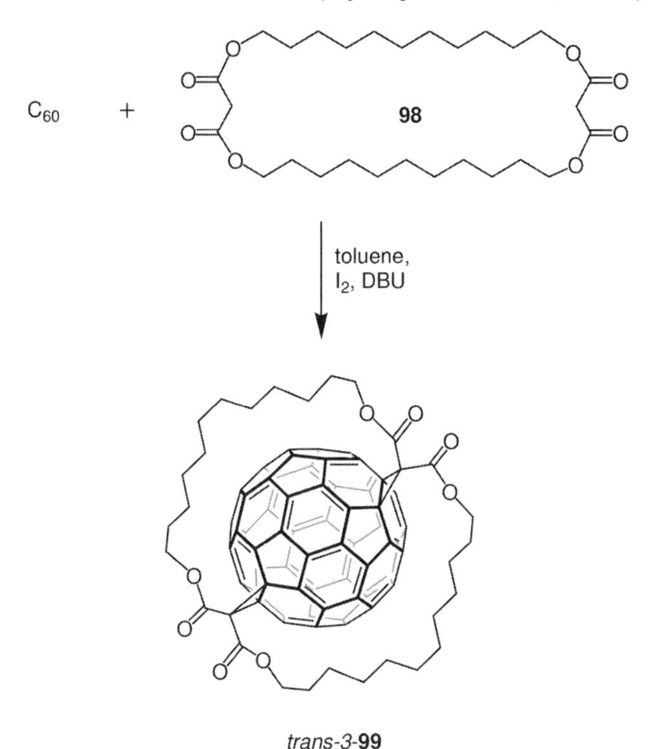

Scheme 10.22 Highly regioselective formation of *trans*-3-**99**.

of identical lengths are forced to form adducts having rotational axes because all distances of adjacent locations are identical. The symmetry of the corresponding adducts can be, for example, C_2, C_3, C_{3v}, D_3 and D_{2h} (Figure 10.9). Only in adducts with such addition patterns is even distribution of strain within the alkyl chains provided. (2) Cyclo-[*n*]-malonates containing alkyl chains of different lengths are selective for the formation of adducts that do not have rotational axes. Typical examples are C_s-symmetrical bisadducts. In these cases the distance between two adjacent addend positions are different. As a consequence, alkyl bridges of different lengths are required to allow for most strain-free arrangements. (3) Within these groups of adducts a given addition pattern can be adjusted by computer-assisted variation of the chain lengths. The corresponding malonates are easily accessible by reaction of malonyl dichloride with alkanediols. Various bis-, tris- and tetrakis-adducts have been synthesized to show the scope of this approach. For example, treatment of cyclo-[2]-dodecylmalonate (**98**) with C_{60} under typical cyclopropanation conditions afforded the *trans*-3 bisadduct **99** (C_2-symmetry) completely regioselective in 56% yield (Scheme 10.22) [100].

When, though, C_{60} was reacted with cyclo-[2]-butyloctylmalonate (**100**) or cyclo-[2]-octyltetradecylmalonate (**101**), C_s-symmetrical bisadducts *cis*-2-**102** and *e*-**103**, respectivley, were formed exclusively (Scheme 10.23) [100].

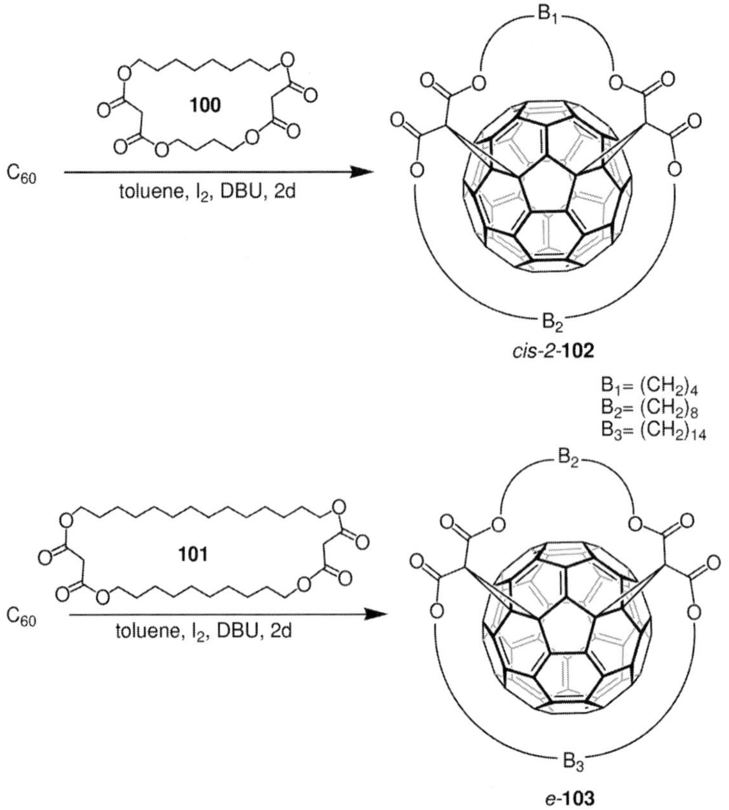

Scheme 10.23 Highly regioselective synthesis of *cis-2*-**102** and *e*-**103**.

Cyclo-[3]-octylmalonate (**104**) reacts with C_{60} to furnish the C_3-symmetrical trisadduct **105** in 94% relative and 42% isolated yield (Scheme 10.24) [100]. As a less polar byproduct, the C_{3v}-symmetrical trisadduct **106** containing the unprecedented *trans*-4,*trans*-4,*trans*-4-addition pattern was formed in 6% relative yield. Solutions of the trisadduct **106** are olive-green. By increasing the length of the alkyl bridges and allowing cyclo-[3]-tetradecylmalonate (**107**) to react with C_{60} under the same conditions, the D_3-symmetrical *trans*-3,*trans*-3,*trans*-3 trisadduct **108**, where the malonate addends are further apart, was formed completely regioselectively. Several tetrakis- and hexakis-adducts have been obtained from the highly regioselective addition of cyclo-[4]-malonates or by successive additions of cyclo-[2]- and cyclo-[3]-malonates [100].

C_{60} + **104**

toluene, I_2, DBU 59%

e,e,e-**105** *trans-4,trans-4,trans-4*-**106**

Scheme 10.24 Synthesis of of trisadducts **105** and **106**.

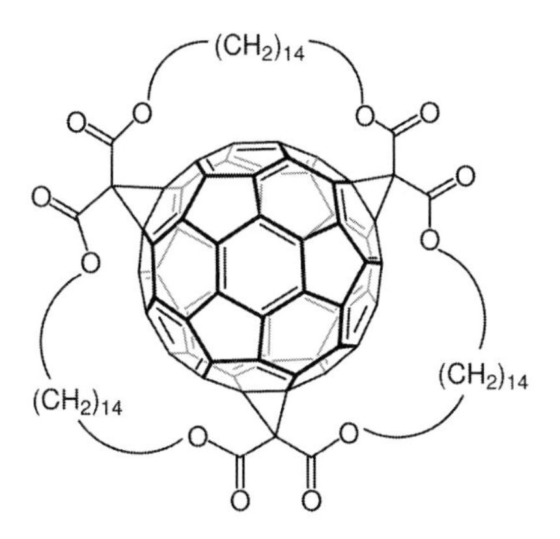

trans-3,trans-3,trans-3-**108**

10.3.3.4 Double Diels–Alder Tethers

Using a double Diels–Alder addition of a tethered bis(buta-1,3-diene) to C_{60} Rubin has achieved the highly regioselective formation of a trans-1 bis-adduct [101–103]. The tethered bis(cyclobutene) **109** was prepared as a mixture of stereoisomers (Scheme 10.25). Upon refluxing **109** with C_{60} in toluene, ring-opening of the cyclobutenes occurred and the resulting 1,3-diene underwent cycloaddition regioselectively to afford the *trans*-1 derivative **110** in 30% yield. This bisadduct is a valuable intermediate for modifications of the fullerene cluster itself and for the highly regioselective synthesis of hexakisadducts with a mixed octahedral addition pattern, which will now be discussed.

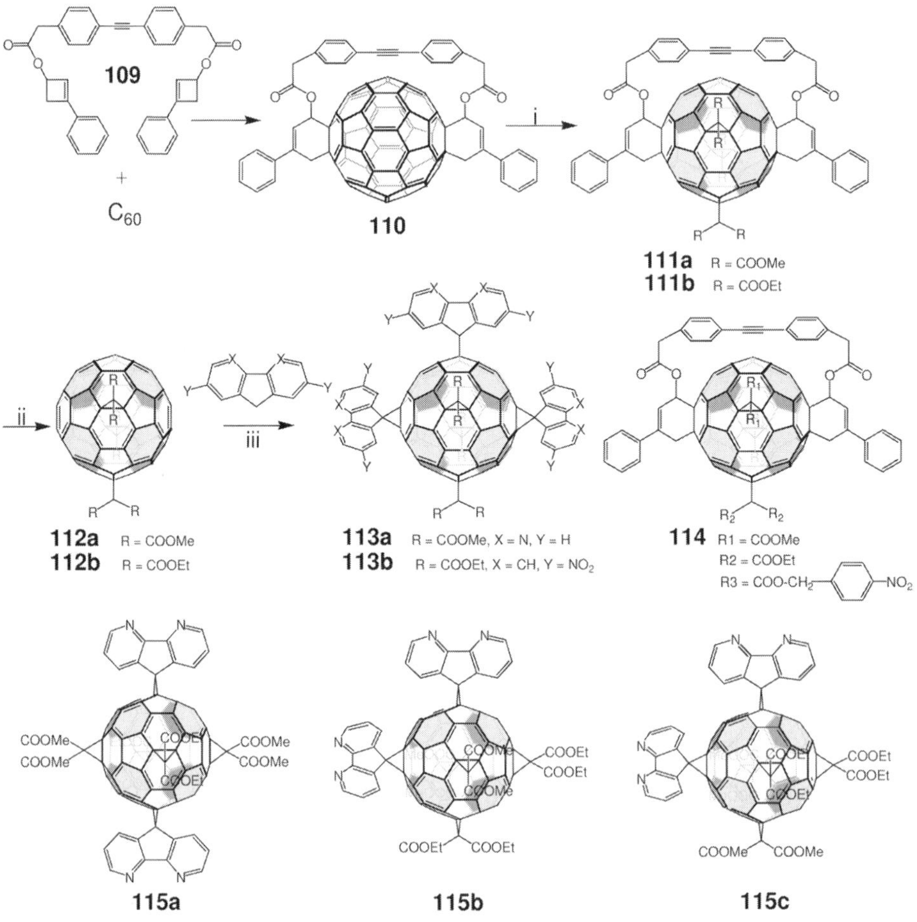

i) methyl and ethyl malonate, respectively, CBr_4,DBU
ii) TosOH, DMAD, toluene, 110 °C
iii) fluorene derivative, CBr_4, DBU

Scheme 10.25 Regioselective synthesis of mixed hexakisadducts of C_{60} starting from the *trans*-1 precursor **110**.

In **110**, thanks to the shielding engendered by the tether, only three of the four reactive e-positions are accessible for further addition reactions. Consequently, treatment with dimethyl or diethyl malonate, under in situ bromination conditions (CBr_4 and DBU), gave mixed pentakisadducts **111a** and **111b**, respectively. The tether was subsequently removed by an elimination/Diels–Alder/retro-Diels–Alder sequence. The resulting C_{2v}-symmetrical trisadducts **112a** and **112b** both underwent highly regioselective threefold addition reactions with 4,5-diazafluorene or 2,7-dinitrofluorene, leading to the mixed [3:3]-hexakisadducts of **113a** and **113b** (type II, Figure 10.12) in 50% and 89% yield, respectively (Scheme 10.25) [102]. Thanks to the trans-1 regiochemistry of the tethered bisadduct **110** and the ensuing single protected *e''*-position, it is possible to address each of the three remaining *e*-positions exclusively, in sequential manner (*e'*, *e''*, and finally *e'*), thus priming this synthesis method for complete differentiation between octahedral addition sites. This synthesis technique is an elegant means of embellishing C_{60} with many different addend types, giving access to multiple fullerene adducts with stereochemically defined geometry. This was demonstrated by the preparation of the [1:1:1:2] pentakisadducts **114**, incorporating three different malonates: methyl, ethyl and *p*-nitrobenzyl, respectively (Scheme 10.25).

Following their initial studies, Rubin and co-workers subsequently succeeded in synthesizing a complete "library" of all-*e* C_{60} [2:2:2] hexakisadducts with three pairs of addends in octahedral sites [104]. Using a "*mer*-313" regiocontrol strategy, involving regiochemically distinct cyclopropanations of *trans*-1 tethered bisadduct **110**, the authors prepared four regioisomeric trismalonates, each with two different ester moieties. This represented a fine-tuning of electronic and steric effects on the surface of C_{60}, enabling consequent, separate addition of further addends. Variation of addition sequences, as well as of the choice of corresponding addends, via [2:2] mixed tetraadducts and [1:2:2] mixed pentaadducts, respectively, was the key to accessing the seven possible regioisomeric [2:2:2] mixed hexakisadducts, such as, for example, **115a–c**. Using a related protocol of reaction sequences starting from **110** allowed complete control over addend permutation at all six pseudooctahedral positions of C_{60} [105]. Two out of 30 possible hexakisadducts with six different addends were selectively synthesized.

In another approach Nishimura and co-workers allowed C_{60} to react with tethers where bis-α,ω-dibromo-*o*-xylene moieties were connected by an oligomethylene chain ($n = 2$–5) (Scheme 10.26) [106]. Only two isomers (*cis*-2- and *cis*-3-isomers) were selectively obtained when $n = 2$ and 3, while another isomer (*e*-isomer) was obtained with $n = 5$. For $n = 4$, a complex mixture of bisadducts was formed which was not separated. Chiral *cis*-3- and *e*-bisadducts were successfully resolved into the respective enantiomers on a chiral HPLC column, although *cis*-2-bisadducts gave a single peak only. Cleavage of the oligomethylene chain afforded corresponding unbridged bis(phenol) adducts. Another templated Diels–Alder bis-functionalization was introduced by Shinkai and co-workers [107, 108]. Using saccharides as templates, two boronic acid groups were regioselectively attached to C_{60}. In this way bis-Diels–Alder adducts involving a *trans*-4 addition pattern were obtained. Removal of the saccharide bridge was possible by treatment with H_2O_2.

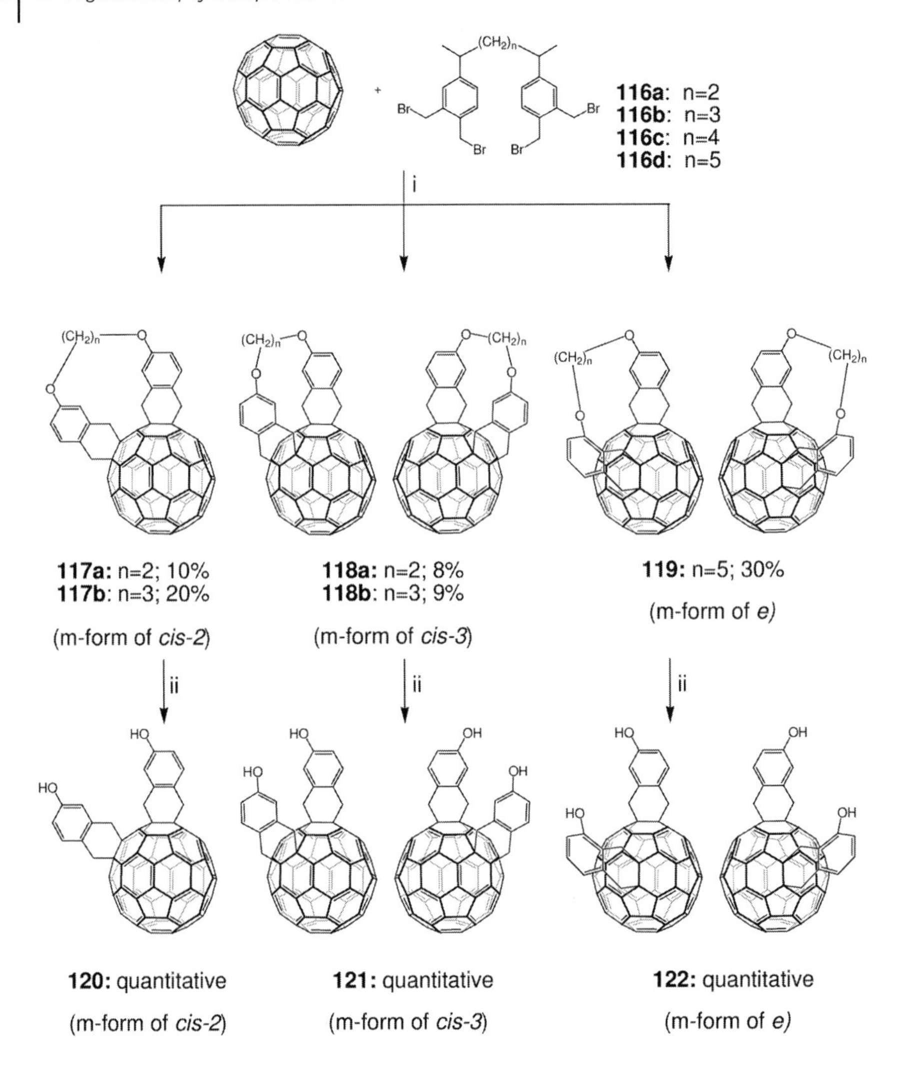

Scheme 10.26 Regioselective tethered double Diels–Alder-additions to C_{60}.

10.3.3.5 **Other Bisfunctional Tethers**

To carry out tethered [3+2]-cycloadditions Nakamura et al. developed the new annulating reagents **123** [109, 110]. The reagents carry, in one molecule, two cyclopropenone acetals that are connected with an *n*-carbon methylene tether. Upon thermolysis in the presence of C_{60}, the reagent undergoes [3+2]-cycloaddition reaction twice in a regio- and stereoselective manner to give C_s- and C_2-symmetrical bisadducts with *cis*-1- and *cis*-3-addition patterns, respectively. The selectivity varies as a function of the tether structure. The experiments have shown that, in each series of different tether lengths, one or two diastereomeric double adducts can be obtained out of several structural possibilities. When using an optically active tether a single, non-racemic cis-3 isomer was obtained in a highly diastereoselective way.

Luh and co-workers showed that the tethered bisazides such as 2,2-dibenzyl-l,3-diazidopropane (**124**) undergo [3+2] cycloadditions to adjacent [6:6]-bonds (*cis*-1 addition). Thermal extrusion of N_2 afforded a mixture of the corresponding *cis-1*-1,2,3,4-bisimino[60]fullerenes and twofold cluster-opened 2,3,4,5-bis-aza-homo[60]-fullerenes [36, 37, 111]. The reaction of C_{60} with optically active bisazides yielded enantiomerically pure bis-azafullerenes with the same addition pattern [111].

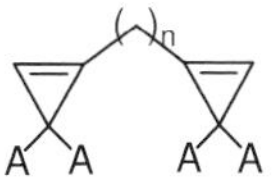

123  **124**

References

1 A. Hirsch, *Top. Curr. Chem.* **1999**, *199*, 1.

2 P. J. Fagan, J. C. Calabrese, B. Malone, *J. Am. Chem. Soc.* **1991**, *113*, 9408.

3 A. Hirsch, I. Lamparth, H. R. Karfunkel, *Angew. Chem.* **1994**, *106*, 453; *Angew. Chem. Int. Ed. Engl.* **1994**, *33*, 437.

4 A. Hirsch, I. Lamparth, G. Schick, *Liebigs Ann.* **1996**, 1725.

5 E. W. Godly, R. Taylor, *Pure Appl. Chem.* **1997**, *69*, 1411.

6 W. H. Powell, F. Cozzi, G. P. Moss, C. Thilgen, R. J. R. Hwu, A. Yerin, *Pure Appl. Chem.* **2002**, *74*, 629.

7 F. Cozzi, C. Thilgen, W. H. Powell, personal communication **2004**.

8 C. Thilgen, A. Herrmann, F. Diederich, *Helv. Chim. Acta* **1997**, *80*, 183.

9 G. Schick, A. Hirsch, H. Mauser, T. Clark, *Chem. Eur. J.* **1996**, *2*, 935.

10 F. Djojo, A. Herzog, I. Lamparth, F. Hampel, A. Hirsch, *Chem. Eur. J.* **1996**, *2*, 1537.

11 K. D. Kampe, N. Egger, M. Vogel, *Angew. Chem.* **1993**, *105*, 1203.

12 J. M. Hawkins, A. Meyer, T. A. Lewis, U. Bunz, R. Nunlist, G. E. Ball, T. W. Ebbesen, K. Tanigaki, *J. Am. Chem. Soc.* **1992**, *114*, 7954.

13 M. Prato, M. Maggini, *Acc. Chem. Res.* **1998**, *31*, 519.

14 C. C. Henderson, C. M. Rohlfing, R. A. Assink, P. A. Cahill, *Angew. Chem.* **1994**, *106*, 803; *Angew. Chem. Int. Ed. Engl.* **1994**, *33*, 786.

15 Y. Nakamura, N. Takano, T. Nishimura, E. Yashima, M. Sato, T. Kudo, J. Nishimura, *Org. Lett.* **2001**, *3*, 1193.

16 A. Hirsch, I. Lamparth, T. Grösser, H. R. Karfunkel, *J. Am. Chem. Soc.* **1994**, *116*, 9385.

17 T. Grösser, M. Prato, V. Lucchini, A. Hirsch, F. Wudl, *Angew. Chem.* **1995**, *107*, 1462; *Angew. Chem. Int. Ed. Engl.* **1995**, *34*, 1343.

18 I. Lamparth, A. Hirsch, *Proc. Electrochem. Soc.* **1995**, *95-10*, 1116.

19 F. Djojo, A. Hirsch, *Chem. Eur. J.* **1998**, *4*, 344.

20 F. Djojo, A. Hirsch, S. Grimme, *Eur. J. Org. Chem.* **1999**, 3027.

21 M. S. Meier, B. R. Weedon, H. P. Spielmann, *J. Am. Chem. Soc.* **1996**, *118*, 11682.

22 B. Kräutler, J. Maynollo, *Angew. Chem. Int. Ed. Engl.* **1995**, *34*, 87.

23 G. Schick, M. Levitus, L. Kvetko, B. A. Johnson, I. Lamparth, R. Lunkwitz, B. Ma, S. I. Khan,

M. A. Garcia-Garibay, Y. Rubin, *J. Am. Chem. Soc.* **1999**, *121*, 3246.
24 K. Hutchison, J. Gao, G. Schick, Y. Rubin, F. Wudl, *J. Am. Chem. Soc.* **1999**, *121*, 5611.
25 I. Lamparth, C. Maichle-Mössmer, A. Hirsch, *Angew. Chem.* **1995**, *107*, 1755; *Angew. Chem. Int. Ed. Engl.* **1995**, *34*, 1607.
26 A. Hirsch, O. Vostrowsky, *Eur. J. Org. Chem.* **2001**, 829.
27 H. Goto, N. Harada, J. Crassous, F. Diederich, *J. Chem. Soc., Perkin Trans. 2* **1998**, 1719.
28 J.-F. Nierengarten, V. Gramlich, F. Cardullo, F. Diederich, *Angew. Chem.* **1996**, *108*, 2242; *Angew. Chem. Int. Ed. Engl.* **1996**, *35*, 2101.
29 J. F. Nierengarten, T. Habicher, R. Kessinger, F. Cardullo, F. Diederich, V. Gramlich, J. P. Gisselbrecht, C. Boudon, M. Gross, *Helv. Chim. Acta* **1997**, *80*, 2238.
30 B. Gross, V. Schurig, I. Lamparth, A. Herzog, F. Djojo, A. Hirsch, *Chem. Commun.* **1997**, 1117.
31 F. Djojo, E. Ravanelli, O. Vostrowsky, A. Hirsch, *Eur. J. Org. Chem.* **2000**, 1051.
32 A. Hirsch, I. Lamparth, T. Grösser, M. Prato, V. Lucchini, F. Wudl, *Phys. Chem. Fullerenes Deriv., Proc. Int. Wintersch. Electron. Prop. Novel Mater.* **1995**, 125.
33 A. Hirsch, I. Lamparth, T. Grösser, M. Prato, V. Lucchini, F. Wudl, *NATO ASI Ser., Ser. E* **1996**, *316*, 267.
34 F. Diederich, L. Isaacs, D. Philp, *Chem. Soc. Rev.* **1994**, *23*, 243.
35 A. Ikeda, C. Fukuhara, S. Shinkai, *Chem. Lett.* **1997**, 407.
36 L.-L. Shiu, K.-M. Chien, T.-Y. Liu, T.-I. Lin, G.-R. Her, T.-Y. Luh, *J. Chem. Soc., Chem. Commun.* **1995**, 1159.
37 C. K. F. Shen, H.-H. Yu, C.-G. Juo, K.-M. Chien, G.-R. Her, T.-Y. Luh, *Chem. Eur. J.* **1997**, *3*, 744.
38 P. J. Krusic, E. Wasserman, P. N. Keizer, J. R. Morton, K. F. Preston, *Science* **1991**, *254*, 1183.
39 P. J. Krusic, E. Wasserman, B. A. Parkinson, B. Malone, E. R. Holler, Jr., P. N. Keizer, J. R. Morton, K. F. Preston, *J. Am. Chem. Soc.* **1991**, *113*, 6274.
40 G. Schick, K.-D. Kampe, A. Hirsch, *J. Chem. Soc., Chem. Commun.* **1995**, 2023.
41 H. Isobe, A. Ohbayashi, M. Sawamura, E. Nakamura, *J. Am. Chem. Soc.* **2000**, *122*, 2669.
42 P. R. Birkett, A. G. Avent, A. D. Darwish, H. W. Kroto, R. Taylor, D. R. M. Walton, *J. Chem. Soc., Chem. Commun.* **1993**, 1230.
43 P. R. Birkett, P. B. Hitchcock, H. W. Kroto, R. Taylor, D. R. M. Walton, *Nature* **1992**, *357*, 479.
44 M. Sawamura, H. Iikura, E. Nakamura, *J. Am. Chem. Soc.* **1996**, *118*, 12850.
45 Y. Murata, M. Shiro, K. Komatsu, *J. Am. Chem. Soc.* **1997**, *119*, 8117.
46 M. Sawamura, Y. Kuninobu, E. Nakamura, *J. Am. Chem. Soc.* **2000**, *122*, 12407.
47 M. Sawamura, Y. Kuninobu, M. Toganoh, Y. Matsuo, M. Yamanaka, E. Nakamura, *J. Am. Chem. Soc.* **2002**, *124*, 9354.
48 E. Nakamura, M. Sawamura, *Pure Appl. Chem.* **2001**, *73*, 355.
49 M. Sawamura, M. Toganoh, K. Suzuki, A. Hirai, H. Iikura, E. Nakamura, *Org. Lett.* **2000**, *2*, 1919.
50 M. Sawamura, M. Toganoh, Y. Kuninobu, S. Kato, E. Nakamura, *Chem. Lett.* **2000**, 270.
51 M. Sawamura, N. Nagahama, M. Toganoh, U. E. Hackler, H. Isobe, E. Nakamura, S.-Q. Zhou, B. Chu, *Chem. Lett.* **2000**, 1098.
52 M. Sawamura, K. Kawai, Y. Matsuo, K. Kanie, T. Kato, E. Nakamura, *Nature* **2002**, *419*, 702.
53 M. Sawamura, N. Nagahama, M. Toganoh, E. Nakamura, *J. Organomet. Chem.* **2002**, *652*, 31.
54 S. Zhou, C. Burger, B. Chu, M. Sawamura, N. Nagahama, M. Toganoh, U. E. Hackler, H. Isobe, E. Nakamura, *Science* **2001**, *291*, 1944.
55 H. Al-Matar, P. B. Hitchcock, A. G. Avent, R. Taylor, *Chem. Commun.* **2000**, 1071.
56 A. a. K. Abdul-Sada, A. G. Avent, P. R. Birkett, H. W. Kroto, R. Taylor, D. R. M. Walton, *J. Chem. Soc., Perkin Trans. 1: Org. Bio-Org. Chem.* **1998**, 393.
57 P. R. Birkett, R. Taylor, N. K. Wachter, M. Carano, F. Paolucci, S. Roffia,

F. Zerbetto, *J. Am. Chem. Soc.* **2000**, *122*, 4209.
58 I. Lamparth, A. Herzog, A. Hirsch, *Tetrahedron* **1996**, *52*, 5065.
59 X. Camps, A. Hirsch, *J. Chem. Soc., Perkin Trans. 1* **1997**, 1595.
60 S. R. Wilson, Q. Lu, *Tetrahedron Lett.* **1995**, *36*, 5707.
61 M. Hetzer, S. Bayerl, X. Camps, O. Vostrowsky, A. Hirsch, T. M. Bayerl, *Adv. Mater.* **1997**, *9*, 913.
62 M. Hetzer, T. Gutberlet, M. F. Brown, X. Camps, O. Vostrowsky, H. Schönberger, A. Hirsch, T. M. Bayerl, *J. Phys. Chem. A* **1999**, *103*, 637.
63 X. Camps, H. Schönberger, A. Hirsch, *Chem. Eur. J.* **1997**, *3*, 561.
64 X. Camps, E. Dietel, A. Hirsch, S. Pyo, L. Echegoyen, S. Hackbarth, B. Röder, *Chem. Eur. J.* **1999**, *5*, 2362.
65 T. Chuard, R. Deschenaux, A. Hirsch, H. Schönberger, *Chem. Commun.* **1999**, 2103.
66 M. Brettreich, S. Burghardt, C. Böttcher, T. Bayerl, S. Bayerl, A. Hirsch, *Angew. Chem.* **2000**, *112*, 1915; *Angew. Chem. Int. Ed. Engl.* **2000**, *39*, 1845.
67 A. Herzog, A. Hirsch, O. Vostrowsky, *Eur. J. Org. Chem.* **2000**, 171.
68 M. Braun, X. Camps, O. Vostrowsky, A. Hirsch, E. Endress, T. M. Bayerl, O. Birkert, G. Gauglitz, *Eur. J. Org. Chem.* **2000**, 1173.
69 P. Timmerman, L. E. Witschel, F. Diederich, C. Boudon, J.-P. Gisselbrecht, M. Gross, *Helv. Chim. Acta* **1996**, *79*, 6.
70 T. Habicher, J.-F. Nierengarten, V. Gramlich, F. Diederich, *Angew. Chem.* **1998**, *110*, 2019; *Angew. Chem. Int. Ed. Engl.* **1998**, *37*, 1916.
71 A. P. Maierhofer, M. Brettreich, S. Burghardt, O. Vostrowsky, A. Hirsch, S. Langridge, T. M. Bayerl, *Langmuir* **2000**, *16*, 8884.
72 B. Kräutler, T. Müller, J. Maynollo, K. Gruber, C. Kratky, P. Ochsenbein, D. Schwarzenbach, H.-B. Bürgi, *Angew. Chem.* **1996**, *108*, 1294; *Angew. Chem. Int. Ed. Engl.* **1996**, *35*, 1204.
73 R. Schwenninger, T. Müller, B. Kräutler, *J. Am. Chem. Soc.* **1997**, *119*, 9317.
74 A. Duarte-Ruiz, K. Wurst, B. Kräutler, *Helv. Chim. Acta* **2001**, *84*, 2167.
75 L. Isaacs, R. F. Haldimann, F. Diederich, *Angew. Chem.* **1994**, *106*, 2434; *Angew. Chem. Int. Ed. Engl.* **1994**, *33*, 2339.
76 F. Diederich, R. Kessinger, in *Templated Organic Synthesis* (Wiley-VCH) **2000**, 189.
77 F. Diederich, R. Kessinger, *Acc. Chem. Res.* **1999**, *32*, 537.
78 P. Seiler, L. Isaacs, F. Diederich, *Helv. Chim. Acta* **1996**, *79*, 1047.
79 L. Isaacs, P. Seiler, F. Diederich, *Angew. Chem.* **1995**, *107*, 1636; *Angew. Chem. Int. Ed. Engl.* **1995**, *34*, 1466.
80 L. Isaacs, F. Diederich, R. F. Haldimann, *Helv. Chim. Acta* **1997**, *80*, 317.
81 R. F. Haldimann, F.-G. Klärner, F. Diederich, *Chem. Commun.* **1997**, 237.
82 E.-U. Wallenborn, R. F. Haldimann, F.-G. Klärner, F. Diederich, *Chem. Eur. J.* **1998**, *4*, 2258.
83 Y.-Z. An, G. A. Ellis, A. L. Viado, Y. Rubin, *J. Org. Chem.* **1995**, *60*, 6353.
84 F. Cardullo, L. Isaacs, F. Diederich, J.-P. Gisselbrecht, C. Boudon, M. Gross, *Chem. Commun.* **1996**, 797.
85 F. Cardullo, P. Seiler, L. Isaacs, J. F. Nierengarten, R. F. Haldimann, F. Diederich, T. Mordasini-Denti, W. Thiel, C. Boudon, J. P. Gisselbrecht, M. Gross, *Helv. Chim. Acta* **1997**, *80*, 343.
86 T. Hamano, K. Okuda, T. Mashino, M. Hirobe, K. Arakane, A. Ryu, S. Mashiko, T. Nagano, *Chem. Commun.* **1997**, 21.
87 D. M. Guldi, K.-D. Asmus, *J. Phys. Chem. A* **1997**, *101*, 1472.
88 J. F. Nierengarten, A. Herrmann, R. R. Tykwinski, M. Rüttimann, F. Diederich, C. Boudon, J. P. Gisselbrecht, M. Gross, *Helv. Chim. Acta* **1997**, *80*, 293.
89 E. Dietel, A. Hirsch, E. Eichhorn, A. Rieker, S. Hackbarth, B. Röder, *Chem. Commun.* **1998**, 1981.
90 A. Soi, A. Hirsch, *New J. Chem.* **1998**, *22*, 1337.
91 G. A. Burley, P. A. Keller, S. G. Pyne, G. E. Ball, *Chem. Commun.* **2000**, 1717.
92 R. W. Alder, S. P. East, *Chem. Rev.* **1996**, *96*, 2097.
93 J.-P. Bourgeois, L. Echegoyen, M. Fibbiroli, E. Pretsch, F. Diederich,

Angew. Chem. **1998**, *110*, 2203; *Angew. Chem. Int. Ed. Engl.* **1998**, *37*, 2118.
94 C. R. Woods, J.-P. Bourgeois, P. Seiler, F. Diederich, *Angew. Chem. Int. Ed. Engl.* **2000**, *39*, 3813.
95 J.-P. Bourgeois, C. R. Woods, F. Cardullo, T. Habicher, J.-F. Nierengarten, R. Gehrig, F. Diederich, *Helv. Chim. Acta* **2001**, *84*, 1207.
96 R. Kessinger, J. Crassous, A. Hermann, M. Rüttimann, L. Echegoyen, F. Diederich, *Angew. Chem.* **1998**, *110*, 2022; *Angew. Chem. Int. Ed. Engl.* **1998**, *37*, 1919.
97 J.-F. Nierengarten, J.-F. Nicoud, *Chem. Commun.* **1998**, 1545.
98 J.-F. Nierengarten, C. Schall, J.-F. Nicoud, *Angew. Chem.* **1998**, *110*, 2037; *Angew. Chem. Int. Ed. Engl.* **1998**, *37*, 1934.
99 G. Rapenne, F. Diederich, J. Crassous, A. Collet, L. Echegoyen, *Chem. Commun.* **1999**, 1121.
100 U. Reuther, T. Brandmüller, W. Donaubauer, F. Hampel, A. Hirsch, *Chem. Eur. J.* **2002**, *8*, 2833.
101 Y. Rubin, *Chem. Eur. J.* **1997**, *3*, 1009.
102 W. Qian, Y. Rubin, *Angew. Chem.* **1999**, *111*, 2504; *Angew. Chem. Int. Ed. Engl.* **1999**, *38*, 2356.
103 W. Qian, Y. Rubin, *J. Org. Chem.* **2002**, *67*, 7683.
104 W. Qian, Y. Rubin, *Angew. Chem.* **2000**, *112*, 3263; *Angew. Chem. Int. Ed. Engl.* **2000**, *39*, 3133.
105 W. Qian, Y. Rubin, *J. Am. Chem. Soc.* **2000**, *122*, 9564.
106 M. Taki, S. Sugita, Y. Nakamura, E. Kasashima, E. Yashima, Y. Okamoto, J. Nishimura, *J. Am. Chem. Soc.* **1997**, *119*, 926.
107 T. Ishi-I, K. Nakashima, S. Shinkai, K. Araki, *Tetrahedron* **1998**, *54*, 8679.
108 T. Ishi-I, S. Shinkai, *Chem. Commun.* **1998**, 1047.
109 E. Nakamura, H. Isobe, H. Tokuyama, M. Sawamura, *Chem. Commun.* **1996**, 1747.
110 H. Isobe, H. Tokuyama, M. Sawamura, E. Nakamura, *J. Org. Chem.* **1997**, *62*, 5034.
111 C. K. F. Shen, K.-M. Chien, C.-G. Juo, G.-R. Her, T.-Y. Luh, *J. Org. Chem.* **1996**, *61*, 9242.

11 Cluster Modified Fullerenes

11.1 Introduction

Reaction sequences leading to an opening of the fullerene cluster [1–3] are not only important in exploring general principles of fullerene chemistry but also provide the basis for the rational synthesis of endohedral fullerenes. The possibility of systematically preparing endohedral fullerenes containing any kind of atom or small molecule would revolutionize molecular materials science. For example, access to new superconductors, magnetic materials and magnetic resonance imaging (MRI) systems by encapsulation of gadolinium ions would be provided. The whole sequence of preparing endohedral fullerenes in such a way would consist of (a) allowing carefully designed exohedral addends to bind to the fullerene core in a very regioselective way, (b) promoting cluster opening within the resulting activated adduct, leading to a hole large enough for the encapsulation of guest atoms or molecules and finally (c) reclosing the cluster while striping off all exohedral addends.

Fullerenes obtained from usual production methods contain five-membered and six-membered rings only. Conceptually, according to Euler's theorem, networks involving specific combinations of hexagons and pentagons with other rings can also form spherical structures. Such fullerene related systems are called *quasi*-fullerenes [4] and are expected to exhibit interesting physical and chemical properties. Synthetic access to *quasi*-fullerenes by chemically modifying conventional fullerenes is challenging.

11.2 Cluster Opened Fullerene Derivatives

11.2.1 "Fulleroids": Bridged Adducts with Open Transannular Bonds

The simplest cluster opened fullerenes are represented by the "fulleroids" (1(6)a-homo[60]fullerenes) [4], the "azafulleroids" (1a-aza-1(6)a-homo[60]fullerenes) [4] and the "oxafulleroids" (1a-oxa-1(6)a-homo[60]fullerenes) [4] (Chapter 4). The

Fullerenes: Chemistry and Reactions. Andreas Hirsch and Michael Brettreich

ISBN: 3-527-30820-2

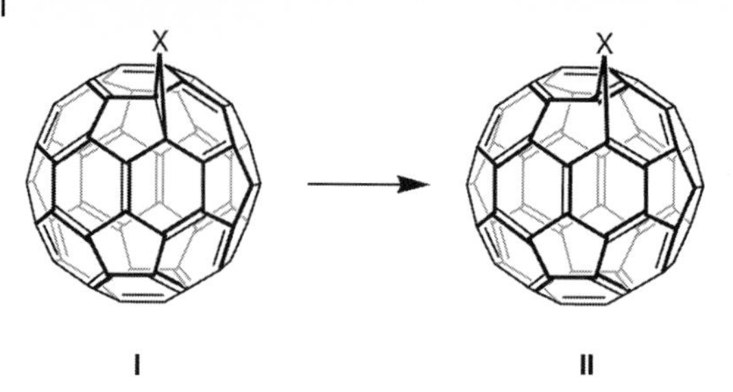

Scheme 11.1 Ring opening reaction from a hypothetical [5,6]-bridged adduct to a fulleroid.

transannular [5,6]-bond in all parent fulleroids **1** [5], **2** [6] and **3** [7] is open and the 60 π-electron system of the fullerene core remains intact. Ring opening from a hypothetical precursor adduct with a closed transannular [5,6]-bond can formally be regarded as a retro-electrocyclic reaction (Scheme 11.1). However, the orbitals that play a role in such a valence isomerization are delocalized over the whole π-system of the fullerene. Consequently, one has to be careful in establishing complete analogy with Woodward–Hoffmann rules obeying cyclohexadiene–hexatriene isomerizations involving six π-electrons only.

In general, open structures **II** are energetically preferred over the closed forms **I** [8, 9]. In the ring closed isomers **I** two unfavorable double bonds within five-membered rings would be required. No monoadduct with such a structure has been observed. Fulleroids such as **1–3** are usually formed via rearrangement of their pyrazoline-, triazoline- or ozonide [6,6]-precursor adducts accompanied by extrusion of N_2 or O_2 (see Chapter 4) [7, 10–12].

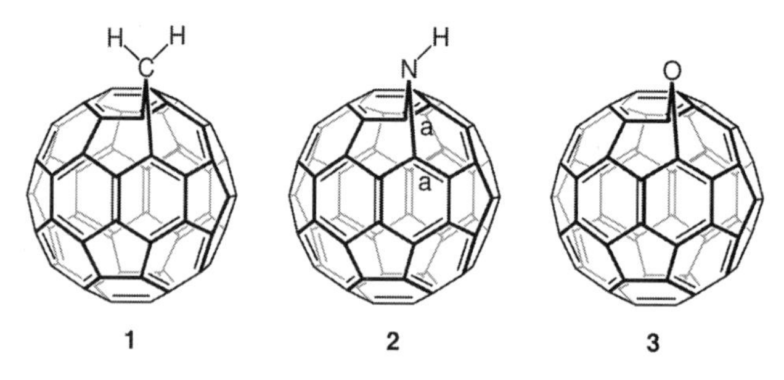

As outlined in Chapter 10, three types of bisazafulleroids, **4** [13], **5** [14–16] and **6** [17], derived from C_{60} involving opened 11-, 13- and 14-membered rings, respectively, have been synthesized.

Whereas in **4** and **5** two [5,6]-bonds are open, the bisazafulleroids **6** represents the first examples of bridged [6,6]-adducts (*cis*-1 addition pattern) with open transannular bonds. In this case three [5,6]-double bonds have to be introduced. This is in contrast to **4** and **5** where no energetically unfavorable [5,6]-double bonds are required. Upon changing the addition pattern, as demonstrated by the investigation of the other possible regioisomers of **6**, regular behavior with closed transannular [6,6]-bonds is observed [17, 18]. The fullerene cage can be closed again

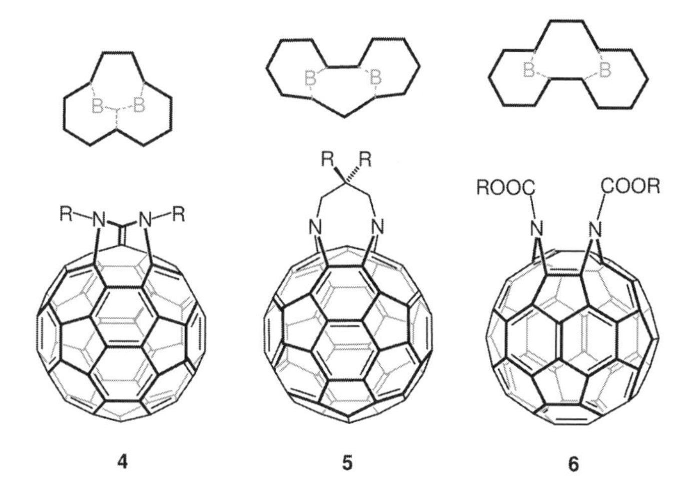

via a 60π electrocyclic reaction by transferring *cis-1*-$C_{60}(NCOOR)_2$ (**6**) into *cis-1*-$C_{60}(NR)_2$ (**7**) as cluster closed valence isomer (Scheme 11.2) [17]. The latter phenomenon clearly demonstrates the role of the addend. An extensive AM1 and DFT study revealed that only in *cis*-1 adducts that prefer planar imino bridges (e.g. carbamates or amides) the open forms are more stable than the closed forms [17].

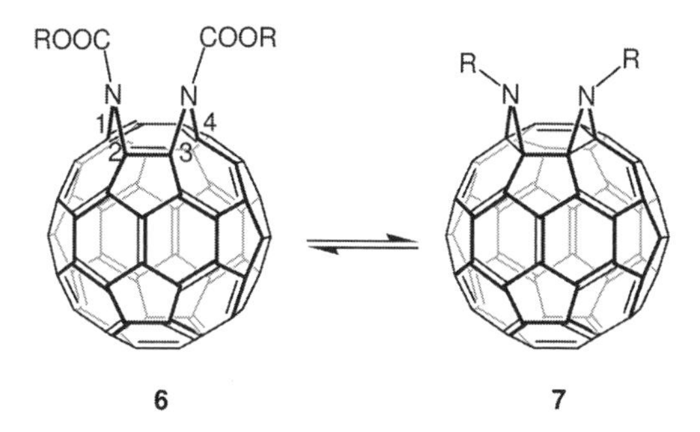

Scheme 11.2 Chemically induced valence tautomerization of *cis*-1 bisazafulleroids and *cis*-1-bisiminofullerenes via formal Diels–Alder/ retro-Diels–Alder reactions (R = e.g. H, alkyl).

The possible opening and closure of transannular *cis*-1 bonds in bisimino-bridged fullerenes such as **6** and **7** is governed by the following factors [17, 18]:

(1) With the exception of a *cis*-1 adduct, where upon the two-fold ring opening due to the location of the imino bridges in the same six-membered ring only three [5,6]-double bonds have to be introduced, six of these energetically unfavorable bonds would be required for the hypothetical open structures of the other seven regioisomers (*trans*-1 to *cis*-2). In the latter regioisomers the transannular [6,6]-bonds are always closed.

(2) In a *cis*-1 adduct a closed valence isomer bears a strained planar cyclohexene ring but the introduction of an unfavorable [5,6]-double bond is avoided, whereas in an open valence isomer no strained planar cyclohexene but three [5,6]-double bonds are present.

(3) For *cis*-1 adducts, open valence isomers are favored for imino addends with planar imino bridges such as carbamates and the closed isomers are favored for imino addends with pyramidalized imino bridges such as alkylimines or NH.

(4) Carbamates or amides prefer planar arrangements of the nitrogen due to resulting favorable conjugation of the free electron pair with the carbonyl group. This has consequences for imino[60]fullerenes, since the planar arrangement of the carbamate N-atoms and the required enlargement of the bond angles C-1–N–C-2 or C-3–N–C-4 are most favorably realized if the transannular [6,6]-bonds are open.

Highly regioselective access to bisazafulleroids **4** is based on the pronounced reactivity of the vinylamine double bonds labeled 'a' within **2** [13]. This addition pattern can only be avoided if sufficiently rigid tethers linking the two binding sites are used. This was demonstrated with the synthesis of **5** where two non-adjacent [5,6]-binding sites within one five-membered ring are bridged by the addend. As well as **4** derived from C_{60}, the two related C_{70} derivatives **8** and **9** have been prepared [19, 20].

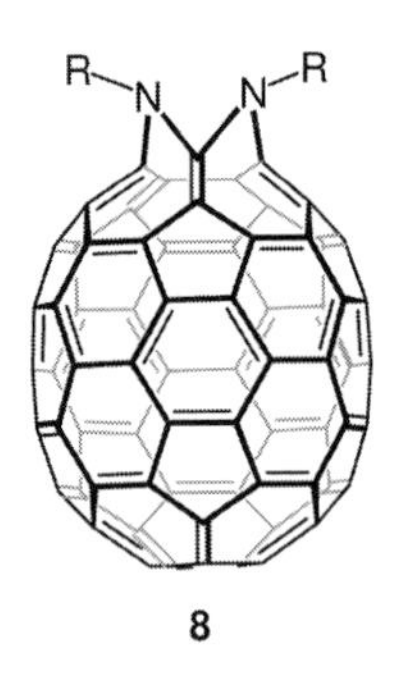

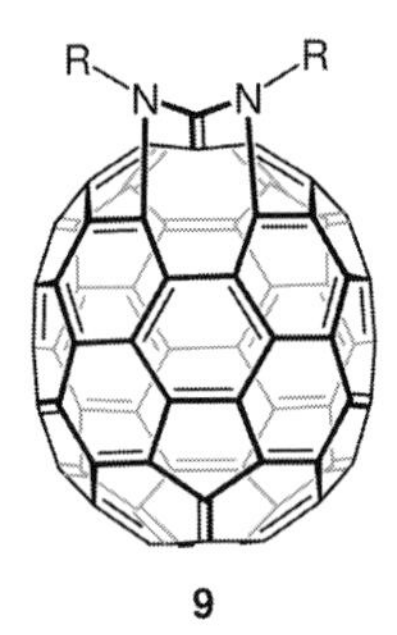

A doubly [5,6]-brigded fulleroid **13** involving another ring opening motif was obtained after photolysis of the precursor adduct **10** under reflux and acidic conditions (TsOH) (Scheme 11.3) [21]. During the reaction sequence the initially formed diene **11** undergoes a photochemically promoted [4+4] cyclization leading to ring closed **12**, which immediately opens up to **13** in a 60π electrocyclic reaction.

Again it has to be noted that the frontier orbitals participating in such a valence isomerization are delocalized over the whole molecule [22]. This has consequences for the orbital symmetry and, thereby, a prior analogy with comparable processes involving 6π-electrons only is not given. However, compared with smaller π-systems the selection rules for orbital symmetry controlled processes in fullerenes seem to be less restrictive, since a large number of π-orbitals with small energy separation are available. Calculations at the AM1 and PM3 level show that the photocyclization

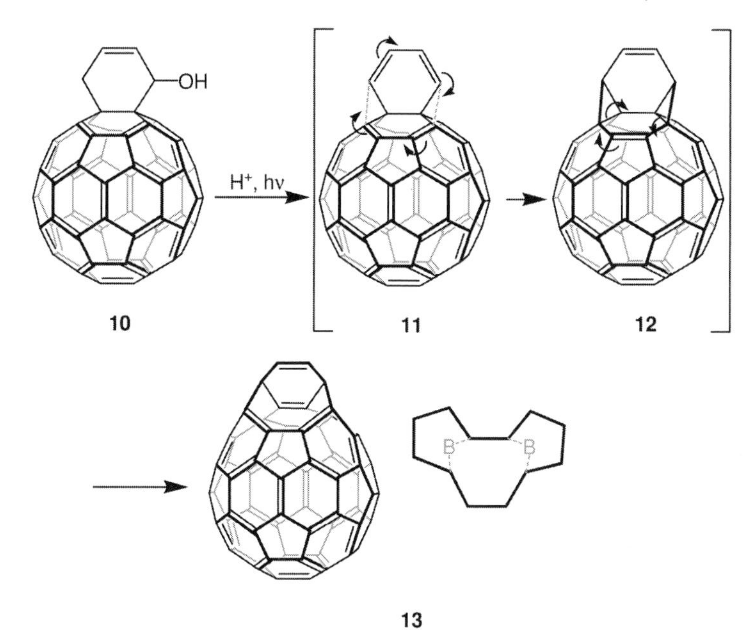

Scheme 11.3 Synthesis of the bisfulleroid **13**.

of **11** leading to **12** goes uphill by about 27 kcal mol^{-1}, and the subsequent ring opening to **13** provides an overall 14–24 kcal mol^{-1} energy gain [21]. Bisfulleroid **13** is purple in solution, which is characteristic for fulleroids.

Subsequently, several related bisfulleroids such as **14–16** have been synthesized [23–28]. In these cases the first reaction step was the cycloaddition of a phthalazine or triazine addend followed by the extrusion of N_2. A sequence of cyclization and ring opening reactions related to that of Scheme 11.3 lead to the corresponding bisfulleroids. The formation of **15** required a final hydrolysis step [24]. The structure of **16** was confirmed by X-ray crystallography [26]. Access to tetrakisfulleroids has been provided via a regioselective tether controlled bis-Diels–Alder addition followed by the removal of the tether and ring opening [29].

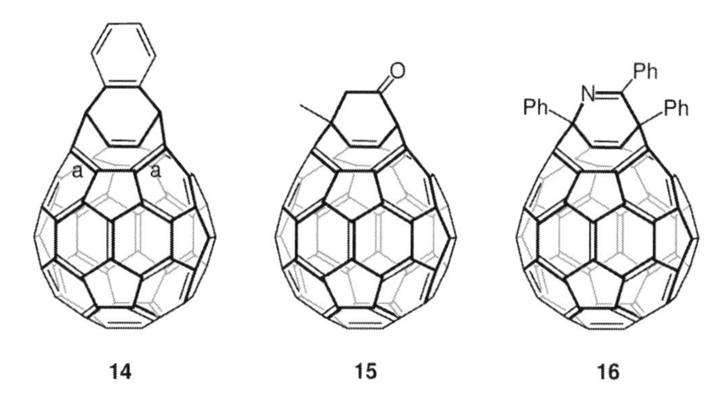

11.2.2
Ring-enlargement Reactions of Bisfulleroids

In all examples presented so far the opened ring fragments within the fullerene framework are singly or doubly bridged. Consequently, successful penetration of guest atoms or molecules to make endohedral derivatives is impossible. However, the cluster-modified fulleroids can serve as starting materials for further ring-enlargement reactions. For example, **13** reacts with $CpCo(CO)_2$ to afford complex **17**, which has been characterized by X-ray crystallography [21]. With the formation of **17**, triple scission of a six-membered ring on the surface of C_{60} leading to an open 15-membered ring was achieved. Several attempts to effect subsequent Co-insertion failed although it was hypothesized that crystal vibrations, in addition to intramolecular vibrations, could promote the metal insertion process at higher temperatures and pressures [2, 30]. As can be seen from space filling representations, the effective size of the ring opening within **17** is still very small [2].

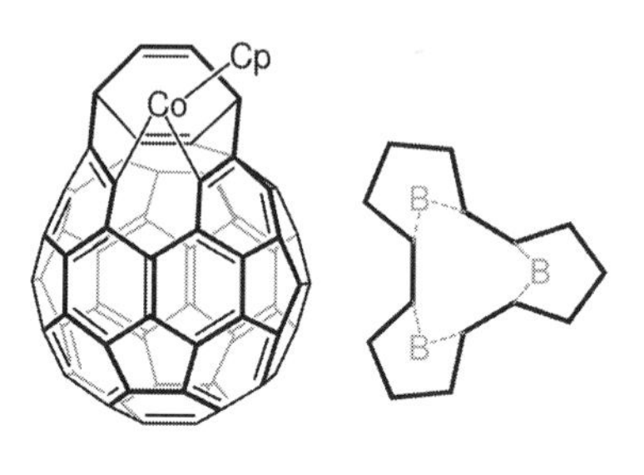

17

The a-type [6,6]-double bonds within bisfulleroids such as **14** are very susceptible to the addition of singlet oxygen. This is due to very pronounced HOMO coefficients at these sites, as demonstrated by quantum mechanical calculations [23]. Indeed, when **14** was photochemically treated with O_2 the open cage fullerene **19**, which contains an enol and a carbonyl group on a 12-membered ring orifice, was obtained (Scheme 11.4) [23]. The proposed reaction intermediate is the dioxolane **18**. The subsequent [2+2]-cycloreversion is accompanied by keto–enol tautomerization, leading to a stable conjugated octatetraene subunit. Altogether, the fullerene core within **19** contains an open doubly bridged 14-membered ring.

Related cluster opened fullerenes have been obtained after the same treatment of other bisfulleroids [24, 26, 27]. A special situation was found for the treatment of **16** with singlet oxygen (Scheme 11.5) [26]. Here, three cluster opened derivatives, **21**, **22** and **24**, were formed. This is because the precursor molecule **16** does not contain a mirror plane and the initial addition of singlet oxygen to the [6,6]-double bonds a and a′ leads to the different dioxetanes **20** and **23**, respectively. The ring opening of the intermediate **20** can proceed via a normal [2+2]-cycloreversion, leading to the expected compound **21**, but also via an alternative ring opening leading to the bis-oxygen-bridged isomer **22**. The related formation of **25** from dioxetane **23** was not observed.

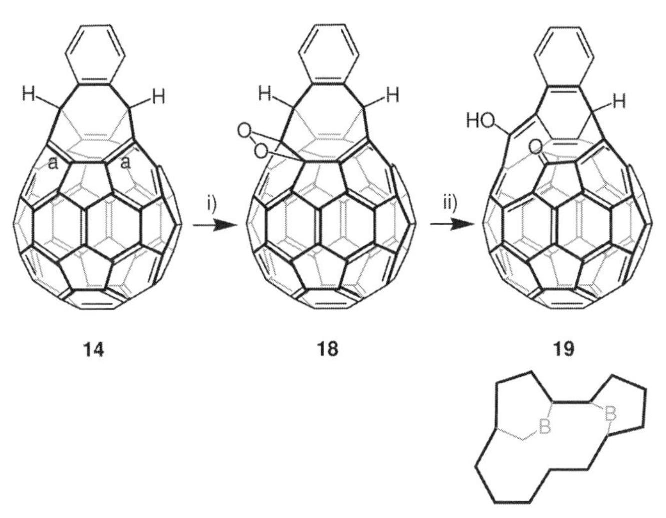

Scheme 11.4 Photochemical reaction of **14** with singlet oxygen, leading to the cluster opened fullerene **19**. (i) O_2, *hv*, benzene; (ii) [2+2] cycloreversion, keto–enol tautomerization.

The thermal reaction of **21** dissolved in ODCB with elemental sulfur in the presence of tetrakis(dimethylamino)ethylene (TDAE) lead to a further increase of the opening of the fullerene cluster to a triply bridged 17-membered ring. This enlargement is caused by the insertion of one sulfur atom into bond b in **21** [26]. As confirmed by X-ray crystallography, the unhindered orifice of **26** consists of a 13-membered ring. This is the largest hole constructed thus far on the surface of C_{60}. The orifice is closer to a circle than to an ellipse as suggested by the schematic representation of **26**.

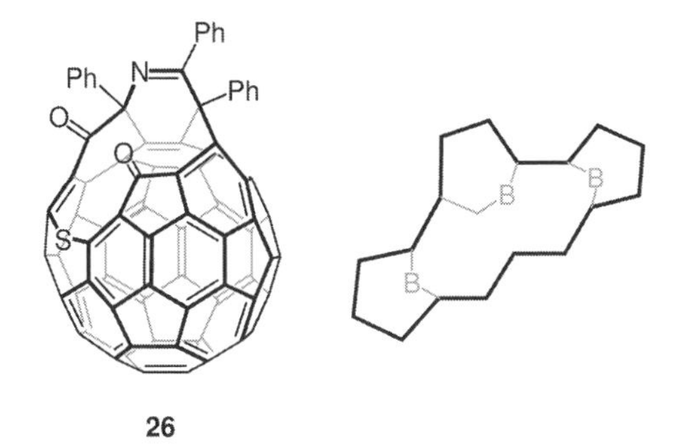

The energies for insertion of He, Ne, H_2 and Ar into **26** were calculated to be 18.9, 26.2, 30.1 and 97.7 kcal mol^{-1}, respectively [31]. These comparatively low energy barriers encouraged experimental investigations on the encapsulation of H_2. Treatment of a powder of **26** with a high pressure of H_2 (800 atm) at 200 °C in an autoclave for 8 h afforded H_2@**26** [31]. The ^1H NMR spectrum showed a sharp

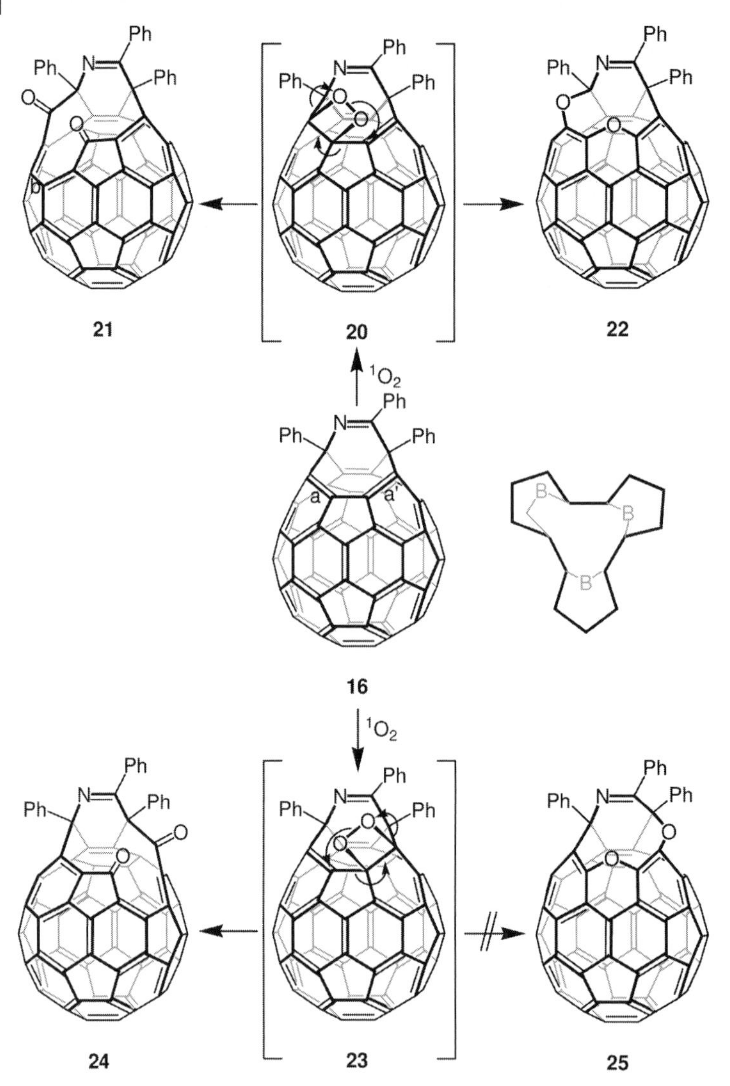

Scheme 11.5 Reaction of **16** with singlet oxygen, and structure motif of the open ring within the cluster framework of **22** and **25**. Curly arrows within the formulas **20** and **23** indicate the cluster opening reaction leading to the valence isomers **22** and **25**.

signal at $\delta = -7.25$ due to the encapsulated H_2 molecule. The integrated relative intensity of this signal demonstrated that 100% encapsulation was achieved. After three months, no escape of the encapsulated H_2 was observed for H_2@**26** dissolved in ODCB at room temperature. However, release of H_2 takes place upon heating above 160 °C. Mass spectrometry investigations on H_2@**26** showed effective fragmentation to $H_2@C_{60}$.

11.2.3
Cluster Opened Lactams, Ketolactams and Lactones

The first cluster opened fullerene containing a non-bridged orifice was obtained when azafulleroid **27** was allowed to react with $^1O_2$ [32]. The regioselective cycloaddition to the vinyl amine bond of **27** leads to the dioxetane intermediate **28**, which, after spontaneous rearrangement, converts into the ketolactam **29** (Scheme 11.6). The 11-membered ring opening in **29** is rather small as compared, for example, to the orifice in **26**. This was demonstrated by the fact that escape of endohedral He in ^{3}He@**29** could not be observed, even after heating up to 200 °C, by ^{3}He NMR [2]. Related C_{70}-based ketolactams could also be synthesized [33, 34]. These cluster opened fullerenes served as starting material for the generation of macroscopic samples of the hetereofullerene dimers $(C_{59}N)_2$ and $(C_{69}N)_2$ (see Chapter 12). The reaction of **29** with phenylhydrazine lead to a further regioselective cage scission to give a 15-membered ring-open fullerene [28].

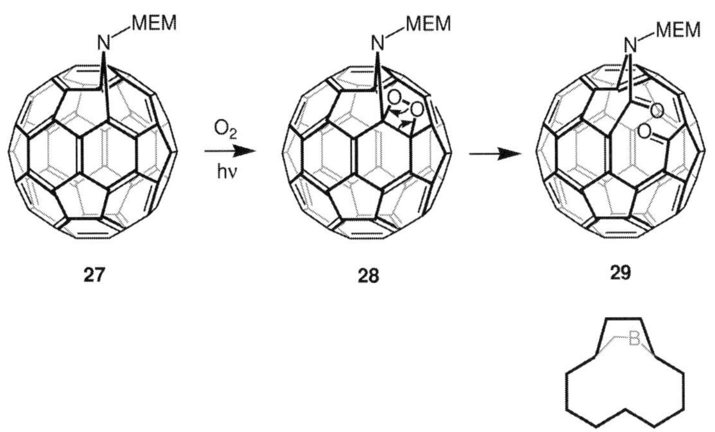

Scheme 11.6 Formation of the cluster opened ketolactam **29**.

Cluster opened bislactams have been obtained in a one-pot reaction by treating diazidobutadiene **30** with C_{60} for four days in ODCB at 55 °C [3, 35]. The first step in the reaction sequence can be assumed to be the formation of the bisazafulleroid **31** (Scheme 11.7). Subsequent addition of $^1O_2$ leads to the endoperoxide intermediate **32**, which isomerizes via a electrocyclic ring opening to **33** [36]. Subsequent oxidation with DDQ or O_2 afforded the dehydrogenated bislactam **34**.

The ring opening within **34** is large enough to allow a partial endohedral filling with He and H_2 [37]. Incorporation of He has been demonstrated by ^{3}He NMR – a host–guest complex He@**34** is formed even at low helium pressure. The degree of filling was determined to be 0.1%. Treatment of **34** with a high pressure of H_2 at 400 °C allowed for the incorporation of up to 5% of molecular hydrogen, as determined by ^{1}H NMR spectroscopy.

The bislactone **36** is the first example of a cluster opened C_{70} derivative [38]. It was formed by the spontaneous oxidation of Ph_8C_{70} (**35**) in air (Scheme 11.8).

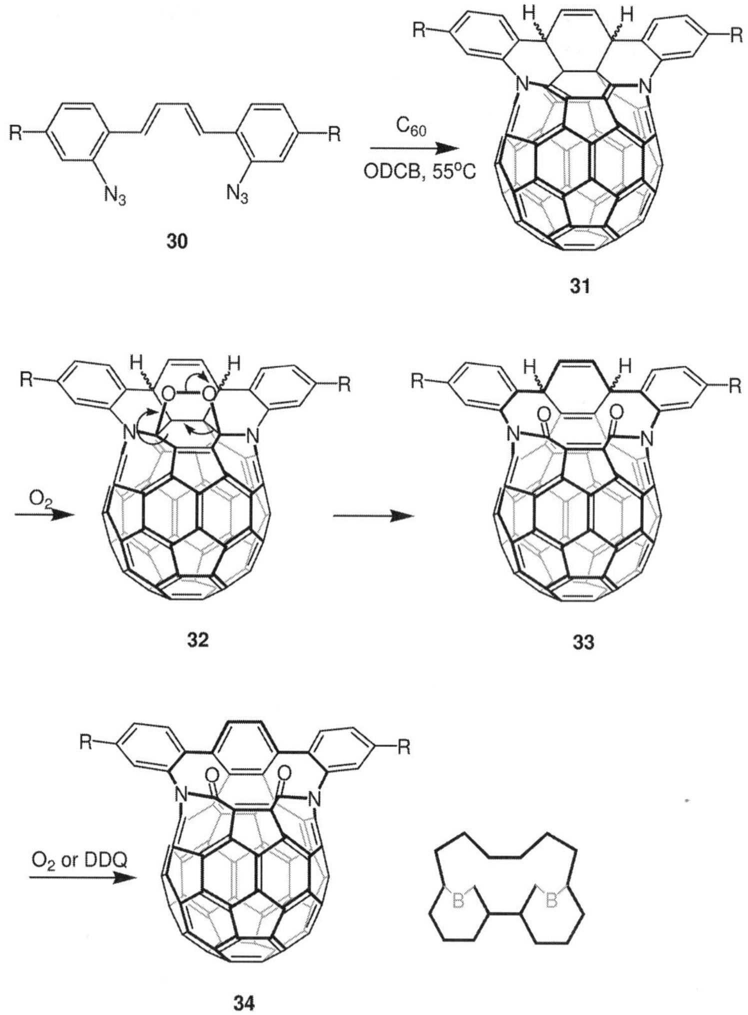

Scheme 11.7 Preparation of the cluster opened bislactam **34**.

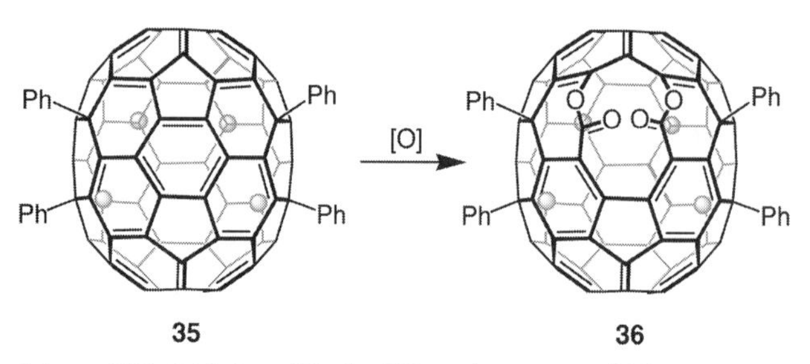

Scheme 11.8 Oxidation of Ph_8C_{70} (**35**) to cluster opened **36**.
The small spheres denote additional locations of the phenyl addends.

Its open structure involving a 13-atom orifice, not counting the bridging –COO– moieties, was assigned on the basis of NMR and IR data. The precursor compound **35** exhibits a reactive [5,6]-double bond, which is very susceptible to O_2 addition. The intermediate dioxetane **36** is probably formed via ring opening and subsequent Baeyer–Villiger-type oxidation [2].

11.3
Quasi-fullerenes

In a first conceptual approach to *quasi*-fullerenes bisfulleroid **37** has been used as precursor (Scheme 11.9) [39]. Mass spectrometric investigations on **37** clearly show efficient fragmentation to C_{62} (**40**) presumably via the intermediates **38** and **39**. It was anticipated that **40** is a C_{2v} symmetric *quasi*-fullerene involving a four-membered ring.

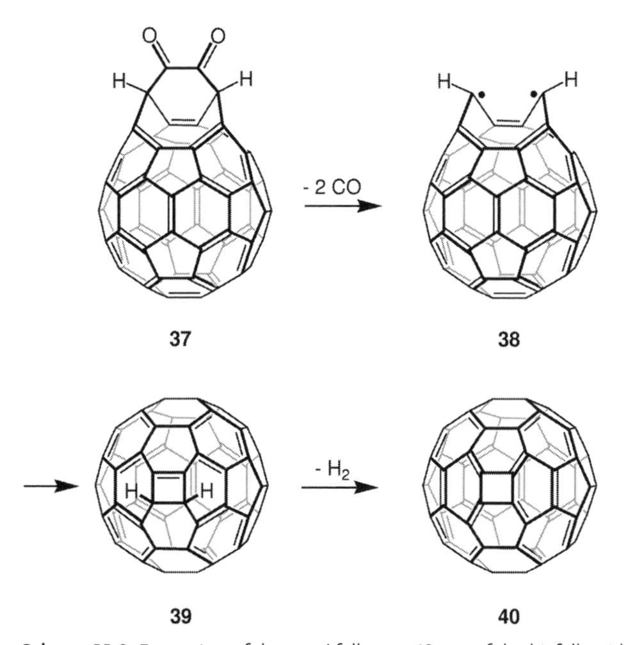

Scheme 11.9 Formation of the *quasi*-fullerene **40** out of the bisfulleroid **37**.

Stable derivatives of the *quasi*-fullerene **40** were obtained in a one-pot reaction upon treating C_{60} with the tetrazine **41** (Scheme 11.10) [40]. The sequential rearrangement and fragmentation of the primary adducts, involving intermediates **42** and **43**, afforded **43** in yields between 14 and 22%. Unambiguous structural proof was obtained by X-ray crystallography.

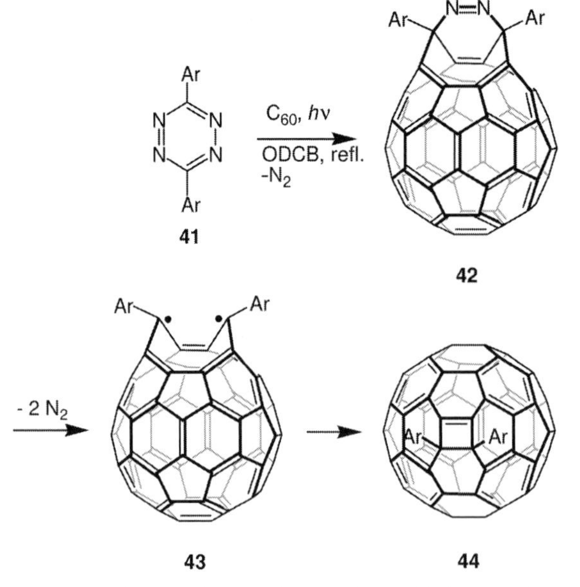

Scheme 11.10 One-pot synthesis of stable *quasi*-fullerene adducts **44**.

11.4 Outlook

The first examples of cluster modified fullerenes presented in this chapter clearly demonstrate the exciting possibilities arising from cage modifications of parent fullerenes. Cluster opened fullerenes represent unprecedented host molecules for supramolecular chemistry. Generating even larger orifices is one of the major challenges for future fullerene chemistry. The electronic and materials properties of the resulting concave PAHs are expected to be outstanding. Moreover, they would be fascinating receptors for small molecules. The latter aspect becomes even more attractive if the spheres can be re-closed once the guest is located in the cage. Whereas the incorporation of small molecules or atoms such as noble gases will lead to neutral systems the encapsulation of ions will give salts. These salts are interesting for themselves but by using electrochemical methods they could also be converted into "molecular atoms", for example with a positively charged core (metal cation) and a negatively charged carbon shell. One can imagine encapsulating all elements of the periodic system in fullerenes. So far, endohedral complexes, in small quantities, have only been prepared either with electropositive metals during fullerene formation or subsequently by the thermal incorporation, for example, of noble gas and nitrogen atoms (Chapter 1). The latter results are already very exciting but the yields of endohedral fullerenes are still low and the method seems to be restricted to the incorporation of noble gases. This could be overcome by using a sequence of mild chemical opening-, encapsulation- and closure reactions leading to defined products. The quantitative encapsulation of H_2 into cluster opened **26** and the fragmentation of H_2@**26** to $H_2@C_{60}$ in a mass spectrometer (Section 11.2.2) is a first encouraging step towards this goal.

References

1 Y. Rubin, *Chem. Eur. J.* **1997**, *3*, 1009.
2 Y. Rubin, *Top. Curr. Chem.* **1999**, *199*, 67.
3 J.-F. Nierengarten, *Angew. Chem.* **2001**, *113*, 3061; *Angew. Chem. Int. Ed. Engl.* **2001**, *40*, 2973.
4 E. W. Godly, R. Taylor, *Pure Appl. Chem.* **1997**, *69*, 1411.
5 T. Suzuki, Q. Li, K. C. Khemani, F. Wudl, *J. Am. Chem. Soc.* **1992**, *114*, 7301.
6 J. Averdung, H. Luftmann, J. Mattay, K.-U. Calus, W. Abraham, *Tetrahedron Lett.* **1995**, *36*, 2957.
7 R. B. Weisman, D. Heymann, S. M. Bachilo, *J. Am. Chem. Soc.* **2001**, *123*, 9720.
8 F. Diederich, L. Isaacs, D. Philp, *Chem. Soc. Rev.* **1994**, *23*, 243.
9 F. Diederich, L. Isaacs, D. Philp, *J. Chem. Soc., Perkin Trans. 2* **1994**, 391.
10 G. Schick, A. Hirsch, *Tetrahedron* **1998**, *54*, 4283.
11 R. F. Haldimann, F.-G. Klärner, F. Diederich, *Chem. Commun.* **1997**, 237.
12 E.-U. Wallenborn, R. F. Haldimann, F.-G. Klärner, F. Diederich, *Chem. Eur. J.* **1998**, *4*, 2258.
13 T. Grösser, M. Prato, V. Lucchini, A. Hirsch, F. Wudl, *Angew. Chem.* **1995**, *107*, 1462; *Angew. Chem. Int. Ed. Engl.* **1995**, *34*, 1343.
14 L.-L. Shiu, K.-M. Chien, T.-Y. Liu, T.-I. Lin, G.-R. Her, T.-Y. Luh, *J. Chem. Soc., Chem. Commun.* **1995**, 1159.
15 P. P. Kanakamma, S.-L. Huang, C.-G. Juo, G.-R. Her, T.-Y. Luh, *Chem. Eur. J.* **1998**, *4*, 2037.
16 C. K. F. Shen, H.-H. Yu, C.-G. Juo, K.-M. Chien, G.-R. Her, T.-Y. Luh, *Chem. Eur. J.* **1997**, *3*, 744.
17 G. Schick, A. Hirsch, H. Mauser, T. Clark, *Chem. Eur. J.* **1996**, *2*, 935.
18 A. Hirsch, *Top. Curr. Chem.* **1999**, *199*, 1.
19 B. Nuber, A. Hirsch, *Fullerene Sci. Technol.* **1996**, *4*, 715.
20 B. Nuber, A. Hirsch, *Chem. Commun.* **1996**, 1421.
21 M.-J. Arce, A. L. Viado, Y.-Z. An, S. I. Khan, Y. Rubin, *J. Am. Chem. Soc.* **1996**, *118*, 3775.
22 A. Hirsch, I. Lamparth, T. Grösser, H. R. Karfunkel, *J. Am. Chem. Soc.* **1994**, *116*, 9385.
23 Y. Murata, K. Komatsu, *Chem. Lett.* **2001**, 896.
24 Y. Murata, M. Murata, K. Komatsu, *J. Org. Chem.* **2001**, *66*, 8187.
25 Y. Murata, N. Kato, K. Komatsu, *J. Org. Chem.* **2001**, *66*, 7235.
26 Y. Murata, M. Murata, K. Komatsu, *Chem. Eur. J.* **2003**, *9*, 1600.
27 H. Inoue, H. Yamaguchi, S. I. Iwamatsu, T. Uozaki, T. Suzuki, T. Akasaka, S. Nagase, S. Murata, *Tetrahedron Lett.* **2001**, *42*, 895.
28 S.-I. Iwamatsu, F. Ono, S. Murata, *Chem. Lett.* **2003**, *32*, 614.
29 W. Qian, Y. Rubin, *J. Org. Chem.* **2002**, *67*, 7683.
30 C. M. Edwards, I. S. Butler, W. Qian, Y. Rubin, *J. Mol. Struct.* **1998**, *442*, 169.
31 Y. Murata, M. Murata, K. Komatsu, *J. Am. Chem. Soc.* **2003**, *125*, 7152.
32 J. C. Hummelen, M. Prato, F. Wudl, *J. Am. Chem. Soc.* **1995**, *117*, 7003.
33 C. Bellavia-Lund, J.-C. Hummelen, M. Keshavarz-K, R. Gonzalez, F. Wudl, *J. Phys. Chem. Solids* **1997**, *58*, 1983.
34 J. C. Hummelen, C. Bellavia-Lund, F. Wudl, *Top. Curr. Chem.* **1999**, *199*, 93.
35 G. Schick, T. Jarrosson, Y. Rubin, *Angew. Chem. Int. Ed. Engl.* **1999**, *38*, 2360.
36 S. Irle, Y. Rubin, K. Morokuma, *J. Phys. Chem. A* **2002**, *106*, 680.
37 Y. Rubin, T. Jarrosson, G.-W. Wang, M. D. Bartberger, K. N. Houk, G. Schick, M. Saunders, R. J. Cross, *Angew. Chem.* **2001**, *113*, 1591; *Angew. Chem. Int. Ed. Engl.* **2001**, *40*, 1543.

38 P. R. Birkett, A. G. Avent, A. D. Darwish, H. W. Kroto, R. Taylor, D. R. M. Walton, *J. Chem. Soc., Chem. Commun.* **1995**, 1869.

39 W. Qian, M. D. Bartberger, S. J. Pastor, K. N. Houk, C. L. Wilkins, Y. Rubin, *J. Am. Chem. Soc.* **2000**, *122*, 8333.

40 W. Qian, S.-C. Chuang, R. B. Amador, T. Jarrosson, M. Sander, S. Pieniazek, S. I. Khan, Y. Rubin, *J. Am. Chem. Soc.* **2003**, *125*, 2066.

12 Heterofullerenes

12.1 Introduction

Apart from exohedral fullerene adducts, endohedral and cluster opened fullerenes, heterofullerenes [1] represent the fourth fundamental group of modified fullerenes. In heterofullerenes, one or more carbon atoms of the cage are substituted by heteroatoms, for example, the trivalent nitrogen or boron atoms. In these cases, substitution of an odd number of C atoms leads to radicals that can be stabilized by dimerization, whereas the replacement of an even number of C atoms would lead directly to closed shell systems. Conceptually, the systematic substitution of cage C-atoms by heteroatoms represents the three-dimensional counterpart of the transition of planar aromatics to heteroaromatics, involving the same gain in structural and functional qualities. The first syntheses of heterofullerenes in bulk quantities were accomplished in 1995, with the preparation of the aza[60]fullerene $C_{59}N$ (**1**) and its dimer $(C_{59}N)_2$ (**2**) [2, 3].

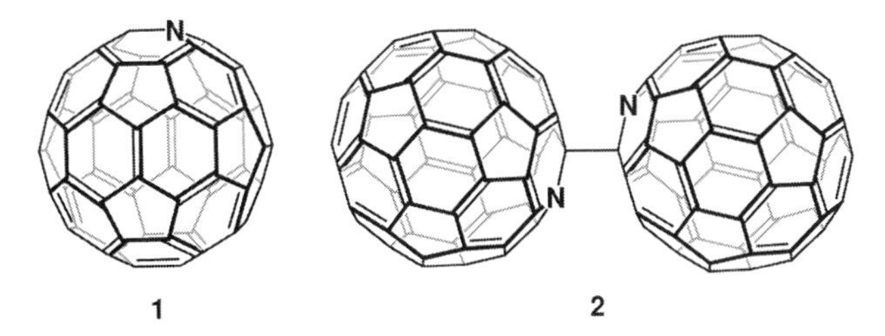

Before that time, experimental evidence for the existence of heterofullerenes was very rare. In 1991 Smalley and co-workers [4, 5] reported on the mass spectrometric detection of borafullerenes $C_{60-n}B_n$ ($n = 1$–6) generated by laser vaporization of graphite–boron nitride composites. Later, in the same year, the preparation of several C_nN_m clusters was reported [6–8]. However, none of these species has been isolated or structurally characterized. In 1994 Clemmer et al. presented evidence for heterofullerenes containing a metal as part of the framework [9]. Muhr et al. reported on the preparation of monoborafullerenes; however, presumably due to the instability of these systems, no pure material could be isolated and completely characterized [10]. Real preparative heterofullerene chemistry began in 1995. The

Fullerenes: Chemistry and Reactions. Andreas Hirsch and Michael Brettreich

ISBN: 3-527-30820-2

groups of Mattay [11] and Hirsch [12] discovered that certain epiminofullerenes and azahomofullerenes (azafulleroids) are suitable precursors for the formation of positively charged heterofullerenes such as $C_{59}N^+$ and $C_{69}N^+$ in the gas phase. Soon thereafter the groups of Wudl [2] and Hirsch [3] published the first syntheses of azafullerenes. With these discoveries, heterofullerene chemistry as a new discipline within the area of fullerene research began [1, 13].

12.2
Synthesis of Nitrogen Heterofullerenes from Exohedral Imino Adducts of C_{60} and C_{70}

So far, all efforts to generate, isolate and characterize heterofullerenes via Krätschmer–Huffman vaporization of graphite in the presence of hetero-element-containing compounds such as boron nitride (BN) or cyanogen $(CN)_2$ have failed. An alternative route for the direct formation of heterofullerenes is cluster rearrangement within exohedral fullerene derivatives such as iminofullerenes and azafulleroids. The first hints of success by this approach were obtained from mass spectrometry investigations of the *cis-1*-diazabishomo[60]fullerene **3** [12], the *n*-butylamine adduct **4** [12] the 1,2-epiminofullerene **5** [11] and the cluster opened ketolactam **6** [2].

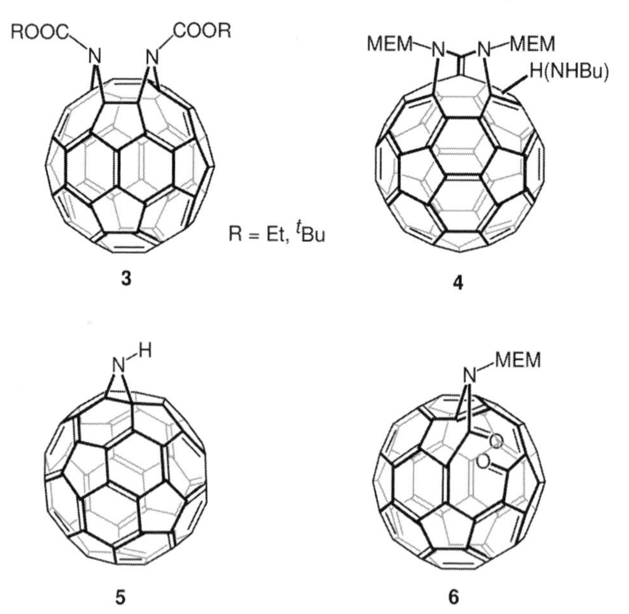

12.2.1
Synthesis of bis(Aza[60]fullerenyl) $(C_{59}N)_2$

FAB mass spectrometry of cluster-opened **3** and the isomeric mixture of *n*-butylamine adducts **4** revealed an unprecedented fragmentation behavior [12]. Whereas the usual exohedral adducts of C_{60} exhibit typical characteristics, namely relatively small M^+ peaks and the peaks of the fullerene fragment ions C_{60}^+ and C_{70}^+ at $m/z = 720$ as the most intense signals, the fragmentation of **3** and **4** in the gas phase leads most efficiently to ions with masses of 722. These peaks are due to the heterofullerenium ion $C_{59}N^+$ since (1) after the first "shrink wrapping" the signals of the fragment ions $C_{57}N^+$ was also of pronounced intensity (under these high energy conditions any exohedrally bound addends would be stripped off); (2) the ^{15}N labeled analogues of **4** gave rise to the most intensive fragmentation signals at $m/z = 723$; and (3) the high-resolution MS signal was consistent with the formula $C_{59}N^{+\bullet}$. The ion $C_{59}N^{+\bullet}$ is isoelectronic with C_{60}. The parent 1,2-epimino[60]fullerene **5**, which was generated via deprotection from the carbamate **7** (R = COOtBu) (Scheme 12.1) under the conditions of DCI (desorptive chemical ionization) mass spectrometry using ammonia as reagent gas, afforded fragmentation signals at $m/z = 723$ and 724 due to $C_{59}NH^+$ and $C_{59}NH_2^+$ [11].

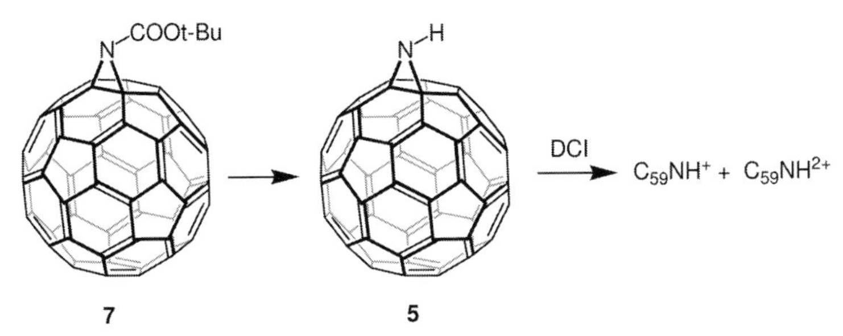

Scheme 12.1 Synthesis of epimino[60]fullerene **5** and formation of nitrogen heterofullerenes in the gas phase by desorptive chemical ionization (DCI) mass spectrometry.

Ketolactam **6** efficiently fragments under FAB mass spectrometric conditions to $C_{59}N^+$ [2]. In this case, no fragmentation to C_{60} was observed. Immediately after this discovery the right synthetic conditions to mimic the fragmentation events in the gas phase were found [2]. Refluxing **6** in *o*-dichlorobenzene (ODCB) in the presence of a 12–20-fold excess of *p*-TsOH, under an atmosphere of nitrogen, leads to the formation of heterofullerene "$C_{59}N$", which exhibits a green solution (Scheme 12.2).

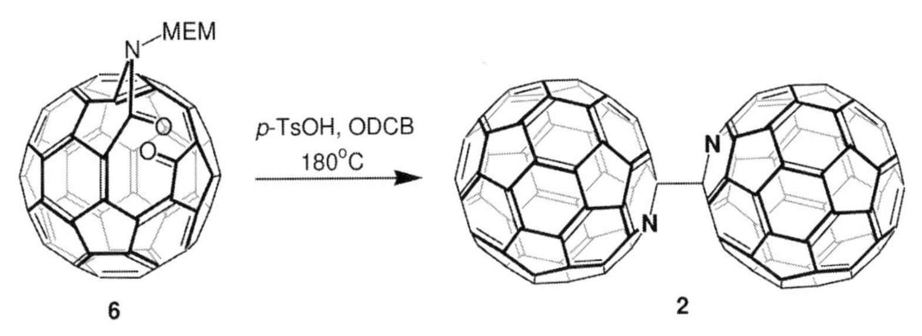

Scheme 12.2 Synthesis of the aza[60]fullerene dimer **2** starting from the cluster opened ketolactam **6**.

As demonstrated by various techniques, including cyclic voltammetry, ESR and ^{13}C NMR spectroscopy, instead of **1**, which is an open shell system, the diamagnetic dimer **2** was isolated. The suggested mechanism for the formation of **2** involves, as first step, the acid-catalyzed cleavage of the MEM group followed by an intramolecular ring formation to an 1,3-oxazetidinium system. This intermediate then loses formaldehyde and CO to form the azafulleronium ion $C_{59}N^+$, which is subsequently reduced to the $C_{59}N$ monomer (**1**). The reducing agent was proposed to be 2-methoxyethanol (the cleavage product of the MEM group) or water. In a final step, dimerization of **1** leads to $(C_{59}N)_2$ (**2**). This mechanism is supported by experiments in which the first cleavage product was trapped, for example, by nucleophiles [14].

The second synthetic approach to heterofullerenes in bulk quantities is based on the fragmentation of **4** in the gas phase (Scheme 12.3) [3, 12]. The reaction of **4** with 20 equiv. *p*-TsOH in refluxing ODCB in an argon atmosphere leads to **2** in an optimized yield of 26%. Interestingly, together with the dimer **2** the alkoxy substituted monomeric compound **8** was formed. This exohedral heterofullerene adduct, however, is not stable in the long term in solution but decomposes to form a cluster opened system exhibiting carbonyl vibrations in the IR spectra. Nevertheless, **8** was the first heterofullerene whose ^{13}C NMR spectrum shows the resonance for sp^3-fullerene C atoms at $\delta = 90.03$ [3] similar to those of the interfullerene bond within **2** [15].

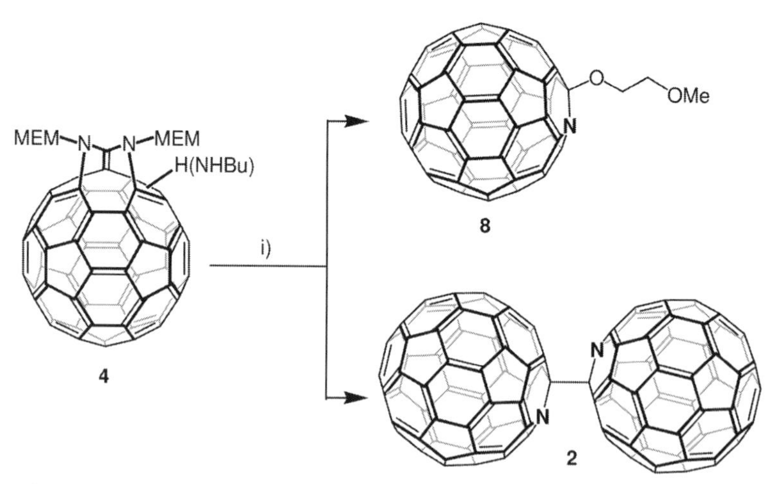

Scheme 12.3 Synthesis of aza[60]fullerenes starting from activated bisazafulleroids. (i) *p*-TsOH, ODCB, heat.

Although the mechanism of the fragmentation reaction is not fully understood, the following considerations seem reasonable: According to the regioselectivity principles pointed out earlier (see Chapter 10), the amine should add preferably to double bonds adjacent to the nitrogen bridges to form the adducts **9** and/or **10**. The central ene-diamine double bond turned out to be not accessible in various addition reactions [2, 12]. Both adducts **9** and **10** contain a BuN-C-N segment that could give rise to an elimination of a carbodiimide moiety, BuN=C=NR. Mass spectrometric investigations clearly show that the nitrogen atom within the azafullerene framework must originate from one of the imino bridges of the diazabishomo[60]fullerene precursor, since **4** with ^{15}N doped imino bridges afforded $C_{59}{}^{15}N^+$ (FAB-MS) [12].

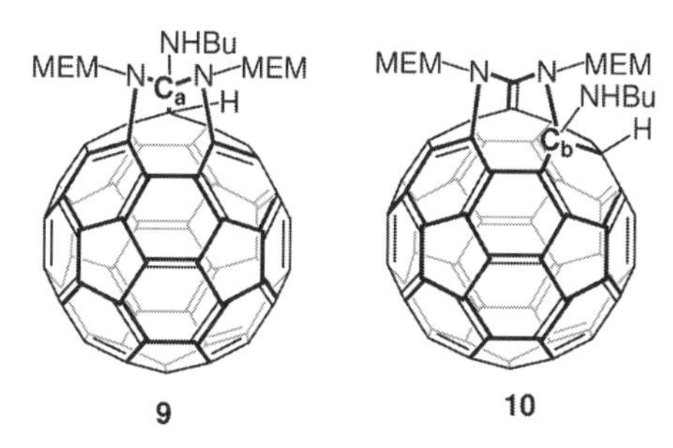

Precursors bearing acid labile groups are required since, for example, the CH_2COOMe substituted disazabishomo[60]fullerene **11** does not lead to heterofullerenes under such synthetic conditions. Trapping the intermediate azafullerenonium ion with 2-methoxymethanol (from cleavage of the MEM groups) explains the formation of **8**, whereas reduction to **1** and subsequent dimerization is the most likely pathway to **2**.

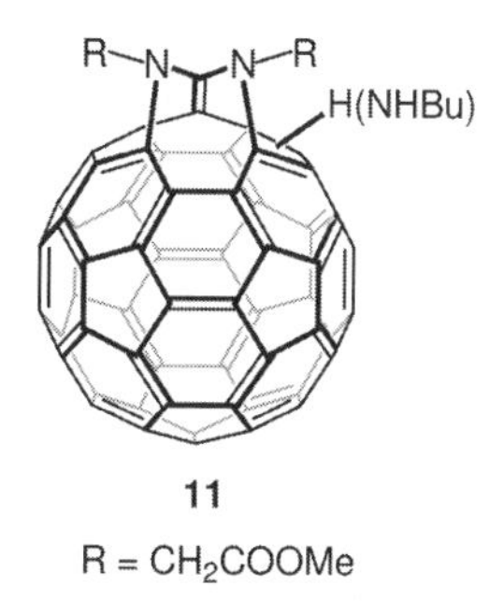

12.2.2
Synthesis of bis(Aza[70]fullerenyl) $(C_{69}N)_2$

The concept of fragmentation of activated diazabishomofullerenes was also applied to the first synthesis of aza[70]fullerenes [3]. Considering the mechanistic investigations described above, the formation of different $(C_{69}N)_2$ isomers is expected, depending on the nature of the bisazafulleroid precursor. Diazabis[70]homofullerene **12** is the precursor of aza[70]heterofullerenes where a C-atom of set **A** is replaced by a N-atom, whereas **13** serves as starting material for heterofullerene isomers containing the N-atom in **B**-position (Scheme 12.4). When a 4 : 1 mixture of **12** and **13** was allowed to form heterofullerenes under the same conditions [3] used for the synthesis of **2** a mixture of all three possible dimers **14–16** involving **AA′**, **AB′** and **BB′** substitution patterns was obtained (Scheme 12.4).

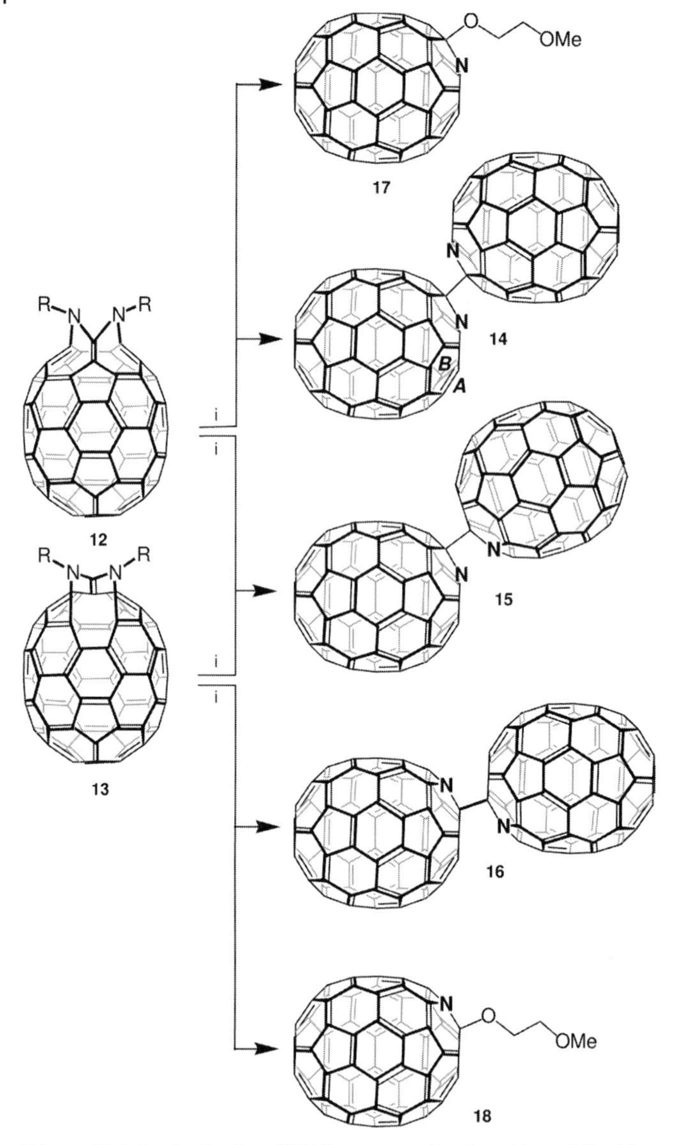

Scheme 12.4 Synthesis of aza[70]fullerenes starting from bisazafulleroids. (i) *n*-Butylamine, *p*-TsOH, ODCB, heat.

The ratio of the constitutional isomers **14–16** as determined by HPLC was 16 : 3 : 1. The fact that mixed dimers are also easily accessible reflects another aspect of the great diversity within heterofullerene chemistry. As with the aza[60]fullerenes, alkoxy substituted monomers **17** and **18** are formed together with the dimers **14–16** (Scheme 12.4) [3]. These two monomers are formed in a ratio of 7 : 1. Also, for these higher heterospheres ^{13}C NMR spectroscopy reveals a closed structure. For example, the signal of the sp^3 C-atom of **17** appears at $\delta = 96.42$.

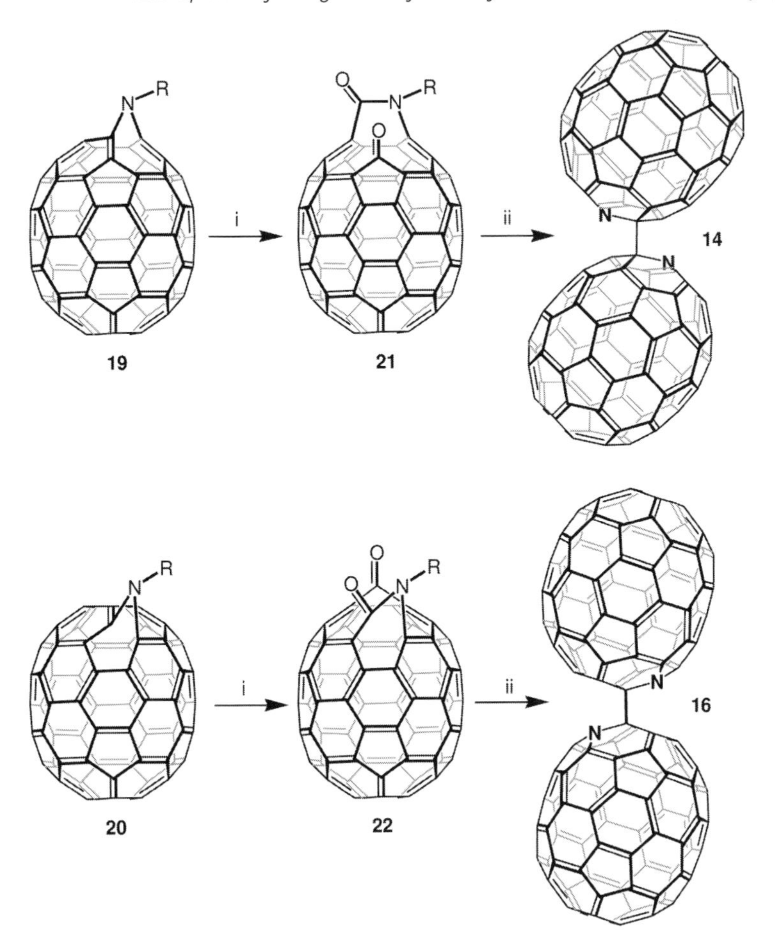

Scheme 12.5 Synthesis of aza[70]fullerenes starting from aza[70]fulleroids. (i) $^1O_2$; (ii) *p*-TsOH, ODCB, heat.

The Wudl group also applied their method of heterofullerene formation to synthesize aza[70]fullerenes [16]. Starting from the [5,6]-bridged azahomo[70]-fullerenes **19** and **20** they selectively synthesized the **AA′** isomer **14** and the **BB′** isomer **16**, respectively (Scheme 12.5).

The first mixed $C_{59}N$-$C_{69}N$ heterodimers, **23** and **24**, have been synthesized in a similar way [1].

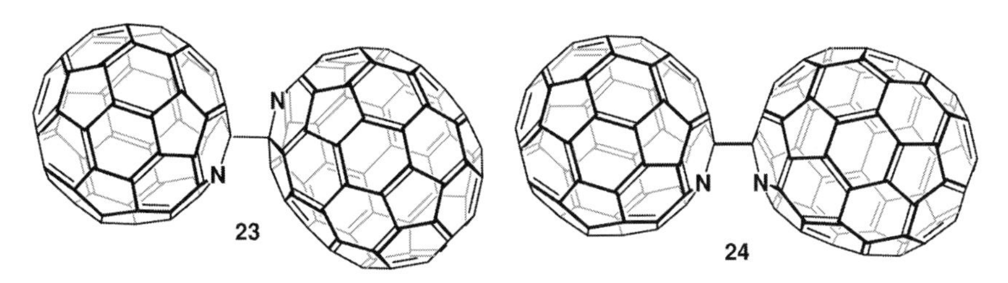

12.3 Chemistry of Azafullerenes

Since heterofullerene chemistry began, as well as **8**, **17** and **18** various further derivatives formed during the formation of azafullerenes or from $(C_{59}N)_2$ (**2**) itself have been synthesized. Heterofullerene transformation of the ketolactam **6** in the presence of a 15-fold excess of hydroquinone leads to the parent hydroaza[60]-fullerene **25** [14]. The hydroquinone is assumed to reduce the $C_{59}N$ radical intermediate (Scheme 12.6).

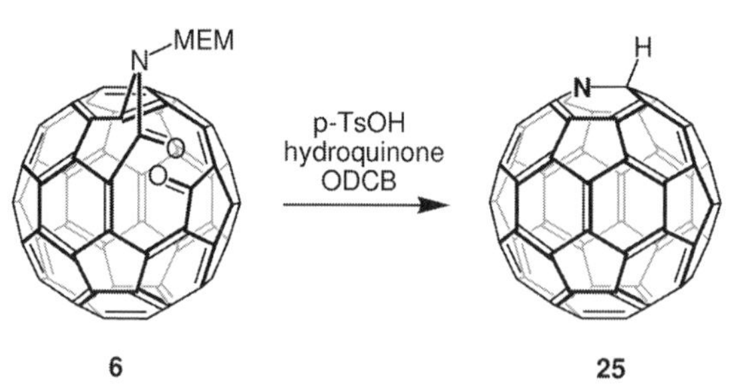

Scheme 12.6 Synthesis of hydroaza[60]fullerene **25**.

The central dimer bond in $(C_{59}N)_2$ was calculated to be 160.9 pm long (average C–C bond 154 pm) with a bond energy of only 18 kcal mol^{-1} [17]. Indeed, this bond can be easily homolytically cleavaged, either thermally or photochemically. Photochemical cleavage allowed for the investigation of $C_{59}N$ (**1**) and of two $C_{69}N$ isomers via light-induced ESR measurements (LESR) [16, 18]. The monomeric radical **1** bears a valuable synthetic potential, as demonstrated by the thermal treatment of **2** in ODCB in the presence of excess diphenylmethane (Scheme 12.7) [19]. The resulting formation of **27** is consistent with a free radical chain mechanism. Similarly, the parent system $C_{59}NH$ (**25**) was obtained by the thermal treatment of **1** in ODCB in the presence of Bu_3SnH. Interestingly, treatment of **2** with diphenylmethane under photolytic conditions does not produce **27**, but the N-oxide **26** [19].

The synthetically most valuable intermediate in heterofullerene chemistry so far has been the aza[60]fulleronium ion $C_{59}N^+$ (**28**). It can be generated in situ by the thermally induced homolytic cleavage of **2** and subsequent oxidation, for example, with O_2 or chloranil [20–24]. The reaction intermediate **28** can subsequently be trapped with various nucleophiles such as electron-rich aromatics, enolizable carbonyl compounds, alkenes and alcohols to form functionalized heterofullerenes **29** (Scheme 12.8). Treatment of **2** with electron-rich aromatics as nucleophilic reagent NuH in the presence of air and excess of *p*-TsOH leads to arylated aza[60]fullerene derivatives **30** in yields up to 90% (Scheme 12.9). A large variety of arylated derivatives **30** have been synthesized, including those containing corannulene, coronene and pyrene addends [20, 22–25].

A special case for such aza[60]fullerene arylation is the reaction of **2** with a 25-fold excess of ferrocenium hexafluorophosphate at 150 °C in an argon atmosphere, which afforded the ferrocenyl-hydroaza[60]fullerene dyad **31** (Scheme 12.10) [23].

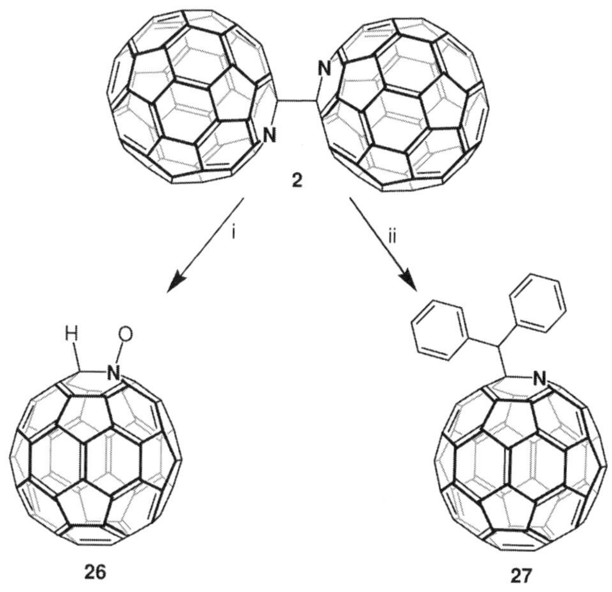

Scheme 12.7 Synthesis of the aza[60]fullerene derivatives **26** and **27**. (i) ODCB, air, *h*ν; (ii) CH_2Ph_2, air, reflux.

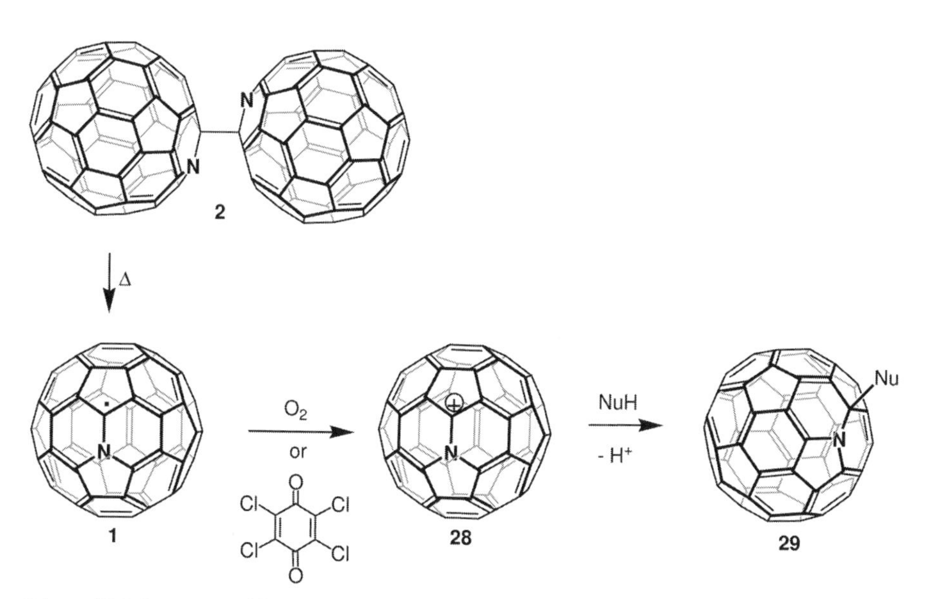

Scheme 12.8 Formation of functionalized heterofullerenes **29** via the azafullerenonium intermediate **28**.

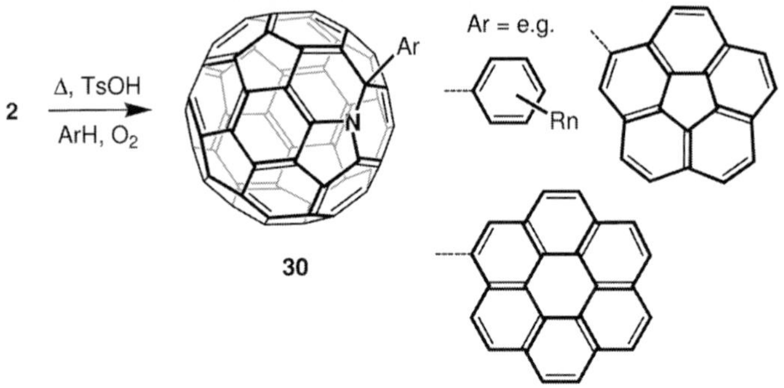

Scheme 12.9 Formation of arylated heterofullerenes **30**.

In this case ferrocenium hexafluorophosphate serves as oxidizing reagent for the generation of **28** and the simultaneously generated ferrocene as its trapping reagent. Compound **31** represents the first example of a fullerene-based dyad where two electroactive groups are connected by only one single σ-bond. Cyclic voltammetry of **31** shows strong electronic coupling between the two chromophores in the ground state. Photophysical studies revealed an intramolecular photoinduced electron transfer with the formation of the charge separated ($C_{59}N^{\bullet -}$)-($Fc^{\bullet +}$) radical pair.

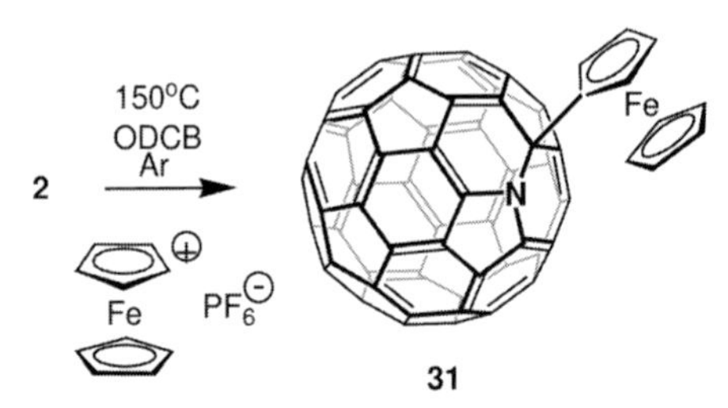

Scheme 12.10 Synthesis of the ferrocenyl hydroazafullerene **31**.

A very versatile tool for the synthesis of monomeric aza[60]fullerene derivatives **32** is the Mannich functionalization (Scheme 12.11), where **2** is typically treated in ODCB with an excess of the carbonyl compound and 40 equiv. TsOH at 150 °C in a constant stream of O_2 [21, 22, 26]. After 15 min the conversion is usually completed and the Mannich bases are formed in good yield and complete regioselectivity.

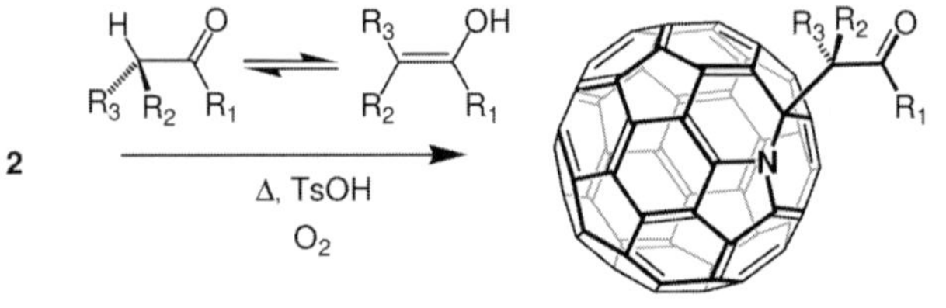

Scheme 12.11 Reaction of **2** with enolizable compounds.

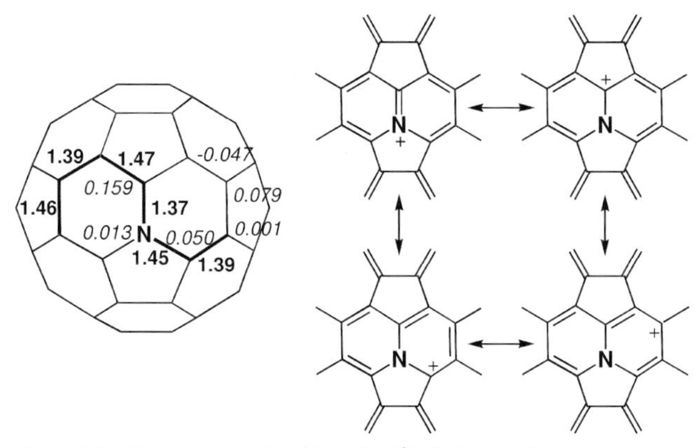

Figure 12.1 Characteristic bond lengths (Å) (bold), Mulliken charges, and the most important resonance structures of $C_{59}N^+$ (**28**) [12].

If 1,3-dicarbonyl compounds are allowed to react with **2** the terminal attack is preferred if possible [22]. This is due to the thermodynamic stability of the corresponding cationic intermediates. α,β-Unsaturated carbonyl compounds also undergo a related reaction with **2** [22].

Aza[60]fulleronium ion **28** has been isolated and characterized by X-ray crystallography via the oxidation of **2** with the radical cation hexabromo(phenyl)carbazole ($HBPC^{\bullet +}$) in dry ODCB [27]. The counter-ion is the silver(I) complex carborane ion $Ag(CB_{11}H_6Cl_6)_2]^-$. The salt was crystallized as dark green crystals by diffusion of hexane vapor. The solid is reasonably air stable. $C_{59}N^+$ has almost the same structure as C_{60}, including the bond length alternation between the shorter [6,6]-bonds and the longer [5,6]-bonds. Semiempirical calculations showed that the positive charge density is predominantly located in the neighborhood of the nitrogen substituent [12]. Therefore, the resonance structures depicted in Figure 12.1 are the most important.

Arylated azafullerenes are also accessible from the heterofullerene precursors **6** and **33** by thermal treatment in the presence of aromatics and air (Scheme 12.12) [3]. Although the yields are lower than those starting from **2**, one reaction step can be saved. Conversely, if the reaction is carried out in 1-chloronaphthalene at 220 °C, **34** is obtained in an isolated yield of 46%.

Since monomeric aza[60]fullerene derivatives carrying an organic addend adjacent to the N-substituent are stable, exhibit favorable solubility properties, and are obtainable in high yields they represent ideal starting materials for the investigation of the behavior of the $C_{59}N$ core towards addition reactions. As an example the chlorination of **30a–c** with ICl in CS_2 at room temperature was investigated [28]. This reaction leads to the tetrachlorinated heterospheres **35a–c**, which were isolated as orange microcrystals in 50–60% yield (Scheme 12.13). Structures of **35a–c** were determined by ^{13}C NMR spectroscopy. They are closely related to the hexachlorides and hexabromides $C_{60}X_6$ (X = Cl, Br) (**36**) previously described by Birkett et al. [29, 30]. In analogy to the cyclopentadiene subunit in **36**, compounds **35** contain an

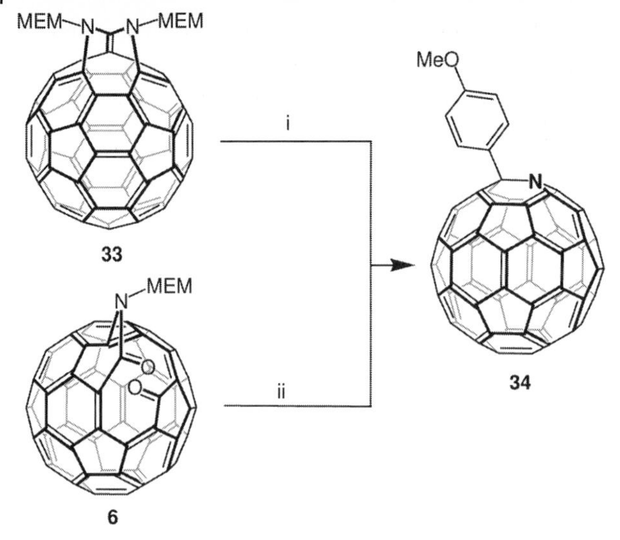

Scheme 12.12 Formation of the arylated heterofullerene **34** from bis-azafulleroid and ketolactam precursors. (i) *n*-Butylamine, ODCB–anisole, *p*-TsOH, air, reflux; (ii) ODCB–anisole, *p*-TsOH, air, reflux.

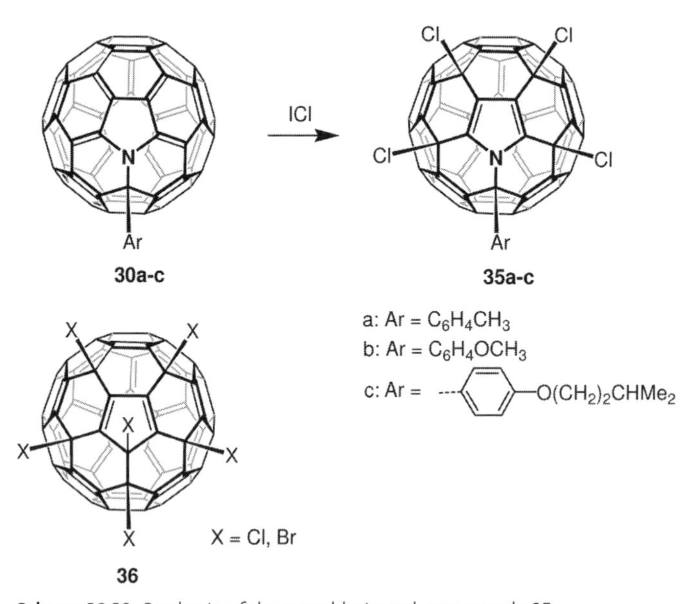

Scheme 12.13 Synthesis of the tetrachlorinated compounds **35** containing an integral pyrrole moiety within the fullerene framework.

integral pyrrole moiety decoupled from the conjugated π-system of the fullerene cage. The chlorine addends can be easily removed from the cage by treating **35** with an excess of PPh_3 at room temperature. The reversible binding could provide a useful strategy for protection of several double bonds of the fullerene core.

The first multiple functionalization of aza[60]fullerene with organic addends was achieved via template mediated addition of malonates into octahedral [6,6]-bonds of the fullerene framework [31]. This method was originally developed for the highly regioselective hexaaddition of C_{60} [32–34]. This approach allows for the synthesis of pentakisadducts derived from $RC_{59}N$ containing a C_s-symmetrical addition pattern. For example, the monoadduct **37** was stirred with a fivefold excess of dimethylanthracene (DMA) in ODCB for three hours and subsequently reacted with a tenfold excess of DBU and diethyl bromomalonate. After chromatography the pentamalonate **38** was isolated in 20% yield (Scheme 12.14).

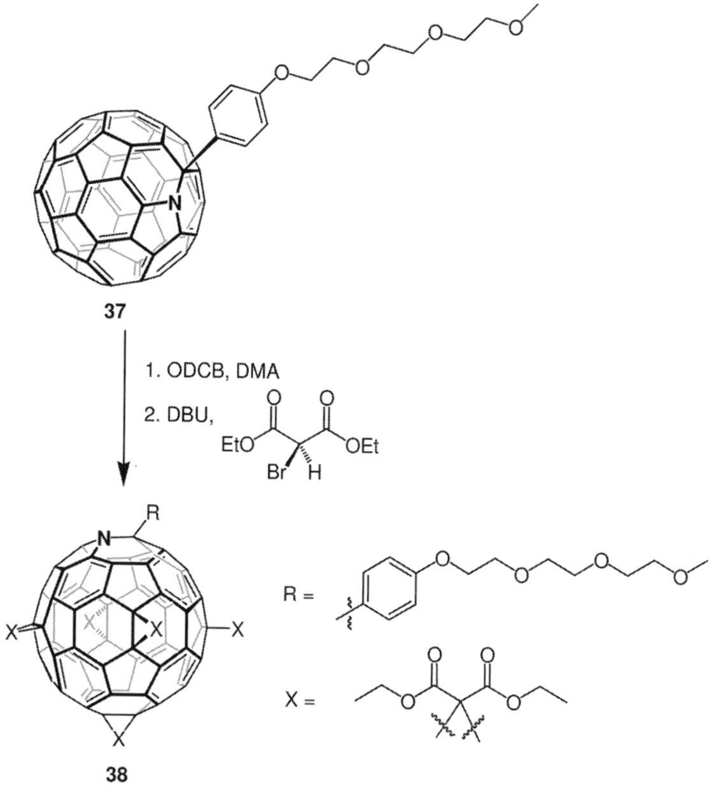

37

1. ODCB, DMA
2. DBU,

38

Scheme 12.14 Highly regioselective formation of the pentamalonate **38**.

12.4 Outlook

Heterofullerene chemistry is still a very young discipline within synthetic organic chemistry and even within fullerene chemistry. So far, it is restricted to mono-azafullerenes. However, the potential of structural diversity within heterofullerenes is enormous (Figure 12.2). Preparative challenges for the future are, for example,

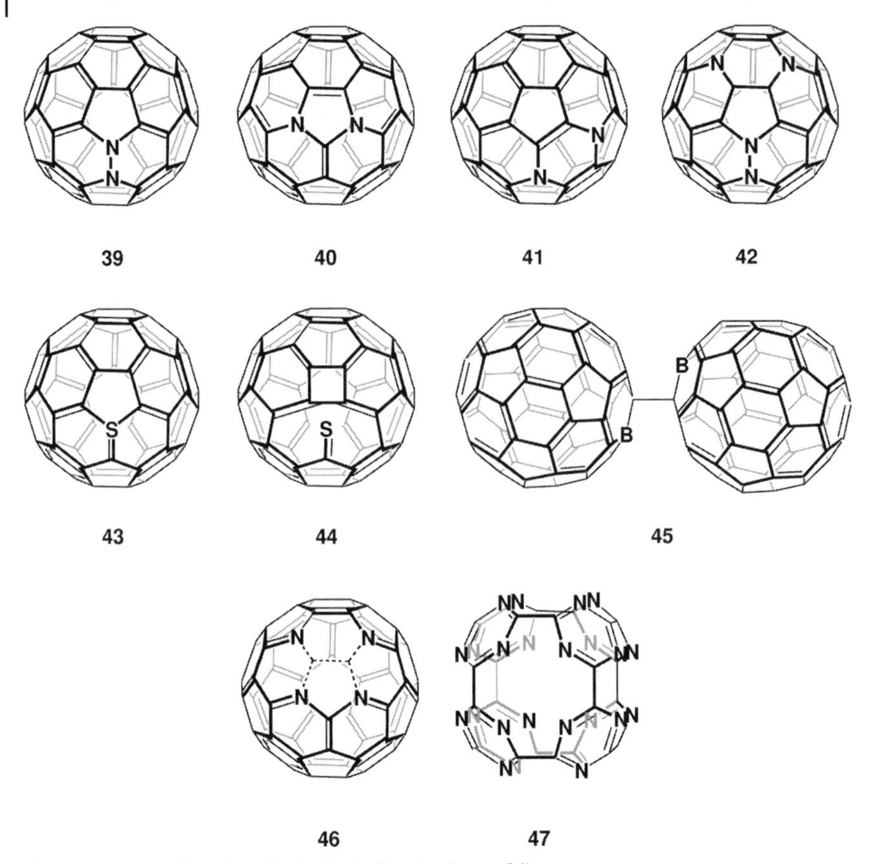

Figure 12.2 Examples of synthetically challenging heterofullerenes.

diazafullerenes $C_{58}N_2$ such as **39–41** as stable diamagnetic monomers, higher substituted azafullerenes $C_{60-n}N_n$ such as **42**, heterofullerenes containing other elements such as thiafullerenes **43** and **44**, which could be either closed or opened molecules [35], bisborafullerenyl **45** and, finally, truncated heterofullerenes such as **46** and **47**. The latter case would also provide the possibility of synthesizing endohedral fullerenes or, in analogy to phthalocyanine chemistry, allow the introduction of a transition metal ion into the window by complexation with the four nitrogen atoms of the reduced truncated system. Various quantum chemical calculations of such structures have been carried out, including investigations on the aromaticity of various heterofullerenes [35–38]. Since our knowledge of the physical and chemical behavior of these fascinating cage molecules and their derivatives is steadily increasing, it is just a question of time as to when these preparative goals will be reached.

References

1 J. C. Hummelen, C. Bellavia-Lund, F. Wudl, *Top. Curr. Chem.* **1999**, *199*, 93.
2 J. C. Hummelen, B. Knight, J. Pavlovich, R. Gonzalez, F. Wudl, *Science* **1995**, *269*, 1554.
3 B. Nuber, A. Hirsch, *Chem. Commun.* **1996**, 1421.
4 T. Guo, C. Jin, R. E. Smalley, *J. Phys. Chem.* **1991**, *95*, 4948.
5 Y. Chai, T. Guo, C. Jin, R. E. Haufler, L. P. F. Chibante, J. Fure, L. Wang, J. M. Alford, R. E. Smalley, *J. Phys. Chem.* **1991**, *95*, 7564.
6 T. Pradeep, V. Vijayakrishnan, A. K. Santra, C. N. R. Rao, *J. Phys. Chem.* **1991**, *95*, 10564.
7 C. N. R. Rao, T. Pradeep, R. Seshadri, A. Govindaraj, *Indian J. Chem. A* **1992**, *31A*, F27.
8 J. F. Christian, Z. Wan, S. L. Anderson, *J. Phys. Chem.* **1992**, *96*, 10597.
9 D. E. Clemmer, J. M. Hunter, K. B. Shelimov, M. F. Jarrold, *Nature* **1994**, *372*, 248.
10 H. J. Muhr, R. Nesper, B. Schnyder, R. Koetz, *Chem. Phys. Lett.* **1996**, *249*, 399.
11 J. Averdung, H. Luftmann, I. Schlachter, J. Mattay, *Tetrahedron* **1995**, *51*, 6977.
12 I. Lamparth, B. Nuber, G. Schick, A. Skiebe, T. Grösser, A. Hirsch, *Angew. Chem.* **1995**, *107*, 2473; *Angew. Chem. Int. Ed. Engl.* **1995**, *34*, 2257.
13 A. Hirsch, B. Nuber, *Acc. Chem. Res.* **1999**, *32*, 795.
14 M. Keshavarz-K, R. Gonzalez, R. G. Hicks, G. Srdanov, V. I. Srdanov, T. G. Collins, J. C. Hummelen, C. Bellavia-Lund, J. Pavlovich, *Nature* **1996**, *383*, 147.
15 C. Bellavia-Lund, M. Keshavarz-K., T. Collins, F. Wudl, *J. Am. Chem. Soc.* **1997**, *119*, 8101.
16 K. Hasharoni, C. Bellavia-Lund, M. Keshavarz-K, G. Srdanov, F. Wudl, *J. Am. Chem. Soc.* **1997**, *119*, 11128.
17 W. Andreoni, A. Curioni, K. Holczer, K. Prassides, M. Keshavarz-K, J.-C. Hummelen, F. Wudl, *J. Am. Chem. Soc.* **1996**, *118*, 11335.
18 A. Gruss, K.-P. Dinse, A. Hirsch, B. Nuber, U. Reuther, *J. Am. Chem. Soc.* **1997**, *119*, 8728.
19 C. Bellavia-Lund, R. Gonzalez, J. C. Hummelen, R. G. Hicks, A. Sastre, F. Wudl, *J. Am. Chem. Soc.* **1997**, *119*, 2946.
20 B. Nuber, A. Hirsch, *Chem. Commun.* **1998**, 405.
21 F. Hauke, A. Hirsch, *Chem. Commun.* **1999**, 2199.
22 F. Hauke, A. Hirsch, *Tetrahedron* **2001**, *57*, 3697.
23 F. Hauke, A. Hirsch, S.-G. Liu, L. Echegoyen, A. Swartz, C. Luo, D. M. Guldi, *Chem. Phys. Chem.* **2002**, *3*, 195.
24 F. Hauke, A. Swartz, D. M. Guldi, A. Hirsch, *J. Mater. Chem.* **2002**, *12*, 2088.
25 F. Hauke, S. Atalick, D. M. Guldi, J. Mack, L. T. Scott, A. Hirsch, *Chem. Commun.* **2004**, 765.
26 F. Hauke, M. A. Herranz, L. Echegoyen, D. Guldi, A. Hirsch, A. Atalik, *Chem. Commun.* **2004**, 600.
27 K.-C. Kim, F. Hauke, A. Hirsch, P. D. W. Boyd, E. Carter, R. S. Armstrong, P. A. Lay, C. A. Reed, *J. Am. Chem. Soc.* **2003**, *125*, 4024.
28 U. Reuther, A. Hirsch, *Chem. Commun.* **1998**, 1401.
29 P. R. Birkett, P. B. Hitchcock, H. W. Kroto, R. Taylor, D. R. M. Walton, *Nature* **1992**, *357*, 479.
30 P. R. Birkett, A. G. Avent, A. D. Darwish, H. W. Kroto, R. Taylor, D. R. M. Walton, *J. Chem. Soc., Chem. Commun.* **1993**, 1230.
31 F. Hauke, A. Hirsch, *Chem. Commun.* **2001**, 1316.
32 I. Lamparth, C. Maichle-Mössmer, A. Hirsch, *Angew. Chem.* **1995**, *107*, 1755; *Angew. Chem. Int. Ed. Engl.* **1995**, *34*, 1607.
33 I. Lamparth, A. Herzog, A. Hirsch, *Tetrahedron* **1996**, *52*, 5065.
34 A. Hirsch, O. Vostrowsky, *Eur. J. Org. Chem.* **2001**, 829.
35 H. Jiao, Z. Chen, A. Hirsch, W. Thiel, *Phys. Chem. Chem. Phys.* **2002**, *4*, 4916.
36 H. R. Karfunkel, T. Dressler, A. Hirsch, *J. Comput.-Aided Mol. Des.* **1992**, *6*, 521.
37 Z. Chen, H. Jiao, A. Hirsch, W. Thiel, *J. Org. Chem.* **2001**, *66*, 3380.
38 Z. Chen, U. Reuther, A. Hirsch, W. Thiel, *J. Phys. Chem. A* **2001**, *105*, 8105.

13
Chemistry of Higher Fullerenes

13.1
Introduction

Most of the properties of the higher fullerenes available in macroscopic quantities (see Chapter 1, Section 1.5.1) are closely related to those of C_{60}. However, there are quantitative differences such as thermodynamic stability, optical properties, aromatic character and, most importantly, structure and symmetry. The development of the chemistry of the higher fullerenes is challenging because they provide a rich potential for the investigation of structure–reactivity relationships. Moreover, their possible use as new building blocks for molecular materials is very attractive. Finally, the inherent chirality of many higher fullerenes and the possibility to generate enantiomerically pure allotropes of carbon using suitable stereochemical approaches represents a fascinating opportunity. Most of the chemical transformations carried out with higher fullerenes were exohedral addition reactions and exciting results have been obtained already. However, compared with C_{60} the chemistry of higher fullerenes is still in its infancy. This is because, except for C_{70}, the quantities of isomerically pure higher fullerenes are still very limited. They are formed as minor products during fullerene generation and their isolation requires tedious HPLC separation protocols. Excellent reviews on the covalent chemistry of higher fullerenes have been provided by Diederich and co-workers [1–3]. The present chapter presents a brief summary of the basis of covalent exohedral functionalization of higher fullerenes.

13.2
Exohedral Reactivity Principles

Like C_{60} most of the characterized higher fullerenes can be considered as electron-deficient polyenes. They exhibit similarities in their chemical behavior [1–3]. Preferred primary addition reactions are, for example, addition of nucleophiles, as well as cycloadditions at the bonds adjacent to two six-membered rings ([6,6]-bonds). The most important difference from the chemistry of C_{60} is due to the less symmetrical cages of the higher fullerenes, which exhibit a broad variety of bond environments with unequal reactivity towards addition reactions. Consequently,

Fullerenes: Chemistry and Reactions. Andreas Hirsch and Michael Brettreich

ISBN: 3-527-30820-2

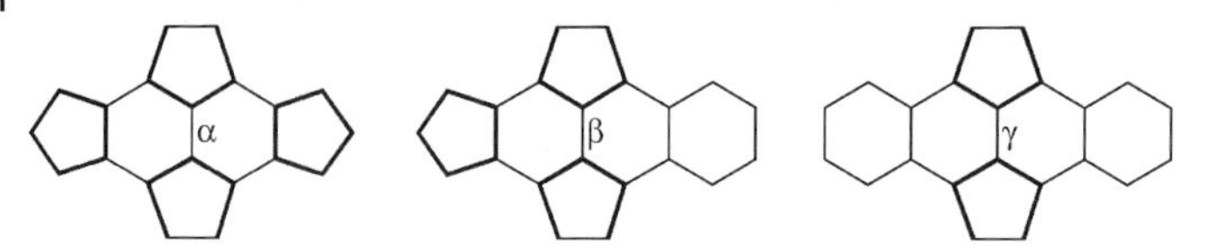

Figure 13.1 The three [6,6]-bonds in fullerenes connecting the C atoms with the highest degree of pyramidalization.

even for monoadduct formation many isomers can be expected. However, it turns out the addition reactions exhibit a remarkable regioselectivity. Most of the additions occur in a 1,2-mode to [6,6]-bonds. Typical exceptions are halogenation, which due to the steric requirement of the addends preferably proceed as intrahexagonal 1,4-additions [4, 5].

A major driving force of fullerene chemistry is relief of strain energy accomplished by rehybridization ($sp^2 \rightarrow sp^3$) due to the covalent binding of the addend [6]. It is therefore reasonable to assume that additions preferably take place at sites with the highest degree of pyramidalization [6–9]. In higher fullerenes, pyramidalization of C atoms depends not only on the overall fullerene size but also on the distribution of the 12 five-membered and the six-membered rings on the fullerene surface. To make *simple* predictions of expected isomeric product distributions, Diederich et al. developed a qualitative model for the evaluation of local curvature [10]. This model predicts an increase in reactivity of a [6,6]-bond the more it is surrounded by five-membered rings, and the shorter its distance is to the latter. According to this description, the most strained fullerene bonds (type α) are located at the center of a pyracylene substructure (Figure 13.1). In C_{70} and C_{76} they are found at the poles (type α, surrounded by type β), whereas the flatter equatorial region contains several different, less curved bonds.

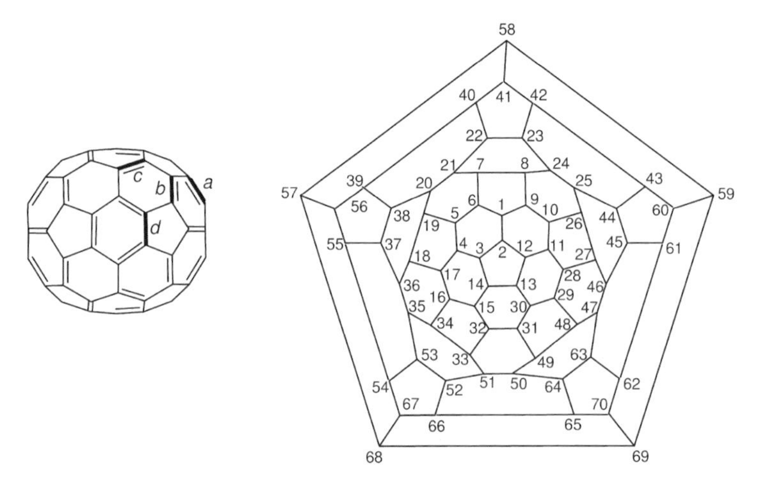

Figure 13.2 The most reactive bonds of C_{70} (*left*) and numbering scheme of C_{70} (*right*). (a) C(1)–C(2) (type α), (b) C(5)–C(6) (type β), (c) C(7)–C(21) and (d) C(7)–C(8) bond. Reactivity decreases in the order a > b > c > d.

This consideration is corroborated by calculations at different levels of theory [11–13], showing that 1,2-additions to C_{70} occur preferentially at α and β bonds between C(1) and C(2) and C(5) and C(6), respectively (Figure 13.2). Also, Mulliken charges on the fullerene core have been used in predicting the regiochemistry of nucleophilic additions to C_{70} [11]. Finally, π-bond orders have also been employed to evaluate the reactivity of different types of bonds [14–16].

13.3 Adducts of C_{70}

13.3.1 Monoadducts

Mono-functionalization of C_{70} affords, preferrably, C(1)-C(2) adducts (type α) (Figure 13.3). In some cases, for example, upon nucleophilic cyclopropanations they even represent the exclusively formed monoadducts [1–3, 17]. Typical examples of addition reactions that afford monoadducts are epoxidations [18, 19], osmylation [9], transition metal complex formations [20, 21], hydrogenation [13, 22], many cycloadditions [1, 2] and additions of nucleophiles [23]. For the formation and the chemical transformation of azahomo[70]fullerenes see also Chapter 12 (Schemes 12.4 and 12.5).

The second most favored bond for the covalent binding of addends is the C(5)-C(6) bond (type β) (Figures 13.2 and 13.3). C(5)-C(6) adducts often accompany C(1)-C(2) adducts in the reaction mixtures. C(7)–C(21) bonds (Figures 13.2 and 13.3) are not among the types α–γ and have a relatively low local curvature [10]. Furthermore, this bond is localized in a six-membered ring with a relatively high benzenoid character [24]. Consequently, it has a low reactivity and few examples of corresponding adducts are known [1, 2]. Owing to the C_1-symmetry of this addition pattern all C(7)-C(21) adducts are chiral, independent of the nature of the addend. The first example was reported by Diederich and co-workers when C_{70} was allowed to undergo Diels–Alder addition of 4,5-dimethoxy-*o*-quinodimethane [10]. It was isolated as a minor product along with the more abundant adducts of types α and β. Similarly, [3+2] cycloaddition of N-methylazomethine ylid to C_{70} yielded a fullero-

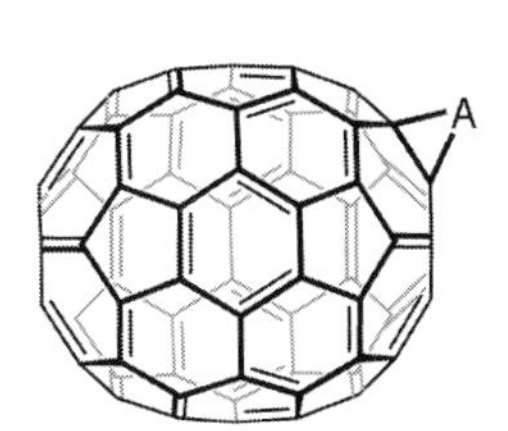

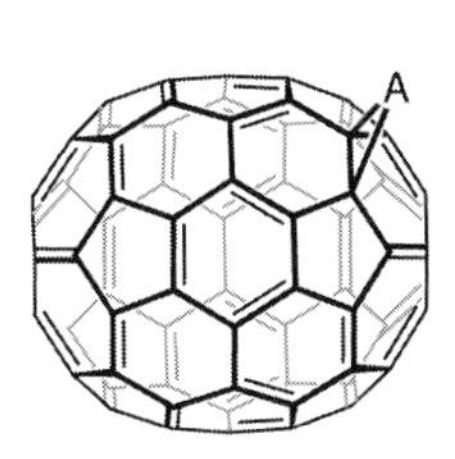

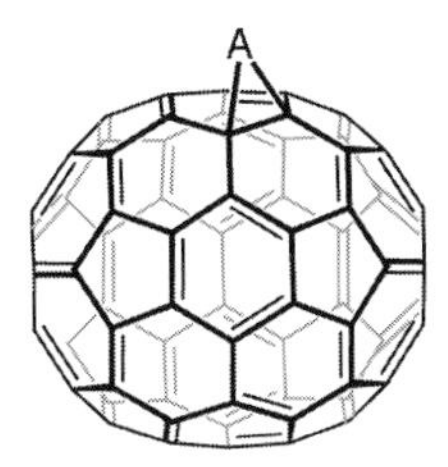

Figure 13.3 Schematic representation of three [6,6]-adducts of C_{70}. *Left:* C(1)-C(2) adduct (type α); *middle:* C(5)-C(6) adduct (type β); *right:* C(7)-C(21) adduct.

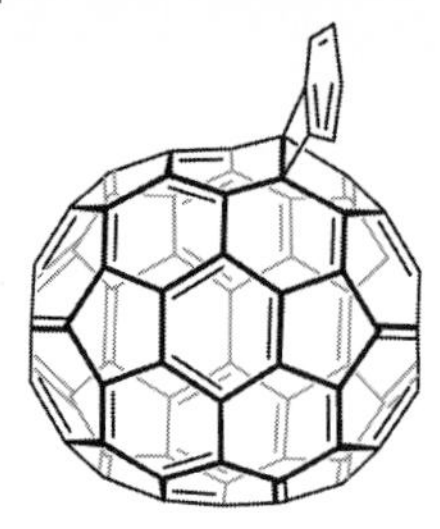
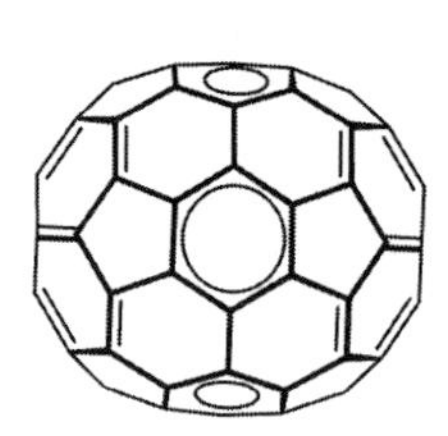

Figure 13.4 C(7)-C(8) isomer resulting from monoaddition of benzyne to C_{70} and Kekulé resonance structure of C_{70}, illustrating the benzenoid character of the equatorial hexagons.

pyrrolidine with the heterocycle fused across the fullerene C(7)–C(21) bond as a minor product [25]. Addition of benzyne as another very hot addend also afforded a C(7)-C(21) adduct, within the mixture of monoadducts [5, 24]. This adduct was also accompanied by another, unusual isomer, resulting from the addition to the [5,6]-bond C(7)–C(8) (Figure13.4) [24].

This isomer represents the first example of a [5,6]-adduct in which the fusion bond remains intact. This addition mode was explained by the relatively high double bond character of the [5,6]-bond C(7)-C(8), which underlines the importance of the Kekulé structure of C_{70} that involves the benzenoid hexagons at the equatorial belt. No such resonance structure is required to describe the proper bonding situation within C_{60} (see also Chapters 1 and 14).

13.3.2
Multiple Adducts

Systematic investigations of twofold additions of malonates to C_{70} revealed that the second addition takes place at one of the five α-bonds of the unfunctionalized pole [17, 26]. With achiral, C_{2v}-symmerical malonate addends, three constitutionally isomeric bisadducts are formed: An achiral one (C_{2v}-symmetrical **1**), and two chiral ones (C_2-symmetrical **2** and **3**), which are obtained as pairs of enantiomers with an inherently chiral addition pattern (Figure 13.5). Twofold addition of chiral malonates leads to the formation of five optically active isomers, two constitutionally isomeric pairs of C_2-symmetrical diastereomers and a third constitutional C_2-symmetrical isomer (Figure 13.5). Twofold additions of azides to C_{70} lead to diazabis[70]homofullerenes, which served as starting material for the synthesis of bis-(aza[70]-fullerenyl) $(C_{69}N)_2$ (Chapter 12) [27]. As further bisadditions, addition reaction to C_{70} [2+2]cycloaddition of electron-rich bis(diethylamino)ethyne and 1-alkylthio-2-(diethylamino)ethynes [28] and the addition of transition metal fragments have been reported [29–32].

Three- and four-fold addition reactions to C_{70} also exhibit a pronounced regioselectivity [17]. Examples for tetrakisadducts with confident structural characterization are shown in Figure 13.6. These tetrakisadducts **4** and **5** were synthesized

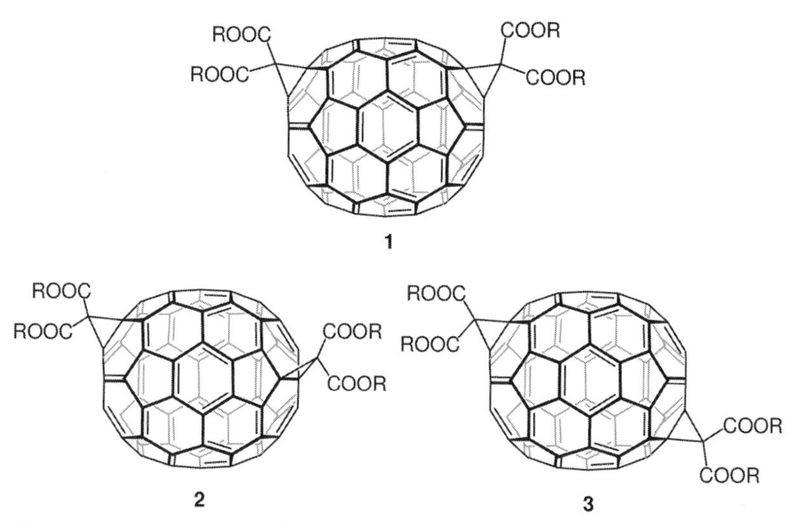

Figure 13.5 Bisadducts of C_{70} obtained from twofold cyclopropanation reactions. Both addends with chiral and achiral substituents have been used. Adducts **2** and **3** exhibit an inherently chiral addition pattern and are formed together with the corresponding enantiomer or diastereomer involving the mirror image addition pattern.

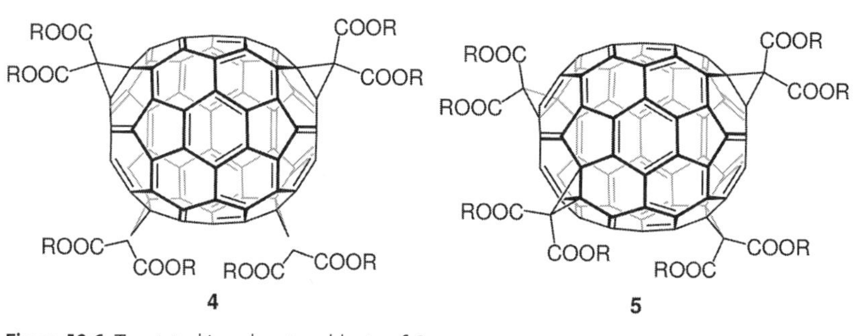

Figure 13.6 Two tetrakismalonate adducts of C_{70}.

from bisadducts such as **1–3** with the two addends bound in opposite hemispheres. Subsequent additions have to take place in a hemisphere that is already functionalized. Significantly, a second attack to another α-type bond does not take place. This is in line with the preferred regiochemistry of twofold cyclopropanations of C_{60}, where an attack of a *cis*-1 bond is sterically forbidden and an attack to a *cis*-2 bond is very unfavorable (see also Chapter 10). The next addition to a bismalonate of C_{70} takes place at a β-type bond. The relative arrangement of the addends in one hemisphere is the same as that of regioselectively preferred equatorial *e*-adducts of C_{60} (see also Chapter 10). Further examples of higher adducts of C_{70} are $C_{70}Cl_{10}$ [33] and its follow up products such as $C_{70}Ph_8$ (see also Chapter 11) [34] and $C_{70}Ph_9OH$ [35]. The C_s-symmetrical $C_{70}Cl_{10}$ exhibits a characteristic intrahexagonal 1,4-addition pattern.

13.4 Adducts of C_{76}, C_{78} and C_{84}

For the higher fullerenes beyond C_{70} only a few examples of adducts with unambiguous structural characterization have been reported [2]. For example, several Diels–Alder and malonate monoadducts of D_2-C_{76} have been isolated [10, 36]. The addends are bound to α-type bonds of the fullerene framework. Nucleophilic cyclopropanations of a 3 : 1 mixture of C_{2v}- and D_3-C_{78} with a slight excess of diethyl bromomalonate afforded a mixture of at least eight trisadducts and two C_1-symmetrical bisadducts. Among these adducts a C_3-symmetrical pure trisadduct of D_3-C_{78} with an *e,e,e*-addition pattern was identified. The first isomerically pure derivative of [84]fullerene was obtained by the crystallization of $[\eta^2\text{-}(D_{2d}\text{-}C_{84})]Ir(CO)Cl(PPh_3)_2$ [37].

The first kinetic resolution of chiral parent fullerenes was achieved by asymmetrical Sharpless oxidation osmylation of racemic D_2-C_{76}, using reagents with enantiomerically pure ligands derived from cinchona alkaloids [7]. Since the reaction with the osmium reagents was incomplete an enantiomeric enrichment of unreacted D_2-C_{76} was obtained. Conversely, the formed adduct fraction was enriched in the other enantiomer, which could be regenerated by reduction of the osmate with $SnCl_2$. Related experiments were carried out with D_3-C_{78} and D_2-C_{84} [38]. The pure enantiomers of D_2-C_{76} were separated by applying a retro-Bingel reaction to each of two optically pure, diastereomeric C_{76} derivatives, which carry chiral malonate addends and whose addition patterns represent a pair of enantiomers [39]. The same approach has been used for the optical resolution of D_2-C_{84} [40].

References

1 C. Thilgen, A. Herrmann, F. Diederich, *Angew. Chem.* **1997**, *109*, 2362; *Angew. Chem. Int. Ed. Engl.* **1997**, *36*, 2268.
2 C. Thilgen, F. Diederich, *Top. Curr. Chem.* **1999**, *199*, 135.
3 C. Thilgen, I. Gosse, F. Diederich, *Top. Stereochem.* **2003**, *23*, 1.
4 S. J. Austin, P. W. Fowler, J. P. B. Sandall, P. R. Birkett, A. G. Avent, A. D. Darwish, H. W. Kroto, R. Taylor, D. R. M. Walton, *J. Chem. Soc., Perkin Trans. 2* **1995**, 1027.
5 A. D. Darwish, A. G. Avent, R. Raylor, D. R. M. Walton, *J. Chem. Soc., Perkin Trans. 2* **1996**, 2079.
6 R. C. Haddon, *Science* **1993**, *261*, 1545.
7 J. M. Hawkins, A. Meyer, *Science* **1993**, *260*, 1918.
8 R. C. Haddon, G. E. Scuseria, R. E. Smalley, *Chem. Phys. Lett.* **1997**, *272*, 38.
9 J. M. Hawkins, A. Meyer, M. A. Solow, *J. Am. Chem. Soc.* **1993**, *115*, 7499.
10 A. Hermann, F. Diederich, C. Thilgen, H.-U. ter Meer, W. H. Müller, *Helv. Chim. Acta* **1994**, *77*, 1689.
11 H. R. Karfunkel, A. Hirsch, *Angew. Chem.* **1992**, *104*, 1529.
12 C. C. Henderson, C. M. Rohlfing, P. A. Cahill, *Chem. Phys. Lett.* **1993**, *213*, 383.
13 C. C. Henderson, C. M. Rohlfing, K. T. Gillen, P. A. Cahill, *Science* **1994**, *264*, 397.
14 J. Baker, P. W. Fowler, P. Lazzeretti, M. Malagoli, R. Zanasi, *Chem. Phys. Lett.* **1991**, *184*, 182.
15 A. Rathna, J. Chandrasekhar, *Fullerene Sci. Technol.* **1995**, *3*, 681.
16 R. Taylor, *J. Chem. Soc., Perkin Trans. 2* **1993**, 813.

17 A. Herrmann, M. Rüttimann, C. Thilgen, F. Diederich, *Helv. Chim. Acta* **1995**, *78*, 1673.
18 A. B. Smith III, R. M. Strongin, L. Brard, G. T. Furst, J. H. Atkins, W. J. Romanow, M. Saunders, H. A. Jimenez-Vazquez, K. G. Owens, R. J. Goldschmidt, *J. Org. Chem.* **1996**, *61*, 1904.
19 V. N. Bezmelnitsin, A. V. Eletskii, N. G. Schepetov, A. G. Avent, R. Taylor, *J. Chem. Soc., Perkin Trans. 2* **1997**, 683.
20 A. L. Balch, V. J. Catalano, J. W. Lee, M. M. Olmstead, S. R. Parkin, *J. Am. Chem. Soc.* **1991**, *113*, 8953.
21 M. Iyoda, Y. Ogawa, H. Matsuyama, H. Ueno, K. Kikuchi, I. Ikemoto, Y. Achiba, *Fullerene Sci. Technol.* **1995**, *3*, 1.
22 A. G. Avent, A. D. Darwish, D. K. Heimbach, H. W. Kroto, M. F. Meidine, J. P. Parsons, C. Remars, R. Roers, O. Ohashi, R. Taylor, D. R. M. Walton, *J. Chem. Soc., Perkin Trans. 2* **1994**, 15.
23 A. Hirsch, T. Grösser, A. Skiebe, A. Soi, *Chem. Ber.* **1993**, *126*, 1061.
24 M. S. Meier, G.-W. Wang, R. C. Haddon, C. P. Brock, M. A. Lloyd, J. P. Selegue, *J. Am. Chem. Soc.* **1998**, *120*, 2337.
25 S. R. Wilson, Q. Lu, *J. Org. Chem.* **1995**, *60*, 6496.
26 C. Bingel, H. Schiffer, *Liebigs Ann.* **1995**, 1551.
27 B. Nuber, A. Hirsch, *Chem. Commun.* **1996**, 1421.
28 X. Zhang, C. S. Foote, *J. Am. Chem. Soc.* **1995**, *117*, 4271.
29 A. L. Balch, V. J. Catalano, D. A. Costa, W. R. Fawcett, M. Federco, A. S. Ginwalla, J. W. Lee, M. M. Olmstead, B. C. Noll, K. Winkler, *J. Phys. Chem. Solids* **1997**, *58*, 1633.
30 A. L. Balch, J. W. Lee, M. M. Olmstead, *Angew. Chem.* **1992**, *104*, 1400.
31 A. L. Balch, L. Hao, M. M. Olmstead, *Angew. Chem. Int. Ed. Engl.* **1996**, *35*, 188.
32 H.-F. Hsu, S. R. Wilson, J. R. Shapley, *Chem. Commun.* **1997**, 1125.
33 P. R. Birkett, A. G. Avent, A. D. Darwish, H. W. Kroto, R. Taylor, D. R. M. Walton, *Chem. Commun.* **1995**, 683.
34 A. G. Avent, P. R. Birkett, A. D. Darwish, H. W. Kroto, R. Taylor, D. R. M. Walton, *Fullerene Sci. Technol.* **1997**, *5*, 643.
35 P. R. Birkett, A. G. Avent, A. D. Darwish, H. W. Kroto, R. Taylor, D. R. M. Walton, *Chem. Commun.* **1996**, 1231.
36 A. Herrmann, F. Diederich, *Helv. Chim. Acta* **1996**, *79*, 1741.
37 A. L. Balch, A. S. Ginwalla, J. W. Lee, B. C. Noll, M. M. Olmstead, *J. Am. Chem. Soc.* **1994**, *116*, 2227.
38 J. M. Hawkins, M. Nambu, A. Meyer, *J. Am. Chem. Soc.* **1994**, *116*, 7642.
39 R. Kessinger, J. Crassous, A. Hermann, M. Rüttimann, L. Echegoyen, F. Diederich, *Angew. Chem.* **1998**, *110*, 2022; *Angew. Chem. Int. Ed. Engl.* **1998**, *37*, 1919.
40 J. Crassous, J. Rivera, N. S. Fender, L. Shu, L. Echegoyen, C. Thilgen, A. Herrmann, F. Diederich, *Angew. Chem.* **1999**, *111*, 1713; *Angew. Chem. Int. Ed. Engl.* **1999**, *38*, 1613.

14
Principles and Perspectives of Fullerene Chemistry

14.1
Introduction

The accessibility of fullerenes in macroscopic quantities [1] opened up the unprecedented opportunity to develop a rich "three-dimensional" chemistry of spherical and polyfunctional all-carbon molecules. Numerous fullerene derivatives, such as covalent addition products, fullerene salts, endohedral fullerenes, heterofullerenes, cluster modified fullerenes and combinations thereof, can be imagined and many examples have already been synthesized. Consequently, new materials with outstanding biological [2–24] or materials properties [25–35] have been discovered. Fullerenes are now established as versatile building blocks in organic chemistry, introducing new chemical, geometric and electronic properties. Most of the chemistry of fullerenes has so far been carried out with C_{60} with little work on higher fullerenes (see Chapter 13), endohedral fullerenes (see Chapter 1) and heterofullerenes (see Chapter 12). This is simply because C_{60} is the most abundant fullerene. The principles of fullerene chemistry can be deduced from the analysis of chemical transformations as well as from theoretical investigations of various fullerenes. This chapter is focused on the principles of C_{60} chemistry [36]. The chemical behavior of other fullerenes tends to be similar, especially in subunits of the molecules closely related to the C_{60} structural elements. To reveal the characteristics of fullerene chemistry [36–55] is not only an academic challenge; an understanding of the chemical behavior and properties of this new class of compounds is an important requirement for the design of fullerene derivatives with technological applications.

14.2
Reactivity

14.2.1
Exohedral Reactivity

Since all the C atoms of fullerenes are quarternary, and therefore contain no hydrogens, substitution reactions characteristic for planar aromatics are not possible

Fullerenes: Chemistry and Reactions. Andreas Hirsch and Michael Brettreich

ISBN: 3-527-30820-2

with fullerenes. Two main types of primary chemical transformations are possible [36]: redox reactions and addition reactions. This simple topological consideration makes it evident that the reactivity of fullerenes differs significantly from that of classical planar aromatics. Redox and addition reactions lead to salts and covalent exohedral adducts, respectively (Figure 14.1). Subsequent transformations of specifically activated adducts pave the way to other classes of fullerene derivatives (Figure 14.1). These are open cage fullerenes (see Chapter 11), *quasi*-fullerenes (see Chapter 11), heterofullerenes (see Chapter 12) and endohedral fullerenes (see Chapter 1).

One example of a challenging yet not completely realized synthetic goal (see Chapter 11) is the introduction of a window into the fullerene framework that is large enough to allow atoms, ions or small molecules to enter the cage followed by reforming, on a preparative scale, the original fullerene cage structure. Such a reaction sequence would provide elegant access to endohedral fullerenes.

Addition reactions have the largest synthetic potential in fullerene chemistry. They also serve as a probe for screening the chemical properties of fullerene surfaces.

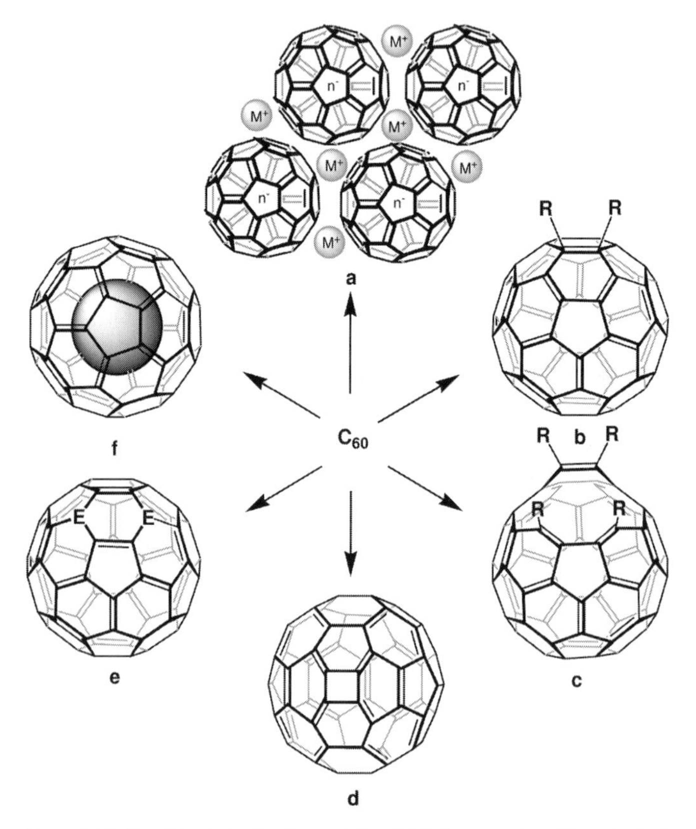

Figure 14.1 Possible derivatizations of C_{60}: (a) fullerene salts, (b) exohedral adducts, (c) open cage fullerenes, (d) *quasi*-fullerenes, (e) heterofullerenes, (f) endohedral fullerenes.

The sp^2 C-atoms in a fullerene are pyramidalized. The deviation from planarity introduces a large strain energy. Of the accessible fullerenes C_{60} is the most strained molecule. Higher fullerenes are thermodynamically more stable but still considerably less stable than graphite. The strain energy of C_{60} (estimated to be about 8 kcal mol^{-1} per carbon) is about 80% of the heat of formation [56]. Heats of formation of C_{60} and C_{70} compared with graphite as reference have been determined experimentally by calorimetry as 10.16 and 9.65 kcal mol^{-1} per C-atom, respectively [57, 58]. The major driving force for addition reactions leading to adducts with a low to moderate number of addends bound to C_{60} is the resulting *relief of strain in the fullerene cage*, which is the *major reactivity principle* of fullerenes. Reactions leading to saturated tetrahedrally hybridized C-atoms are strongly assisted by the local strain of pyramidalization present in the fullerene (Figure 14.2) [56].

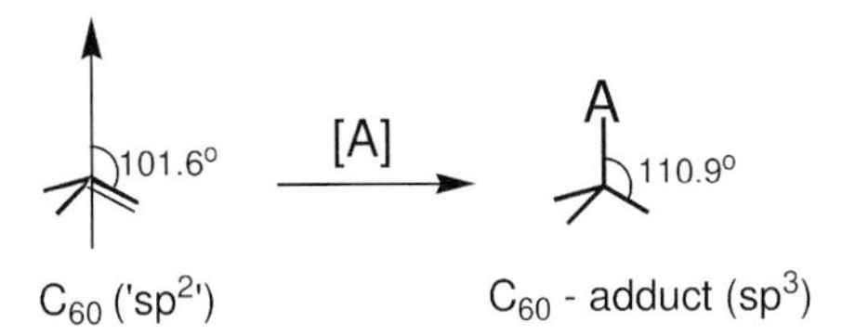

Figure 14.2 Strain-assisted addition of an addend **A** to a pyramidalized C-atom of a fullerene.

Therefore, most additions to C_{60} are exothermic reactions. The exothermicity of subsequent additions decreases with an increasing number of addends already bound to C_{60}. This holds also for the formation of η^2-complexes. For example, the addition of one Pt atom to C_{60} releases 17 kcal mol^{-1} of global strain energy, whereas in the hexakisadduct less than 10 kcal mol^{-1} are released per Pt atom [56].

The conjugated C-atoms of a fullerene respond to the deviation from planarity by rehybridization of the sp^2 σ and the p π orbitals, since pure p character of π orbitals is only possible in strictly planar situations [56]. The electronic structure of non-planar organic molecules has been analyzed by Haddon using the π orbital axis vector (POAV) analysis. For C_{60} an average σ bond hybridization of $sp^{2.278}$ and a fractional s character of 0.085 (POAV1) or 0.081 (POAV2) was found [59–63]. Consequently, the π orbitals extend further beyond the outer surface than into the interior of C_{60}. This consideration implies, moreover, that fullerenes and, in particular, C_{60} are fairly electronegative molecules [64, 65] since, due to the rehybridization, low-lying π^* orbitals also exhibit considerable s character.

Indeed, compared with the chemical behavior of other classes of compounds the reactivity of C_{60} is that of a *fairly localized* and *electron-deficient* polyolefin. The electrophilicity per se is especially reflected by the ease of electrochemical and chemical reductions as well as by nucleophilic additions (Scheme 14.1). In reactions with nucleophiles, the initially formed intermediates $Nu_nC_{60}{}^{n-}$ can be stabilized by (1) the addition of electrophiles E^+, e.g. H^+, or carbocations to give $C_{60}(ENu)_n$ [66]; (2) the addition of neutral electrophiles E-X such as alkyl halogenides to give $C_{60}(ENu)_n$ [67]; (3) an S_{Ni} or internal addition reaction to give methanofullerenes [68, 69], and fullereno-cyclohexanones [70]; or (4) by an oxidation (air) to give, for example, $C_{60}Nu_2$ (Scheme 14.1) [71, 72].

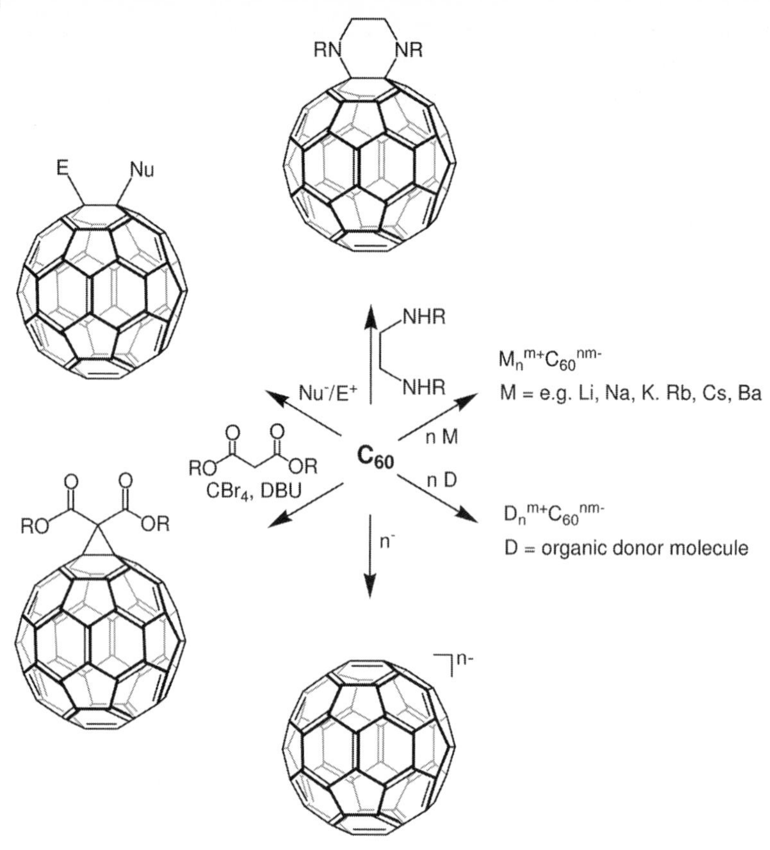

Scheme 14.1 Example reductions and nucleophilic additions to C_{60}.

The electrophilicity and the ease of reduction is associated by the fact that, because of their considerable s character, the π^* orbitals are low lying. Conversely, the reduction of C_{60} and other fullerenes is also supported by strain-relief [56], because many carbanions prefer pyramidal geometries. Reductions, for example up to the dodecaanion (see Chapter 2) as well as multiple nucleophilic additions (see Chapter 3) are possible or unavoidable if an excess of reagent is used. The number of reduction steps or nucleophilic additions can be controlled by the stoichiometry of the reagent or sometimes, for nucleophilic additions, by the size of the addend. The more addends already bound to C_{60} the less nucleophilic and electronegative becomes the fullerene. For example, the longer reaction times required to synthesize the T_h-symmetrical hexakisadduct $C_{66}(COOEt)_{12}$ (see Chapter 10) by nucleophilic addition of diethyl bromomalonate starting from the pentaadduct compared with that of the monoadduct $C_{61}(COOEt)_2$ from C_{60} are a consequence of both decreasing nucleophilicity of the fullerene cage and decreasing exothermicity of the addition [50, 73, 74]. The formation of a heptakisadduct $C_{67}(COOEt)_{14}$ from T_h-$C_{66}(COOEt)_{12}$ is impossible under the same reaction conditions. In this case steric hindrance of the addends also becomes important. Along the same lines, the electrochemical reductions of a series of mono- through to hexakisadducts of C_{60} become increasingly difficult and more irreversible with increasing number of addends [75]. Compared with a monoadduct the first reduction of a hexakisadduct with T_h-symmetrical addition pattern is typically shifted by about 0.8 V to more negative potentials. With incremental functionalization of the fullerene, the LUMO of the remaining conjugated framework is raised in energy.

With transition metal reagents, C_{60} undergoes either hydrometalations or forms η^2- rather than η^5- or η^6-π-complexes (see Chapter 7). The latter would be typical reactions for planar aromatics (Scheme 14.2). Conversely, η^5-π-complexes can be obtained with fullerene derivatives, where due to a specific addition pattern an isolated cyclopentadienide substructure is present (see Chapter 7) [76–78].

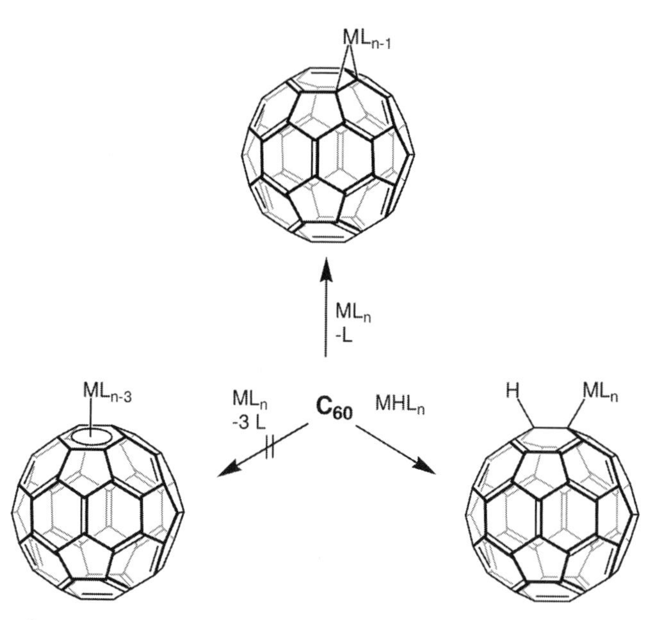

Scheme 14.2 Typical reactions of C_{60} with transition metal complexes.

In analogy to olefins, C_{60} undergoes a broad variety of cycloadditions (see Chapter 4 and Scheme 14.3). In many cases cycloadducts of C_{60} exhibit the same stability as the corresponding non-fullerene based adducts. These reactions are very useful for the introduction of functional groups. Among the most important cycloadditions are [4+2] cycloadditions such as Diels–Alder and hetero-Diels–Alder reactions, where C_{60} reacts always as dienophile, [3+2] cycloadditions with 1,3 dipoles, thermal or photochemical [2+2] cycloadditions, [2+1] cycloadditions and others, for example, [8+2] cycloadditions. Among these general reactions several examples deserve special attention, since they reflect characteristic chemical properties of C_{60} [36]:

(1) *Reversibility of Diels–Alder reactions:* Some [4+2] cycloadditions, for example, the reaction of C_{60} with anthracenes [79–83] or cyclopentadiene [80, 84] are reversible. The parent components are obtained from the adduct upon heating. For 9,10-dimethylanthracene (DMA) as diene the Diels–Alder reaction is already reversible at room temperature, making it difficult to isolate the adduct. Such facile retro-reactions have been used for regioselective formations of stereochemically defined multiple adducts such as the template mediated syntheses that use DMA as a reversibly binding diene [82] and the topochemically controlled solid-state reaction of C_{60} with C_{60}(anthracene) to give *trans-1-*C_{60}(anthracene)$_2$ [83].

(2) *[2+2] Reactivity with benzyne:* Even with the good dienophile benzyne C_{60} does not react as diene but rather forms a [2+2] cycloadduct [85–87]. One reason for such reactivity is certainly due to the fact that in a hypothetical [4+2]-adduct one unfavorable [5,6]-double bond in the lowest energy VB structure is required.

(3) *Formation of cluster opened methano- and imino[60]fullerenes (fulleroids and azafulleroids):* Thermal [3+2]-cycloadditions of diazo compounds or azides leads to the formation of fulleropyrazolines or fullerotriazolines. Thermolysis of such adducts after extrusion of N_2 affords as kinetic products the corresponding [5,6]-bridged methano and iminofullerenes with an intact 60 π-electron system and an open transannular bond (see Chapter 4) [88–91]. The corresponding [6,6]-bridged structures with 58 π-electrons and a closed transannular bond are formed only in traces.

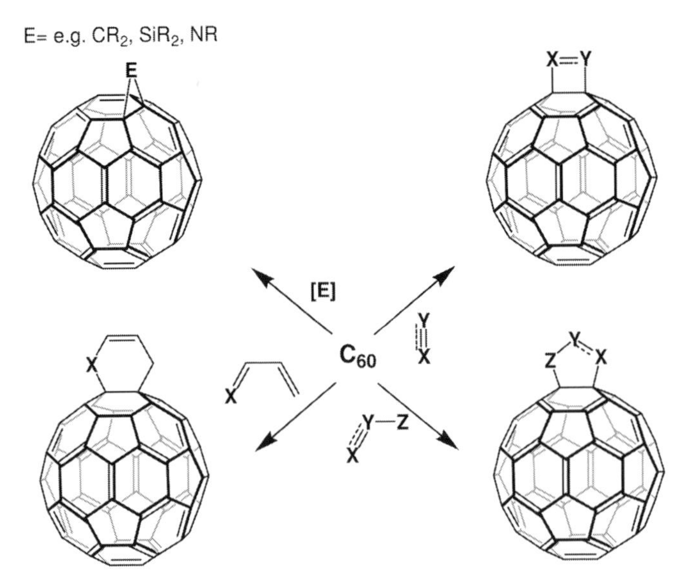

Scheme 14.3 Typical cycloaddition reactions of C_{60}.

C_{60} is a radical sponge (Scheme 14.4) and readily adds organic as well as inorganic radicals (see Chapter 6). As with nucleophilic additions, multiple additions take place if an access of radicals is allowed to react with the fullerene. The pronounced reactivity towards radicals is also preserved in fullerene adducts. A water-soluble C_3-symmetrical trisadduct of C_{60} showed excellent radical scavenging properties in vitro and in vivo and exhibits remarkable neuroprotective properties [7, 8]. It is a drug candidate for the prevention of ALS and Parkinson's disease. Concerning the reaction mechanism, nucleophilic additions and radical additions are closely related and in some cases it is difficult to decide which mechanism actually operates [92]. For example, the first step in the reaction of C_{60} with amines is a single electron transfer (SET) from the amine to the fullerene. The resulting amines are finally formed via a complex sequence of radical recombinations, deprotonations and redox reactions [36].

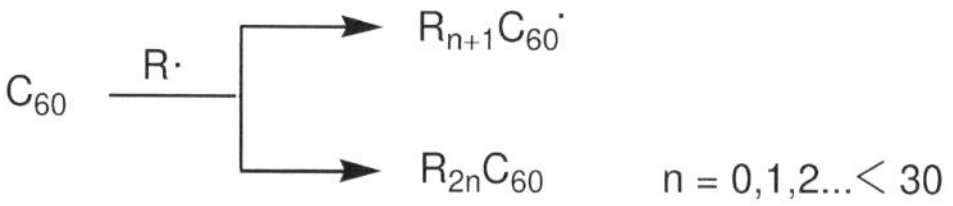

C_{60} —R·→ $R_{n+1}C_{60}^{\cdot}$

C_{60} —R·→ $R_{2n}C_{60}$ $n = 0,1,2... < 30$

Scheme 14.4 Reactions of C_{60} with free radicals.

Similar to usual olefins, C_{60} undergoes osmylations, epoxidations, and additions of Lewis acids (see Chapter 9). Here, C_{60} acts as electron-donating species (Scheme 14.5).

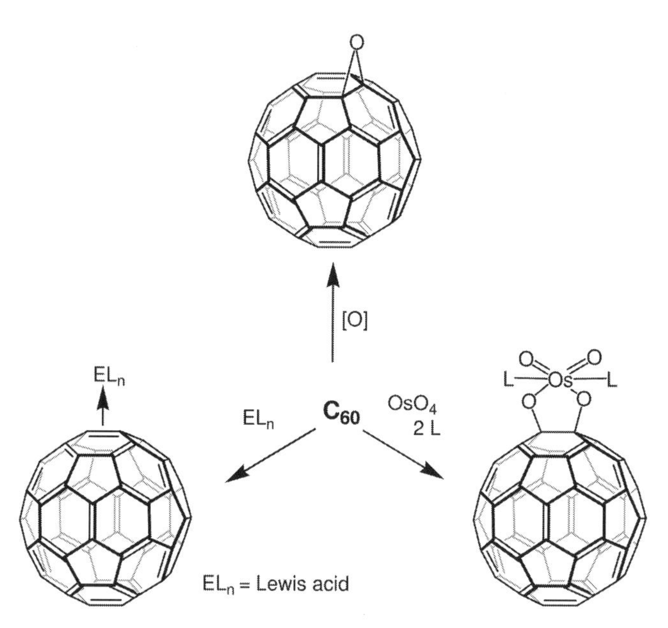

Scheme 14.5 Reactions of C_{60} with electrophiles.

All attempts to isolate the completely hydrogenated (see Chapter 5) or halogenated (see Chapter 9) icosahedral $C_{60}X_{60}$ have failed. If formed at all, $C_{60}X_{60}$ is rather unstable or it is a ring-opened degraded product. This instability is due to another driving force in fullerene chemistry, namely the *reduction of strain* in polyadducts due to *elimination*. These new types of strains are caused by the increasing deviation from tetrahedral angles of sp^3 carbons (e.g. planar cyclohexane rings) as well as accumulating eclipsing interactions between the addends X bound in 1,2-positions. Eclipsing interactions between X are unfavorable, and increase with increasing size of X. Whereas, for a small number of addends, the strain relief in the parent fullerene cage caused by the addition is predominant, the opposite behavior is valid for polyadducts, because new types of strain are increasingly built up. Therefore, the degree of addition is a consequence of the balance of these two opposing effects. In general, exhaustive hydrogenations and halogenations lead to adducts $C_{60}X_n$ with an intermediate number of addends bound to the fullerene, e.g. $n = 24$ and 36. These adducts, however, are thermally and chemically not very stable and tend to revert to C_{60}. Several polyfluorofullerenes such as $C_{60}F_{18}$ are more stable [93]. This is due to the comparatively strong C–F bonds and the presence of aromatic substructures within the fullerene framework [44]. Chemical oxidation of C_{60} leading to free cationic species is possible under dramatic conditions, e.g. in superacidic media (Chapter 8). Mono-cations of fullerenes such as C_{60}^{+} and C_{70}^{+} and $C_{59}N^{+}$ can be stabilized and characterized in the presence of very non-nucleophilic hexahalocarborane anions [94, 95].

In general, addition products of C_{60} tend to revert to the parent fullerene by eliminating the addends. This aspect will be discussed in more detail in Section 14.4, since the retention of the structural type has been frequently used as aromaticity criterion. However, in many cases exohedral fullerene adducts exhibit thermal and chemical stability that is high enough to allow their use as building blocks in synthetic chemistry and technological applications.

14.2.2
Endohedral Reactivity

The spherical structure of fullerenes allows exohedral and endohedral chemistries to be distinguished. As demonstrated above, the covalent exohedral chemistry of C_{60} has been well established over the last few years. For such chemistry, which is mainly addition chemistry, fullerenes are certainly more reactive than planar aromatics because an important driving force for such addition reactions is the reduction of strain, which results from pyramidalization in the sp^2-carbon network. To investigate the potential covalent chemistry taking place in the interior of a fullerene, access to suitable endohedral model systems is required. Most known endohedral fullerenes known are either metallofullerenes where electropositive metals are encapsulated or endohedral complexes involving a noble gas guest. However, these prototypes do not serve as model systems for investigations on endohedral covalent bond formation since the electropositive metals transfer electrons to the fullerene shell, leading to components with ionic character, and due to their inert nature the noble gas hosts do not react with the fullerene guest at all. In contrast to these guests reactive non-metal atoms or a reactive molecular species such as a methyl radical would be more instructive probes for screening the chemical properties of the inner concave surface of C_{60}.

The first example of such an endohedral fullerene was N@C_{60} [96], where the encapsulated N-atom is in its atomic $^4S_{3/2}$ ground state and undergoes no charge transfer with the fullerene cage [36]. This result reveals an astonishing and unprecedented situation: The N atom does not form a covalent bond with C atoms of the fullerene cage, showing that, in contrast to the outer, the inner concave surface is extremely inert. In other words, the reactivity of a carbon network depends on its shape. This conclusion was corroborated by various semi-empirical (PM3-UHF/RHF) and density functional calculations (UB3LYP/D95//PM3) on the system N/C_{60} [97]. The structure of N@C_{60} with nitrogen in the center of the cage is the global minimum of the endohedral complexation. The formation of N@C_{60} from the free compounds is more or less thermoneutral. The Coulson and Mulliken charges (PM3 and DFT) of the nitrogen are both zero. The spin density is exclusively localized at the nitrogen. No charge is transferred from the nitrogen to the fullerene because of the low, even slightly negative, electron affinity of N ($E_A = -0.32$ eV), which is comparable to that of He ($E_A = -0.59$ eV). With a fixed cage-geometry the energy increases continuously when the N-atom moves from the center to the cage, independent of the way of approach. In contrast, there are strong attractive interactions if the N-atom approaches a C-atom, a [6,6]-bond or a [5,6]-bond from the outside. This can be explained with the pyramidalization of the C-atoms of C_{60} and the reduced p-character of the π-orbital. Owing to electron pair repulsion the charge density on the outside of the fullerenes is higher than on the inside. This implies that, in contrast to the reactive exterior, the orbital overlap with an N-atom is essentially unfavorable inside C_{60} and at the same time there is a repulsion of the valence electron pairs (Figure 14.3). If the fullerene cage is allowed to relax during the various approaches of the N-atom to the cage, there are local minima in the endohedral case corresponding to the covalently bound structures with a C–N bond, an aza-bridge over a closed [6,6]-bond and an aza-bridge over a closed

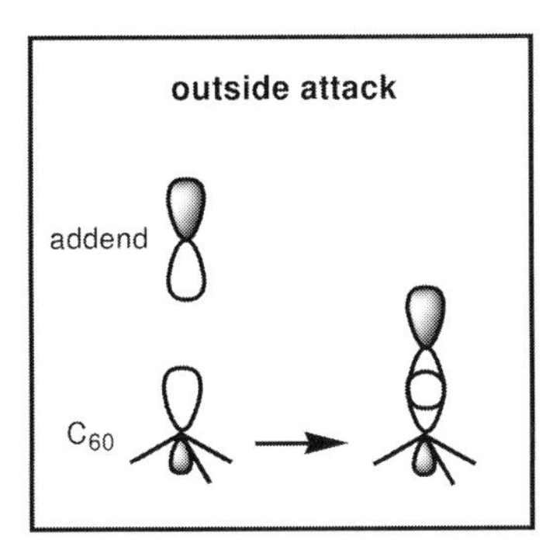

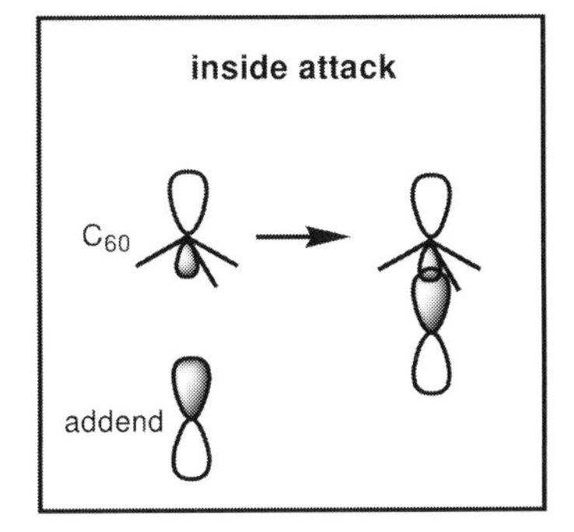

Figure 14.3 Schematic representation of an exohedral and an endohedral attack of an addend, represented by a p orbital, to the fullerene with fixed cage geometry.

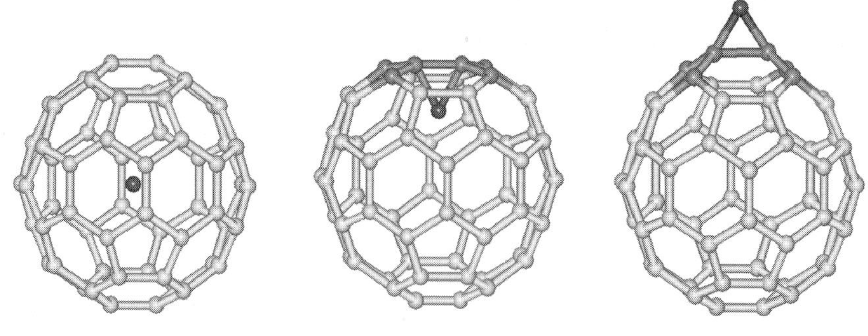

Figure 14.4 Calculated structures of N@C_{60} (*left*) (endohedral minimum), C_{60} with an endohedral [6,6]-aza-bridge (*middle*), and C_{60} with an exohedral [6,6]-aza-bridge (*right*). The relative PM3 energies are 0, +33.3 and –48.4 kcal mol^{-1}, respectively.

[5,6]-bond (Figure 14.4). These local minima are about 20–50 kcal mol^{-1} higher than the global minimum of N@C_{60}. In contrast to the corresponding exohedral binding, no energy gain is accompanied by covalent bond formation and subsequent relaxation of the cage geometry. This is because the endohedral addition at the relatively rigid cage structure of C_{60} leads to the introduction of additional strain. The opposite effect is operative for the corresponding exohedral adducts. Here, the cage geometry supports the formation of normal, almost strain-free geometries in the region of the addend. For endohedral additions the considerable σ-strain of the fullerene cage would be enlarged, but if the addition is exohedral the strain of the cage would decrease. The tendency of the outside surface of fullerenes to undergo addition reactions easily is confirmed by many experimental examples of exohedral adduct formation, as demonstrated in the preceding paragraphs.

The corresponding calculations on the endohedral and exohedral complexes of C_{60} with H, F or the methyl radical as guest predict the same behavior [98]. In all cases the formation of endohedral covalent bonds is energetically unfavorable due to the analogous strain arguments, even if such reactive species as F atoms are exposed to the inner surface of C_{60}.

These investigations focused for the first time on a *new aspect of topicity* that takes into account the influence of the shape of a bent structure on its reactivity. The remarkable inertness of the inner surface contrasts with the pronounced reactivity of the outer concave surface of C_{60}. Almost unperturbed, atomic species or reactive molecules can be studied at ambient conditions once they are encapsulated by C_{60}. Moreover, the wavefunction of the guest atom can be influenced by a permanent distortion of the C_{60} cage. This was demonstrated by exohedral derivatization of N@C_{60}, leading to a lowering of the I_h symmetry, which influences the ESR spectra of the paramagnetic guest [99–103].

14.3 Regiochemistry of Addition Reactions

14.3.1 Bond Length Alternation – Preferred Additions to [6,6]-Double Bonds

It has been calculated that there are 12 500 Kekulé resonance structures possible for C_{60} [104]. However, the two types of bonds in C_{60} have different lengths, with the [6,6]-bonds being shorter than the [5,6]-bonds (see Chapter 1). Thus, the lowest energy Kekulé structure of C_{60} is that with all the double bonds located at the junctions of the hexagons ([6,6]-double bonds) (Figure 14.5).

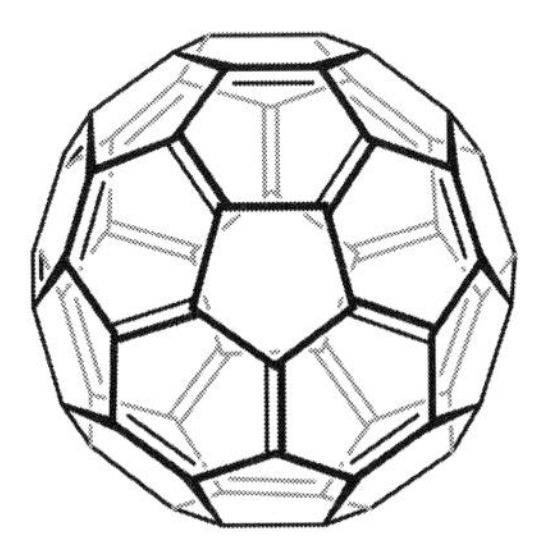

Figure 14.5 Lowest energy Kekulé structure of C_{60}.

The bond length alternation in C_{60} is due to symmetry and the occupation of its π-orbitals [36, 48, 105–108]. In a first approximation the π-electron system of an icosahedral fullerene such as C_{20}, C_{60} or C_{80} can be described with a spherical electron gas incasing the σ framework in a double skin. The wavefunctions of this electron gas are characterized by the angular momentum quantum numbers $l = 0$, 1, 2, 3, etc. corresponding to s-, p-, d-, and f-π-shells. These shells are analogs of atomic orbitals. The most significant difference is that the sphere defined by the σ framework represents a nodal plane and the electron density in the center of the sphere approaches zero. The irreducible representations of the icosahedral group can be found using group theory by lowering the symmetry from full rotational symmetry to icosahedral symmetry treated as a perturbation (Table 14.1) [36, 48]. Considering the Pauli principle, it can be seen that upon occupation with $2(N+1)^2$ electrons all π-shells are completely filled. The shape of the molecular s-, p-, d-, and g- etc. orbitals is still very reminiscent of that of the corresponding atomic H-like orbitals. The charged fullerene $C_{60}{}^{10+}$ [48, 107], for example, represents a closed shell system with $n = 50$ π electrons where all π orbitals up to $l = 4$ (g shell) are completely occupied (Figure 14.6). Since the angular momenta are symmetrically distributed no distortion from spherical or icosahedral symmetry is expected in cases where all states are completely filled. Hence, no significant bond length alternation is expected for $C_{60}{}^{10+}$. The calculated lengths of the [6,6]- and [5,6]-bonds of $C_{60}{}^{10+}$ (I_h symmetry) are 1.48 and 1.44 Å [48, 107]. Therefore, the bond length alternation is lower than that of neutral C_{60} and comparable with that in T_h symmetrical hexakisadducts [82] containing benzenoid substructures (see Chapter 10).

Table 14.1 Electron ground-state configurations[17] of the fully (bold) and partially filled π shells of icosahedral fullerenes.

l [a]	*shell*	*electrons/shell*	n_c [b]	*HOMO (I_h symmetry)* [c]
0	s	2	**2**	$\boldsymbol{a_g^{2}}$
1	p	6	**8**	$\boldsymbol{t_{1u}^{6}}$
2	d	10	**18**	$\boldsymbol{h_g^{10}}$
3	f	14	24 26 **32**	t_{2u}^{6} g_u^{8} $\boldsymbol{t_{2u}^{6}g_u^{8}}$
4	g	18	40 42 **50**	g_g^{8} h_g^{10} $\boldsymbol{g_g^{8}h_g^{10}}$
5	h	22	56 60 62 66 **72**	t_{1u}^{6} or t_{2u}^{6} h_u^{10} $t_{1u}^{6}t_{2u}^{6}$ $t_{1u}^{6}h_u^{10}$ or $t_{2u}^{6}h_u^{10}$ $\boldsymbol{t_{1u}^{6}t_{2u}^{6}h_u^{10}}$

a) Angular momentum quantum number for a spherical shell of π electrons.
b) Number of π electrons for the ground state configurations with closed-shell or inner-shell.
c) HOMO-symmetries for all levels of given *l*; the superscripted digit shows the number of electrons for complete occupation of the shell.

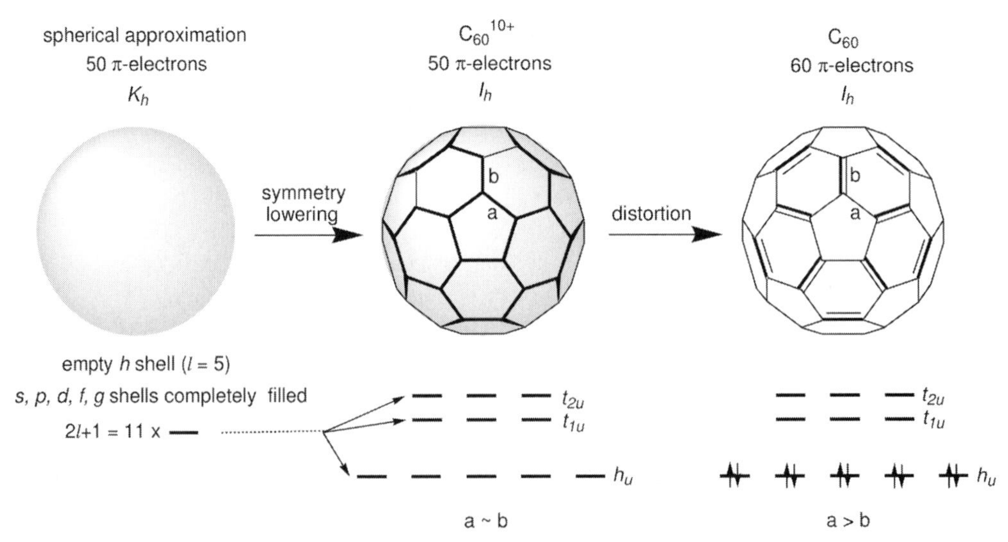

Figure 14.6 Incomplete filling of the h shell deduced from the spherical approximation and the resulting bond length alternation in neutral C_{60}.

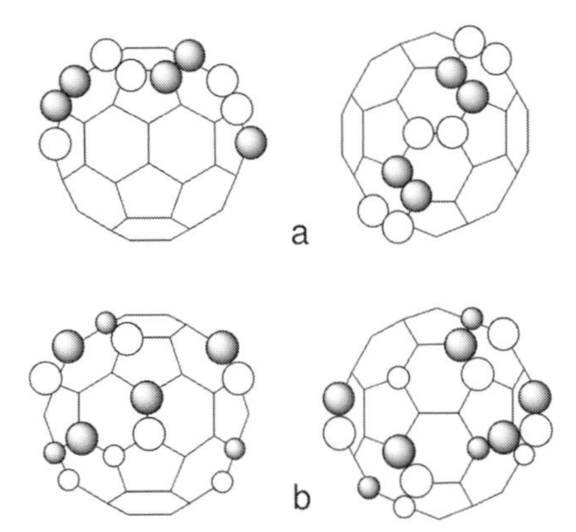

Figure 14.7 Schematic representation of one of the five degenerate HOMOs (a) and one of the three degenerate LUMOs (b) of C_{60}. Each orbital is represented by a front and side view. Only the exohedral part of the orbitals projected to the corresponding C-atoms is shown for clarity.

What are the consequences of filling the $l = 5$ (h) shell on the geometry of C_{60}? The complete filling of the $l = 5$ state would lead to an accumulation of 72 π electrons, assuming that all the $l = 5$ states are filled, before any $l = 6$ level is occupied. In icosahedral symmetry the $l = 5$ states split into the $h_u + t_{1u} + t_{2u}$ irreducible representations (Table 14.1) [36, 48]. The lowest energy level is the h_u level, which in neutral C_{60} is completely filled with the ten available electrons. Looking at the symmetries of the h_u orbitals reveals bonding interactions with the [6,6]-bonds and antibonding interactions in the [5,6]-bonds (Figure 14.7) [36, 48, 107]. Consequently, filling of the h_u orbitals causes a shortening of the [6,6]-bonds, which leads to an increase in bond energy and at the same time favors longer [5,6]-bonds by reducing antibonding interactions [48].

The bond length alternation in C_{60} is obviously not only caused by the nature of the σ framework. The difference in bond lengths between [6,6]-bonds and [5,6]-bonds is rather directly connected with the symmetry and the occupation of its molecular π-orbitals. The shortening of one type of bonds is equivalent to a partial localization of the π-orbitals into localized bonds. It is the filling of the h_u, t_{1u} and t_{1g} orbitals that is responsible for the degree of bond length alternation. This is closely related to the situation in the annulenes where, depending on the filling of the molecular orbitals with either $4N$ or $4N$+2 π electrons, a distortion from the ideal D_{nh} takes place or not. Calculations on the icosahedral homologues C_{20} and C_{80} revealed the same behavior [107]. Only the closed-shell species C_{20}^{2+} (closed $l = 2$ shell, $n_c = 18$) and C_{80}^{8+} (closed $l = 5$ shell, $n_c = 72$) exhibit ideal I_h symmetry, whereas symmetry lowering to C_2 is predicted for C_{20} and to D_{5d} for C_{80}.

Comparable alternations between the [6,6]- and [5,6]- bond lengths are observed in many fullerene adducts and higher fullerenes. Moreover, from the evaluation of many exohedral transformation of fullerenes, in particular C_{60}, an important principle of fullerene chemistry can be deduced, namely *minimization of [5,6]-double bonds* in the *lowest energy Kekulé structure.* Minimization of [5,6]-double bonds is

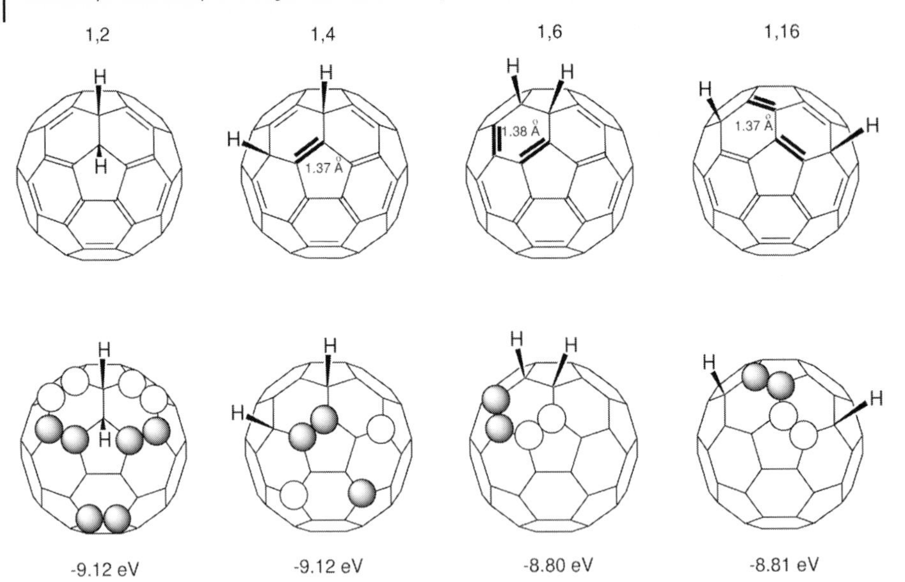

Figure 14.8 Lowest energy VB structures, PM3 calculated lengths of [5,6]-"double" bonds, HOMO coefficients and energies of different dihydro[60]fullerenes.

one factor that dictates the regiochemistry of addition reactions. Although many isomers are, in principle, possible for additions of segregated addends, the preferred mode of addition is 1,2. For a combination of sterically demanding addends, 1,4 additions [36, 72] and even 1,6 additions [109] (to the positions 1 and 16) can take place alternatively or exclusively (Figure 14.8) [36].

If the size of two addends in a substituted dihydro[60]fullerene does not cause steric constraints 1,2-dihydro[60]fullerenes are formed as the thermodynamically most stable structures after addition of the two segregated addends to a [6,6]-double bond. This addition pattern represents the only cluster closed dihydro[60]fullerene, whose lowest energy Kekulé structure does not have double bonds placed in a five-membered ring. This VB consideration is confirmed by various X-ray single-crystal structure investigations and computations showing that in 1,2-adducts the bond length alternation between [6,6]-bonds and [5,6]-bonds is totally preserved, with significant deviations from the values of the parent C_{60} only in the direct neighborhood of the addends [110]. Moreover, as in the parent C_{60}, the bonding interactions in the occupied frontier orbitals are located predominantly at [6,6]- and the nodes at the [5,6]-sites (Figure 14.9) [36].

It is therefore useful to look at a 1,2-dihydro[60]fullerene as a stereoelectronically slightly perturbed C_{60}. Introduction of a double bond into a five-membered ring costs about 8.5 kcal mol^{-1} (Figure 14.10) [111]. In a 1,4-adduct (1,4-dihydro[60]-fullerene) one, and in a 1,6-adduct (1,16-dihydro[60]fullerene or 1,6- dihydro[60]-fullerene) two, double bonds in five-membered rings are required for the corresponding lowest-energy Kekulé structure. This VB consideration is also confirmed experimentally and by computations (Figure 14.8) [36].

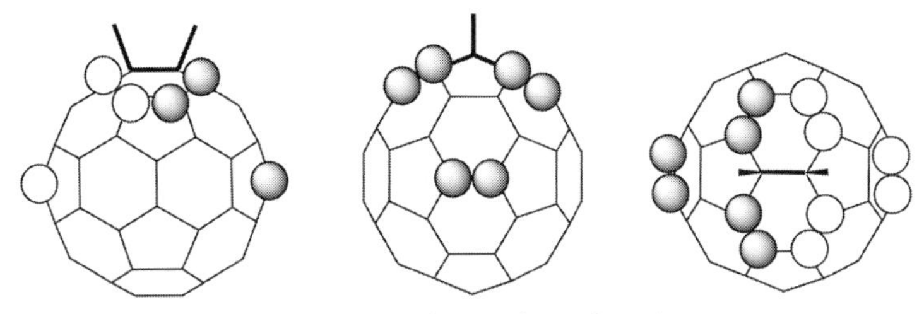

Figure 14.9 PM3 calculated HOMO of $C_{60}H_2$ shown from three perspectives.

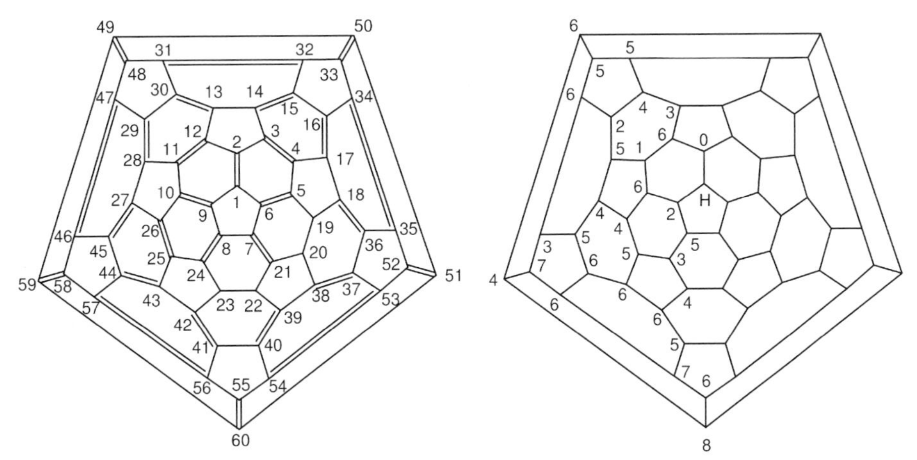

Figure 14.10 Numbering system within the lowest energy VB structure of C_{60} and dependence of the topologically counted minimum number of [5,6]-double bonds that have to be introduced by the addition of two H-atoms on the positions of the H-atoms.

The 1,4 and 1,16 addition patterns are only preferred for sterically demanding addends, since in the corresponding 1,2-adducts, or even in some 1,4-adducts, the eclipsing interactions become predominant (Table 14.2). A 1,6-addition pattern with a cluster closed structure is unfavorable and has never been isolated due to both introduction of two [5,6]-double bonds and eclipsing interactions. In 1,2-$C_{60}H_2$, the eclipsing interaction was estimated to be about 3–5 kcal mol^{-1}. This increases with increasing size of the addend. The relative product distribution is balanced by minimizing [5,6]-double bonds, on the one hand, and eclipsing interactions on the other hand. Dihydro[60]fullerenes having the addends attached further apart than those described above would require the introduction of even more [5,6]-double bonds and have not been isolated.

The charge or spin density of primary adducts RC_{60}^- or $RC_{60}^{\bullet}$ formed by nucleophilic or radical additions is highest at position 2, followed by 4 (11), for nucleophilic additions and at position 2, followed by 4 (11) and 6 (9), for radical additions (Figure 14.11). This is also a result of avoiding formal [5,6]-double bonds. From this viewpoint, an electrophilic attack or a radical recombination process at the position 2 is also favored, neglecting eclipsing interactions.

Table 14.2 Calculated relative PM3 energies (kcal mol^{-1}) of dihydro[60]fullerene derivatives with addends of different steric requirement [36].

	1,2-Adduct	*1,4-Adduct*	*1,6-Adduct*	*1,16-Adduct*
$C_{60}H_2$	0	3.8	18.3	15.5
$C_{60}H^tBu$	0	3.1	17.8	15.0
$C_{60}(^tBu)_2$	31.3	0	40.6	9.2
$C_{60}[Si(^tBu)_3]_2$	–	13.3	–	0

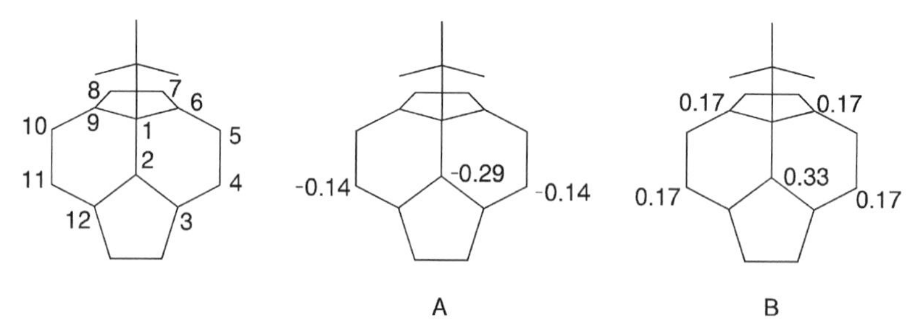

Figure 14.11 Charge distribution (A) (Mulliken charges) and spin distribution (B) in the intermediates $^tBuC_{60}^-$ and $^tBuC_{60}^{\bullet}$.

With methano- and azirenofullerenes and the corresponding homofullerenes, either ring-closed 1,2-bridged structures or ring-open 1,6-bridged structures have been observed (Figure 14.12) [89, 90, 112–114]. In both cases no [5,6]- double bond has to be introduced into the lowest energy Kekulé structure. This is not valid for hypothetical open 1,2-bridged or closed 1,6-bridged structures. In both cases, two [5,6]-double bonds would be required. The ring-open 1,6-methanofullerenes (fulleroids) are formed upon diazomethane additions to C_{60} after a rearrangement of 1,2-bridged intermediates. The only exception so far are specific bis-azahomofullerenes where the imino-bridges are located at [6,6]-bonds in *cis*-1 positions (see Chapter 11) [115].

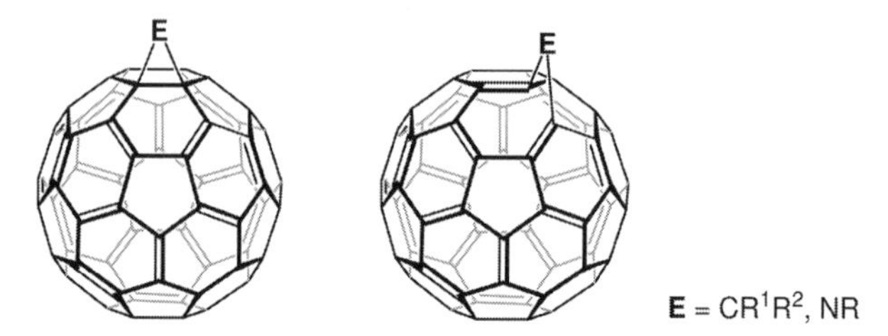

Figure 14.12 Stable ring-closed methano- and azirenofullerenes and the corresponding ring-open homofullerenes with no [5,6]-double bonds.

14.3.2
1,2-Additions with Preferred *e*- and *cis*-1 Modes: the *trans*-1 Effect

Although, usually, a mixture of many regioisomers is formed, successive 1,2-additions to [6,6]-bonds of C_{60} exhibit an inherent regioselectivity (see Chapter10). For a second 1,2-addition to a [6,6]-monoadduct the *cis*-1 positions are the preferred reactions sites. If the addends are too bulky, *cis*-1 positions are inaccessible for a second attack. In this case e- followed by *trans*-3 positions are the preferred for a second attack (Scheme 14.6). *Trans*-1 additions belong to the least preferred reactions.

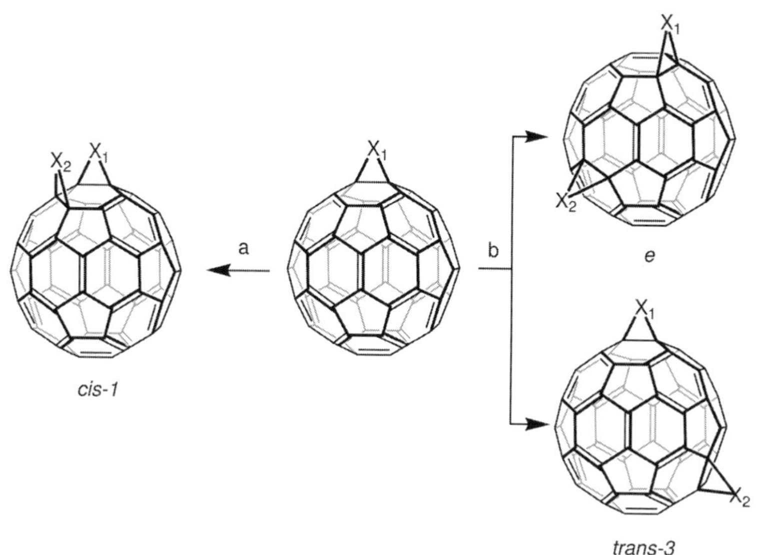

Scheme 14.6 Preferred sites for a second 1,2-addition to C_{60} of (a) sterically non-demanding addends and (b) sterically demanding addends.

This inherent regioselectivity is qualitatively independent of the nature of the addends. It is governed by a combination of thermodynamic stability of the bisadducts and orbital coefficients at the corresponding sites [36]. Frontier orbitals with high coefficients in *cis*-1-, *e*- and *trans*-3-sites are energetically accessible whereas frontier orbitals with high coefficients in other sites, especially *trans*-1, are energetically less accessible. The regioselectivity of additions into *e*-positions according to Scheme 14.6 increases with the number of addends already bound. Both pronounced frontier orbital coefficients at the corresponding reaction sites in the precursor adducts as well as an increasing relative thermodynamic stability of the adducts with addends bound in all e-sites are the reasons for the regioselectivity depicted in Scheme 14.7 [36]. Different regioselectivities can be achieved by applying specific synthesis concepts such as tether controlled reactions (see Chapter 10). The formation of hexakisadducts with a T_h-symmetrical addition pattern is unlikely for irreversibly binding and sterically non-demanding addends (preferred *cis*-1 additions).

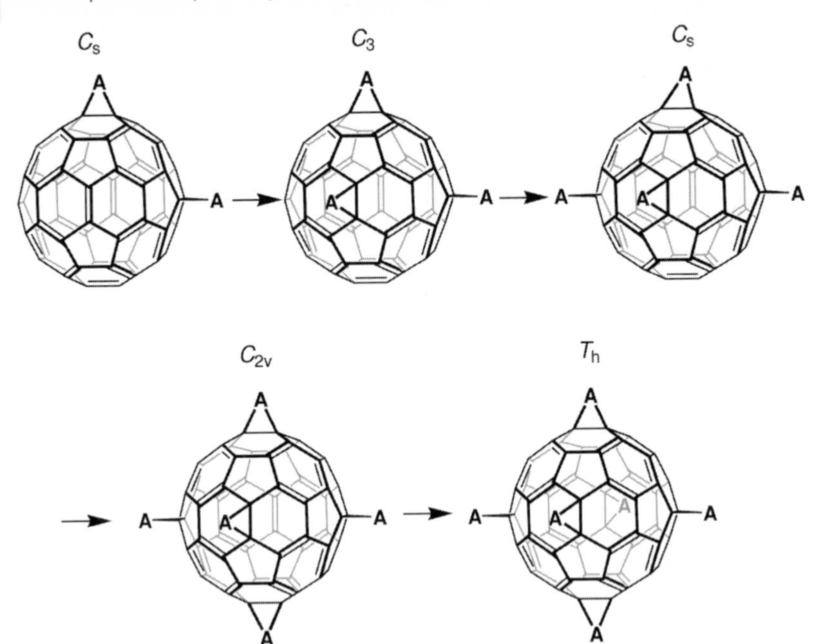

Scheme 14.7 Regioselective formation of hexakisadducts of C_{60} with a T_h-symmetrical addition pattern.

14.3.3 Vinyl Amine Mode

The very pronounced inherent regioselectivity observed for the formation of bis-azahomofullerenes where the two aza-bridges are located at two adjacent [5,6]-sites within the same five-membered ring is due to the vinyl amine character of bond α within the precursor azahomofullerene (Figure 14.13) (see Chapter 10).

Bond α is the most polarized double bond within the fullerene framework and the preferred reaction site for a subsequent attack.

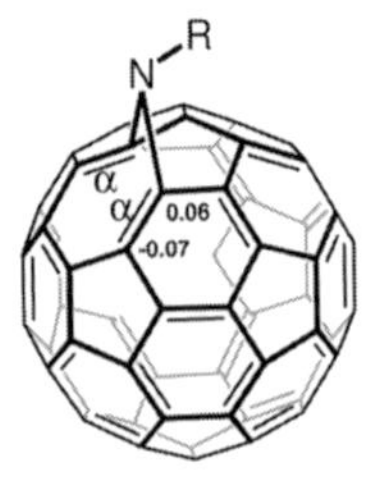

Figure 14.13 Preferred addition sites (α) in an azahomofullerene and Mulliken charges at the vinyl amine bond.

14.3.4
Cyclopentadiene Mode

The regiochemistry of subsequent additions of bulky segregated groups, such as free radicals, amines in the presence of O_2, iodine chloride, bromine and specific organometallic reagents in the presence of O_2 is largely influenced by the possibility of electronic delocalization in intermediates such as the open shell systems shown in Figure 14.14 (see Chapter 10). The stable reaction product of such reactions is a 1,4,11,14,15,30-hexahydro[60]fullerene. In these hexa- or pentaadducts of C_{60} the addends are bound in five successive 1,4- and one 1,2-position. The corresponding corannulene substructure of the fullerene core contains an integral cyclopentadiene moiety whose two [5,6]-double bonds are decoupled from the remaining conjugated π-electron system of the C_{60} core.

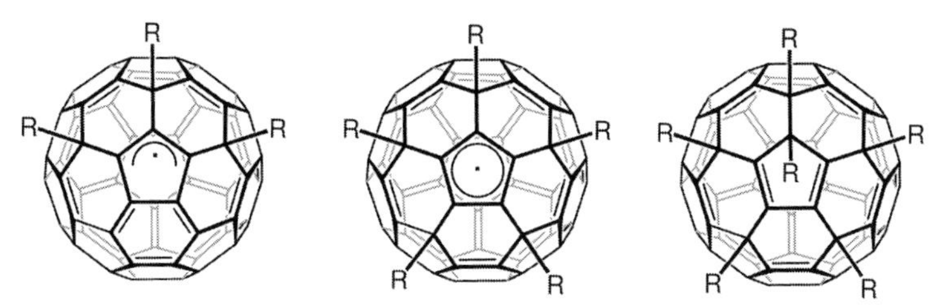

Figure 14.14 Reaction intermediates and reaction products found upon subsequent additions to C_{60} via the cyclopentadiene mode.

14.3.5
Further Modes

Further addition modes are found for the polybromination and polyfluorination of C_{60} and higher fullerenes such as C_{70} (see Chapter 9). Formation of the highly symmetrical addition pattern of T_h-$C_{60}Br_{24}$, for example, is probably due to thermodynamic control, including crystal packing effects, since bromination of C_{60} exhibits reversible character. The regioselective formation of many polyfluorofullerenes such as $C_{60}F_{18}$ seems to be governed by the formation of aromatic substructures within the intact sp^2-fullerene framework.

14.4
Aromaticity of Fullerenes

Whether or not fullerenes have to be considered aromatic has been debated since their discovery [48]. This is because the definition of aromaticity [116–118] is controversial and has changed many times over the last 175 years. The structure of fullerenes, especially that of the icosahedral representatives such as C_{60}, implies their consideration as three-dimensional analogues of benzene and other planar aromatics. In contrast to such classical systems, however, the sp^2-networks of fullerenes have no boundaries that are saturated by hydrogen atoms [56]. Consequently, the characteristic aromatic substitution reactions retaining the conjugated π-system are not possible for fullerenes. Compared with benzene the discussion of aromaticity in fullerenes must take into account the strain provided by the pyramidalization of the C atoms. The rich exohedral chemistry of fullerenes, which is basically addition to the conjugated π-system, is to a great extent driven by the reduction of strain (Section 14.2.1). Analysis of reactivity and regiochemistry of addition reactions reveals a behavior reminiscent of electron-deficient olefins. However, especially, the magnetic properties of fullerenes clearly reflect delocalized character of the conjugated π-system, which, depending on the number of π-electrons, can cause the occurrence of diamagnetic or paramagnetic ring currents within the loops of the hexagons and pentagons. Here, the aromaticity of fullerenes will be evaluated according to the classical criteria of structure, energy, reactivity and magnetisms. Finally, the nature of spherical aromaticity of fullerenes on the basis of the $2(N+1)^2$ count rule considering the whole π-system will be discussed.

14.4.1 Structural Criteria

An important structural aspect of neutral C_{60} is the bond length alternation (shorter [6,6]-"double bonds" and longer [5,6]-"single bonds") that is due to the incomplete filling of the outer $l = 5$ π-shell (Section 14.3.1). This leads to a clear distinction from the archetypal aromatic benzene [119, 120]. In benzene the fully symmetric charge density requires a VB description where at least two D_{3h} symmetric VB structures are mixed, leading to the equivalence of all six bonds. The VB structure of C_{60}, however, (Figure 14.5) exhibits already the full I_h symmetry of the carbon framework. This VB structure is the only one of the 12 500 to satisfy the symmetry criterion. In C_{70} two equivalent VB structures per equatorial hexagon are required to describe its structure and reactivity properties (Figure 14.15).

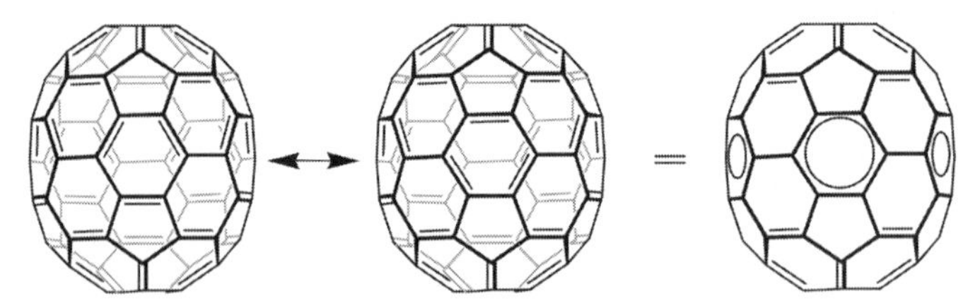

Figure 14.15 Lowest energy Kekulé resonance structures of C_{70}.

Charged C_{60} systems with completely filled shells, especially $C_{60}{}^{10+}$, are much more aromatic and exhibit less pronounced bond length alternation [107]. The same holds true for the icosahedral fullerenes C_{20} and C_{80} [107]. Application of structural aromaticity criteria such as HOMA [121] indices reveal ambiguous aromatic character, namely aromatic hexagons and antiaromatic pentagons within neutral C_{60}. The geometry of the polar caps of C_{70} is comparable to C_{60}. The geometry of the cyclic phenylene-type belt within C_{70} suggests a much more pronounced aromatic character of this substructure. Also here, the pentagons are less aromatic than the hexagons. Bond length alternation between [6,6]- and [5,6]-bonds is also a typical phenomenon of higher fullerenes.

14.4.2 Energetic Criteria

The aromatic stabilization of a molecule is the energy contribution due to the cyclic bond delocalization. This contribution is defined as the resonance energy (RE) [119, 120]. The problem in determining resonance energies is to single out of the total energy of the molecule the contribution from the cyclic bond delocalization. In other words, model reference structures have to be defined whose energy would differ from that of the molecule under consideration precisely by the component corresponding to the delocalization in question. Approaches to evaluate the amount of aromatic stabilization in C_{60} were based on isodesmic calculations, and on estimations of the heat of hydrogenation [48, 119, 120]. Based on these energetic criteria of aromaticity, C_{60} clearly exhibits an aromatic stability, which is significantly inferior to that of benzene. Upon increasing the cluster size to giant fullerenes the resonance energies approach the value for graphite.

14.4.3 Reactivity Criteria

The lack of substitution reactions, which are characteristic for planar aromatics, cannot be taken as a reactivity criterion of fullerenes, since they contain no hydrogen atoms that could be substituted. Only addition reactions can take place as primary steps on the fullerene surface. The reversibility of several addition reactions (Section 14.2.1) could be an indication of the propensity of the retention of the structural type that is considered to be a reactivity criterion of aromaticity. The regioselectivity of two-step additions, such as addition of a nucleophile, followed by the trapping of the fullerenide intermediate with an electrophile, on the other hand, shows that the negative charge is not delocalized over the whole fullerenide (Section 14.2.1). This speaks against a pronounced aromatic character. Also, other reactions, such as the additions of transition metal fragments are reminiscent of olefins rather than aromatics (Section 14.2.1). In many cases an important driving force for the regioselectivity of multiple addition reactions is the formation of substructures that are more aromatic than the substructures of the parent fullerene.

14.4.4 Magnetic Criteria

An aromatic molecule tends to sustain a diamagnetic ring current while an antiaromatic molecule tends to sustain a paramagnetic ring current. Fullerenes show evidence for ring currents on the length scale of individual rings (segregated ring currents) (Figure 14.16) [122–124]. The strength of the ring currents is

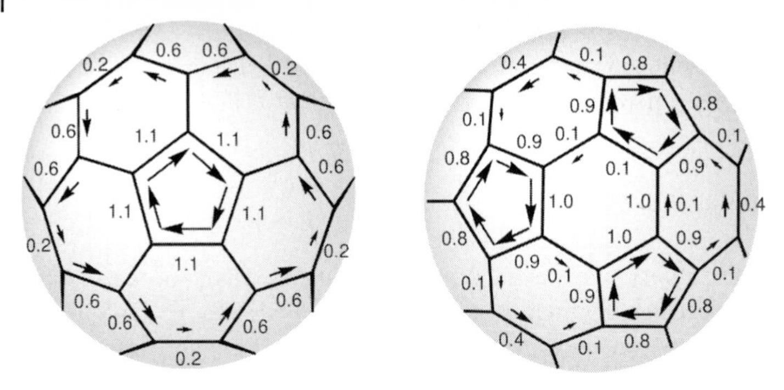

Figure 14.16 π-Electron ring currents in C_{60} for a magnetic field oriented perpendicular to a five-membered ring (5-MR) (left) and a six-membered ring (6-MR) (right), calculated with the London theory. Currents in the 5-MRs are paramagnetic (clockwise), whereas those in the 6-MRs are mostly diamagnetic. The current strength is given with respect to that in benzene [36, 122].

comparable to those in benzenoid hydrocarbons. Of particular importance is the finding of paramagnetic ring currents in the pentagons. These currents are found to be sufficiently large that their sum over the twelve pentagons almost exactly cancelled the diamagnetic currents in the twenty six-membered rings.

The cancellation of ring currents in neutral C_{60} is a direct consequence of the fullerene topology, with the diamagnetic term of the magnetic susceptibility largest in the hexagons and its paramagnetic Van Vleck term predominant in the pentagons in such a way that the overall effect of the π electrons on the magnetic susceptibility is negligible. These findings explain the vanishingly small ring-current contribution to the total magnetic susceptibility of C_{60} determined experimentally. A pronounced diamagnetic ring current gives rise to magnetic shielding in the center of the ring. The opposite is true for paramagnetic ring currents. A theoretical probe to evaluate magnetic shielding is the nucleus-independent chemical shift (NICS) [125]. Calculation of NICSs in the center of individual rings of neutral fullerenes that have been isolated revealed a general trend of diatropic hexagons and paratropic pentagons [48].

A powerful tool for measuring the magnetic shielding in the interior of a fullerene is ^3He NMR spectroscopy on endohedral fullerenes encapsulating the nucleus ^3He [126, 127]. Whereas the chemical shift of He in $^3He@C_{60}$ is comparatively small ($\sigma = -6.3$) a large additional shielding by more than 40 ppm is observed in $^3He@C_{60}{}^{6-}$, with C_{60} being in the hexaanionic state with the t_{1u} orbitals occupied [127]. The opposite behavior was found for C_{70}. The endohedral shielding within $^3He@C_{70}$ is −28.8, whereas reduction to the hexaanion cause considerable deshielding, with $\delta(^3He)$ for $^3He@C_{70}{}^{6-}$ being +8.3. Endohedral chemical shifts of higher fullerenes such as C_{76}, C_{78} and C_{84} are in between the values of C_{60} and C_{70} [48]. Consequently, a simple correlation between aromaticity based on magnetic criteria and the size of a fullerene cannot be found. Moreover, there is no correlation between the thermodynamic stability of a fullerene and its magnetic properties [48]. The considerable strain within the fullerene framework and its variation between isomers clearly dominates effects associated with aromatic stabilization. Based on the magnetic criterion, neutral C_{60}, for example, containing diatropic hexagons and paratropic pentagons can be labeled "ambiguously aromatic" [56].

14.4.5
$2(N+1)^2$-Rule for Spherical Aromaticity

The aromaticity of annulenes and heteroannulenes can be described with the Hückel rule. Due to their closed-shell structures, annulenes with $4N+2$ electrons are not distorted (D_{nh}-symmetry) and show strong diamagnetic ring currents, while singlet $4N$ annulenes are often distorted and have paratropic character. However, the Hückel rule cannot be applied in polycyclic systems, where, for example, benzenoid rings are joined by four- or five-membered rings. Thus, neither biphenylene nor fluoranthene are antiaromatic, although they represent 12π and 16π systems, respectively. Spherical fullerenes represent a special group of polycyclic π systems. To explain the nature of aromaticity of fullerenes and to deduce a count rule for their aromatic character it is important to choose the most suitable and unambiguous criterion. For the reasons outlined above, structural, energetic, and reactivity criteria alone are insufficient to properly describe the aromaticity of fullerenes. The magnetic behavior however, evaluated, for example, by ^{3}He NMR spectroscopy or by NICS calculations seems to be the most independent criterion.

It was discovered, however, that the spherical aromaticity of the icosahedral fullerenes C_{20}, C_{60} and C_{80} depends on the filling of the π-shells with electrons [107]. As pointed out in Section 14.3.1 no distortion of the cage structure is expected in these fullerenes if their shells are fully filled. Closed-shell situations are realized if the fullerene contains $2(N+1)^2$ π electrons. This is closely related to the stable noble-gas configuration of atoms or atomic ions [108]. In this case the electron distribution is spherical and all angular momenta are symmetrically distributed. Correlation of the aromatic character determined by the magnetic properties is shown in Table 14.3.

Table 14.3 Calculated endohedral ^{3}He chemical shifts and symmetry of He@C_{20}, He@C_{60} and He@C_{80} species with different filling degree of the π electron system [107].

Species	*Symmetry*	*$\delta(^3He)$* [a]
C_{20}^{2+} (closed $l = 2$ shell)	I_h	−66.2
C_{20}	C_2	−31.7
C_{60}^{+10} (closed $l = 4$ shell)	I_h	−81.4
C_{60}	I_h	−8.0
C_{80}^{8+} (closed $l = 5$ shell)	I_h	−82.9
C_{80}^{6+}	D_{5d}	−70.0
C_{80}^{2+}	D_{5d}	−54.5
C_{80} (triplet)	D_{5d}	−8.4
C_{80}^{2-}	D_{5d}	+78.6
C_{80}^{6-}	D_{5d}	−32.8

a) GIAO-SCH/3-21G//HF/6-31G.

The closed shell systems C_{20}^{2+} ($N = 2$), C_{60}^{10+} ($N = 4$), and C_{80}^{8+} ($N = 5$) exhibit very pronounced ^{3}He chemical shifts in the center of the cage. Both the pentagons and hexagons are very paratropic in all cases. Therefore, icosahedral fullerenes that contain $2(N + 1)^2$ π-electrons show the maximum degree of spherical aromaticity. Exceptions are possible if levels of the most outer shell are occupied before all "inner" shells are completely filled. This is the case for C_{60}^{12-}. Here, the t_{1g} levels of the $l = 6$ shell are filled before the t_{2u} levels of the $l = 5$ shell. This is in analogy to the filling of the atomic 4s levels prior to the 3d levels. Compared with the cyclic annulenes, which follow the $4N + 2$ rule, the spherical fullerenes show the maximum diatropicity more rarely, and there are numerous intermediate situations, including molecules with both aromatic and antiaromatic regions.

One key conclusion is that the entire molecule must be taken into account to understand the aromatic properties of icosahedral fullerenes. The $2(N + 1)^2$ rule of spherical aromaticity also sufficiently describes the magnetic behavior of non-icosahedral fullerenes [128], homoaromatic cage molecules [129], and inorganic cage molecules [130].

Clearly, for fullerenes with a lower symmetry, local substructures, such as the cyclic phenylene belt of C_{70}, can also largely determine the magnetic behavior.

14.5 Seven Principles of Fullerene Chemistry: a Short Summary

(1) The spherical shape causes a pyramidalization of the C atoms and, therefore, a large amount of strain energy. Addition chemistry is driven by the strain relief introduced by the formation of almost strain-free sp^3-C atoms. At a certain degree of addition this strain relief mechanism has to compete with new strain-introducing processes such as the increasing introduction of eclipsing interactions and the formation of planar cyclohexane substructures.

(2) Due to the s character of the π-orbitals, caused by the pyramidalization and the resulting repulsion of valence electron pairs (rehybridization), C_{60} is a comparatively electronegative molecule that is easily accessible for reductions and for the addition of nucleophiles.

(3) In neutral C_{60} the h shell is incompletely filled, resulting in a distortion corresponding to the only internal freedom that the C_{60} molecule has without breaking the I_h symmetry. This is the bond length alternation between [6,6]- and [5,6]-bonds. Since in the occupied h_u states (HOMOs) bonding interactions are predominantly located at the [6,6]-sites and the antibonding interactions (nodes) at the [5,6]-sites a contraction of the [6,6]-bonds causes an energy lowering. Hence, the lowest energy VB structure of neutral C_{60} contains only [6,6]-double bonds and [5,6]-single bonds.

(4) The regiochemistry of additions to C_{60} is driven by the maintenance of the MO structure and the minimization of energetically unfavorable [5,6]-double bonds.

(5) The most easy accessible sites in the HOMOs and LUMOs of [6,6]-adducts of C_{60} are the *e*- and *cis*-1 positions. Subsequent attacks occur preferably into the *e*- and *trans*-3 positions for sterically demanding addends and the *cis*-1 position for sterically less demanding ones. The regioselectivity of additions into *equatorial* sites increases with number of addends already bound.

(6) Due to the pyramidalization of the C atoms and the rigid cage structure of C_{60} the outer convex surface is very reactive towards addition reactions but at the same time the inner concave surface is inert (*chemical Faraday cage*). This allows the encapsulation, observation and tuning of the wavefunction of extremely reactive species that otherwise would immediately form covalent bonds with the outer surface.

(7) Icosahedral fullerenes that contain $2(N + 1)^2$ π-electrons are closed shell systems and are spherically aromatic.

14.6 The Future of Fullerene Chemistry

Although comparatively young, fullerene chemistry is already a mature discipline. Fullerenes are organic building blocks with unprecedented properties that chemists can utilize routinely in their syntheses. Ten years ago four unsolved synthesis problems were identified as major topics for future fullerene chemistry [131]:

(a) Multiple addition products with a defined three-dimensional structure.
(b) Heterofullerenes.
(c) Ring-opening and degradation reactions – synthesis of endohedrals.
(d) Chemistry with higher fullerenes.

Considerable progress in all these fields has been achieved, as demonstrated with the inclusion of the four new Chapters 10–13. Problems (a) and (d) can now be considered as well elaborated although a lot of room for new investigations is left such as: development of new addition patterns, combination of different types of addition reactions, addition chemistry of cluster modified fullerenes such as homofullerenes (fulleroids). The chemistry of higher fullerenes will benefit from the fact that larger quantities will be available due to the possibility of producing fullerenes on the ton scale (Frontier Carbon Cooperation, Tokyo). This will allow, for example, the application of procedures developed for C_{60} that provide access to cluster modified species and heterofullerenes. For C_{60}, as the basic fullerene, encouraging breakthroughs have been made towards these two synthesis goals. However, the potential of elusive cluster modified fullerenes and heterofullerenes with outstanding physical and chemical properties is enormous.

As a third group of fullerene-like systems whose synthetic development is identified as a major future challenge are structurally defined single-walled carbon nanotubes (SWNTs) with uniform helicity and length, including their covalent and non-covalent derivatives. SWNTs can be considered as elongated giant fullerenes.

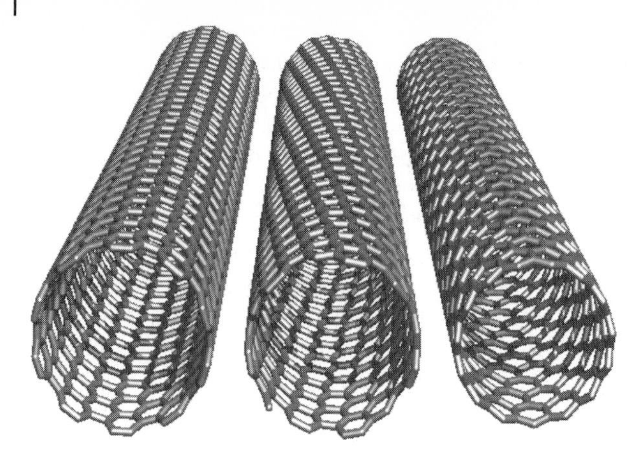

Figure 14.17 Idealized representation of defect-free (n,m)-SWNTs with open ends. *Left:* a metallic conducting (10,10)-tube (armchair); *middle:* a chiral, semiconducting (12,7)-tube; *right:* a conducting (15,0)-tube (zigzag). Armchair and zigzag tubes are achiral. All (n,n)-armchair-tubes are metallic, whilst this is only the case with chiral or zigzag tubes if $(n - m)/3$ is a whole number. Otherwise, they are semiconducting [132].

In most cases the caps at the end of the tubes are open. Carbon nanotubes exhibit outstanding electronic and mechanical properties. They are considered to be among the most promising materials for future nanotechnology. A problem with SWNTs is that they come as impossible, or difficult, to isolate mixtures of different helicities (Figure 14.17) and lengths and different amount of defects when prepared and purified according to standard procedures [132]. If it were possible, for example, to use cyclic organic molecules as templates for the growth of SWNTs then isomerically pure SWNTs with one specific helicity only could be accessible. This would revolutionize carbon nanotube research and would also form the basis for the preparation of stereochemically defined derivatives of SWNTs such as covalent adducts [132]. In this way SWNTs with tunable properties and improved solubility and processablility would be available.

In conclusion the main challenges for future fullerene chemistry are considered to be:

(a) Heterofullerenes.
(b) Cluster modified fullerenes.
(c) Structurally defined and isomerically pure SWNTs and their derivatives.
(d) Use of fullerenes as building blocks for molecular engineering and practical applications.

14.7
Fullerenes as Building Blocks for Molecular Engineering (Nanotechnology) and Practical Applications

Due to their unique electronic and chemical properties fullerenes have a tremendous potential as building blocks for molecular engineering, new molecular materials and supramolecular chemistry [54, 133]. Many examples of fullerene derivatives (Section 14.1), which are promising candidates for nanotechnological or medical applications, have been synthesized already and even more exciting developments are expected. A detailed description of the potential of fullerene derivatives for technological applications would require an extra monograph. Since this book focuses on the chemical properties and the synthetic potential of fullerenes only a few concepts for fullerene based materials will be briefly presented.

An approach is based on the remarkable electronic and photophysical properties of fullerenes (see Chapter 1, Section 1.5.2) that form the basis for the development of photosynthetic antenna and reaction center models as well as building blocks for solar cell devices. This subject has been extensively reviewed [27, 29, 30, 134–141]. For this purpose many molecular architectures have been synthesized with fullerenes, serving as electron acceptor, linked either covalently or supramolecularly to electron-donor moieties such as porphyrins, phthalocyanines, tetrathiafulvalenes, carotenes, Ru-bipy and Ru-terpy as well as ferrocene. As well as donor–acceptor dyads, oligoads involving a redox-gradient of the aligned electrophores have also been designed (Figure 14.18). Depending, for example, on the redox-potentials, the bridging moieties, the relative orientations and the solvent, photoinduced energy- or electron-transfer processes can be observed (Figure 14.19). The pronounced electron delocalization in negatively charged C_{60} together with its very rigid structure (low reorganization energy) offers unique opportunities to slow down back electron transfer in the charge-separated states. For example, the photoinduced multi-step electron transfer in the tetrad (Fc-ZnP-H_2P-C_{60}) **1** leads to the final state (Fc^+-ZnP-H_2P-C_{60}^-) in which the charges are separated at a long distance (4.9 nm) [142]. The lifetime of the resulting charge-separated state in frozen benzonitrile is 0.38 s and is comparable to that of observed for the bacterial photosynthetic reaction center. Related fullerene-based oligoads within a semibiological liposomal system have been successfully used to mimic the photosynthetic solar energy transduction, including ATP synthesis [138].

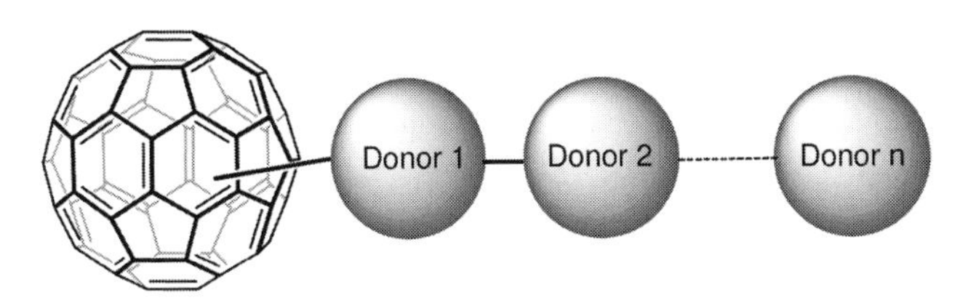

Figure 14.18 Schematic representation of fullerene based donor–acceptor oligoads suitable for photoinduced energy- or electron transfer.

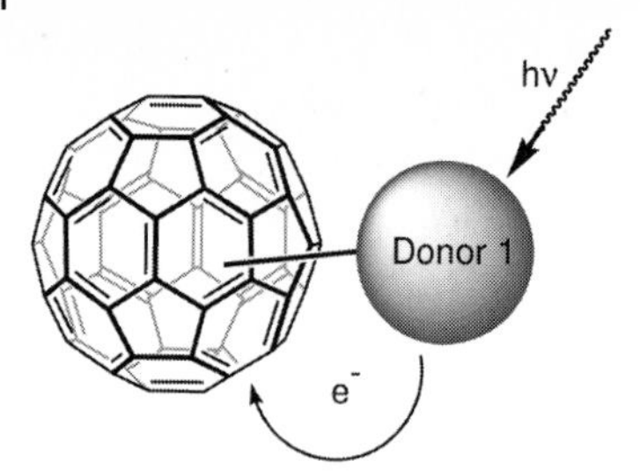

Figure 14.19 Schematic representation of a photoinduced electron transfer in a fullerene-based donor–acceptor dyad.

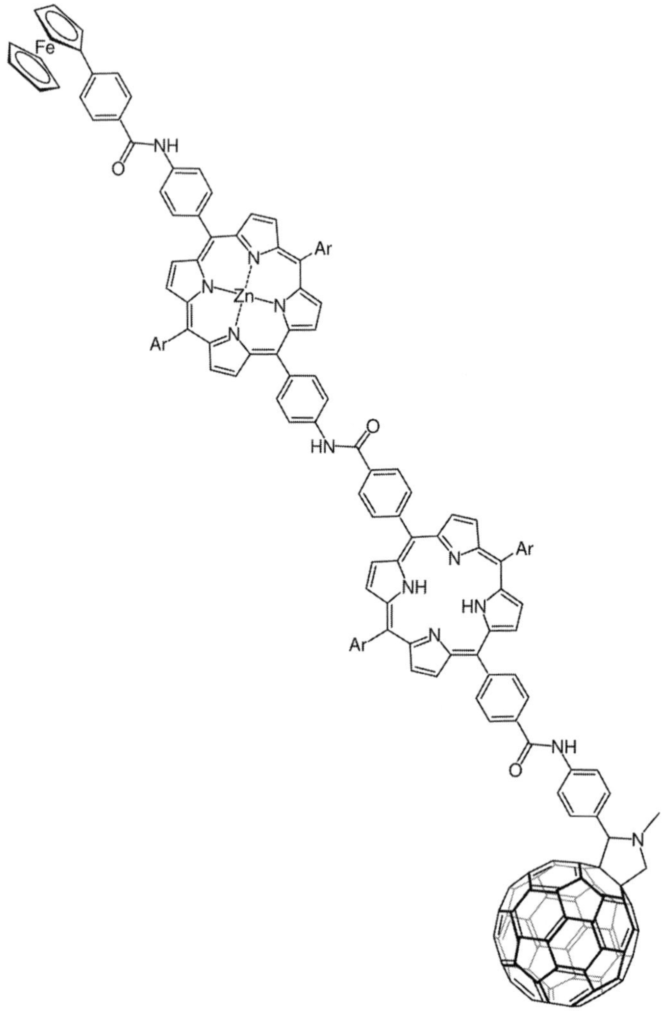

1

Heterogeneous mixing of fullerenes and fullerene derivatives with π-conjugated polymers has been used to produce excellent materials for photovoltaic devices [141]. Upon irradiation of fullerene/polymer blends, charge transfer from the polymer to C_{60} occurs, resulting in efficient photoconductivities. Better behavior of fullerene derivatives than with pristine C_{60} has been observed, and attributed to the improved miscibility of the functionalized species.

Thin fullerene films for laser protection can be obtained by incorporation or covalent attachment of fullerenes to transparent solid matrices. The optical limiting (OL) properties of C_{60} originally detected in toluene solutions can be transferred to solid substrates without significant activity loss [27, 143, 144].

Various covalent and non-covalent approaches for the incorporation of fullerene building blocks into liquid crystalline mesomorphic materials have been developed [35]. Lyotropic mesophases (buckysomes) were obtained upon dissolving globular amphiphilic fullerene dendrimers in water [145].

Biological investigations on many water-soluble fullerene derivatives have revealed a significant potential for medical applications [2–24]. Examples are: enzyme inhibition (HIV-protease and reverse transcriptase), anticancer activity, DNA cleavage, photodynamic therapy, radiotherapy, medical imaging and antioxidant properties. This latter aspect is based on the pronounced propensity of fullerenes to act as scavenger for radicals and lead to the development of drug candidates for neurodegenerative disorders. The water-soluble trismalonic acid **2** [8], for which an efficient large-scale synthesis has been developed [146], is a lead compound within the drug development program of C Sixty Inc., Houston (www.CSixty.com). It shows significant activity against a spectrum of neurodegenerative disorders in animal models that replicate many of the features of important human neurodegenerative diseases, including amyotrophic lateral sclerosis (ALS) and Parkinson's disease [147].

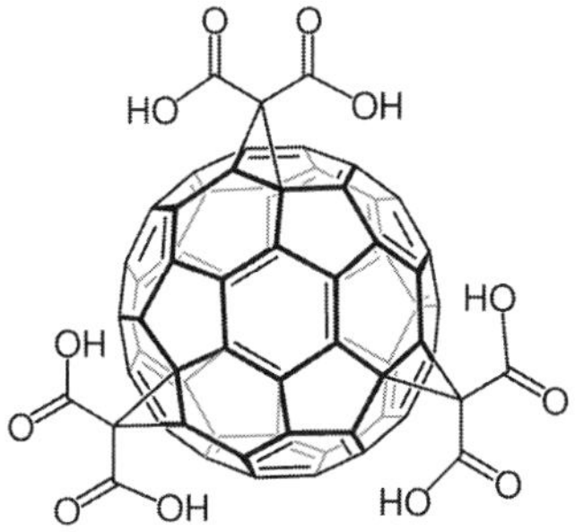

References

1 V. Alcazar Montero, L. Tomlinson, K. N. Houk, F. Diederich, *Tetrahedron Lett.* **1991**, *32*, 5309.
2 R. Sijbesma, G. Srdanov, F. Wudl, J. A. Castoro, C. Wilkins, S. H. Friedman, D. L. DeCamp, G. L. Kenyon, *J. Am. Chem. Soc.* **1993**, *115*, 6510.
3 S. H. Friedman, D. L. DeCamp, R. P. Sijbesma, G. Srdanov, F. Wudl, G. L. Kenyon, *J. Am. Chem. Soc.* **1993**, *115*, 6506.
4 G. Schick, T. Jarrosson, Y. Rubin, *Angew. Chem.* **1999**, *111*, 2508; *Angew. Chem. Int. Ed. Engl.* **1999**, *38*, 2360.
5 S. H. Friedman, G. L. Kenyon, *J. Am. Chem. Soc.* **1997**, *119*, 447.
6 S. H. Friedman, P. S. Ganapathi, Y. Rubin, G. L. Kenyon, *J. Med. Chem.* **1998**, *41*, 2424.
7 L. L. Dugan, D. M. Turetsky, C. Du, D. Lobner, M. Wheeler, C. R. Almli, C. K. F. Shen, T.-Y. Luh, D. W. Choi, T.-S. Lin, *Proc. Natl. Acad. Sci.* **1997**, *94*, 9434.
8 I. Lamparth, A. Hirsch, *J. Chem. Soc., Chem. Commun.* **1994**, 1727.
9 U. Reuther, T. Brandmüller, W. Donaubauer, F. Hampel, A. Hirsch, *Chem. Eur. J.* **2002**, *8*, 2261.
10 A. W. Jensen, S. R. Wilson, D. I. Schuster, *Bioorg. Med. Chem.* **1996**, *4*, 767.
11 D. I. Schuster, S. R. Wilson, R. F. Schinazi, *Bioorg. Med. Chem. Lett.* **1996**, *6*, 1253.
12 S. R. Wilson, *Proc. – Electrochem. Soc.* **1997**, *97-42*, 322.
13 B. X. Chen, S. R. Wilson, M. Das, D. J. Coughlin, B. F. Erlanger, *Proc. Natl. Acad. Sci.* **1998**, *95*, 10809.
14 D. I. Schuster, S. R. Wilson, A. N. Kirschner, R. F. Schinazi, S. Schlüter-Wirtz, P. Tharnish, T. Barnett, J. Ermolieff, J. Tang, M. Brettreich, A. Hirsch, *Proc. – Electrochem. Soc.* **2000**, *2000-11*, 267.
15 B. C. Braden, B. F. Erlanger, B. X. Chen, A. N. Kirschner, S. R. Wilson, *Proc. – Electrochem. Soc.* **2000**, *2000-11*, 233.
16 B. C. Braden, F. A. Goldbaum, B.-X. Chen, A. N. Kirschner, S. R. Wilson, B. F. Erlanger, *Proc. Natl. Acad. Sci.* **2000**, *97*, 12193.
17 D. J. Wolff, A. D. P. Papoiu, K. Mialkowski, C. F. Richardson, D. I. Schuster, S. R. Wilson, *Arch. Biochem. Biophys.* **2000**, *378*, 216.
18 D. J. Wolff, C. M. Barbieri, C. F. Richardson, D. I. Schuster, S. R. Wilson, *Arch. Biochem. Biophys.* **2002**, *399*, 130.
19 C. Cusan, T. Da Ros, G. Spalluto, S. Foley, J.-M. Janot, P. Seta, C. Larroque, M. C. Tomasini, T. Antonelli, L. Ferraro, M. Prato, *Eur. J. Org. Chem.* **2002**, 2928.
20 N. Gharbi, M. Pressac, V. Tomberli, T. Da Ros, M. Brettreich, M. Hadchouel, B. Arbeille, F. Trivin, R. Ceolin, A. Hirsch, M. Prato, H. Szwarc, R. V. Bensasson, F. Moussa, *Proc. – Electrochem. Soc.* **2000**, *2000-11*, 240.
21 T. Da Ros, M. Prato, *Chem. Commun.* **1999**, 663.
22 S. Bosi, T. Da Ros, S. Castellano, E. Banfi, M. Prato, *Bioorg. Med. Chem. Lett.* **2000**, *10*, 1043.
23 T. Da Ros, G. Spalluto, M. Prato, *Croat. Chem. Acta* **2001**, *74*, 743.
24 T. Da Ros, G. Spalluto, A. S. Boutorine, R. V. Bensasson, M. Prato, *Curr. Pharm. Design* **2001**, *7*, 1781.
25 H. W. Kroto, E. Fischer John, D. E. Cox, Eds., *The Fullerenes*, Pergamon Press Oxford **1993**.
26 M. Prato, *J. Mater. Chem.* **1997**, *7*, 1097.
27 M. Prato, *Top. Curr. Chem.* **1999**, *199*, 173.
28 M. Carano, P. Ceroni, F. Paolucci, S. Roffia, T. Da Ros, M. Prato, M. I. Sluch, C. Pearson, M. C. Petty, M. R. Bryce, *J. Mater. Chem.* **2000**, *10*, 269.
29 D. M. Guldi, *Chem. Soc. Rev.* **2002**, *31*, 22.
30 D. M. Guldi, N. Martin, *J. Mater. Chem.* **2002**, *12*, 1978.
31 M. V. Martinez-Diaz, N. S. Fender, M. S. Rodriguez-Morgade, M. Gomez-Lopez, F. Diederich, L. Echegoyen, J. F. Stoddart, T. Torres, *J. Mater. Chem.* **2002**, *12*, 2095.
32 D. Bonifazi, A. Salomon, O. Enger, F. Diederich, D. Cahen, *Adv. Mater.* **2002**, *14*, 802.
33 F. Diederich, *Abstr. Pap. – Am. Chem. Soc.* **2001**, *221*.

34 N. Tirelli, F. Cardullo, T. Habicher, U. W. Suter, F. Diederich, *J. Chem. Soc., Perkin Trans. 2* **2000**, 193.
35 T. Chuard, R. Deschenaux, *J. Mater. Chem.* **2002**, *12*, 1944.
36 A. Hirsch, *Top. Curr. Chem.* **1999**, *199*, 1.
37 F. Wudl, *Acc. Chem. Res.* **1992**, *25*, 157.
38 F. Wudl, *Buckminsterfullerenes* **1993**, 317.
39 P. J. Fagan, B. Chase, J. C. Calabrese, D. A. Dixon, R. Harlow, P. J. Krusic, N. Matsuzawa, F. N. Tebbe, D. L. Thorn, E. Wasserman, *Carbon* **1992**, *30*, 1213.
40 P. J. Fagan, J. C. Calabrese, B. Malone, *Acc. Chem. Res.* **1992**, *25*, 134.
41 G. A. Olah, I. Bucsi, R. Aniszfeld, G. K. S. Prakash, *Carbon* **1992**, *30*, 1203.
42 J. M. Hawkins, *Acc. Chem. Res.* **1992**, *25*, 150.
43 R. Taylor, D. R. Walton, *Nature* **1993**, *363*, 685.
44 R. Taylor, *Lecture Notes on Fullerene Chem.: A Handbook for Chemists*, Imperial College Press, London **1999**.
45 R. Taylor, *Synlett* **2000**, 776.
46 A. Hirsch, *Angew. Chem.* **1993**, *105*, 1189; *Angew. Chem. Int. Ed. Engl.* **1993**, *32*, 1138.
47 A. Hirsch, *Synthesis* **1995**, 895.
48 M. Bühl, A. Hirsch, *Chem. Rev.* **2001**, *101*, 1153.
49 A. Hirsch, *Angew. Chem.* **2001**, *113*, 1235; *Angew. Chem. Int. Ed. Engl.* **2001**, *40*, 1195.
50 A. Hirsch, O. Vostrowsky, *Eur. J. Org. Chem.* **2001**, 829.
51 F. Diederich, L. Isaacs, D. Philp, *Chem. Soc. Rev.* **1994**, *23*, 243.
52 F. Diederich, C. Thilgen, *Science* **1996**, *271*, 317.
53 F. Diederich, *Pure Appl. Chem.* **1997**, *69*, 395.
54 F. Diederich, M. Gomez-Lopez, *Chem. Soc. Rev.* **1999**, *28*, 263.
55 F. Diederich, R. Kessinger, in *Templated Organic Synthesis* (Wiley-VCH) **2000**, 189.
56 R. C. Haddon, *Science* **1993**, *261*, 1545.
57 H. D. Beckhaus, C. Rüchardt, M. Kao, F. Diederich, C. S. Foote, *Angew. Chem.* **1992**, *104*, 69; *Angew. Chem. int. Ed. Engl.* **1992**, *31*, 63.
58 H. D. Beckhaus, S. Verevkin, C. Rüchardt, F. Diederich, C. Thilgen, H. U. ter Meer, H. Mohn, W. Müller, *Angew. Chem.* **1994**, *106*, 1033; *Angew. Chem. Int. Ed. Engl.* **1994**, *33*, 996.
59 R. C. Haddon, *J. Am. Chem. Soc.* **1986**, *108*, 2837.
60 R. C. Haddon, *J. Am. Chem. Soc.* **1987**, *109*, 1676.
61 R. C. Haddon, *Acc. Chem. Res.* **1988**, *21*, 243.
62 R. C. Haddon, *J. Am. Chem. Soc.* **1990**, *112*, 3385.
63 R. C. Haddon, *Acc. Chem. Res.* **1992**, *25*, 127.
64 R. C. Haddon, L. E. Brus, K. Raghavachari, *Chem. Phys. Lett.* **1986**, *125*, 459.
65 R. C. Haddon, L. E. Brus, K. Raghavachari, *Chem. Phys. Lett.* **1986**, *131*, 165.
66 A. Hirsch, T. Grösser, A. Skiebe, A. Soi, *Chem. Ber.* **1993**, *126*, 1061.
67 H. Okamura, Y. Murata, M. Minoda, K. Komatsu, T. Miyamoto, T. S. M. Wan, *J. Org. Chem.* **1996**, *61*, 8500.
68 C. Bingel, *Chem. Ber.* **1993**, *126*, 1957.
69 X. Camps, A. Hirsch, *J. Chem. Soc., Perkin Trans. 1* **1997**, 1595.
70 P. S. Ganapathi, Y. Rubin, *J. Org. Chem.* **1995**, *60*, 2954.
71 K. D. Kampe, N. Egger, M. Vogel, *Angew. Chem.* **1993**, *105*, 1203.
72 G. Schick, K.-D. Kampe, A. Hirsch, *J. Chem. Soc., Chem. Commun.* **1995**, 2023.
73 A. Hirsch, I. Lamparth, H. R. Karfunkel, *Angew. Chem.* **1994**, *106*, 453; *Angew. Chem. int. Ed. Engl.* **1994**, *33*, 437.
74 A. Hirsch, I. Lamparth, T. Grösser, H. R. Karfunkel, *J. Am. Chem. Soc.* **1994**, *116*, 9385.
75 C. Boudon, J.-P. Gisselbrecht, M. Gross, L. Isaacs, H. L. Anderson, R. Faust, F. Diederich, *Helv. Chim. Acta* **1995**, *78*, 1334.
76 M. Sawamura, H. Iikura, E. Nakamura, *J. Am. Chem. Soc.* **1996**, *118*, 12850.
77 M. Sawamura, Y. Kuninobu, E. Nakamura, *J. Am. Chem. Soc.* **2000**, *122*, 12407.
78 M. Sawamura, Y. Kuninobu, M. Toganoh, Y. Matsuo, M. Yamanaka, E. Nakamura, *J. Am. Chem. Soc.* **2002**, *124*, 9354.
79 M. Tsuda, T. Ishida, T. Nogami, S. Kurono, M. Ohashi, *J. Chem. Soc., Chem. Commun.* **1993**, 1296.
80 J. A. Schlueter, J. M. Seaman, S. Taha, H. Cohen, K. R. Lykke, H. H. Wang, J. M. Williams, *J. Chem. Soc., Chem. Commun.* **1993**, 972.

81 K. Komatsu, Y. Murata, N. Sugita, K. i. Takeuchi, T. S. M. Wan, *Tetrahedron Lett.* **1993**, *34*, 8473.
82 I. Lamparth, C. Maichle-Mössmer, A. Hirsch, *Angew. Chem.* **1995**, *107*, 1755; *Angew. Chem. Int. Ed. Engl.* **1995**, *34*, 1607.
83 B. Kräutler, T. Müller, J. Maynollo, K. Gruber, C. Kratky, P. Ochsenbein, D. Schwarzenbach, H.-B. Bürgi, *Angew. Chem.* **1996**, *108*, 1294; *Angew. Chem. Int. Ed. Engl.* **1996**, *35*, 1204.
84 V. M. Rotello, J. B. Howard, T. Yadav, M. M. Conn, E. Viani, L. M. Giovane, A. L. Lafleur, *Tetrahedron Lett.* **1993**, *34*, 1561.
85 S. H. Hoke II, J. Molstad, D. Dilettato, M. J. Jay, D. Carlson, B. Kahr, R. G. Cooks, *J. Org. Chem.* **1992**, *57*, 5069.
86 M. Tsuda, T. Ishida, T. Nogami, S. Kurono, M. Ohashi, *Chem. Lett.* **1992**, 2333.
87 Y. Nakamura, N. Takano, T. Nishimura, E. Yashima, M. Sato, T. Kudo, J. Nishimura, *Org. Lett.* **2001**, *3*, 1193.
88 T. Suzuki, Q. Li, K. C. Khemani, F. Wudl, *J. Am. Chem. Soc.* **1992**, *114*, 7301.
89 M. Prato, Q. C. Li, F. Wudl, V. Lucchini, *J. Am. Chem. Soc.* **1993**, *115*, 1148.
90 L. Isaacs, A. Wehrsig, F. Diederich, *Helv. Chim. Acta* **1993**, *76*, 1231.
91 B. Nuber, F. Hampel, A. Hirsch, *Chem. Commun.* **1996**, 1799.
92 A. Skiebe, A. Hirsch, H. Klos, B. Gotschy, *Chem. Phys. Lett.* **1994**, *220*, 138.
93 I. S. Neretin, K. A. Lyssenko, M. Y. Antipin, Y. L. Slovokhotov, O. V. Boltalina, P. A. Troshin, A. Y. Lukonin, L. N. Sidorov, R. Taylor, *Angew. Chem.* **2000**, *112*, 3411; *Angew. Chem. Int. Ed. Engl.* **2000**, *39*, 3273.
94 C. A. Reed, K.-C. Kim, R. D. Bolskar, L. J. Müller, *Science* **2000**, *289*, 101.
95 K.-C. Kim, F. Hauke, A. Hirsch, P. D. W. Boyd, E. Carter, R. S. Armstrong, P. A. Lay, C. A. Reed, *J. Am. Chem. Soc.* **2003**, *125*, 4024.
96 T. A. Murphy, T. Pawlik, A. Weidinger, M. Höhne, R. Alcala, J. M. Späth, *Phys. Rev. Lett.* **1996**, *77*, 1075.
97 H. Mauser, N. J. R. van Eikema Hommes, T. Clark, A. Hirsch, B. Pietzak, A. Weidinger, L. Dunsch, *Angew. Chem.* **1997**, *109*, 2858; *Angew. Chem. Int. Ed. Engl.* **1997**, *36*, 2835.
98 H. Mauser, A. Hirsch, N. J. R. van Eikema Hommes, T. Clark, *J. Mol. Model.* **1997**, *3*, 415.
99 E. Dietel, A. Hirsch, B. Pietzak, M. Waiblinger, K. Lips, A. Weidinger, A. Gruss, K.-P. Dinse, *J. Am. Chem. Soc.* **1999**, *121*, 2432.
100 B. Pietzak, M. Waiblinger, T. Almeida Murphy, A. Weidinger, M. Hohne, E. Dietel, A. Hirsch, *Chem. Phys. Lett.* **1997**, *279*, 259.
101 M. Waiblinger, K. Lips, W. Harneit, A. Weidinger, E. Dietel, A. Hirsch, *Phys. Rev. B: Condens. Matter Mater. Phys.* **2001**, *64*, 159901/1.
102 B. Pietzak, M. Waiblinger, T. A. Murphy, A. Weidinger, M. Hohne, E. Dietel, A. Hirsch, *Carbon* **1998**, *36*, 613.
103 M. Waiblinger, K. Lips, W. Harneit, A. Weidinger, E. Dietel, A. Hirsch, *Phys. Rev. B: Condens. Matter Mater. Phys.* **2001**, *63*, 045421/1.
104 D. J. Klein, T. G. Schmalz, G. E. Hite, W. A. Seitz, *J. Am. Chem. Soc.* **1986**, *108*, 1301.
105 G. Stollhoff, *Phys. Rev. B: Condens. Matter Mat. Phys.* **1991**, *44*, 10998.
106 G. Stollhoff, H. Scherrer, *Mater. Sci. Forum* **1995**, *191*, 81.
107 A. Hirsch, Z. Chen, H. Jiao, *Angew. Chem.* **2000**, *112*, 4079; *Angew. Chem. Int. Ed. Engl.* **2000**, *39*, 3915.
108 M. Reiher, A. Hirsch, *Chem. Eur. J.* **2003**, *9*, 5442.
109 T. Kusukawa, W. Ando, *Angew. Chem.* **1996**, *108*, 1416; *Angew. Chem. Int. Ed. Engl.* **1996**, *35*, 1315.
110 F. Djojo, A. Herzog, I. Lamparth, F. Hampel, A. Hirsch, *Chem. Eur. J.* **1996**, *2*, 1537.
111 N. Matsuzawa, D. A. Dixon, T. Fukunaga, *J. Phys. Chem.* **1992**, *96*, 7594.
112 M. Prato, V. Lucchini, M. Maggini, E. Stimpfl, G. Scorrano, M. Eiermann, T. Suzuki, F. Wudl, *J. Am. Chem. Soc.* **1993**, *115*, 8479.
113 L. Isaacs, F. Diederich, *Helv. Chim. Acta* **1993**, *76*, 2454.
114 G. Schick, T. Grösser, A. Hirsch, *J. Chem. Soc., Chem. Commun.* **1995**, 2289.

115 G. Schick, A. Hirsch, H. Mauser, T. Clark, *Chem. Eur. J.* **1996**, *2*, 935.

116 V. I. Minkin, M. N. Glukhovtsev, B. Y. Simkin, *Aromaticity and Antiaromaticity: Electronic and Structural Aspects*, Wiley, New York **1994**.

117 P. v. R. Schleyer, H. Jiao, *Pure Appl. Chem.* **1996**, *68*, 209.

118 T. M. Krygowski, M. K. Cyranski, Z. Czarnocki, G. Hafelinger, A. R. Katritzky, *Tetrahedron* **2000**, *56*, 1783.

119 P. W. Fowler, D. J. Collins, S. J. Austin, *J. Chem. Soc., Perkin Trans. 2* **1993**, 275.

120 S. J. Austin, P. W. Fowler, P. Hansen, D. E. Manolopoulos, M. Zheng, *Chem. Phys. Lett.* **1994**, *228*, 478.

121 T. M. Krygowski, A. Ciesielski, *J. Chem. Inf. Comp. Sci.* **1995**, *35*, 1001.

122 R. C. Haddon, *Nature* **1995**, *378*, 249.

123 V. Elser, R. C. Haddon, *Nature* **1987**, *325*, 792.

124 M. Bühl, *Chem. Eur. J.* **1998**, *4*, 734.

125 P. v. R. Schleyer, C. Maerker, A. Dransfeld, H. Jiao, N. J. R. van Eikema Hommes, *J. Am. Chem. Soc.* **1996**, *118*, 6317.

126 M. Saunders, H. A. Jimenez-Vazquez, R. J. Cross, S. Mroczkowski, D. I. Freedberg, F. A. L. Anet, *Nature* **1994**, *367*, 256.

127 M. Saunders, R. J. Cross, H. A. Jimenez-Vazquez, R. Shimshi, A. Khong, *Science* **1996**, *271*, 1693.

128 Z. Chen, H. Jiao, A. Hirsch, W. Thiel, *J. Mol. Model.* **2001**, *7*, 161.

129 Z. Chen, H. Jiao, A. Hirsch, P. von Rague Schleyer, *Angew. Chem.* **2002**, *114*, 4485; *Angew. Chem. Int. Ed. Engl.* **2002**, *41*, 4309.

130 A. Hirsch, Z. Chen, H. Jiao, *Angew. Chem.* **2001**, *113*, 2916; *Angew. Chem. Int. Ed. Engl.* **2001**, *40*, 2834.

131 A. Hirsch, *Chemistry of the Fullerenes*, Georg Thieme Verlag, Stuttgart, Stuttgart, **1994**.

132 A. Hirsch, *Angew. Chem.* **2002**, *114*, 1933; *Angew. Chem. Int. Ed. Engl.* **2002**, *41*, 1853.

133 F. Diederich, M. Gomez-Lopez, *Chimia* **1998**, *52*, 551.

134 H. Imahori, Y. Sakata, *Adv. Mater.* **1997**, *9*, 537.

135 N. Martin, L. Sanchez, B. Illescas, I. Perez, *Chem. Rev.* **1998**, *98*, 2527.

136 H. Imahori, Y. Sakata, *Eur. J. Org. Chem.* **1999**, 2445.

137 D. M. Guldi, M. Prato, *Acc. Chem. Res.* **2000**, *33*, 695.

138 D. Gust, T. A. Moore, A. L. Moore, *Acc. Chem. Res.* **2001**, *34*, 40.

139 S. Fukuzumi, *Org. Biomol. Chem.* **2003**, *1*, 609.

140 J.-F. Nierengarten, *Top. Curr. Chem.* **2003**, *228*, 87.

141 A. Cravino, N. S. Sariciftci, *J. Mater. Chem.* **2002**, *12*, 1931.

142 H. Imahori, D. M. Guldi, K. Tamaki, Y. Yoshida, C. Luo, Y. Sakata, S. Fukuzumi, *J. Am. Chem. Soc.* **2001**, *123*, 6617.

143 P. Innocenzi, G. Brusatin, M. Guglielmi, R. Signorini, M. Meneghetti, R. Bozio, M. Maggini, G. Scorrano, M. Prato, *J. Sol-Gel Sci. Technol.* **2000**, *19*, 263.

144 R. Signorini, A. Tonellato, M. Meneghetti, R. Bozio, M. Prato, M. Maggini, G. Scorrano, G. Brusatin, P. Innocenzi, M. Guglielmi, *MCLC S&T, Sect. B: Nonlinear Opt.* **2001**, *27*, 193.

145 M. Brettreich, S. Burghardt, C. Böttcher, T. Bayerl, S. Bayerl, A. Hirsch, *Angew. Chem.* **2000**, *112*, 1915; *Angew. Chem. Int. Ed. Engl.* **2000**, *39*, 1845.

146 U. Reuther, T. Brandmüller, W. Donaubauer, F. Hampel, A. Hirsch, *Chem. Eur. J.* **2002**, *8*, 2833.

147 L. L. Dugan, E. Lovett, S. Cuddihy, B.-W. Ma, T.-S. Lin, D. W. Choi, *Fullerene: Chem., Phys. Technol.* **2000**, 467.

Subject Index

Fullerenes: Chemistry and Reactions. Andreas Hirsch and Michael Brettreich

ISBN: 3-527-30820-2